Power Systems

Electrical power has been the technological foundation of industrial societies for many years. Although the systems designed to provide and apply electrical energy have reached a high degree of maturity, unforeseen problems are constantly encountered, necessitating the design of more efficient and reliable systems based on novel technologies. The book series Power Systems is aimed at providing detailed, accurate and sound technical information about these new developments in electrical power engineering. It includes topics on power generation, storage and transmission as well as electrical machines. The monographs and advanced textbooks in this series address researchers, lecturers, industrial engineers and senior students in electrical engineering.

Power Systems is indexed in Scopus

Farhad Shahnia · Josep M. Guerrero
Editors

Microgrids and Virtual Power Plants

Springer

Editors
Farhad Shahnia
School of Engineering and Energy
Murdoch University
Perth, WA, Australia

Josep M. Guerrero
Department of Energy Technology
Aalborg University
Aalborg, Denmark

ISSN 1612-1287 ISSN 1860-4676 (electronic)
Power Systems
ISBN 978-981-97-6622-2 ISBN 978-981-97-6623-9 (eBook)
https://doi.org/10.1007/978-981-97-6623-9

This Springer imprint is published by the registered company Springer Nature Singapore Pte Ltd.
The registered company address is: 152 Beach Road, #21-01/04 Gateway East, Singapore 189721, Singapore

Preface

Microgrids and virtual power plants are integral to our future electrical energy systems. Over the past decade, substantial theoretical and practical research in this field has resulted in the development of technologically advanced products. Numerous manufacturers are adopting these technologies, and utilities worldwide have initiated and developed many pilot projects. Consequently, microgrids and virtual power plants are rapidly gaining traction in the global energy industry.

A microgrid is a network of interconnected loads powered by distributed energy resources, functioning as a single controllable unit. These energy resources, ranging from a few kilowatts to megawatts, are predominantly driven by renewable energy, are mostly interfaced with power electronics converter systems, and are distributed within the network.

Microgrids can connect to or disconnect from the electricity grid, enabling them to operate in grid-connected or island mode. In grid-connected mode, the grid determines the microgrid's voltage and frequency, acting as a power exchange backbone. It facilitates power import or export to the microgrid during power generation-demand imbalances, making it more economical for the distributed energy resources to operate at maximum power tracking mode. Conversely, a stand-alone or isolated microgrid operates independently of the grid, in which the energy resources must meet the loads' active and reactive demand and also regulate the voltage and frequency.

A virtual power plant is a microgrid system that emphasizes energy economics and the roles and benefits of the participating loads and energy resources. The aim is to enhance the financial benefits for customers and prosumers. Furthermore, virtual power plants play a crucial role in properly aggregating multiple loads and energy resources, offering numerous flexibilities and technical ancillary services that can benefit the electricity grid and network operators.

This book aims to provide a research-oriented and comprehensive platform for microgrids and virtual power plants. It covers their features, designs, technologies, control methods, and applications, making it a valuable resource for postgraduate students and researchers. The book, which benefits from the contributions and feedback of 46 researchers and experts from academia and industry from 14 countries, comprises 17 chapters on various aspects such as planning, control, modelling, energy transaction, stability, protection, communications, cyber-security, and regulation. The breakdown of the chapters is as follows:

- Chapter "Holistic Data-Driven Approach for Sizing and Energy Management of an Urban Islanded Microgrid" presents a data-driven approach for optimal sizing and energy management of an islanded microgrid,
- Chapter "Probabilistic Microgrid Investment Planning with Integrated Game-Theoretic Demand Response Management" formalizes a probabilistic microgrid investment planning with integrated game-theoretic demand response management,
- Chapter "Design and Modelling of Microgrids Operated at Constant Frequency and with a Power Level of Megawatts" discusses the design and modelling of microgrids operated at constant frequency and with a power level of megawatts,
- Chapter "Design, Sizing, and Simulation of a DC Microgrid for Real Implementation" focuses on the design, sizing, and simulation of dc microgrids,
- Chapter "Microgrid Control Assessment Using Advanced Hardware in the Loop Technologies" introduces microgrid control assessment using various hardware-in-loop technologies,
- Chapter "Stability and Control of Hybrid AC/DC Microgrids" highlights the stability and control aspects of hybrid ac/dc microgrids,
- Chapter "Quantifying Transient Dynamics for Microgrid's Inverter-Based Resources" quantifies the transient dynamics of microgrid's inverter-based resources using entropy and complexity metrics,
- Chapter "Machine Learning and Internet-of-Things Solutions for Microgrid Resilient Operation" uses machine learning and Internet-of-Things solutions for microgrid's resilient operation,
- Chapter "Deep Learning-Based Microgrid Protection" presents deep learning-based microgrid protection,
- Chapter "Cyber-Attacks Detection and Mitigation in Microgrids" focuses on mixed cyber attack detection and mitigation in microgrids,
- Chapter "Peer-To-Peer Trading Among Microgrid Prosumers in Local Energy Markets" introduces peer-to-peer trading among microgrid prosumers in local energy markets,
- Chapter "Embedding Regulatory Frameworks in Microgrids Management" discusses the need to embed regulatory frameworks in the management of microgrids,
- Chapter "Virtual Power Plant Participation in Australian Wholesale Electricity Markets" elaborates on the virtual power plants' participation in the Australian wholesale electricity markets,
- Chapter "Interconnected Microgrid Clusters Through Various Provisional Power Exchange Links" presents interconnected microgrid clusters through various provisional power exchange links,
- Chapter "Cooperative and Transactive Integration of Multiple Microgrids" formulates the cooperative and transactive integration of multiple microgrids,

- Chapter "Practical Aspects of Pre-engineering Design of Clustered Microgrids" highlights the practical aspects of the pre-engineering design of clustered microgrids, and
- Chapter "Military Microgrids with Renewable Energy and 5G Communication" focuses on military microgrids with renewable energy and 5G communication applications.

As the book's editors, we thank all the contributors for their support and hard work. We also thank the reviewers who provided valuable comments to improve the book's quality. Also, we are grateful to Springer Nature, the publisher, for agreeing to publish this book. Last but not least, we would like to thank our families—Farhad thanks his parents (Ali and Nahideh) and his spouse (Negar), and Josep thanks his family for their continuous encouragement and support.

Perth, Australia
Aalborg, Denmark
May 2024

Farhad Shahnia
Josep M. Guerrero

Contents

Editors and Contributors

About the Editors

Farhad Shahnia received his Ph.D. in Electrical Engineering from Queensland University of Technology, Brisbane, Australia in 2011. Currently, he is a Professor of Electrical Power Engineering at Murdoch University, Perth, Australia. His professional experience includes three years at the Research Office-Eastern Azarbayjan Electric Power Distribution Company, Tabriz, Iran. Prior to joining Murdoch University, he was a Research Fellow at the Queensland University of Technology, Brisbane, Australia, and a Lecturer at Curtin University, Perth, Australia. He has published eight books, 13 book chapters, and over 200 research articles on microgrids and renewable energy integration.

Josep M. Guerrero received his B.Sc. degree in telecommunication engineering, M.Sc. degree in electronics engineering, and Ph.D. degree from the Technical University of Catalonia, Barcelona. Since 2011, he has been a Full Professor at Aalborg University, Denmark. He is the founder and Director of the Center for Research on Microgrids (CROM) at Aalborg University. From 2019, he became a Villum Investigator by the Villum Fonden, which supports the CROM. In 2023, he joined the Technical University of Catalonia as an ICREA Research Professor. He has published eight books, 36 book chapters and, over 1500 research papers on microgrids and renewable energy systems.

Contributors

Zulfiqar Ali Department of Electrical Engineering, National Kaohsiung University of Science and Technology, Kaohsiung City, Taiwan

Hassan Bevrani Smart/Micro Grids Research Center, Department of Electrical Engineering, University of Kurdistan, Sanandaj, Iran

Alan C. Brent Wellington Faculty of Engineering, Sustainable Energy Systems, Victoria University of Wellington, Wellington, New Zealand;
Department of Industrial Engineering and the Centre for Renewable and Sustainable Energy Studies, Stellenbosch University, Stellenbosch, South Africa

Siqi Bu Department of Electrical and Electronics Engineering, Hong Kong Polytechnic University, Hong Kong, China

Syed Basit Ali Bukhari Department of Electrical Engineering, University of Azad Jammu and Kashmir, Muzaffarabad, Pakistan

Jordi de la Hoz Department of Electrical Engineering, Barcelona East School of Engineering (EEBE), Polytechnic University of Catalonia (UPC), Barcelona, Spain

Jan L. Díaz University of Puerto Rico, Mayaguez, Puerto Rico

Nelson L. Díaz Universidad Distrital Francisco José de Caldas, Bogotá, Colombia

Chris S. Edrington College of Engineering, Clemson University, Clemson, USA

Mahmoud Elsisi Department of Electrical Engineering, National Kaohsiung University of Science and Technology, Kaohsiung City, Taiwan

Xue Feng Engineering Cluster, Singapore Institute of Technology, Singapore, Singapore

S. M. Ferdous School of Engineering and Energy, Murdoch University, Perth, Australia

Farzad Ferdowsi College of Engineering, University of Louisiana at Lafayette, Lafayette, USA

Deepak Fulwani Departmentof Electrical Engineering, Indian Institute of Technology Jodhpur (IITJ), Rajasthan, India

Yuly V. García University of Puerto Rico, Mayaguez, Puerto Rico

Mojgan Hojabri Competence Center of Digital Energy and Electric Power, Institute of Electrical Engineering, Lucerne, Switzerland

Qian Hu Department of Physics, Hong Kong Baptist University, Hong Kong, China

Mohit Kachhwaha Departmentof Electrical Engineering, Indian Institute of Technology Jodhpur (IITJ), Rajasthan, India

Takeyoshi Kato Institute of Materials and Systems for Sustainability, Nagoya University, Nagoya, Japan

Hoang Le Quang Nhat Department of Electrical Engineering, National Kaohsiung University of Science and Technology, Kaohsiung City, Taiwan

Li Li School of Electrical and Data Engineering, University of Technology Sydney, Ultimo, Australia

Chen-Ching Liu Bradley Department of Electrical and Computer Engineering, Virginia Tech, Blacksburg, VA, USA

Adriana C. Luna University of Puerto Rico, Mayaguez, Puerto Rico

Helena Martín Department of Electrical Engineering, Barcelona East School of Engineering (EEBE), Polytechnic University of Catalonia (UPC), Barcelona, Spain

Moudud Ahmed Masum School of Engineering, RMIT University, Melbourne, Australia

Lasantha Meegahapola School of Engineering, RMIT University, Melbourne, Australia

Ali Mehrizi-Sani Bradley Department of Electrical and Computer Engineering, Virginia Tech, Blacksburg, VA, USA

Touria El Mezyani School of Engineering, University of North Florida, Jacksonville, USA

Himani Modi Departmentof Electrical Engineering, Indian Institute of Technology Jodhpur (IITJ), Rajasthan, India

Soheil Mohseni Wellington Faculty of Engineering, Sustainable Energy Systems, Victoria University of Wellington, Wellington, New Zealand;
Institute for Sustainable Futures, University of Technology Sydney, Sydney, NSW, Australia

Alexandre Nassif LUMA Energy, San Juan, PR, USA

Jeffery H. Reed Bradley Department of Electrical and Computer Engineering, Virginia Tech, Blacksburg, VA, USA

Sharara Rehimi Institute of Materials and Systems for Sustainability, Nagoya University, Nagoya, Japan

Seyed Hossein Rouhani Department of Electrical Engineering, National Kaohsiung University of Science and Technology, Kaohsiung City, Taiwan

Muhammad Sadiq Department of Electrical Engineering, National Kaohsiung University of Science and Technology, Kaohsiung City, Taiwan

Mohammad Seyedi College of Engineering, University of Louisiana at Lafayette, Lafayette, USA

G. M. Shafiullah School of Engineering and Energy, Murdoch University, Perth, Australia

Farhad Shahnia School of Engineering and Energy, Murdoch University, Perth, Australia

Chun-Lien Su Department of Electrical Engineering, National Kaohsiung University of Science and Technology, Kaohsiung City, Taiwan

Julius Susanto School of Engineering and Energy, Murdoch University, Perth, Australia

Chiyori T. Urabe Institute of Materials and Systems for Sustainability, Nagoya University, Nagoya, Japan

Muhammad Uzair School of Electrical and Data Engineering, University of Technology Sydney, Ultimo, Australia

Arash Vahidnia School of Engineering, RMIT University, Melbourne, Australia

Yan Xu Center for Power Engineering, School of Electrical and Electronic Engineering, Nanyang Technological University, Singapore, Singapore

Daming Zhang School of Electrical Engineering and Telecommunication, University of New South Wales, Sydney, Australia

Ziqing Zhu Department of Electrical and Electronics Engineering, Hong Kong Polytechnic University, Hong Kong, China

Abbreviations

2-DOF	Two-degree-of-freedom
AES	All-electric ship
AI	Artificial intelligence
ANFIS	Adaptive neuro-fuzzy inference system
AoI	Age of information
ApEn	Approximate entropy
BAU	Business-as-usual
BESS	Battery energy storage system
BRP	Balancing responsible party
CCI	Commonwealth cyber initiative
CCMA-DDPG	CVaR-constrained multi-agent deep deterministic policy gradient
CEAM	Carbon emission auction market
CEQ	Carbon emission quota
C-HIL	Controller hardware-in-the-loop
CMG	Coupled microgrids
CNN	Convolutional neural network
CRM	Capacity remuneration mechanism
CSC	Current source converter
C-SPAACE	Coordinated set-point automatic adjustment with correction enabled
CVaR	Conditional value-at-risk
DCPH	Dc-powered home
DER	Distributed energy resource
DFIG	Doubly-fed induction generator
DG	Distributed generator
DNSP	Distribution network service provider
DoD	Depth of discharge
DOE	Dynamic operating envelope
DOL	Direct-on-line
DoS	Denial-of-service
DR	Demand response

DRM	Demand response mode
DSCA	Defense support of civil authorities
DSM	Demand-side management
DSO	Distribution system operator
DUT	Device under test
DVCC	Dual vector current control algorithm
DWT	Discrete wavelet transform
ELM	Extreme learning machine
EM	Electricity market
eMBB	Enhanced mobile broad band
ESC	Electricity supply cost
ESS	Energy storage system
EV	Electrical vehicle
FCEV	Fuel cell electric vehicle
FCEV2G	Fuel cell electric vehicle-to-grid
FDI	False data injection
FMG	Floating microgrid
FPGA	Field-programmable gate array
GCC	Grid-connected converter
GFM	Grid forming
GFOV	Ground fault overvoltage
GOOSE	Generic object oriented substation event
HHT	Hilbert-huang transform
HIL	Hardware-in-the-loop
HMG	Healthy microgrid
IBR	Inverter based resource
ICT	Information and communication technologies
IED	Intelligent electronic device
IGBT	Insulated-gate bipolar transistor
ILC	Inter-linking converter
IM	Induction motor
IoT	Internet-of-things
LEM	Local energy market
LFO	Low-frequency oscillation
LSC	Line-side converter
LSTM	Layered long short-term memory
LUIFO	Linear unknown input functional observer
MAD	Maximal age difference
MADDPG	Multi-agent deep deterministic policy gradient
MDPUR	Markov decision process with uncertain reward
MEC	Mobile edge computing
MF	Membership function
MFOA	Moth-flame optimization algorithm
MGCS	Microgrid control system
MIL	Model-in-the-loop

ML	Machine learning
mMTC	Machine-to-machine communication
MP	Marginal price
MPPT	Maximum power point tracking
MRA	Multiresolution analysis
MSC	Microgrid-side converter
MSE	Mean square error
MV	Medium voltage
NEM	National electricity market
NEP	Nash equilibrium point
NGC	Next generation core network
NG-RAN	Next generation radio access network
NPC	Net present cost
NRMSE	Normalized root mean square error
OGFC	Over-generation frequency controller
OLFC	Overload frequency controller
OPPEn	Ordinal pattern and permutation entropy
P&O	Perturbation and observation
P2P	Peer-to-peer
PAB	Pay-as-bid
PAO	Pay as the opportunity cost
PCC	Point of common coupling
PCS	Power converter system
PDF	Probability density function
PEC	Power electronic converter
PF	Proportional fairness
P-HIL	Power hardware-in-the-loop
PI	Proportional-integral
PLL	Phase-locked loops
PMG	Problem microgrid
PMS	Predictive maintenance as a service
PMSG	Permanent magnet synchronous generator
POD	Power oscillation damping
PSH	Peak-sun-hour
pu	Per unit
PV	Photovoltaic
PWM	Pulse width modulation
RAN	Radio access network
RAS	Remedial action scheme
RB	Radio resources
REC	Regulated electricity cost
RES	Renewable energy resource
RL	Reinforcement learning
RR	Round robin
RUL	Remaining useful life

SCADA	Supervisory control, and data acquisition
SIL	Software-in-the-loop
SMG	Shipboard microgrid
SoC	State-of-charge
SPE	Shipboard power equipment
SSR	Sub-synchronous (or torsional) resonance
SST	Solid state transformer
SVM	Support vector machine
TCPD	Total cost per day
TE	Transactive energy
THD	Total harmonic distortion
TQWT	Tunable q-factor wavelet transform
TSO	Transmission system operator
UI	User interface
URLLC	Ultra-reliable and low latency communication
VFD	Variable frequency drive
VFT	Variable frequency transformer
V-HIL	Virtual hardware-in-the-loop
VPP	Virtual power plant
VSC	Voltage source converter
VSM	Virtual synchronous machine
WEM	Wholesale electricity market

Holistic Data-Driven Approach for Sizing and Energy Management of an Urban Islanded Microgrid

Xue Feng and Yan Xu

Abstract This chapter presents a data-driven approach for optimal sizing and operation of islanded microgrids within an urban context. The study employs a building-level urban islanded microgrid as a testbed for case studies. A randomized learning model is introduced for supply and demand forecasting, and based on these results, a data-driven approach is employed to generate uncertainty scenarios characterizing uncertain supply and demand characteristics. Optimal sizing is conducted considering operational constraints to determine the sizes of different components, particularly the energy storage system and distributed generators, with a focus on minimizing capital and operational costs. A two-stage coordinated energy management strategy is used to minimize operating costs while meeting various system constraints.

Keywords Data-driven · Optimal sizing · Energy management

1 Introduction

The adoption of islanded or standalone microgrids has gained widespread traction as a strategic measure to reduce dependence on conventional, carbon-intensive power generation sources and centralized power grids [1]. Notably, in urban landscapes like Singapore, microgrids have emerged as effective solutions for advancing the utilization of renewable energy. Existing research on standalone microgrids tends to fall into two broad categories: those primarily focused on practical implementation [1] and those academically oriented, addressing optimal system design and sizing [2], as well as energy management strategies under predefined design parameters [3]. This chapter endeavors to bridge the existing gap between theoretical research and applied

X. Feng
Engineering Cluster, Singapore Institute of Technology, Singapore, Singapore
e-mail: Elsa.Feng@singaporetech.edu.sg

Y. Xu (✉)
Center for Power Engineering, School of Electrical and Electronic Engineering, Nanyang Technological University, Singapore, Singapore
e-mail: xuyan@ntu.edu.sg

F. Shahnia and J. M. Guerrero (eds.), *Microgrids and Virtual Power Plants*, Power Systems, https://doi.org/10.1007/978-981-97-6623-9_1

projects in islanded microgrids. The introduced approach encompasses comprehensive aspects such as data analysis and forecasting for renewable energy generation and load demand, data-driven system uncertainty modeling, optimal sizing, and coordinated energy management strategies in the face of uncertainties. The ultimate goal is to implement this holistic framework in real-world scenarios.

In the realm of microgrid applications, operational strategies strive for optimal economic benefits while maintaining system stability amidst intermittent renewable resources and uncertainties [4]. Microgrid optimization studies are broadly categorized into three groups [4]: operational strategies under specified or assumed system designs [5–7], optimal design including sizing and siting under various constraints [8, 9], and a combined approach addressing both operational strategies and optimal design for global optimal solutions [4]. The prevailing argument emphasizes the interdependence of system sizing and operational strategies, asserting that an optimal sizing foundation facilitates the development of operational strategies, ensuring practicality and superior outcomes. Despite extensive research in this domain, most works tend to focus on either the operational or sizing phase, with fewer adopting a holistic approach that coordinates both [10]. Additionally, the challenge of uncertainty, particularly from renewable energy sources like solar and wind power, is a significant concern in microgrid optimization. Existing approaches, such as stochastic programming and robust optimization, address uncertainty, but each has its limitations and trade-offs [7, 11–13].

This chapter aims to present a comprehensive, data-driven approach for optimal sizing and operation of islanded microgrids within an urban context. The study employs a building-level urban islanded microgrid as a testbed for case studies. Building load data and photovoltaic irradiance data are collected, and an energy monitoring dashboard is developed for real-time visualization and data collection from various sensors. A randomized learning model is introduced for supply and demand forecasting, and based on these results, a data-driven approach is employed to generate uncertainty scenarios characterizing uncertain supply and demand characteristics. Optimal sizing is conducted considering operational constraints to determine the sizes of different components, particularly the energy storage system and distributed generators, with a focus on minimizing capital and operational costs. Building upon the optimal sizing results, a two-stage coordinated energy management strategy is used to minimize operating costs while meeting various system constraints. The main contributions of this chapter include data analysis on field-collected historical data, data-driven uncertainty modeling, an optimal sizing method considering operational constraints, and a temporal two-stage coordinated energy management strategy. The subsequent sections delve into the detailed structure of the islanded microgrid, component modeling, forecasting, data-driven uncertainty modeling, optimal sizing methodology, coordinated energy management strategy, and showcase the results of the methods. The chapter concludes with a comprehensive summary and insights gained from the study.

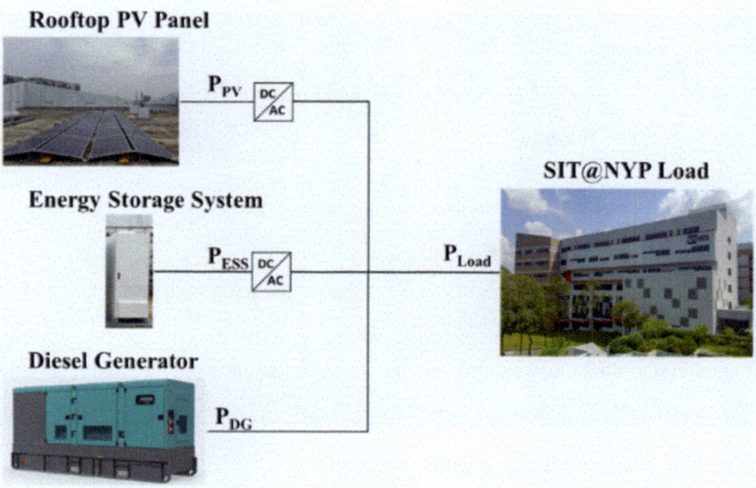

Fig. 1 Structure of the islanded microgrid at SIT@NYP

2 Microgrid Modelling

2.1 *Structure of the Islanded Microgrid*

The schematic structure of the islanded microgrid, depicted in Fig. 1, showcases its key components, namely the rooftop photovoltaic (PV) panel, energy storage system (ESS), distriubuted generator (DG), and building load. The system incorporates meters strategically placed within the building to measure specific load consumption, PV irradiance, and temperature. Figure 2 displays a real-time energy monitoring dashboard, providing a dynamic overview of the microgrid's energy metrics. All data referenced in this chapter originate from the comprehensive monitoring system depicted in Fig. 2.

2.2 *Modelling of Microgrid Components*

This section delves into the mathematical modeling of various components within the islanded microgrid.

The rooftop PV panel is characterized by a mathematical model that translates irradiance into power. The conversion equation is expressed as:

$$P_{PV}(t) = Y_{PV}\left(\frac{G(t)}{G_{STC}}\right)[1 + \alpha_p(T(t) - T_{STC})] \tag{1}$$

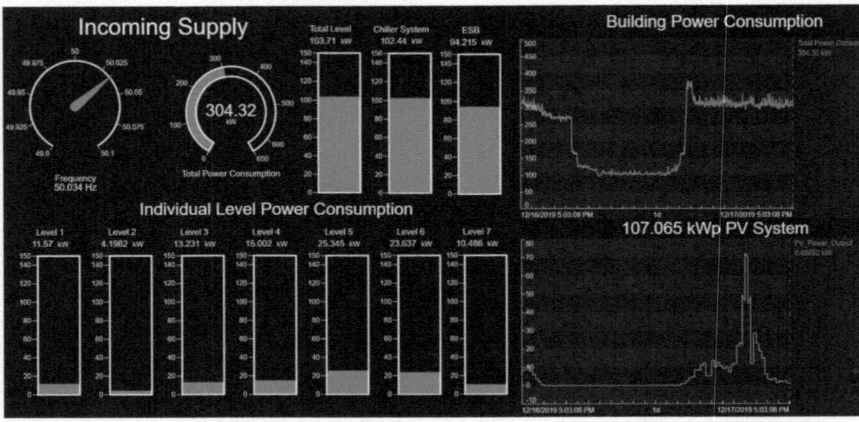

Fig. 2 Real time energy consumption monitoring platform

where P_{PV} is the PV output power, Y_{PV} is the total rated capacity of the PV array, G is the current irradiance, G_{STC} is the irradiance at standard test conditions, α_p is the temperature coefficient of power, T is the PV cell temperature, and T_{STC} is the PV cell temperature at standard test conditions.

Energy storage is characterized in terms of power, energy ratings, and efficiencies.

$$0 \leq P_c(t) \leq \overline{P_c} \tag{2}$$

$$0 \leq P_d(t) \leq \overline{P}_d \tag{3}$$

$$SoC(t) = SoC(t-1) + \frac{P_{ESS}(t)\Delta t}{E_{max}} \tag{4}$$

$$P_{ESS}(t) = P_C(t) - P_D(t) \tag{5}$$

$$P_C(t) = \frac{p_c(t)}{\eta_c} \tag{6}$$

$$P_D(t) = p_d(t)\eta_d \tag{7}$$

where $\overline{P_c}$ and $\overline{P}_d$ are the maximum charging and discharging rate of the ESS; $P_{ESS}(t)$, Δt and E_{max} are the power, time interval and maximum capacity of the ESS, respectively; p, η and P are DC power, efficiency and AC power respectively; subscripts c and d denotes charging and discharging. The energy losses during conversion between DC/AC and AC/DC are considered.

The diesel generator's output power is characterized by output power limits and ramp rates:

$$P_{DG}(t) = p_{DG}(t)\eta_{DG} \tag{8}$$

$$\underline{P_{DG}} \le P_{DG}(t) \le \overline{P_{DG}} \tag{9}$$

$$R_{DG}^{down}\Delta t \le P_{DG}(t) - P_{DG}(t-1) \le R_{DG}^{up}\Delta t \tag{10}$$

where p_{DG} and η_{DG} are the output power and efficiency of the diesel generator, $\underline{P_{DG}}$ and $\overline{P_{DG}}$ are the lower and upper output power limits of the diesel generator, R_{DG}^{down} and R_{DG}^{up} are the ramp down and ramp up limits of the diesel generator, and R_{DG}^{up} is a positive value and R_{DG}^{down} is a negative value.

2.3 Modeling of Microgrid Operational Uncertainties

The unpredictable and intermittent nature of power output from renewable energy sources, such as solar PV, poses a considerable challenge to the reliability of operational and planning solutions. This chapter addresses this uncertainty through the application of extreme learning machine (ELM), a type of artificial neural network, for PV irradiance and load forecasting. The structure of the ELM, illustrated in Fig. 3, embodies a single-layer feed-forward network [14].

The ELM's output function is formulated based on its fundamental structure:

$$\mathrm{f_N}(x_j) = \sum\nolimits_{i=1}^{N} \beta_i \cdot g(\boldsymbol{w}_i \cdot \boldsymbol{x}_j + b_i) = t_j, j = 1,2, \ldots, N \tag{11}$$

where N is the number of hidden layer nodes, g is the activation function, wi ∈RN represents the input weight vectors connecting the input layer and hidden layer, βi∈RN represents the output weights connecting the hidden layer and output layer, and bi is the input bias at the hidden layer nodes.

Fig. 3 Single layer feed-forward network

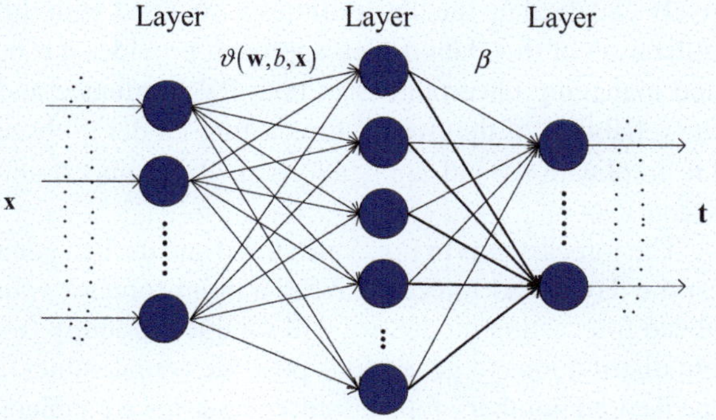

During the training stage of the ELM, the input weights and biases are randomly selected, leaving the output weights β to be analytically obtained by a direct matrix calculation. The output weight vector β* can be estimated by using the minimal norm least square method:

$$\beta* = \mathrm{H}^{\dagger}\mathrm{T} \tag{12}$$

where $\mathrm{H}^{\dagger}$ represents the Moor-Penrose generalized inverse of H, and H is referred to as the hidden layer output matrix. The generalized inverse is calculated as $\mathrm{H}^{\dagger}$=HT(HHT)-1.

This chapter addresses the uncertainty modeling for both PV irradiance and load forecasting, employing a data-driven method that leverages historical data distribution. The forecasting models for PV irradiance and load consider specific inputs and outputs. For PV irradiance forecasting, the inputs encompass PV irradiance, temperature, and the hour of the day, with the output being the PV irradiance. Conversely, the load forecasting model considers load power, the hour of the day, and the day of the week as inputs, producing load power as the output.

To capture and represent uncertainties, forecast bins are introduced, offering an approximation of uncertainty levels associated with forecasting values [15]. The methodology begins by forming a data pool of point forecasting values using the forecasting model over an extensive period. Each point forecast corresponds to a set of actual measurement values, facilitating the creation of a comprehensive historical dataset.

Figure 4 illustrates a day-ahead forecast of PV irradiance along with actual values for a selected day. The forecast results are organized into equidistant bins, categorizing forecast values into distinct uncertainty levels. Taking the highlighted forecast bin ranging from 930 W/m^2 to 1033 W/m^2 as an example, the actual values corresponding to historical point forecasts within this range form a distribution. This distribution serves as an approximation of uncertainties for the [930-1033 W/m^2] forecast bin. Applying the same methodology, uncertainty distributions for other forecast bins are derived.

Figure 5 presents uncertainty distributions for all forecast bins, derived from data spanning from October 8, 2020, to February 4, 2020. These distributions effectively encapsulate the uncertainties associated with different forecasted ranges. The utilization of this data-driven approach provides a robust method for understanding and managing uncertainties in both PV irradiance and load forecasting, enhancing the reliability of the overall forecasting model. Subsequent sections will delve into the implications and applications of these uncertainty distributions in real-world scenarios.

The forecast bins in Fig. 5 are stored and used to generate different scenarios based on the day-ahead forecasts. After a point forecast value is generated from the day-ahead forecasting model, we will sample randomly from the corresponding forecast bin distributions to get various possible actual values. This process is repeated until the desired number of uncertainty scenarios are generated. Figure 6 shows three PV scenarios generated from the forecast bin approach. The method can generate any

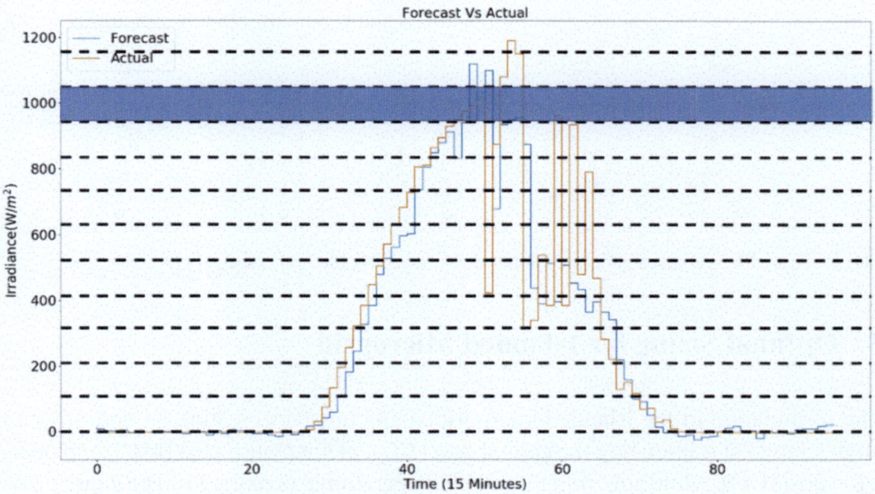

Fig. 4 Equidistant bins of PV irradiance forecast

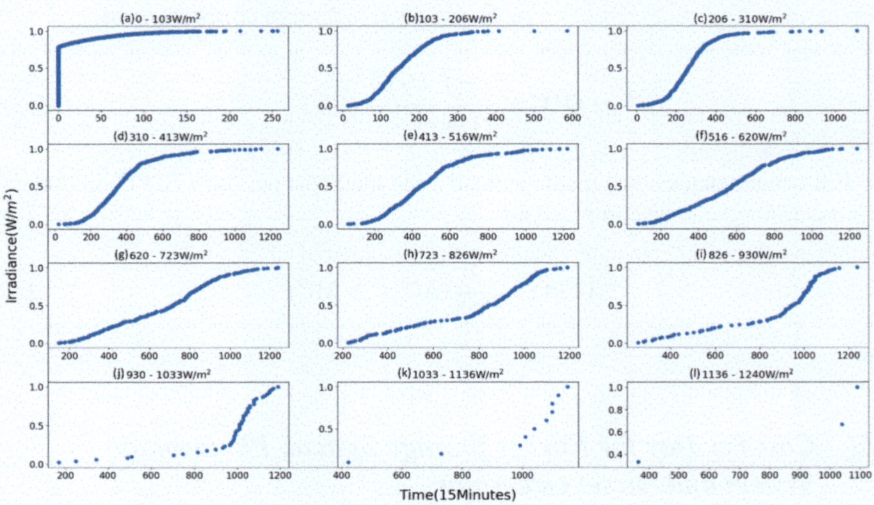

Fig. 5 Forecast bins for PV irradiance and their associated uncertainty distribution

number of scenarios based on a single day-ahead forecast. A similar approach is adopted to generate uncertainty scenarios for the load distributions. The combined uncertainty scenarios can then be generated by combining the two.

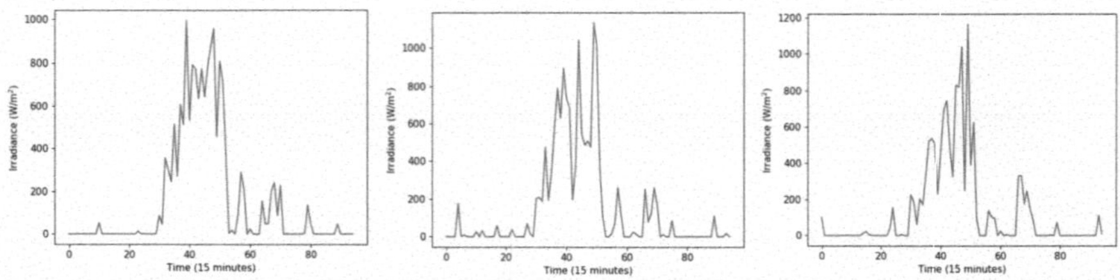

Fig. 6 Three PV scenarios generated using the forecasting bin approach

3 Optimal Sizing for Islanded Microgrid

The formulation of the islanded microgrid sizing involves creating an optimization model aimed at minimizing the capital cost (CC), maintenance cost (MC), and operation cost (OC), while adhering to specified operational constraints. The capital cost represents a singular investment made in the initial year, and assuming a fixed interest rate of r and a project lifespan of l years, the annualized one-time cost ($AOTC$) in dollars per year ($/year) can be computed using the formula outlined in reference [9].

$$\mathrm{OTC} = \frac{r(1+r)^l}{(1+r)^l - 1}(CC) \tag{13}$$

If the maintenance cost recurs annually, the total cost per day ($TCPD$) in dollars per day ($/day) can be computed as:

$$\mathrm{TCPD} = \frac{1}{365}(\mathrm{AOTC} + \mathrm{MC}) \tag{14}$$

3.1 Cost Per Day for Energy Storage System, Photovoltaic System and Diesel Generator

The ESS comprises two primary components: the battery array and the power converter system. The costs associated with batteries and their installation are proportionate to the system's size and are measured in dollars per kilowatt-hour ($/kWh). On the other hand, the cost of the power converter system (PCS), responsible for linking the ESS to the islanded microgrid, is expressed in dollars per kilowatt ($/kW). The cost of batteries is denoted as SC, and it is calculated as:

$$\mathrm{SC} = \mathrm{ESS_{bat}}\,\mathrm{E_{max}} \tag{15}$$

where ESS_{bat} is the cost of batteries in \$/kWh and E_{max} is the maximum capacity of the battery in kWh.

The cost of the power converter system denoted as PCS_{ess} is:

$$\mathrm{PCS_{ess}} = \mathrm{ESS_{pcs}}\overline{P_c} \tag{16}$$

where ESSpcs is the cost of the PCS in \$/kW and $\overline{P_c}$ is the maximum power output of the ESS.

The maintenance cost of the ESS denoted as MC_{ess} is considered in this project as shown in (17)

$$\mathrm{MC_{ess}} = \mathrm{MC}_{EC} \times E_{max} + \mathrm{MC_{pcs}} \times \overline{P_c} \tag{17}$$

where MC_{EC} and MC_{pcs} are the maintenance cost of the batteries and PCS in \$/kWh and \$/kW respectively.

$$\mathrm{AOTC_{ess}} = \frac{r(1+r)^l}{(1+r)^l - 1}(\mathrm{SC} + \mathrm{PCS_{ess}}) \tag{18}$$

The TCPD of the ESS in \$/day can be calculated as

$$\mathrm{TCPD_{ess}} = \frac{1}{365}(\mathrm{AOTC_{ess}} + \mathrm{MC_{ess}}) \tag{19}$$

The PV system comprises the PV array and the power converter system. The installation cost is directly proportional to the system's size and is measured in dollars per kilowatt-peak (\$/kWp). The PV array cost is calculated by:

$$\mathrm{PAC} = \mathrm{PV_{cost}}\,\mathrm{PV_{max}} \tag{20}$$

where PAC represents the PV array cost, PV_{max} denotes the maximum PV output power, and PV_{cost} signifies the cost in kWp (\$/kWp).

The cost of the Power Converter System for PV (PCS_PV) is computed as:

$$\mathrm{PCS_{PV}} = \mathrm{PV_{pcs}}\,\mathrm{PV_{max}} \tag{21}$$

where PCS_{PV} represents the cost of the power converter system, and PV_{pcs} signifies the cost of the PCS in dollars per kWp (\$/kWp).

The maintenance cost of the PV system (MC_{PV}) is given by:

$$\mathrm{MC_{PV}} = \mathrm{PV_{max}}\left(\mathrm{MC_{pac}} + \mathrm{MC_{pcs}}\right) \tag{22}$$

where MC_{pac} and MC_{pcs} represent the maintenance cost of the PV and PCS in dollars per kWp (\$/kWp).

The Annualized One-Time Cost (*AOTC*) of the PV in dollars per year ($/year) is calculated using the formula:

$$\mathrm{AOTC_{pv}} = \frac{r(1+r)^l}{(1+r)^l - 1}\left(\mathrm{PAC} + \mathrm{PCS_{pv}}\right) \tag{23}$$

where $\mathrm{AOTC_{pv}}$ represents the annualized one-time cost of the PV system, r denotes the fixed interest rate, and l signifies the number of years.

The Total Cost per Day (TCPD) of the PV system in dollars per day ($/day) is computed as:

$$\mathrm{TCPD_{pv}} = \frac{1}{365}(\mathrm{AOTC_{pv}} + \mathrm{MC_{pv}}) \tag{24}$$

where $\mathrm{TCPD_{pv}}$ represents the total cost per day of the PV system, incorporating the annualized one-time cost and the maintenance cost.

The capital cost of the diesel generator, denoted as DG, is determined by:

$$\mathrm{DG} = \mathrm{DG_P}\overline{P_{DG}} \tag{25}$$

where $\mathrm{DG_P}$ denotes the cost per unit of power output of the diesel generator in dollars per kilowatt ($/kW), and $\overline{P_{DG}}$ signifies the peak power output of the diesel generator in kilowatts (kW).

The maintenance cost of the diesel generator, denoted as $\mathrm{MC_{DG}}$, is computed as:

$$\mathrm{MC_{DG}} = \mathrm{UMC_{DG}} \times \overline{P_{DG}} \tag{26}$$

Here, $\mathrm{UMC_{DG}}$ signifies the unit maintenance cost of the diesel generator in dollars per kilowatt ($/kW).

The capital cost of the diesel generator is also annualized using the formula:

$$\mathrm{AOTC_{DG}} = \frac{r(1+r)^l}{(1+r)^l - 1}(\mathrm{DG}) \tag{27}$$

where $\mathrm{AOTC_{DG}}$ represents the annualized one-time cost of the diesel generator, considering the fixed interest rate (r) and the project lifespan (l).

The Total Cost per Day (TCPD) of the diesel generator is calculated as:

$$\mathrm{TCPD_{DG}} = \frac{1}{365}(\mathrm{AOTC_{DG}} + \mathrm{MC_{DG}}) \tag{28}$$

Here, $\mathrm{TCPD_{DG}}$ represents the total cost per day of the diesel generator, encompassing the annualized one-time cost and the maintenance cost.

The total TCPD of the ESS PV, and DG can be computed by adding the respective TCPD values:

$$CPD = TCPD_{ESS} + TCPD_{PV} + TCPD_{DG} \tag{29}$$

3.2 Operating Cost for Energy Storage System, Photovoltaic System and Diesel Generator

The operating cost of the diesel generator in dollars per kilowatt-hour ($/kWh) is determined by:

$$C_{DG} = \sum\nolimits_{t=1}^{T} P_{DG}(t) F_{DG} \tag{30}$$

where C_{DG} represents the total operating cost of the diesel generator, and F_{DG} signifies the total fuel cost associated with the diesel generator.

The operating cost of the ESS is computed as:

$$C_{ESS} = \sum\nolimits_{t=1}^{T} P_{ESS}(t) F_{ESS} \tag{31}$$

where F_{ESS} denotes the unit operating cost of the ESS.

In the event of power deficiency in an islanded microgrid, leading to load curtailment, the cost of power curtailment is calculated as:

$$C_{Curtail} = \sum\nolimits_{t=1}^{T} P_{curtail}(t) F_{curtail} \tag{32}$$

where $P_{curtail}$ represents the curtailed power, and $F_{curtail}$ signifies the cost coefficient of curtailment.

The overall total operating cost of the islanded microgrid is given by:

$$OC = C_{DG} + C_{ESS} + C_{Curtail} \tag{33}$$

where OC represents the sum of the operating costs of the diesel generator, ESS, and power curtailment cost, constituting the total operating cost of the islanded microgrid.

3.3 Optimal Sizing of Islanded Microgrid

The optimal sizing is formulated to minimize the total cost of the microgrid comprising of TCPD (Eq. (29)) and OC (Eq.(33)).

$$\min TCPD + OC \tag{34}$$

subject to the following constraints

Power balance:

$$P_{DG}(t) + P_{PV}(t) + P_D(t) = P_{Load}(t) + P_C(t) + P_{curtail}(t) \forall t \in T \tag{35}$$

Power limit of ESS:

$$0 \leq P_c(t) \leq \overline{P_c} \tag{36}$$

$$0 \leq P_d(t) \leq \overline{P}_d \tag{37}$$

Power limit of the diesel generator:

$$\underline{P_{DG}} \leq P_{DG}(t, s) \leq \overline{P_{DG}} \forall t \in T, \forall s \in S \tag{38}$$

Ramping rate limit of the diesel generator:

$$R_{DG}^{down} \Delta t \leq P_{DG}(t, s) - P_{DG}(t - 1, s) \leq R_{DG}^{up} \Delta t \forall t \in T, \forall s \in S \tag{39}$$

4 Two-Stage Coordinated Energy Management Under Operational Uncertainties

The two-stage coordinated energy management strategy, illustrated in Fig. 7, addresses uncertainties in supply and demand within an islanded microgrid.

In this strategy, the optimization process occurs in two stages:

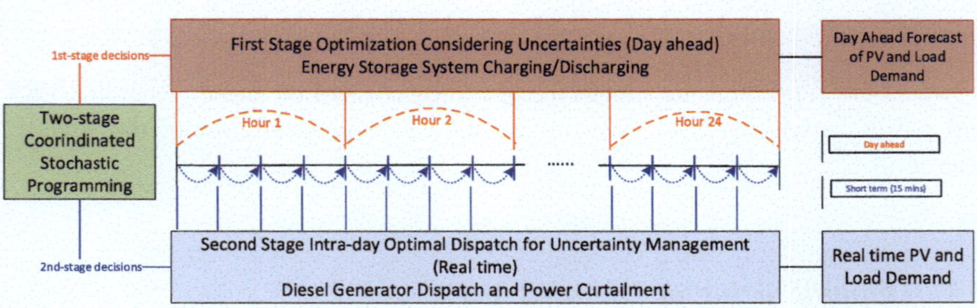

Fig. 7 Two-stage coordinated energy management strategy

- First Stage: Optimization of day-ahead ESS power outputs, which remain fixed for the subsequent day's operation (second stage). Meanwhile, real-time adjustments are made to DG and curtailment power to compensate for forecast inaccuracies.
- Second Stage: Re-dispatching of DG and power curtailment in real-time to accommodate the actual realized uncertainties. This stage involves decisions on DG dispatch and power curtailment, while the ESS power output remains consistent with the day-ahead optimization.

This approach is formulated as a two-stage stochastic programming model, represented as:

$$\min\{f(x) + E\big[Q(x, \zeta)\big]\} \tag{40}$$

Here, $f(x)$ denotes the first-stage problem (i.e., the operating cost of the ESS), x represents the first-stage decision variables related to day-ahead charging/discharging for the ESS, $E\big[Q(x, \zeta)\big]$ signifies the expected cost of the second-stage problem involving the operating cost of the DG and power curtailment, and $Q(x, \zeta)$ reflects the intra-day operation objective, considering scenarios generated from day-ahead forecasts. It is equal to $\min\limits_{y \in \Omega(x,\xi)} g(y)$, where y is the second stage decision variables, i.e. the DG output and power curtailment. The formulation encompasses uncertainties in electrical load demands and PV power outputs ξ, represented by a finite number of scenarios $\xi_1, \xi_2, \ldots \xi_n$ with respective probabilities $\omega_{1,}\omega_2, \ldots \omega_{n.}$

The first-stage optimization with uncertainties is expressed as:

$$\min \sum_{t=1}^{T} C_{ESS}(t) + \sum_{t=1}^{T} \sum_{s=1}^{S} \omega_s [C_{DG}(t, s) + C_{curtail}(t, s)] \tag{41}$$

s.t. Eqs. (35–39), detailing the constraints on the system.

During the second-stage real-time operation, the ESS adheres to first-stage decisions, while the DG output and power curtailment are adjusted in response to realized uncertainties in the system. This approach ensures a coordinated response to fluctuations in supply and demand within the islanded microgrid.

Table 1 Performance of day-ahead 15-min forecasting

	NRMSE (%)	
	Train	Test
PV	12.0	15.3
Load	8.85	15.3

Table 2 Performance of day-ahead hourly forecasting

	NRMSE (%)	
	Train	Test
PV	12.9	15.5
Load	13.3	19.0

5 Simulation Results

5.1 *Forecasting Results*

Assessing the precision of our forecasts, we rely on the Normalized Root Mean Square Error (NRMSE) as a key measure. This metric evaluates the accuracy of our predictions by calculating the discrepancy between the forecasted and actual values. The NRMSE is determined using the following formulas:

$$\mathrm{NRMSE} = \frac{RMSE}{\max(Y) - \min(\mathrm{Y})} \times 100\% \tag{43}$$

$$\mathrm{RMSE} = \sqrt{\frac{\sum_{t=0}^{t=T} (\widehat{Y}(\mathrm{t}) - \mathrm{Y(t)})^2}{T}} \times 100\% \tag{44}$$

For our analysis, we utilize data collected from our PV sensors and building load systems, spanning specific periods: PV irradiance data from October 8, 2019, to February 4, 2020, and load data from November 27, 2019, to February 4, 2020. To ensure robustness, the dataset is split into training and testing sets, maintaining a 7:3 ratio. The forecasting accuracy is depicted in Tables 1 and 2, showcasing the performance of our ELM models for day-ahead predictions at both 15-minute and hourly intervals.

Illustrations of the forecasting results are shown in Fig. 8 and 9.

5.2 *Optimal Sizing Results*

In the optimization process for sizing the islanded microgrid, we utilize typical load and PV irradiance profiles extracted from historical data, as depicted in Fig. 10.

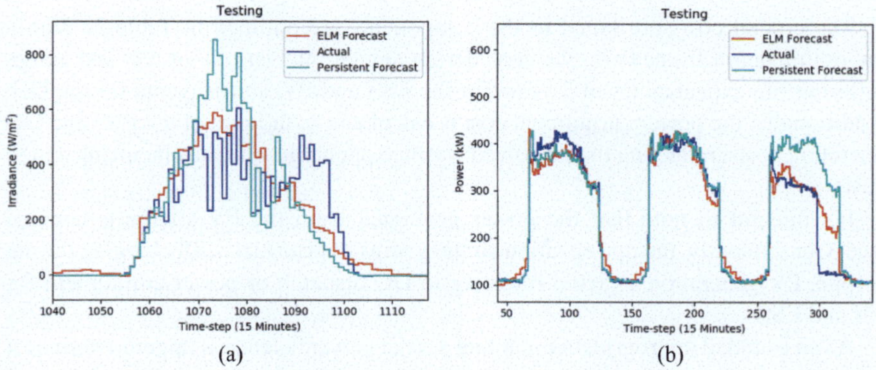

Fig. 8 Day-ahead 15 min forecast of PV (**a**) and load (**b**)

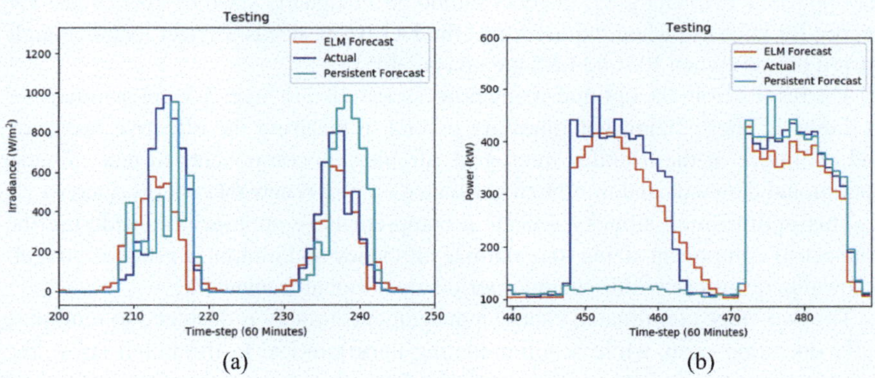

Fig. 9 Day-ahead hourly forecast of PV (**a**) and load (**b**)

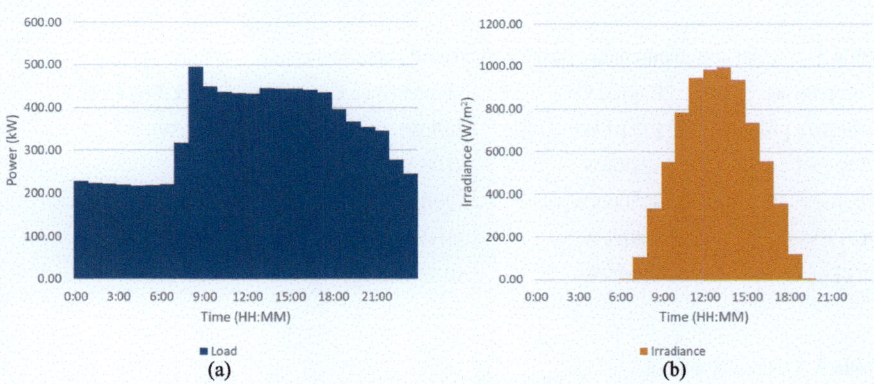

Fig. 10 A typical building load (**a**) and PV irradiance (**b**)

The crucial costs attributed to the components are outlined in Table 3. Within the optimization framework, we consider an annual interest rate of 6% and assign differing life expectancies of 3 years for the ESS and PV, and 10 years for the DG. Additionally, the power curtailment cost is calculated as the sum of the ESS and DG operating costs, reflecting the perceived wastage of energy, and thus incurring a high cost.

It's integral to note that the power generated from the PV source is deemed free. Consequently, the optimization strategy aims to prioritize ESS charging during surplus PV generation to avoid resorting to DG dispatch or power curtailment for PV and ESS.

As an islanded microgrid, maintaining a strict power balance is imperative since it operates independently of the main grid. The sizing of each component holds pivotal importance: the PV system must possess ample capacity to generate surplus power for opportune ESS charging; the ESS should be adequately sized to store excess PV energy for later use when required; and the DG sizing needs to meet load demands during periods when PV and ESS are unavailable.

The results from the optimal sizing process, detailed in Sect. 3.3, are summarized in Table 4. These sizing outcomes are pivotal in ensuring the effective operation and resilience of the islanded microgrid, aligning its components to meet diverse operational demands and minimizing reliance on non-renewable energy sources.

This optimization strategy reflects a comprehensive approach, considering the nuances of component sizing to maximize efficiency and minimize reliance on non-renewable resources within the microgrid's operational framework.

The two-stage coordinated energy management strategy is designed to minimize daily operating costs while accommodating uncertainties. In the initial stage, the optimization of ESS outputs aligns with the day-ahead forecast, considering uncertainty scenarios related to both PV and load. Decisions regarding DG power output

Table 3 Capital and maintenance cost for microgrid components

Components	Capital Cost	Maintenance Cost	Operating Cost
ESS Size [16]	$271/kWh	3¢/kWh	3¢/kWh
PCS ESS	$288/kW	$10/kW	
PV Array [17]	$2500/kWp	$20/kW [18]	$0
PCS PV	$288/kW	$10/kW	
DG Power	$100/kW	$0.75/kW [19]	$0.12299

Table 4 Optimal sizing of islanded microgrid components

Components	Size
ESS PCS, $\overline{P_c}$	391 kW
ESS Size, $Emax$	2284 kWh
PV PCS/Size, PV_{max}	795 kWp
Diesel Generator, $\overline{P_{DG}}$	1056 kW

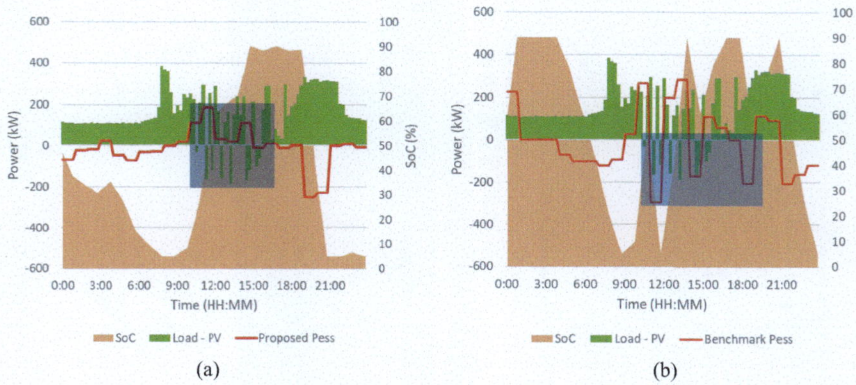

Fig. 11 First stage day ahead ESS Schedule using (**a**) this method and (**b**) the benchmark deterministic method for case study on 05 Feb 2020

and PV power curtailment are omitted. Subsequently, in the second stage, DG power output and PV power curtailment are readjusted to counterbalance forecast errors and real-time uncertainties.

To illustrate the advantages of our approach, we compare it with a benchmark method using a two-stage deterministic energy management strategy. In the benchmark method's initial stage, the ESS schedule is based solely on the day-ahead forecast, aiming to minimize daily operating costs without factoring in uncertainties. Subsequently, in the second stage, DG and power curtailment are recalibrated when uncertainties materialize.

Figure 11 provides a comparative view of the day-ahead ESS schedules between this method (Fig. 11a) and the benchmark deterministic method (Fig. 11b) for the case study on February 5, 2020.

In the benchmark method (Fig. 11b), the ESS discharges during a period of excess PV generation, displaying a failure to fully capitalize on the surplus PV energy, potentially leading to unnecessary power curtailment. This outcome is attributed to the absence of considerations for forecasting uncertainties. Conversely, this method (Fig. 11a) optimally charges the ESS during excessive PV generation periods and discharges it when needed to meet the load demand, showcasing its ability to make informed decisions considering forecasting uncertainties.

Figures 12(a) and (b) depict the DG power output and power curtailment under ten selected uncertainty scenarios out of a total of 100. These decisions are preliminary and will be discarded, as the second-stage online optimization will readjust the DG output and power curtailment based on realized uncertainties.

Moving to Fig. 13, it showcases a comparison of the second-stage real-time dispatch for DG and power curtailment between this method and other benchmark methods.

Moreover, Fig. 14 presents detailed insights into the accumulated energy consumption, standard deviation of ESS operating profiles, and cost comparisons for the case study on February 5, 2020. These comparative analyses vividly illustrate

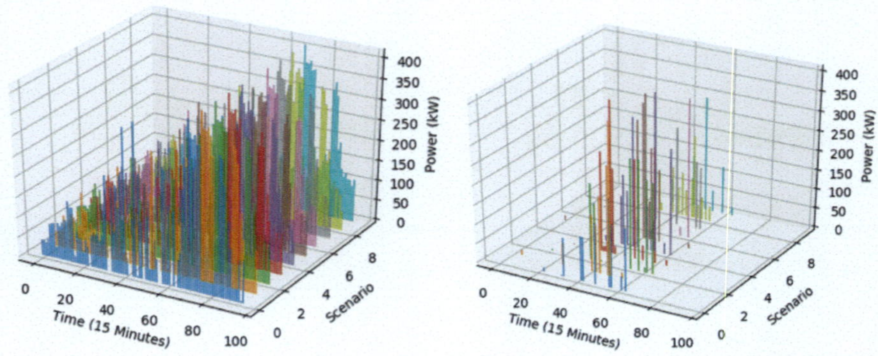

(a) (b)

Fig. 12 (**a**) DG power output and (**b**) Power curtailment under uncertainty scenarios in the first stage for case study on 05 Feb 2020 (selected 10 scenarios)

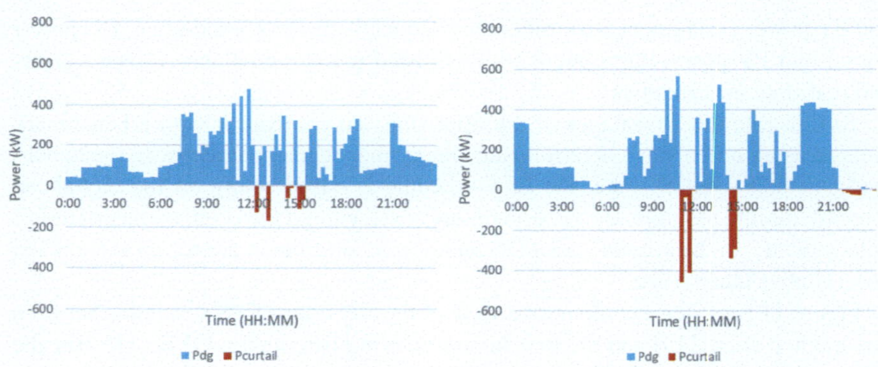

(a) (b)

Fig. 13 Second stage real-time DG re-dispatch and power curtailment for (**a**) this method and (**b**) the benchmark method in case study on 05 Feb 2020

the effectiveness of the two-stage coordinated energy management strategy in minimizing power curtailment and optimizing the utilization of excess PV generation, thus significantly reducing operating costs. The strategy's incorporation of forecasting uncertainties demonstrates its potential to enhance the microgrid's efficiency and adaptability in real-time operational scenarios.

Figure 14 presents a comparative analysis between the benchmark and the suggested methods. In Fig. 14a, it's evident that this method consumes less energy from both ESS and DG and experiences reduced power curtailment. This reduction is attributed to the method's consideration of uncertainties. Additionally, Fig. 14b demonstrates that this method generates smoother ESS and DG operating profiles, potentially contributing to a longer ESS lifespan. Finally, Fig. 14c reveals that the

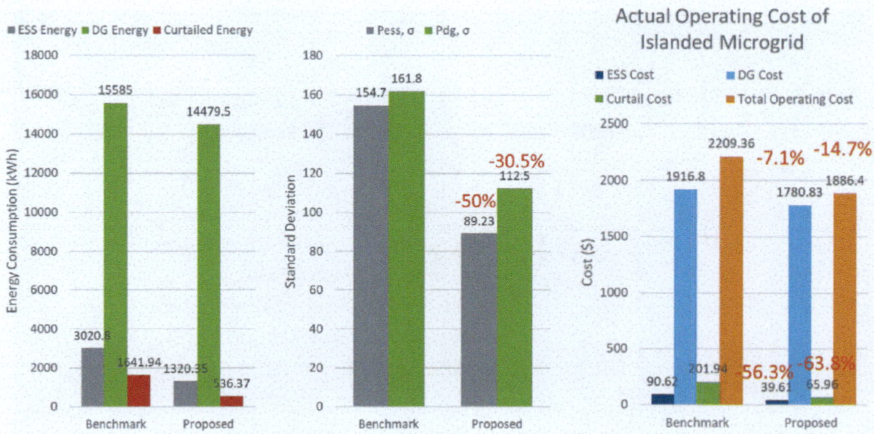

(a) (b) (c)

Fig. 14 (**a**) Accumulated energy consumption for ESS output, DG output and power curtailment (**b**) Standard deviation of ESS operating profile (**c**) Cost comparison for case study on 05 Feb 2020

total operating cost using this method is 14.7% lower than that of the benchmark method.

For a subsequent case study using data from February 7, 2020, comparative results are depicted in Fig. 15. Similarly, this method demonstrates smoother fluctuations in ESS and DG operating profiles, as seen in Fig. 15b. Overall, an 18% cost reduction is achieved using this method.

Moving to Fig. 16, it illustrates a dashboard showcasing control decisions generated by the two-stage coordinated energy management strategies. The real-time data

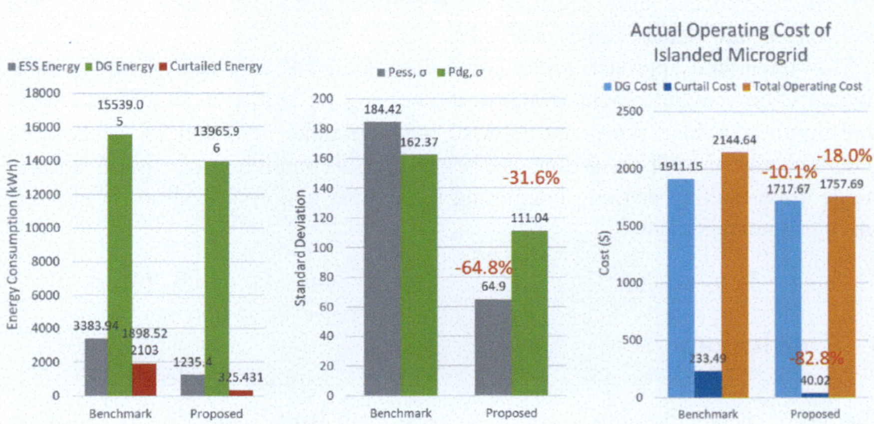

(a) (b) (c)

Fig. 15 (**a**) Accumulated energy consumption for ESS ouptut, DG output and power curtailment (**b**) Standard deviation of ESS operating profile (**c**) Cost comparison for case study on 07 Feb 2020

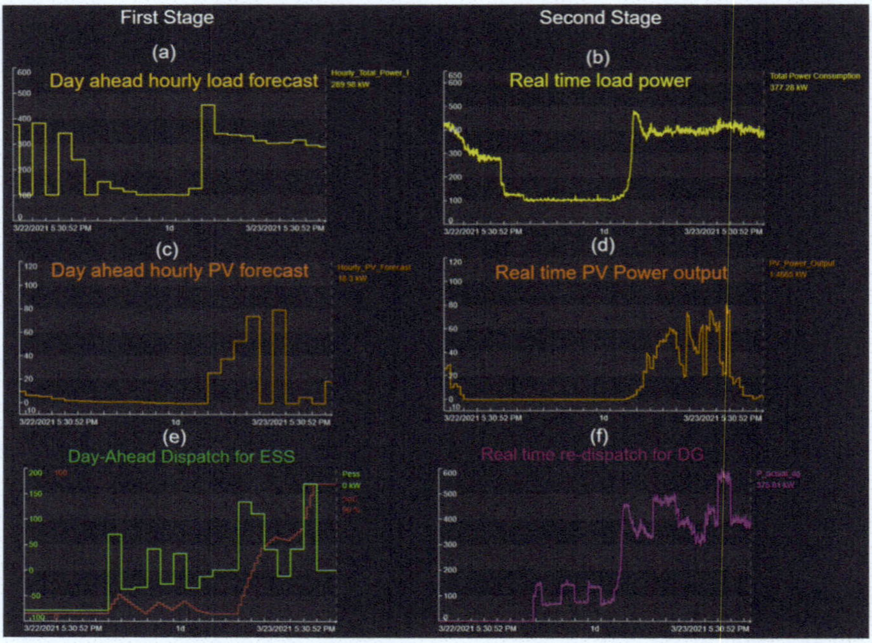

Fig. 16 Dashboard that demonstrates control decisions for the two stage coordinated energy management strategies

from the energy monitoring system, as depicted in Fig. 2, feeds into the forecasting, scenario generation, and optimization engine. The resulting decisions are displayed on this dashboard. Data extraction, forecasting, scenario generation, and optimization processes occur daily, updating the ESS's day-ahead schedules and real-time decisions as new data becomes available.

This integrated approach underscores the effectiveness of the two-stage coordinated energy management strategy, showcasing its ability to minimize energy consumption, reduce power curtailment, and optimize ESS and DG operations under uncertainties. The real-time dashboard offers a comprehensive visualization of the decision-making process, integrating forecasting and optimization for efficient microgrid management.

6 Conclusion

This book chapter presents a comprehensive framework addressing energy forecasting, uncertainty analysis, optimal system sizing, and coordinated energy management strategies tailored for an urban-based islanded microgrid. It commences with an online energy monitoring system, capturing diverse data sources crucial for analysis and forecasting. Employing a data-driven methodology, this approach models and generates various scenarios to account for forecasting uncertainties.

Leveraging this collected field data, the chapter navigates through optimal sizing procedures, ensuring precise microgrid dimensions to minimize investment and operational costs while upholding system stability. Introducing a two-stage coordinated energy management method, specifically designed to mitigate uncertainties, the chapter focuses on minimizing operational costs. Comparative analysis against a benchmark deterministic approach demonstrates the efficacy of this method in achieving reduced operational costs and enhancing the longevity of the ESS.

Additionally, a dedicated software platform has been developed, serving as a hub to consolidate collected data, exhibit forecasting outcomes, and visually represent operational suggestions. This platform acts as a pivotal tool for stakeholders, offering insights into the microgrid's functioning and optimization potential.

References

1. Lancel G, et al. (2017) Energy storage systems (ESS) and microgrids in Brittany islands. CIRED-Open Access Proc J 10(1):1741–1744
2. Li P, Li R, Cao Y, Li D, Xie G (2018) Multiobjective sizing optimization for Island microgrids using a triangular aggregation model and the levy-harmony algorithm. IEEE Trans Industr Inf 14(8):3495–3505
3. Kong X, Liu X, Ma L, Lee KY (2019) Hierarchical distributed model predictive control of standalone wind/solar/battery power system. IEEE Trans Syst, Man, Cybern: Syst 49(8):1570–1581
4. Moradi MH, Eskandari M, Mahdi Hosseinian S (2015) Operational strategy optimization in an optimal sized smart microgrid. IEEE Trans Smart Grid 6(3):1087–1095
5. Li Z, Xu Y (2018) Optimal coordinated energy dispatch of a multi-energy microgrid in grid-connected and islanded modes. Appl Energy 210:974–986
6. Chen Y, Xu Y, Li Z, Feng X (2019) Optimally coordinated dispatch of combined-heat-and-electrical network with demand response. IET Gener Transm Distrib 13(11):2216–2225
7. Li Z, Xu Y, Feng X, Wu Q Optimal stochastic deployment of heterogeneous energy storage in a residential multi-energy microgrid with demand-side management. IEEE Trans Ind Inform
8. Li Z, Xu Y, Fang S, Mazzoni S (2019) Optimal placement of heterogeneous distributed generators in a grid-connected multi-energy microgrid under uncertainties. IET Renew Power Gener 13(14):2623–2633
9. Chen SX, Gooi HB, Wang MQ (2012) Sizing of energy storage for microgrids. IEEE Trans Smart Grid 3(1):142–151
10. Ashfaq S, Zhang D, Zhang C, Dong ZY (2020) Regionalisation of islanded microgrid considering planning and operation stages. IET Renew Power Gener 14(1):145–153
11. Zhang C, Xu Y, Dong ZY (2020) Robustly coordinated operation of a multi-energy microgrid in grid-connected and Islanded modes under uncertainties. IEEE Trans Sustain Energy 11(2):640–651
12. Zhang C, Xu Y, Li Z, Dong ZY (2019) Robustly coordinated operation of a multi-energy microgrid with flexible electric and thermal loads. IEEE Trans Smart Grid 10(3):2765–2775
13. Zhang C, Xu Y, Dong ZY, Ma J (2017) Robust operation of microgrids via two-stage coordinated energy storage and direct load control. IEEE Trans Power Syst 32(4):2858–2868
14. Huang G-B, Zhu Q-Y et al (2006) Extreme learning machine: theory and applications. Neurocomputing 70(1):489–501

15. Ma XY, Sun YZ, Fang HL (2013) Scenario generation of wind power based on statistical uncertainty and variability. IEEE Trans Sustain Energy 4(4):894–904
16. Mongird K, Fotedar V, Viswanathan V, Koritarov V, Balducci P, Hadjerioua B, Alam J (2019) Energy Storage Technology and Cost Characterization Report. U.S. Department of Energy
17. Ran Fu DFRM (2018) U.S. Solar Photovoltaic System Cost Benchmark: Q1 2018. NREL
18. Enbar N, Weng D, Klise G (2015) Budgeting for solar PV plant operations & maintenance: Practices and pricing. Electric Power Research Institute, U.S
19. Logenthiran T, Srinivasan D, Khambadkone AM, Raj TS (2010) Optimal sizing of an islanded microgrid using evolutionary strategy. In: 2010 IEEE 11th International Conference on Probabilistic Methods Applied to Power Systems, Singapore, Singapore

Probabilistic Microgrid Investment Planning with Integrated Game-Theoretic Demand Response Management

Soheil Mohseni and Alan C. Brent

Abstract This Chapter presents an innovative modelling framework for advancing integrated metaheuristic-based energy planning optimization, with a focus on resilient, renewable community microgrids. The study brings attention to a critical aspect of long-term sustainable energy system planning by illuminating biases intrinsic to consumer preference-based demand response projections. Central to the framework is the emphasis on market-driven sectoral flexibility procurement, achieved through sealed-bid auctions, thereby optimizing social welfare, improving demand response capacity liquidity, and promoting sectoral stability. To this end, the integration of game theory is highlighted as a means to leverage demand-side flexibility for community-level clean energy projects. Further, navigating the uncertainties stemming from non-dispatchable sources and smart grid interventions, the Chapter advocates for a comprehensive evaluation involving holistic uncertainty-aware models for renewable system planning. In this context, a novel probabilistic investment planning framework is introduced, which incorporates climatological, demand, and price uncertainties within a stochastic microgrid sizing model integrated with demand response solutions. The significance of this framework lies in its capacity to address gaps in uncertainty-aware methodologies and the integration of operational and investment planning. By synergizing intricate modelling with practical implementation considerations, the Chapter establishes a guiding framework that not only enriches scholarly deduction, but also aids practical, sustainable energy decision-making. The efficacy of the approach is substantiated through numerical simulations of a case study in Aotearoa New Zealand, to validate the applicability of the results.

S. Mohseni (✉) · A. C. Brent
Wellington Faculty of Engineering, Sustainable Energy Systems, Victoria University of Wellington, Wellington 6140, New Zealand
e-mail: Soheil.Mohseni@vuw.ac.nz

S. Mohseni
Institute for Sustainable Futures, University of Technology Sydney, Sydney, NSW 2007, Australia

A. C. Brent
Department of Industrial Engineering and the Centre for Renewable and Sustainable Energy Studies, Stellenbosch University, Stellenbosch 7600, South Africa

F. Shahnia and J. M. Guerrero (eds.), *Microgrids and Virtual Power Plants*, Power Systems, https://doi.org/10.1007/978-981-97-6623-9_2

Keywords Distributed energy resources · Renewable energy sources · Metaheuristics · Optimization · Demand response · Uncertainty

1 Introduction

1.1 Motivation

As the world transitions towards sustainable and renewable energy systems, microgrids have emerged as a vital component of this transformation. Microgrids offer the promise of localized, reliable, and resilient energy supply, often powered by renewable sources and incorporating demand-side management [1, 2]. However, the planning and operation of these complex systems are associated with uncertainties arising from various sources, such as weather conditions, load demand fluctuations, and electricity price volatility. Addressing these uncertainties is paramount for realizing the full potential of microgrids and achieving their sustainable integration into the broader energy landscape [3–5].

While microgrids hold immense promise, realizing their benefits requires a thorough understanding of how uncertainties impact their design and operation. Neglecting these uncertainties can lead to sub-optimal decisions, misalignment between simulated and real-world outcomes, and potential financial risks. Therefore, a robust framework that can comprehensively consider multi-dimensional uncertainties while optimizing both the capacity planning and real-time operation of microgrids is crucial [6–8].

1.2 Literature Review: Knowledge Gaps

In recent years, considerable research efforts have been directed towards developing 'uncertainty-aware' microgrid investment planning modelling frameworks [9–13]. These approaches aim to incorporate uncertainty quantification into the decision-making process for designing and operating microgrids. Table 1 provides an overview of the prevailing methodologies within the realm of microgrid capacity planning optimization that have demonstrated robustness and rigor in their approaches. This compilation synthesizes the landscape of existing research efforts, presenting a chronological sequence of studies that have delved into the intricate task of addressing uncertainties in microgrid planning. The table serves as a comprehensive repository of these methodologies, positioning them within the trajectory of microgrid research and highlighting their respective strengths, limitations, and contributions.

The selected studies encompass various dimensions of uncertainty, showcasing the diversity of approaches taken to model and mitigate these complex factors. From the table, a continuum of research efforts aimed at refining methodologies, expanding

Table 1 Overview of key prior research on stochastic investment planning for renewable and sustainable energy systems

Reference	System configuration	Uncertainty characterization technique	Random parameter(s)	Dispatch management (with arbitrage)	Optimization algorithm
[17]	A microgrid connected to the grid, integrating solar photovoltaic, fuel cell, battery energy storage system, and super-capacitor	Employing MATLAB's white noise block for quantifying uncertainties	Wind speed, solar irradiance, and load demand forecasts	× (×)	Particle swarm optimization
[18]	A microgrid connected to the grid, comprising photovoltaic, wind turbines, and battery energy storage system	Monte Carlo simulation	Wind speed and solar irradiance forecasts	× (×)	Mixed-integer linear programming
[19]	An energy hub incorporating wind turbines, combined heat and power units, boilers, battery energy storage system, and thermal energy storage system	Monte Carlo simulation	Wind speed, load demand, and electricity price forecasts	✓(×)	Mixed-integer nonlinear programming
[20]	A distribution grid with significant distributed energy resources	Monte Carlo simulation	Demand response events	× (×)	Simple matching of supply and demand
[21]	A grid incorporating combined heat and power units, distributed generators, and boilers	Scenario analysis	Load demand forecasts	✓ (×)	Particle swarm optimization
[22]	A network integrating coal, gas, wind, solar, and biopower plants	Monte Carlo simulation	Load demand forecasts, techno-economic variables, electric vehicle charging	× (×)	Robust flexible programming
[23]	A renewable distribution network	Monte Carlo simulation	Wind speed and load demand forecasts	× (×)	Mixed-integer programming

(continued)

Table 1 (continued)

Reference	System configuration	Uncertainty characterization technique	Random parameter(s)	Dispatch management (with arbitrage)	Optimization algorithm
[24]	A standard IEEE 33-bus distribution network	Monte Carlo simulation	Wind speed and load demand forecasts	× (×)	Mixed-integer nonlinear programming
[25]	A photovoltaic/wind turbine/battery energy storage system microgrid	Monte Carlo simulation	Wind speed, solar irradiance, and load demand forecasts	× (×)	Mixed-integer linear programming
[26]	A refined IEEE 33-bus distribution network	Monte Carlo simulation	Wind speed and load demand forecasts	✓ (×)	Metamodel-based global optimization
[27]	A microgrid integrating wind turbines, photovoltaic panels, a biopower plant, a battery energy storage system, a boiler, and a thermal energy storage system	Monte Carlo simulation	Wind speed and solar irradiance forecasts, techno-economic variables	✓ (×)	Linear programming
[28]	A microgrid integrating photovoltaic panels, a biopower plant, and a boiler	Monte Carlo simulation	Wind speed and solar irradiance forecasts, techno-economic variables	✓ (×)	Linear programming
[29]	A stand-alone solar photovoltaic/battery energy storage system/diesel microgrid	Chance-constrained programming	Wind speed, solar irradiance, and load demand forecasts	✓ (×)	Mixed-integer nonlinear programming
[30]	An energy hub integrating combined heat and power units, solar photovoltaic panels, battery energy storage system, boilers, and chillers	Monte Carlo simulation	Solar irradiance and load demand forecasts	✓ (×)	Particle swarm optimization
[31]	A microgrid integrating wind, solar photovoltaic, and battery	Artificial neural network	Wind speed, solar irradiance, and load demand forecasts	× (×)	Particle swarm optimization

(continued)

Table 1 (continued)

Reference	System configuration	Uncertainty characterization technique	Random parameter(s)	Dispatch management (with arbitrage)	Optimization algorithm
[32]	A microgrid integrating solar photovoltaic, thermal storage, combined heat and power, and battery	Monte Carlo simulation	Wind speed, solar irradiance, and load demand forecasts	✓ (×)	Genetic algorithm
[33]	A microgrid integrating solar photovoltaic and battery	Generative adversarial networks	Solar irradiance and load demand forecasts	✓ (×)	Stochastic optimization
[34]	A renewable, battery-backed cloud energy storage system	Monte Carlo simulation	Solar irradiance and load demand forecasts	✓ (×)	Artificial ecosystem optimization
This study	A grid-connected microgrid integrating solar photovoltaic, wind, micro-hydro, biopower plant, fuel cell, battery and super-capacitors	Monte Carlo simulation	Wind speed, solar irradiance, load demand, electricity price, ambient temperature, and river streamflow forecasts	✓(✓)	Moth-flame optimization algorithm

the scope of uncertainties considered, and integrating dynamic dispatch strategies can be observed. A notable feature of this compilation is the identification of whether each approach incorporates an optimization-based operational scheduling model. This delineation provides insights into the extent of integration between the strategic capacity planning and the operational dispatch decisions. Moreover, the table reflects the presence or absence of energy arbitrage considerations within these methodologies, shedding light on the varying degrees of attention given to the economic implications of optimal storage utilization.

By juxtaposing these methodologies and identifying their distinctive characteristics, Table 1 facilitates the identification of gaps and opportunities in the existing literature. The selected studies collectively represent a composite of innovative strategies, yet they also reveal areas that warrant further exploration and refinement. For instance, the scarcity of methodologies that systematically account for the intricate interplay of uncertainty sources or that comprehensively integrate game-theoretic demand response (DR) strategies underscores the potential for future advancements in these areas [14–16].

More specifically, despite these advancements, there are several critical knowledge gaps in the existing literature, namely:

- *Incomplete handling of multi-dimensional uncertainty*: While the literature has made progress in addressing uncertainty, many studies still limit their scope to a few uncertain parameters. The inherent complexity of microgrid systems, coupled with the multitude of potential uncertainty sources, often results in an incomplete understanding of the broader uncertainties influencing microgrid performance. Therefore, there is a need for an approach that can simultaneously handle a wide range of uncertain inputs, providing a more comprehensive and realistic assessment of microgrid feasibility, reliability, and cost-effectiveness.
- *Lack of systematic long-term energy planning under uncertainty*: Despite the recognition of uncertainties, the majority of existing studies focus on the deterministic aspect of microgrid planning and sizing, neglecting the impact of uncertainties on long-term energy planning optimization decisions. Few studies systematically analyze how varying levels of uncertainty affect key design metrics such as reliability and total discounted cost. This omission leads to sub-optimal decisions and undermines the feasibility of implementing microgrids in dynamic real-world environments.
- *Limited integration of optimal dispatch strategies*: While some studies address uncertainties in microgrid planning, there is a dearth of research that effectively integrates optimal dispatch strategies into the decision-making process. The potential for energy storage arbitrage, which can significantly impact microgrid profitability, has been largely neglected. Consequently, an approach that seamlessly integrates uncertainty-aware design with optimal dispatch scheduling is lacking in the literature.
- *Underutilization of advanced metaheuristics*: While many studies have employed metaheuristics in microgrid planning, there is an opportunity to leverage advanced metaheuristic optimization techniques to improve solution quality and efficiency.

Many existing approaches rely on simplistic metaheuristic optimization methods, potentially overlooking opportunities for achieving more optimal system designs and schedules.

- *Lack of comprehensive game theory integration in DR solutions*: Despite the recognized importance of DR solutions in microgrid operation, the majority of studies do not fully harness the power of game theory to model the interaction between consumers, prosumers, aggregators, and utilities. Incorporating game-theoretic strategies could lead to more accurate representation of real-world behavior and, therefore, improved DR outcomes.

1.3 Novel Contributions

This section outlines the substantial contributions of this research, addressing critical knowledge gaps that currently impede the advancement of uncertainty-aware microgrid capacity planning optimization. By introducing a holistic framework that transcends existing limitations, this Chapter endeavors to reshape the landscape of microgrid planning and operation in the face of intricate uncertainties.

Accordingly, in response to the knowledge gaps identified, the Chapter makes the following novel contributions:

- *Comprehensive handling of multi-dimensional uncertainty*: Addressing the limitation of incomplete uncertainty consideration, this Chapter introduces an innovative approach capable of accommodating a wide spectrum of uncertain parameters. By simultaneously addressing multiple sources of uncertainty, including meteorological data, power load demand, wholesale electricity prices, solar irradiance, ambient temperature, wind speed, and river streamflow, this research advances the understanding of the intricate uncertainties that influence microgrid performance.
- *Systematic long-term energy planning under uncertainty*: In response to the gap in systematic long-term energy planning, this Chapter presents a methodological advancement that quantifies the impact of uncertainties on critical microgrid design metrics. By comprehensively analyzing the effects of varying uncertainty levels on the total discounted cost, the Chapter bridges the gap between deterministic planning and real-world dynamic conditions, contributing to more robust and informed microgrid deployment decisions.
- *Seamless integration of optimal dispatch strategies*: Building on the overlooked potential of energy storage arbitrage, this Chapter seamlessly integrates uncertainty-aware design with optimal dispatch scheduling. By considering daily arbitrage opportunities through storage utilization, the research introduces a holistic framework that not only maximizes microgrid profitability, but also mitigates the impact of uncertainties on operational decisions, addressing the gap in integrated design and dispatch solutions.
- *Leveraging advanced metaheuristics*: Recognizing the untapped potential of advanced metaheuristic optimization techniques, this Chapter enriches the microgrid planning landscape. By harnessing the power of these techniques, the

research improves solution quality and computational efficiency. The integration of advanced metaheuristics addresses the gap in the utilization of state-of-the-art optimization methods, contributing to optimal system designs.

- *Comprehensive game theory integration in DR solutions*: In response to the lack of comprehensive game theory integration in DR solutions, this Chapter pioneers the incorporation of game-theoretic strategies. By modelling the intricate interactions between consumers, prosumers, aggregators, and utilities using game theory principles, the research offers a more accurate representation of real-world behavior. This integration elevates DR outcomes and fills the gap in the holistic modelling of microgrid operation.

2 Test-Case System

This section introduces a grid-connected, DC-coupled, multiple energy carrier microgrid test-case system, as illustrated by the configuration and power flow diagram depicted in Fig. 1.

The notional microgrid is designed to provide green power and transportation fuel to semi-urban communities, normally accommodating around 1,000 people. The electrical consumption patterns of these communities experience significant seasonality, often influenced by factors like tourist flows. The energy demands are categorized into five types, namely: (1) residential, (2) agricultural, (3) commercial, and (4) industrial power loads, along with (5) the demand for hydrogen serving fuel cell electric vehicles (FCEVs). This test-case microgrid system serves as a validation platform for the DR-integrated energy planning optimization model. To synthesize the electrical load profile, local Aotearoa New Zealand representative load profiles [35] were scaled in accordance with real-world consumption data [36].

Key features of the microgrid include: (i) leveraging a diverse portfolio of variable renewable energy sources with complementary characteristics alongside sustainable agricultural and woody biomass resources, (ii) utilizing a sophisticated three-timescale hybrid energy storage system, and (iii) harnessing vehicle-to-grid technologies and FCEV charging/discharging coordination through demand-side management (DSM) mechanisms. The system primarily relies on onsite renewable resources, such as solar photovoltaic (PV), wind turbines, micro-hydro turbines, and biomass power. It incorporates three complementary energy storage mediums—super-capacitors, battery packs, and a hydrogen-based energy storage system. Additionally, the main grid is considered as an auxiliary storage system with the potential for energy exports. For detailed mathematical modelling of the system components, refer to [37–41].

The leading equipment brands from the Australian and New Zealand renewable energy asset markets were selected based on viability and prevalence (see Table 2). The costs and efficiencies of power inverters/converters within dashed boxes in Fig. 1, as well as those of unshown intermediate-pressure hydrogen compressors,

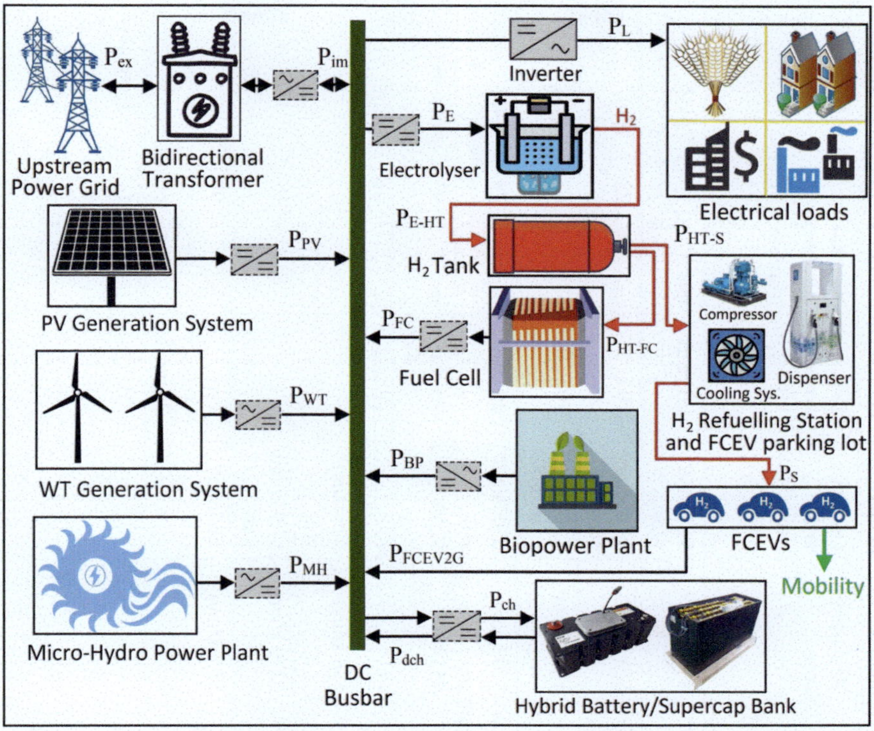

Fig. 1 Schematic architecture and energy flow pathways of the DC-linked, grid-connected community microgrid system, catering to residential, commercial, industrial, agricultural, and FCEV-refueling loads

are included in the corresponding main components. Also note that throughout this Chapter, all costs are cited in the Aotearoa New Zealand currency.[1]

3 Methodology

The methodology employed in this study is designed to comprehensively address the identified knowledge gaps and provide a robust approach for uncertainty-aware microgrid capacity planning optimization. The methodology consists of several key steps, each aimed at improving the accuracy, efficiency, and practicality of the optimization process while accounting for multiple dimensions of uncertainty.

[1] Mean NZD to USD exchange rate in 2022 [42]: 1 NZD = 0.64 USD.

Table 2 Data values and sources for techno-economic specifications of the components of test-case microgrid

Component	Manufacturer part number	Nameplate rating	Capital cost		Replacement cost	Operation and maintenance cost	Expected service life	Nominal efficiency		Source
			Per unit	Per standard unit				Notation	Value [%]	
PV module	CS6K-280P	280 W	$437/unit	$1.6 k/kW	$350/unit	$1.9/unit/year	20 years	η_{PV}	17.11	[43]
Biopower unit	PP30 Cogen-CS	25 kW	$32 k/unit	$1.28 k/kW	$23 k/unit	$0.01/unit/hour	10 k hours	η_{BP}	23	[44]
Wind turbine	ECO 48/750	750 kW	$1.096 m/unit	$1.46 k/kW	$822 k/unit	$21 k/unit/year	20 years	N/A	N/A	[45]
Micro-hydro turbine	XJ50-100SCTF6-Z	100 kW	$56 k/unit	$560/kW	$50 k/unit	$500/unit/year	20 years	η_{MH}	78	[46]
Transformer	Generic	N/A	N/A	$165/kVA	$55/kVA	$2/kVA/year	30 years	η_T	93	[47, 48]
Inverter	GTP-501	33 kW	$12 k/unit	$364/kW	$12 k/unit	$85/unit/year	15 years	η_I	95	[49]
	GTP-506	115 kW	$38 k/unit	$330/kW	$38 k/unit	$250/unit/year				
	GTP-519-S	900 kW	$270 k/unit	$300/kW	$270 k/unit	$1.9 k/unit/year				
Super-capacitor module	XLR-48R6167-R	166 F, 48 V ≡ 0.054 kWh	$1.3 k/unit	$24 k/kWh	$1.3 k/unit	$3.3/unit/year	1 m cycles or 20 years	η_{SC}	97	[50]

(continued)

Table 2 (continued)

Component	Manufacturer part number	Nameplate rating	Capital cost		Replacement cost	Operation and maintenance cost	Expected service life	Nominal efficiency		Source
			Per unit	Per standard unit				Notation	Value [%]	
Battery pack	FB 10-100	100 kWh	\$110 k/unit	\$1.1 k/kWh	\$110 k/unit	\$220/unit/year	12 k cycles or 15 years	η_B	92	[51]
	FB 200-400	400 kWh	\$400 k/unit	\$1 k/kWh	\$400 k/unit	\$840/unit/year				
	FB 400-1600	1600 kWh	\$1.442 m/unit	\$901/kWh	\$1.442 m/unit	\$4 k/unit/year				
Electrolyzer stack	H-TEC S 30/50	5 kW	\$6 k/unit	\$1.2 k/kW	\$5.5 k/unit	\$20/unit/year	20 years	η_E	75	[52]
Hydrogen tank	Generic	N/A	N/A	\$500/kg	\$500/kg	\$1/kg/year	20 years	η_{HT}	95	[53]
Fuel cell stack	FCgen-1020ACS	3.3 kW	\$5 k/unit	\$1.5 k/kW	\$4 k/unit	\$0.02/unit/hour	10 k hours	η_{FC}	40	[54]
Hydrogen station—refueling unit	Generic (Pure Energy Centre)	N/A	N/A	\$10 k/(kg-H_2/h)	\$5 k/(kg- H_2/h)	\$350/(kg-H_2/h)/year	20 years	η_S	95	[38, 55, 56]

3.1 Probabilistic Parametric Uncertainty Characterization

To systematically address the inherent uncertainties within the microgrid capacity planning optimization framework, a comprehensive uncertainty characterization approach is employed. This approach encompasses the probabilistic quantification of model-specific, case-specific parametric uncertainties associated with various hourly-based input variables. These uncertainties include solar irradiance, ambient temperature, wind speed, river streamflow, wholesale electricity prices, and power load demand input data. The deterministic model is transformed into a stochastic formulation using a Monte Carlo simulation-based approach, comprising the following steps.

3.1.1 Probability Density Function (PDF) Construction

The uncertainty quantification process commences with constructing probability density functions (PDFs) for each uncertain variable based on real or synthetic historical data. Distributions including: (i) beta, (ii) two-parameter Weibull, (iii) normal, and (iv) two-parameter gamma are selected as best-fitting distributions for specific uncertain variables, including (i) solar irradiance, (ii) wind speed, (iii) power load demand, ambient temperature, wholesale electricity prices, and (iv) river streamflow data, respectively [57–61]. Notably, time-dependent dummy variables are introduced to differentiate between weekday and weekend data points for load demand and wholesale price data.

3.1.2 PDF Discretization

To facilitate the solution of continuous probability distribution-driven stochastic optimization problems, the continuous distributions are discretized into a finite set of scenarios with associated probabilities. Each uncertain variable's PDF is divided into equal-width segments, and these segments are represented by means and probabilities of occurrence. Increasing granularity improves the accuracy of the approximations, but incurs greater computational complexity. Figure 2 portrays the process of discretizing a common representative normal PDF through the application of equal-width intervals.

3.1.3 Scenario Vector Generation

For each uncertain input, scenario vectors are generated by systematically combining the uncertain values according to the discretized PDFs. Each scenario represents a specific realization of uncertain variables, and the probabilities of occurrence are calculated based on the PDF approximations.

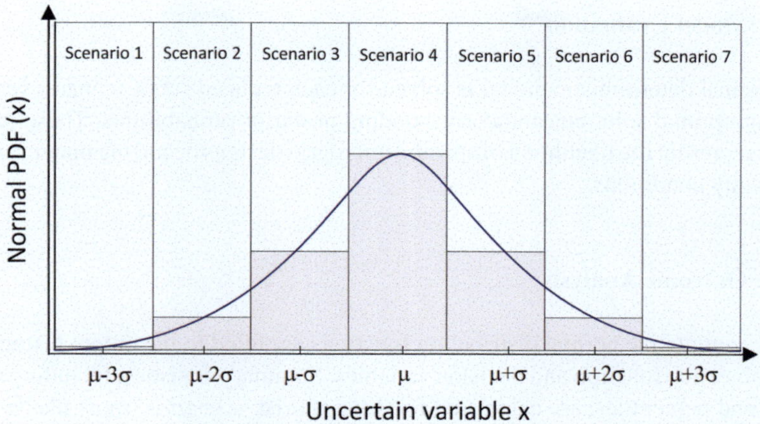

Fig. 2 Illustration of discretizing a representative normal PDF using equal-width intervals

In this setting, the probabilities of occurrence and the associated parameter values (probability-value pairs) obtained from a discrete approximation of the corresponding PDFs (representing a scenario for the realization of the uncertain variable x) are given in Eqs. 1 and 2, respectively.

$$p_i = \int_{R_i} PDF(x)dx \;\; for \;\; i = 1, 2, \ldots, 7 \tag{1}$$

$$x_i = \int_{R_i} x \frac{PDF(x)}{p_i} dx \;\; for \; i = 1, 2, \ldots, 7 \tag{2}$$

where x denotes an uncertain parameter, R_i is the ith region on the respective PDF's value axis (conforming to the predetermined count of equal-width intervals), p is the associated probability density, and the numerical value 7 represents the aggregate count of scenarios generated for uncertain variable x as a result of the discretization process applied to its PDF.

3.1.4 Scenario Vector Reduction

Considering computational constraints, a scenario reduction algorithm is employed to select a minimal subset of scenarios while preserving key statistical characteristics. A mixed-integer linear programming approach is utilized to find an optimal set of reduced scenarios that represent the original distribution's behavior [62].

3.1.5 Model Evaluation

The original deterministic model is solved for each reduced set of scenario vectors, yielding optimal solutions and corresponding posterior probabilities. These probabilities represent the likelihood of specific microgrid designs occurring under varying uncertainty conditions.

3.1.6 Outcome Analysis

Finally, cumulative normal distribution functions are fitted to the output histograms of the objective function and decision variables. Summary statistics, including mean values and percentiles, are calculated for different risk scenarios (most likely, best-case, worst-case). This analysis offers insights into decision-making under different risk preferences, accounting for the multi-dimensional parametric uncertainties affecting the microgrid capacity planning process.

In this context, given that the modelling results are independent and identically distributed, the average values of the model outputs in different hourly scenarios (probability-value pairs) converge to the expected values, according to the strong law of large numbers. Accordingly, the stochastic modelling results for each output variable x in the most likely case are also frequently referred to as realistic-uncertainty-adjusted results. This can be expressed mathematically as [63]

$$E_{rs} = \sum\nolimits_{rs \in RS} \gamma_{rs} Y_{rs} \tag{3}$$

where E_{rs} represents the vector of the expected values of the uncertain variables and the returned values of the fitness function (Y_{rs}) optimized for the reduced input probability-value pairs ($p_{rs}(t), X^{rs}(t)$) with the same index number over the year-long, hourly-basis microgrid operation, γ_{rs} is the posterior probability of the associated microgrid design, while RS represents the set of reduced scenario vectors.

3.2 *Investment Planning Under Uncertainty*

3.2.1 Objective Function

The fundamental objective revolves around minimizing the project cost encompassing the lifetime expenses of system components, coupled with the net cost of electricity transactions with the utility grid. This is quantified in present value terms as [64, 65]

$$NPC_c = N_c \times \left(CC + RC \times SPPW + \frac{O\&M}{CRF(ir, PL)} - SV \right) \tag{4}$$

where N_c denotes the optimal capacity/quantity of component c; CC and RC respectively represent the capital cost and replacement cost of the component; $SPPW$ stands for the single-payment present-worth factor, which is defined in Eq. 5 [64]; $O\&M$ indicates the operation and maintenance cost of the component; CRF stands for the capital recovery factor that is a function of the real interest rate, ir (2.45% [66]), and the expected service life of the project, PL (20 years), as expressed in Eq. 7 [64]; and SV is the salvage value of the component, which is expressed in Eq. 8 [67]. Any additional residual value, other than what is reflected in the equipment salvage value, is assumed to be counterbalanced by the costs associated with the equipment recycling or disposal

$$SPPW = \sum_{n=1}^{N} \frac{1}{(1+ir)^{CL\times n}} \tag{5}$$

where CL denotes the component lifetime and N can be determined by the following equation

$$N = \begin{cases} \left\lfloor \frac{PL}{CL} \right\rfloor - 1 \; if \; PL \mod CL = 0 \\ \left\lfloor \frac{PL}{CL} \right\rfloor \quad otherwise \end{cases} \tag{6}$$

$$CRF(ir, PL) = \frac{ir(1+ir)^{PL}}{(1+ir)^{PL} - 1} \tag{7}$$

$$SV = RC \times \frac{CL - \left(PL - CL \times \left\lfloor \frac{PL}{CL} \right\rfloor\right)}{CL} \tag{8}$$

3.2.2 Planning-Level Constraints

At the investment planning level, constraints focus on reliability, self-sufficiency, resilience, and decision variable limits.

- Reliability: Constraints regulate power supply reliability using the loss of power supply probability index and hydrogen supply reliability alike—zero loss of load.
- Self-sufficiency: A minimum self-sufficiency ratio enforces a high percentage (85%) of local demand coverage by distributed energy resources.
- Resilience: Constraints include autonomy hours (8 h) and grid outage survivability ratio (0.9), indicating the system's resilience to outages.
- Initial and terminal constraints: Constraints ensure realistic operation at the beginning and end of the simulation horizon.

3.3 Integration of Optimal Dispatch

The overarching goal of the day-ahead energy management optimization problem, solved using linear programming, is to maximize the microgrid's overall profit from power exchanges with the grid. Mathematically, the optimal scheduling problem can be expressed as [68–70]

$$\min \boldsymbol{OPEX} = \boldsymbol{P_{im}}\boldsymbol{\pi}^T \Delta t - \boldsymbol{P_{ex}}\boldsymbol{\pi}^T \Delta t + 10^{-6}\|\boldsymbol{u}\|_1 \tag{9}$$

subject to

$$P_{im} - P_{ex} = \left(\frac{P_L}{\eta_I}\right) + \left(\frac{P_S}{\eta_S}\right) + P_E - P_{PV} - P_{WT} - P_{MH} - P_{BP} - P_{FC} - P_{FCEV2G} + P_{ch} - P_{dch} - \left(\frac{Q_L}{\eta_I}\right) - \left(\frac{Q_{H_2}}{\eta_S}\right) \tag{10}$$

where $\boldsymbol{OPEX}$ denotes the 24 h column vector of the daily operational expenditure; $\boldsymbol{P_{im}}$ and $\boldsymbol{P_{ex}}$ respectively denote the 24 h column vectors of imported power and exported power; $\boldsymbol{\pi}$ is the 24 h column vector of wholesale electricity price; $\boldsymbol{P_L}$ and $\boldsymbol{P_S}$ respectively represent the 24 h column vectors of power and hydrogen loads; $\boldsymbol{P_E}$ denotes the 24 h column vector of power supplied to the electrolyzer; $\boldsymbol{P_{PV}}$, $\boldsymbol{P_{WT}}$, $\boldsymbol{P_{MH}}$, and $\boldsymbol{P_{BP}}$ respectively denote the 24 h column vectors of solar PV, wind turbine, micro-hydro, and biopower plant generation; $\boldsymbol{P_{FC}}$ is the 24 h column vector of fuel cell power generation; $\boldsymbol{P_{FCEV2G}}$ is the 24 h column vector of FCEV-to-grid (FCEV2G) provisions; $\boldsymbol{P_{ch}}$ and $\boldsymbol{P_{dch}}$ are the 24 h column vectors of the hybrid battery/super-capacitor storage system's charging power and discharging power, respectively; while $\boldsymbol{Q_L}$ and $\boldsymbol{Q_{H_2}}$ respectively denote the 24 h column vectors of unmet power and hydrogen loads. Also, the term $10^{-6}\|\boldsymbol{u}\|_1$ in Eq. 9 represents the L1-norm of the energy storage schedules over the 24 h operational horizon that is included to penalize any needless cycling of the storage devices, the intended operational timescales of which can be accommodated within the daily time horizon of the optimization-based dispatch strategy, namely the battery and super-capacitor banks.

3.4 Integration of Game-Theoretic Demand Response

This sub-section formulates the Stackelberg [71] strategic game framework mathematically, outlining the objective functions and constraints of the players in the game. In accordance with Fig. 3, the game involves three key sets of players, namely: the microgrid operator, DR aggregators, and end-consumers. Each set of players is motivated by utility maximization, and they interact to optimize their objectives while considering the operational constraints.

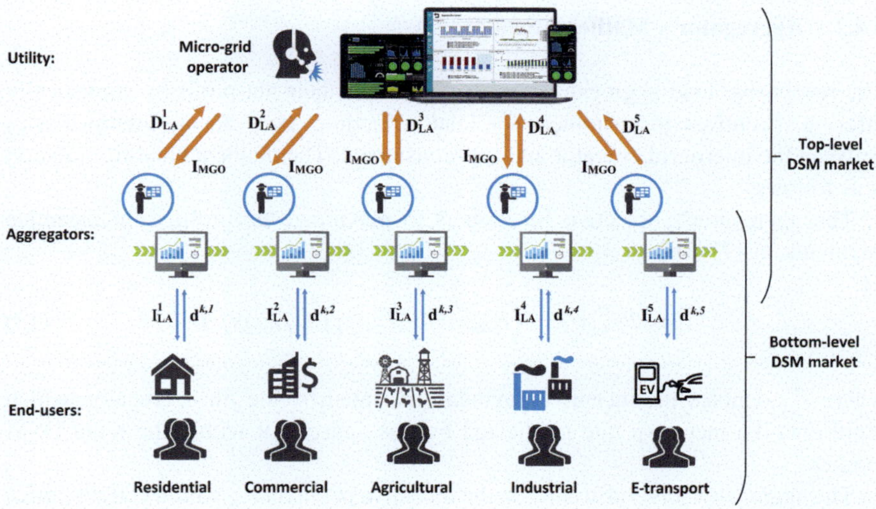

Fig. 3 Overview of the interactions of the players of the DR provisioning game-theoretic framework

3.4.1 Microgrid Operator's Mathematical Model

The microgrid operator aims to minimize operational costs while ensuring power supply–demand balance. The operational cost includes the cost of imported power and incentive payments to DR aggregators for load reduction. The incentive offered by the microgrid operator is based on the forecasted net energy deficit. The mathematical formulation is as follows:

The microgrid operator's objective function is to minimize the operational cost, as [72–76]

$$\min OC_{MG}(t) = cost_{im}(t) + I_{MGO}(t).\sum_{j\in J} D_{LA}^{j}(t)\forall t \tag{11}$$

where OC_{MG} stands for the operational cost of the microgrid, defined through the combination of the cost of power imported from the national grid ($cost_{im}$) and the overall incentive payments for load reduction. The term $I_{MGO}(t)$ signifies the rate of incentive payment for load reduction by the utility, pertaining to time-step t within the day-ahead microgrid scheduling. This corresponds to the incentive price signal posted by the microgrid operator to the wholesale DSM market. Furthermore, D_{LA}^{j} denotes the total load reduction contributed by the jth aggregator (participating in DR aggregation) within the set of aggregators with |J| members.

The microgrid operator sets constraints on the incentive payment I_{MGO} within certain limits, as [72–76]

$$I_{MGO}^{min} \leq I_{MGO}(t) \leq I_{MGO}^{max}\forall t \tag{12}$$

3.4.2 Aggregator's Mathematical Model

The responsive load aggregators aim to maximize their net profit by strategically offering incentives to customers for load reduction. They act as intermediaries between the microgrid operator and end-consumers. The mathematical formulation is as follows:

The aggregator's objective function is to maximize the profit from incentive payments, as [72–76]

$$\max\ Pr_{LA}^{j}(t) = \left(I_{MGO}(t) - I_{LA}^{j}(t)\right).D_{LA}^{j}(t)\forall j, t \quad (13)$$

where I_{LA}^{j} denotes the incentive payment presented by the jth aggregator, which represents the incentive rate publicized by this aggregator within the retail DSM market.

The aggregator sets constraints on the incentive payment I_{LA}^{j} within certain limits, as

$$I_{LA}^{j,min} \leq I_{LA}^{j}(t) \leq I_{LA}^{j,max}\forall j, t \quad (14)$$

3.4.3 End-Consumer's Mathematical Model

End-consumers seek to maximize their utility from financial incentives while maintaining a certain comfort level. They decide on the amount of load reduction based on the offered incentives. The mathematical formulation is as follows:

The end-consumer's objective function is to maximize utility from incentive payments minus discomfort cost, as [72–76]

$$\max\ U^{k,j}(t) = d^{k,j}(t).I_{LA}^{j}(t) - dis^{k,j}(t)\forall k, t \quad (15)$$

where notation $d^{k,j}(t).I_{LA}^{j}(t)$ signifies the compensation received by the kth customer of the jth aggregator in exchange for tolerating a specific discomfort level denoted as $dis^{k,j}(t)$ related to load reductions, serving as an indicator of the electricity value.

The end-consumer sets constraints on the amount of load reduction within non-critical range, as

$$0 \leq d^{k,j}(t) \leq d_{ncr}^{k,j}(t)\forall k, t \quad (16)$$

where $d_{ncr}^{k,j}$ represents the non-critical (dispatchable) segment of the original load.

The mathematical models of the microgrid operator, aggregators, and end-consumers collectively form the foundation of the proposed game-theoretic framework. It provides a systematic way to understand the decisions and interactions among different players in the DR scheduling process.

3.5 Overview of the Approach

Figure 4 portrays the overarching framework of the proposed algorithm to address the probabilistic optimization model for microgrid capacity planning. In this algorithm, the model receives a range of input data encompassing historical load profiles, meteorological conditions, wholesale power prices, equipment specifications, meta-heuristic parameterizations, as well as project lifespan and interest rate. As delineated in the illustration, the optimization process for microgrid planning unfolds for each annual collection of clustered hourly scenario vectors. It is worth noting that these clustered scenarios arise from the resolution of a mixed-integer linear programming problem linked to hour-specific scenario vectors, which are themselves generated via the Monte Carlo simulation technique.

The Monte Carlo simulation procedure captures all feasible combinations of parameter-specific realizations, underpinned by discretized PDFs tailored for each uncertain parameter. Following this, the optimization outputs considering game-theoretic DR and optimal scheduling are utilized to construct normal distribution curves that effectively approximate the corresponding histograms of optimized decision variables and the associated overall microgrid cost over its operational lifetime. Notably, the solutions pertaining to the best-case, most likely case, and worst-case scenarios are computed, leveraging the 5th percentile, expected values, and 95th percentile values, respectively.

Also, the planning-level optimization is conducted using the moth-flame optimization algorithm (MFOA) [77]. This choice is motivated by its demonstrated efficiency in solving the microgrid planning problem, as evidenced in the literature [78–81].

4 Numeric Simulation Results and Discussion

4.1 Simulation Setup

All numerical simulations presented in this Chapter were conducted by implementing the model through MATLAB software (version 9.5, R2018b) [82]. The simulations were executed on a desktop computer operating on the 64-bit Windows 10 platform, equipped with an Intel® CoreTM i7-8700 @ 3.20 GHz processor and 16 GB of RAM.

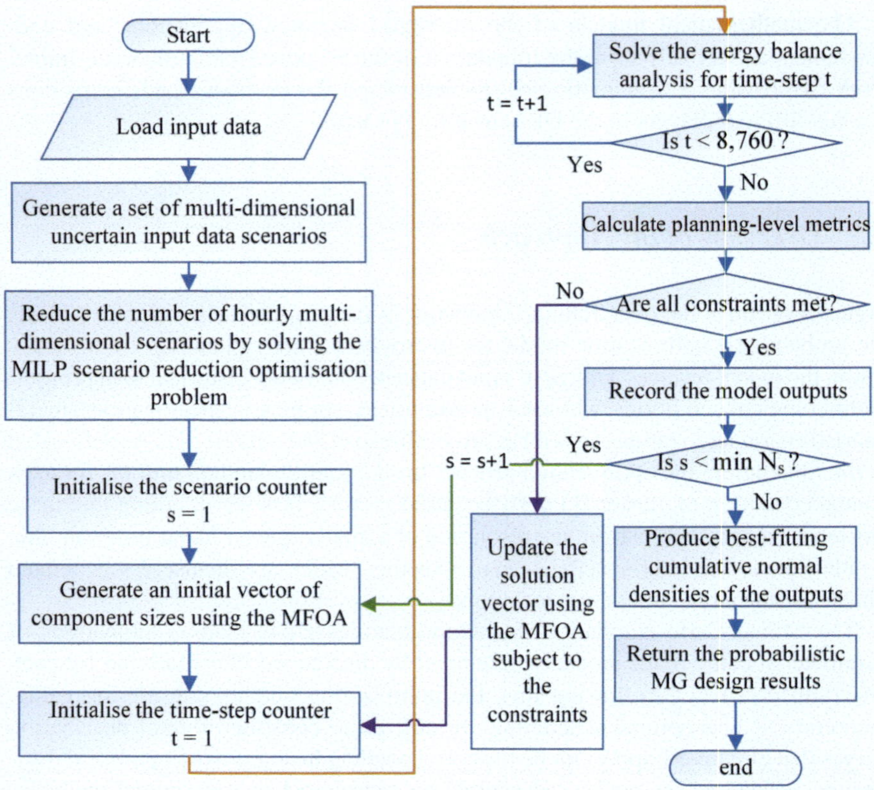

Fig. 4 Overview of the overall microgrid investment planning framework

4.2 *Case Study*

To showcase the effectiveness of the Stackelberg, non-cooperative game-theoretic strategy for optimizing the distribution of sectoral DR resources and its value in reducing capital costs within microgrid development analyses under uncertainty, a case study is conducted for a conceptual community microgrid scheme in Ohakune, New Zealand (latitude 39.4180°S, longitude 175.3985°E). Situated in the central region of the North Island, Ohakune serves as the 'snow season' hub of the area, acting as the gateway to the Turoa Ski Area and holding the title of the 'Carrot Capital' of the country. Its permanent population of approximately 1,000 residents swells to 7,000–10,000 individuals during the winter ski season [36]. Given this, and the significant demand for low-temperature heating on its distribution network, the electrical load experiences pronounced seasonality. The town's current electricity demand is met through a distribution network solely fed by the national grid. However, the residential community grapples with elevated winter bills due to congestion-induced capacity shortages in the transmission line/transformer connecting the local distribution network to the national grid.

In response, the conceptualized microgrid emerges as a robust energy system that fosters democratic energy independence while safeguarding the region's natural environmental assets. Moreover, Ohakune boasts an abundance of renewable energy resources, both dispatchable and non-dispatchable: it has ample untapped solar potential with an average annual total of about 1,400 kWh/m^2 [83], possesses significant untapped wind resources (averaging around 6 m/s yearly at a 10 m height [83]) and could harness power from the Mangawhero River through micro-hydro (with a yearly average streamflow of 2.9 m^3/s [83]), alongside substantial high-quality biomass from discarded carrot crops and indigenous forests in the form of foliage and woody biomass. It is important to highlight that pumped hydropower is not a viable option in the local energy mix due to its inability to meet criteria outlined in Aotearoa New Zealand's National Water Conservation Order 1988, ensuring minimal environmental impact of storage on river flow. Consequently, the conceptualized microgrid takes shape as a pathway to decarbonize Ohakune's energy landscape.

Furthermore, aligning with the objective of decarbonizing the transportation sector, a fleet comprising 180 light-duty personal passenger FCEVs is considered for integration into the system through the strategic utilization of the refueling infrastructure.

4.3 Optimal Planning Results

4.3.1 Impact of Uncertainty and Optimal Scheduling

Figure 5 illustrates the optimal cumulative normal distribution function fitted to the dataset of total discounted system costs, accompanied by superimposed values for the deterministic case, expected outcomes, as well as the 5th and 95th percentiles.

Figure 6 visually portrays the outlined procedure using the finest-fit normal distribution curve for the expanded collection of total system net present cost (NPC) values, acquired through the optimal cost solutions derived for the seven annual sets of diminished scenario vectors. These diverse uncertainty budgets (quantities of interest) are superimposed for clarity. The calculated whole-life cost values under the three prominent probabilistic scenarios exhibit exceedingly small margin of error as follows: best-case scenario, 0.157%; most likely case, 0.351%; and worst-case scenario, 0.161%.

Tables 3 and 4 provide a comprehensive overview of the microgrid investment planning model outcomes, contrasting the deterministic and probabilistic scenarios while delineating the distribution of total discounted system costs across system components and cost/income parameters. Specifically, the probabilistic scenario delves into three distinct levels of accepted risk, aligned with separate uncertainty

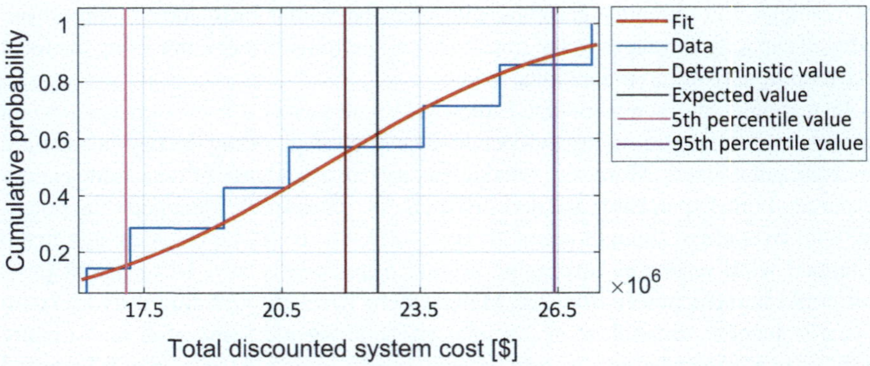

Fig. 5 Cumulative normal distribution fitting for total NPC outputs from stochastic sub-models

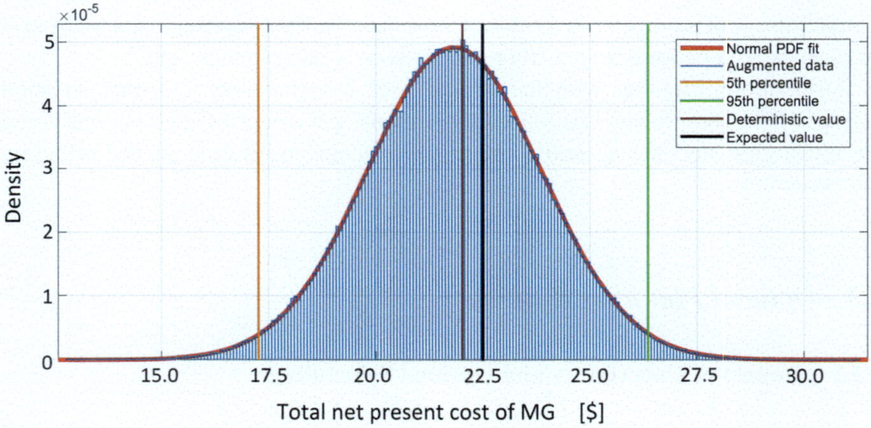

Fig. 6 Fitting a normal PDF to the augmented optimal microgrid life-cycle cost histograms with overlaid uncertainty budgets of interest

budgets. These encompass two scenarios characterized by low probabilities but high impacts, and one scenario marked by high probabilities and low impacts.

Table 3 provides a detailed breakdown of the optimal cost and income components integral to the life-cycle analysis of the microgrid system. Also, Table 4 presents the optimal capacities of microgrid equipment, concurrently presenting the proportional contribution of each component to the total discounted equipment-related cost. The percentages enclosed in parentheses signify the relative variations between the outcomes of the probabilistic cases and those of the deterministic case. This comparative analysis provides valuable insights into the effects of varying levels of risk acceptance on system optimization results.

The key insights collectively emerging from Tables 3 and 4 and Figs. 5 and 6 are as follows:

Table 3 Breakdown of the total discounted system cost under deterministic and budget-constrained probabilistic microgrid planning simulation cases

Cost component	Cost sub-component	Simulation case			
		Deterministic	Best case	Most likely case	Worst case
Total discounted equipment-related costs $((\sum_{c\in C} \underset{20-yr_c}{NPC}) + \underset{20-yr}{NPC_I})$ [\$]	–	18.25 m	17.38 m (−5%)	20.16 m (+10%)	26.59 m (+46%)
Total discounted microgrid operational costs $(\underset{20-yr}{NPV}(\sum_{t=1}^{8760} OC_{MG}(t)))$	Total discounted incentive payment to the aggregators $(\underset{20-yr}{NPV}(\sum_{t=1}^{8760} I_{MGO}(t) \sum_{j\in J} D_{LA}^{j}(t)))$ [\$]	3.99 m	2.84 m (−29%)	3.13 m (−22%)	1.12 m (−72%)
	Total discounted cost of electricity imports $(\underset{20-yr}{NPV}\left(\sum_{t=1}^{8760} cost_{im}(t)\right))$ [\$]	0.46 m	0.57 m (+24%)	0.38 m (−17%)	0.11 m (−76%)
Total discounted FCEV2G electricity provision costs $(\underset{20-yr}{NPV}(\sum_{t=1}^{8760} \pi_{FCEV2G} P_{FCEV2G}(t)))$ [\$]	–	0.42 m	0.35 m (−17%)	0.39 m (−7%)	0.30 m (−29%)
Total discounted operating costs of the biopower plant	Total discounted emission credits $(\underset{20-yr}{NPV}\left(\sum_{t=1}^{8760} cost_{em}(t)\right))$ [\$]	0.52 m	0.52 m (0%)	0.52 m (0%)	0.52 m (0%)
	Total discounted biomass feedstock costs $(\underset{20-yr}{NPV}\left(0.098\,[\$/kg] \times \sum_{t=1}^{8760} M_{BP}(t)[kg]\right))$ [\$]	0.49 m	0.49 m (0%)	0.49 m (0%)	0.49 m (0%)

(continued)

Table 3 (continued)

Cost component	Cost sub-component	Simulation case			
		Deterministic	Best case	Most likely case	Worst case
Total discounted income derived from electricity exports $(-\underset{20-yr}{NPV}\left(\sum_{t=1}^{8760} income_{ex}(t)\right))$ [$]	–	−2.41 m	−5.08 m (−111%)	−2.56 m (−6%)	−2.96 m (−23%)
Whole-life cost of the system (WLC) [$]	–	21.72 m	17.07 m (−21%)	22.51 m (+4%)	26.17 m (+20%)
CPU usage time [s]		82,920	690,950 (+733%)		

Table 4 Comparative deterministic and stochastic (under different uncertainty budgets) optimal microgrid equipment capacity solution sets

Component		Simulation case			
		Deterministic	Best case	Most likely case	Worst case
PV plant	N_{PV} [no.]	3,594	3,824 (+6%)	3,801 (+6%)	4,315 (+20%)
	STDEC* [%]	9.3	10.4 (+1.1)	8.7 (−0.6)	7.9 (−1.4)
Wind plant	N_{WT} [no.]	4	5 (+25%)	5 (+25%)	8 (+100%)
	STDEC* [%]	19.5	31.0 (+11.5)	26.0 (+6.5)	29.0 (+9.5)
Micro-hydro power plant	N_{MH} [no.]	6	7 (+17%)	7 (+17%)	8 (+33%)
	STDEC* [%]	2.2	2.6 (+0.4)	2.2 (0)	2.0 (−0.2)
Biopower plant	N_{BP} [no.]	4	4 (0%)	4 (0%)	4 (0%)
	STDEC* [%]	0.8	0.9 (+0.1)	0.7 (−0.1)	0.6 (−0.2)
Transformer	N_T [kVA]	310	415 (+34%)	362 (+17%)	391 (+26%)
	STDEC* [%]	0.3	0.4 (+0.1)	0.3 (0)	0.2 (−0.1)
Hydrogen tank	N_{HT} [kg]	6,079	3,742 (−38%)	6,179 (+2%)	7,452 (+23%)
	STDEC* [%]	16.7	8.0 (−8.7)	15.4 (−1.3)	15.0 (−1.7)
Electrolyzer	N_E [no.]	122	75 (−39%)	124 (+2%)	149 (+22%)
	STDEC* [%]	3.7	2.0 (−1.7)	3.5 (−0.2)	3.3 (−0.4)
Fuel cell	N_{FC} [no.]	238	148 (−38%)	240 (+1%)	288 (+21%)
	STDEC* [%]	7.1	4.1 (−3.0)	6.4 (−0.7)	6.3 (−0.8)
Battery bank	N_{1600} [no.]	2	2 (−3%)†	2 (+3%)†	2 (+15%)†
	N_{400} [no.]	0	0	0	1
	N_{100} [no.]	2	1	3	3
	STDEC* [%]	15.3	14.8 (−0.5)	13.9 (−1.4)	13.8 (−1.5)
Super-capacitor bank	N_{SC} [no.]	1,982	1,922 (−3%)	1,998 (+1%)	2,094 (+6%)
	STDEC* [%]	13.8	14.5 (+0.7)	12.5 (−1.3)	12.0 (−1.8)
FCEV2G setup	N_{FCEV2G} [kW]	504	462 (−8%)	489 (−3%)	431 (−14%)
	STDEC* [%]	2.1	2.1 (0)	1.9 (−0.2)	1.3 (−0.8)
Hydrogen station	N_S [kg-H_2/h]	6.14	6.14 (0%)	6.14 (0%)	6.14 (0%)
	STDEC* [%]	0.5	0.6 (+0.1)	0.5 (0)	0.4 (−0.1)

(continued)

Table 4 (continued)

Component		Simulation case			
		Deterministic	Best case	Most likely case	Worst case
Inverter	N_{900} [no.]	5	5 (−6%)†	5 (+1%)†	5 (+6%)†
	N_{115} [no.]	2	0	2	4
	N_{33} [no.]	1	0	2	2
	STDEC* [%]	8.7	8.6 (−0.1)	7.9 (−0.8)	8.0 (−0.7)

* STDEC stands for the share of the total discounted equipment-related costs, which can be expressed explicitly in mathematical terms as $((\sum_{c \in C} NPC_{20-yr_c}) + NPC_{I_{20-yr}})$. Also, given that STDEC is measured in percentage points, the associated deviations of the probabilistically optimized decision variables from the corresponding deterministic values are represented by absolute changes, rather than relative changes—to make clear comparisons

† For the battery bank and multi-mode inverter, the percentage of stochastic results' deviation from the deterministic results refers to the associated optimal overall capacity, rather than the constituent elements alone

- The probabilistic model effectively supports decision-making in designing cost-optimal microgrids under various parametric uncertainties according to the investment planning budget. By accounting for forecast uncertainties, the most likely total NPC of the system indicates an approximately 4% uncertainty cost premium (equivalent to about NZ$0.79 million) compared to deterministic modelling results. This means the life cycle cost of the microgrid could have been underestimated by up to 4% without factoring variability in forecast inputs. The lower (best-case) and upper (worst-case) extremes show total NPC deviations of around 21% lower and 20% higher than the deterministic case, respectively. This suggests that characterizing parametric uncertainties could increase or decrease the system's total NPC based on the decision-maker's risk attitude. The additional cost from strongly risk-averse decisions is roughly equivalent to the expected savings from risk-seeking scenarios.
- The computational cost of the probabilistic model is about 8.3 times higher than the deterministic model. However, given the offline nature of microgrid sizing, these costs are acceptable for metaheuristic-based microgrid capacity planning optimization. Parallelization can, therefore, further manage computational costs.
- The deterministic model's computational time is dominated by running the non-cooperative game-theoretic DR-integrated model. Each iteration for a search agent determines hourly energy dispatch decisions in approximately 4.77 s. The remaining computational time for probabilistic sub-problems accounts for the execution of the optimization algorithm.
- Comparing discounted equipment-related costs and whole-life costs offers insights into capital expenditure and operational expenditure trade-offs. Best-case, most likely-case, and worst-case scenarios lead to different changes in these costs.

The most likely scenario incurs a capital cost increase of about 10% compared to deterministic results, highlighting the need for reliable and resilient operations.

- Comparing the optimal microgrid configurations between deterministic and stochastic simulations provides insights into balancing supply and demand under uncertainty. In the best-case scenario, overbuilding renewables for grid export is cost-optimal due to highly favorable conditions. The worst-case scenario increases non-dispatchable renewables and energy storage due to less favorable conditions. Trade-offs occur among various decision variables.
- The optimal generation mix ensures energy security through diversification, leveraging temporal complementarities of renewables. Component size variables do not reach saturation points, reaffirming the validity of energy storage power allocation strategies and the impact of temporal complementarities.

The contributions of energy storage to system-wide uncertainty management decrease as uncertainty reduction increases, due to the sensitivity of storage capacity to different risk attitudes.

4.3.2 Impact of Game-Theoretic DR Management

To assess the impact of DR scheduling framework on microgrid cost reduction, a comparative analysis was conducted for equipment capacity planning in the proposed microgrid. Three scenarios were examined, namely: (1) employing a business-as-usual (BAU) static interruptible load strategy for managing smaller DR resources, (2) utilizing the market-based dynamic integration of aggregator-mediated interruptible responsive loads, and (3) not implementing any DSM strategies.

The outcomes of these scenarios, denoted as 'BAU-DR', 'MBI-DR', and 'NO-DR', are presented in Table 5, which provides a breakdown of optimized cost components considered in the life-cycle analysis of the microgrid system. These results account for the value of biomass feedstock, with the total cost associated with pelletizing blended agricultural and woody biomass feedstocks set at $98 per ton [84]. The site's abundant forest biomass and conducive climate for agriculture mitigate feedstock supply uncertainty, supporting the approach of incorporating biomass feedstock costs exogenously in the post-optimization phase.

It is important to note that the reported results in the table represent the best-case outcomes of the MFOA algorithm after 30 independent trials. Furthermore, to demonstrate the effectiveness of the chosen number of iterations (n = 300) and search agents (n = 50), Fig. 7 illustrates the convergence curves of the MFOA in its best and worst performances across the three scenarios.

Table 5 illustrates the compelling results of the market-based interruptible DSM approach when integrated into the microgrid planning framework. This innovative approach effectively reduces the estimated whole-life cost of the microgrid system by a significant margin. Specifically, it achieves cost reductions ranging from 21% to as much as 32%, as compared to the cases of BAU-integrated interruptible DR and non-DR-integrated microgrid planning scenarios.

Table 5 Breakdown of the total discounted system cost under different DR provision strategies

Cost component	Cost sub-component	Simulation case		
		MBI-DR	BAU-DR	NO-DR
Total discounted equipment-related costs $((\sum_{c\in C} \underset{20-yr}{NPC}_c) + \underset{20-yr}{NPC_I})$ [$]	–	18.25 m	21.88 m	25.62 m
Total discounted microgrid operational costs $(\underset{20-yr}{NPV}(\sum_{t=1}^{8760} OC_{MG}(t)))$	Total discounted incentive payment to the aggregators $(\underset{20-yr}{NPV}(\sum_{t=1}^{8760} I_{MGO}(t) \sum_{j\in J} D_{LA}^{j}(t)))$ [$]	3.99 m	3.48 m	–
	Total discounted cost of electricity imports $(\underset{20-yr}{NPV}\left(\sum_{t=1}^{8760} cost_{im}(t)\right))$ [$]	0.46 m	2.76 m	7.46 m
Total discounted FCEV2G electricity provision costs $(\underset{20-yr}{NPV}(\sum_{t=1}^{8760} \pi_{FCEV2G} P_{FCEV2G}(t)))$ [$]	–	0.42 m	0.49 m	0.50 m
Total discounted operating costs of the biopower plant	Total discounted emission credits $(\underset{20-yr}{NPV}\left(\sum_{t=1}^{8760} cost_{em}(t)\right))$ [$]	0.52 m	0.57 m	0.62 m
	Total discounted biomass feedstock costs $(\underset{20-yr}{NPV}\left(0.098\,[\$/kg] \times \sum_{t=1}^{8760} M_{BP}(t)[kg]\right))$ [$]	0.49 m	0.54 m	0.58 m
Total discounted income derived from electricity exports $(-\underset{20-yr}{NPV}\left(\sum_{t=1}^{8760} income_{ex}(t)\right))$ [$]	–	−2.41 m	−2.42 m	−2.97 m
Whole-life cost of the system (WLC) [$]	–	21.72 m	27.30 m	31.81 m

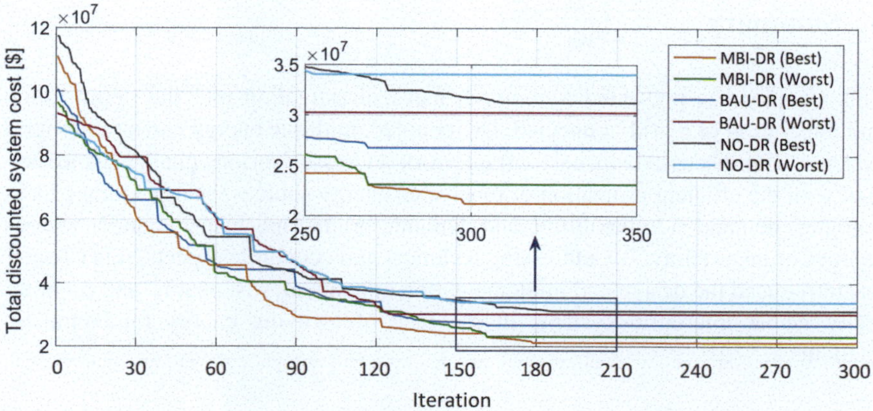

Fig. 7 Convergence process of the MFOA in its best and worst runs throughout 30 simulation cases

The summarized outcomes in Table 5 underscore the efficiency of the model, showcasing the following key findings:

- Under the BAU case, the total discounted equipment-related costs are approximately 20% higher than those under the MBI case. Notably, the total discounted income generated from electricity exports remains nearly constant. This is primarily due to the allocation of extra cost towards backup power equipment, the energy output of which, guided by the microgrid's hourly operational strategy, cannot be sold to the main grid for energy efficiency reasons.
- To verify this assumption's robustness, a separate unreported model was run for both the MBI and BAU scenarios, where the sale of backup power to the grid was not allowed. The results show that this alteration increases the relative cost difference between MBI and BAU cases to around 20% from the original 15%. The change stems from the unreported model's inability to exploit intertemporal price variations, leading to a reduced proportion of total nominal storage to generation capacity, with a corresponding 76% to 429% decrease in net income from energy exchange with the utility grid. Consequently, while resulting in slightly lower energy conversion losses due to the lack of arbitrage operations, the microgrid's whole-life cost increases by approximately 3% to 5%.

5 Summary

This Chapter has presented a comprehensive investigation into the optimization and integration of a grid-connected, DC-coupled, multiple energy carrier microgrid systems. The discussed stochastic framework addresses various challenges associated with the efficient utilization of renewable energy sources, energy storage technologies, demand-side flexibility, and dynamic pricing mechanisms under various sources of uncertainty. By addressing technical and economic aspects, this Chapter has introduced the design and implementation of resilient, sustainable, and economically viable microgrid systems in the face of evolving energy (and climate) challenges.

References

1. Mohseni S, Brent AC, Kelly S, Browne WN (2022) Demand response-integrated investment and operational planning of renewable and sustainable energy systems considering forecast uncertainties: A systematic review. Renew Sustain Energy Rev 158:112095
2. Mansouri SA, Rezaee Jordehi A, Marzband M, Tostado-Véliz M, Jurado F, Aguado JA (2023) An IoT-enabled hierarchical decentralized framework for multi-energy microgrids market management in the presence of smart prosumers using a deep learning-based forecaster. Appl. Energy, 333:120560
3. Taheri B, Foroud AA, Jabari F (2023) Design and performance optimization of a tri-generation energy hub considering demand response programs. Energy Sci. Eng. 11(2):675–684
4. Mobtahej M, Barzegaran M, Esapour K (2023) A novel Three-Stage demand side management framework for stochastic energy scheduling of renewable microgrids. Sol Energy 256:32–43
5. Mohseni S, Brent AC, Burmester D (2021) Off-Grid Multi-Carrier Microgrid Design Optimisation: The Case of Rakiura-Stewart Island, Aotearoa-New Zealand. Energies 14(20):6522
6. Moretti L, Astolfi M, Vergara C, Macchi E, Pérez-Arriaga JI, Manzolini G (2019) A design and dispatch optimization algorithm based on mixed integer linear programming for rural electrification. Appl Energy 233–234:1104–1121
7. Castillo-Calzadilla T, Cuesta MA, Olivares-Rodriguez C, Macarulla AM, Legarda J, Borges CE (2022) Is it feasible a massive deployment of low voltage direct current microgrids renewable-based? A technical and social sight. Renew Sustain Energy Rev 161:112198
8. Mohseni S, Brent AC, Burmester D (2020) Community Resilience-Oriented Optimal Micro-Grid Capacity Expansion Planning: The Case of Totarabank Eco-Village, New Zealand. Energies 13(15):3970
9. Scheller F, Bruckner T (2019) Energy system optimization at the municipal level: An analysis of modeling approaches and challenges. Renew Sustain Energy Rev 105:444–461
10. Tostado-Véliz M, Rezaee Jordehi A, Icaza D, Mansouri SA, Jurado F (2023) Optimal participation of prosumers in energy communities through a novel stochastic-robust day-ahead scheduling model. Int. J. Electr. Power Energy Syst 147:108854
11. Kumar KP, Saravanan B (2017) Recent techniques to model uncertainties in power generation from renewable energy sources and loads in microgrids—A review. Renew Sustain Energy Rev 71:348–358
12. Mohseni S, Brent AC, Burmester D, Browne WN, Kelly S (2021) Adding a computationally-tractable probabilistic dimension to meta-heuristic-based microgrid sizing. In: TENCON 2021 - 2021 IEEE Region 10 Conference (TENCON), pp 464–469

13. Mohseni S, Brent AC (2022) Risk-based dispatch optimization of microgrids considering the uncertainty in EV driving patterns. In: 2022 17th International Conference on Probabilistic Methods Applied to Power Systems (PMAPS), pp 1–6
14. Mohseni S, Brent AC (2022) Game-theoretic sectoral demand response procurement in multi-energy microgrid planning. In: 2022 IEEE Power & Energy Society General Meeting (PESGM), pp 1–5
15. Jasim AM, Jasim BH, Mohseni S, Brent AC (2023) Consensus-based dispatch optimization of a microgrid considering meta-heuristic-based demand response scheduling and network packet loss characterization. Energy AI 11:100212
16. Fan W et al (2023) A two-stage optimal scheduling model of integrated energy system based on CVaR theory implementing integrated demand response. Energy 263:125783
17. Hassanzadehfard H, Moghaddas-Tafreshi SM, Hakimi SM (2015) Optimization of grid-connected microgrid consisting of PV/FC/UC with considered frequency control. Turkish J Electr Eng Comput Sci 23(1):1–16
18. Atia R, Yamada N (2016) Sizing and analysis of renewable energy and battery systems in residential microgrids. IEEE Trans Smart Grid 7(3):1204–1213
19. Pazouki S, Haghifam MR (2016) Optimal planning and scheduling of energy hub in presence of wind, storage and demand response under uncertainty. Int J Electr Power Energy Syst 80:219–239
20. Schachter JA, Mancarella P, Moriarty J, Shaw R (2016) Flexible investment under uncertainty in smart distribution networks with demand side response: Assessment framework and practical implementation. Energy Policy 97:439–449
21. Hussain A, Arif SM, Aslam M, Danial S, Shah A (2017) Optimal siting and sizing of Tri-generation equipment for developing an autonomous community microgrid considering uncertainties. Sustain Cities Soc 32:318–330
22. Yu L, Li YP, Huang GH, An CJ (2017) A robust flexible-probabilistic programming method for planning municipal energy system with considering peak-electricity price and electric vehicle. Energy Convers Manag 137:97–112
23. Nojavan S, Majidi M, Zare K (2017) Stochastic multi-objective model for optimal sizing of energy storage system in a microgrid under demand response program considering reliability: A weighted sum method and fuzzy satisfying approach. J Energy Manag Technol 1(1):61–70
24. Nojavan S, Majidi M, Esfetanaj NN (2017) An efficient cost-reliability optimization model for optimal siting and sizing of energy storage system in a microgrid in the presence of responsible load management. Energy 139:89–97
25. Chen J et al (2018) Optimal sizing for grid-tied microgrids with consideration of joint optimization of planning and operation. IEEE Trans Sustain Energy 9(1):237–248
26. Xiao H, Pei W, Dong Z, Kong L (2018) Bi-level planning for integrated energy systems incorporating demand response and energy storage under uncertain environments using novel metamodel. CSEE J power energy Syst 4(2):155–167
27. Zheng Y, Jenkins BM, Kornbluth K, Træholt C (2018) Optimization under uncertainty of a biomass-integrated renewable energy microgrid with energy storage. Renew Energy 123:204–217
28. Zheng Y, Jenkins BM, Kornbluth K, Kendall A, Træholt C (2018) Optimization of a biomass-integrated renewable energy microgrid with demand side management under uncertainty. Appl Energy 230:836–844
29. Prathapaneni DR, Detroja KP (2019) An integrated framework for optimal planning and operation schedule of microgrid under uncertainty. Sustain Energy, Grids Networks 19:100232
30. Mansouri SA, Ahmarinejad A, Javadi MS, Catalão JPS (2020) Two-stage stochastic framework for energy hubs planning considering demand response programs. Energy 206:118124
31. Masoumi A, Ghassem-zadeh S, Hosseini SH, Ghavidel BZ (2020) Application of neural network and weighted improved PSO for uncertainty modeling and optimal allocating of renewable energies along with battery energy storage. Appl Soft Comput 88:105979
32. Amir V, Azimian M (2020) Dynamic multi-carrier microgrid deployment under uncertainty. Appl Energy 260:114293

33. Bagheri F, Dagdougui H, Gendreau M (2022) Stochastic optimization and scenario generation for peak load shaving in Smart District microgrid: sizing and operation. Energy Build 275:112426
34. Saini VK, Kumar R, Al-Sumaiti AS, Panigrahi BK (2023) Uncertainty aware optimal battery sizing for cloud energy storage in community microgrid. Electr Power Syst Res 222:109482
35. Stephenson J, Ford R, Nair NK, Watson N, Wood A, Miller A (2018) Smart grid research in New Zealand—A review from the GREEN Grid research programme. Renew Sustain Energy Rev 82(2017):1636–1645
36. Anderson J (2009) Pulling the plug on network congestion. In: B.A. dissertation, Department of Economics, University of Otago, Dunedin, New Zealand
37. Mohseni S, Brent AC, Kelly S, Browne WN, Burmester D (2021) Strategic design optimisation of multi-energy-storage-technology micro-grids considering a two-stage game-theoretic market for demand response aggregation. Appl Energy 287:116563
38. Mohseni S, Brent AC (2020) Economic viability assessment of sustainable hydrogen production, storage, and utilisation technologies integrated into on- and off-grid micro-grids: A performance comparison of different meta-heuristics. Int J Hydrogen Energy 45(59):34412–34436
39. Mohseni S, Brent AC, Burmester D (2020) A comparison of metaheuristics for the optimal capacity planning of an isolated, battery-less, hydrogen-based micro-grid. Appl Energy 259:114224
40. Mohseni S, Brent AC, Burmester D (2019) A demand response-centred approach to the long-term equipment capacity planning of grid-independent micro-grids optimized by the moth-flame optimization algorithm. Energy Convers Manag 200:112105
41. Mohseni S, Brent AC (2022) A metaheuristic-based micro-grid sizing model with integrated arbitrage-aware multi-day battery dispatching. Sustainability 14(19)
42. English Rates UK. New Zealand Dollar to US Dollar Spot Exchange Rates for 2022. [Online]. Available: https://www.exchangerates.org.uk/NZD-USD-spot-exchange-rates-history-2022.html
43. Canadian Solar Inc. CS6K-270|275|280P, 2017, PV Module Product Datasheet V5.552_EN. [Online]. Available: https://www.collectiu-solar.cat/pdf/2-Panel-Canadian_Solar-Datasheet-CS6K.pdf
44. All Power Labs. Old PP20 vs New PP30 Cogen-CS: Technical Specifications & Comparisons (2019) Rev 03. [Online]. Available: http://www.allpowerlabs.com/wp-content/uploads/2019/07/PP30-vs-PP20-Spec-Sheet-Public-2019-Rev-03-current-July-2019.pdf
45. Alstom (2014) Ecotècnia ECO 48/750 Datasheet. [Online]. Available: https://en.wind-turbine-models.com/turbines/791-ecot-cnia-eco-48-750#datasheet/
46. Sunceo Hydro (2015) XJ50–100SCTF6-Z. [Online]. Available: https://www.micro-hydro-power.com/100kw-hydro-turbine-generator/
47. Qin H, Kimball JW (2013) Solid-state transformer architecture using AC-AC dual-active-bridge converter. IEEE Trans Ind Electron 60(9):3720–3730
48. Nerini FF, Broad O, Mentis D, Welsch M, Bazilian M, Howells M (2016) A cost comparison of technology approaches for improving access to electricity services. Energy 95:255–265
49. Gandini D, de Almeida AT (2017) Direct current microgrids based on solar power systems and storage optimization, as a tool for cost-effective rural electrification. Renew Energy 111:275–283
50. Eaton Corporation (2019) XLR-48 Supercapacitor. Technical Data 10510. [Online]. Available: https://datasheet.octopart.com/XLR-48R6167-R-Eaton-datasheet-130052459.pdf
51. My Solar Quotes. Solar Battery Storage—Product Comparison (2020) [Online]. Available: https://www.mysolarquotes.co.nz/about-solar-power/residential/solar-battery-storage---product-comparison/
52. H-TEC Systems (2019) The core of electrolysis: PEM electrolyser stacks—Designed for ideal integration in systems. H-TEC Series-S: S 30/30. [Online]. Available: https://www.h-tec-systems.com/fileadmin/Content/PDFs/19022019/H-TEC_SYSTEMS_Datenblatt_Stacks_SE30_30_EN.PDF

53. Duman AC, Güler Ö (2018) Techno-economic analysis of off-grid PV/wind/fuel cell hybrid system combinations with a comparison of regularly and seasonally occupied households. Sustain Cities Soc 42:107–126
54. Ballard Power Systems Inc. FCgen–1020ACS, 2015, SPC5101559–0H. [Online]. Available: https://www.ballard.com/docs/default-source/spec-sheets/fcgen-1020-acs-v2.pdf?sfvrsn=c3ebc380_4
55. Graham JDT, Mulvenna AJ, Mufford WE, Borck JG, Ko J, Harper MAM (2004) Hydrogen fueling station. Google PatS
56. Pure Energy Centre Limited. Hydrogen fueling station (2019) [Online]. Available: https://pureenergycentre.com/hydrogen-fueling-station/
57. Huang L, Liao Q, Qiu R, Liang Y, Long Y (2021) Prediction-based analysis on power consumption gap under long-term emergency: A case in China under COVID-19. Appl Energy 283:116339
58. Khmaladze E (2007) Statistical analysis of electricity prices. J. Data Sci. 5(1):103–129
59. Li H, Liu P, Guo S, Ming B, Cheng L, Yang Z (2019) Long-term complementary operation of a large-scale hydro-photovoltaic hybrid power plant using explicit stochastic optimization. Appl Energy 238:863–875
60. Borowy BS, Salameh ZM (1994) Optimum photovoltaic array size for a hybrid wind/PV system. IEEE Trans Energy Convers 9(3):482–488
61. Lu N, Diao R, Hafen RP, Samaan N, Makarov YV (2013) A comparison of forecast error generators for modeling wind and load uncertainty. In: 2013 IEEE Power & Energy Society General Meeting, IEEE, pp 1–5
62. Karuppiah R, Martín M, Grossmann IE (2010) A simple heuristic for reducing the number of scenarios in two-stage stochastic programming. Comput Chem Eng 34(8):1246–1255
63. Ahmadi SE, Rezaei N (2020) A new isolated renewable based multi microgrid optimal energy management system considering uncertainty and demand response. Int J Electr Power Energy Syst 118:105760
64. Hakimi SM, Moghaddas-Tafreshi SM (2009) Optimal sizing of a stand-alone hybrid power system via particle swarm optimization for Kahnouj area in south-east of Iran. Renew Energy 34(7):1855–1862
65. Mohseni S, Brent AC, Burmester D, Chatterjee A (2018) Optimal sizing of an Islanded microgrid using meta-heuristic optimization algorithms considering demand-side management. In: 2018 Australasian Universities Power Engineering Conference (AUPEC), IEEE, pp 1–6
66. Trading Economics. New Zealand – Real Interest Rate. [Online]. Available: https://tradingeconomics.com/new-zealand/real-interest-rate-percent-wb-data.html/
67. HOMER Energy, *Salvage Value.* [Online]. Available: https://www.homerenergy.com/products/pro/docs/latest/salvage_value.html
68. Pimm AJ, Palczewski J, Morris R, Cockerill TT, Taylor PG (2020) Community energy storage: A case study in the UK using a linear programming method. Energy Convers Manag 205:112388
69. Mohseni S, Brent AC (2022) Quantifying the effects of forecast uncertainty on the role of different battery technologies in grid-connected solar photovoltaic/wind/micro-hydro microgrids: An optimal planning study. J Energy Storage 51:104412
70. Mohseni S, Brent AC (2022) Co-optimization of the sizing and dispatch of microgrids considering vehicle-to-grid and arbitrage. In: 2022 IEEE PES 14th Asia-Pacific Power and Energy Engineering Conference (APPEEC), pp 1–7
71. Jørgensen S, Sigue S-P, Zaccour G (2001) Stackelberg leadership in a marketing channel. Int Game Theory Rev 3(01):13–26
72. Mohseni S, Brent AC, Kelly S (2020) A hierarchical, market-based, non-cooperative game-theoretic approach to projecting flexible demand-side resources: Towards more realistic demand response-integrated, long-term energy planning models. In: 2020 17th International Conference on the European Energy Market (EEM), IEEE, pp 1–6
73. Mohseni S, Brent AC, Burmester D, Browne WN (2020) A game-theoretic approach to model interruptible loads: Application to micro-grid planning. In: 2020 IEEE Power & Energy Society General Meeting (PESGM), IEEE, pp 1–5

74. Mohseni S, Brent AC, Kelly S, Browne WN, Burmester D (2021) Modelling utility-aggregator-customer interactions in interruptible load programmes using non-cooperative game theory. Int J Electr Power Energy Syst 133:107183
75. Fahrioglu M, Alvarado FL (2001) Using utility information to calibrate customer demand management behavior models. IEEE Trans Power Syst 16(2):317–322
76. Fahrioglu M, Alvarado FL (2000) Designing incentive compatible contracts for effective demand management. IEEE Trans Power Syst 15(4):1255–1260
77. Mirjalili S (2015) Moth-flame optimization algorithm: A novel nature-inspired heuristic paradigm. Knowledge-Based Syst 89:228–249
78. Mohseni S, Brent AC, Burmester D (2019) A sustainable energy investment planning model based on the micro-grid concept using recent metaheuristic optimization algorithms. In: 2019 IEEE Congress on Evolutionary Computation (CEC), IEEE, pp 219–226
79. Mohseni S, Brent AC, Burmester D, Chatterjee A (2019) Stochastic optimal sizing of micro-grids using the moth-flame optimization algorithm. In: 2019 IEEE Power & Energy Society General Meeting (PESGM), IEEE, pp 1–5
80. Mohseni S, Brent AC, Burmester D, Browne WN (2021) Lévy-flight moth-flame optimisation algorithm-based micro-grid equipment sizing: An integrated investment and operational planning approach. Energy AI 3:100047
81. Mohseni S, Khalid R, Brent AC (2022) Metaheuristic-based isolated microgrid sizing and uncertainty quantification considering EVs as shiftable loads. Energy Rep 8:11288–11308
82. MATLAB (2018) Natick, MA: The MathWorks Inc. R2018b, Version 9.5
83. CliFlo: New Zealand's national climate database. [Online]. Available: http://cliflo.niwa.co.nz/
84. Karkania V, Fanara E, Zabaniotou A (2012) Review of sustainable biomass pellets production—A study for agricultural residues pellets' market in Greece. Renew Sustain Energy Rev 16(3):1426–1436

Design and Modelling of Microgrids Operated at Constant Frequency and with a Power Level of Megawatts

Daming Zhang

Abstract Hardware prototyping of large-scale autonomous microgrids is costly and dangerous. To reduce such risks, this chapter adopts a fixed-time-step multi-rate modelling in Matlab/Simulink to reduce the differences between modelling and hardware implementation. For a comparison purpose, single-rate modelling has also been conducted. Furthermore, phasor-domain analytical solutions of the voltages at each bus in the microgrid have also been sought and are compared with results from both single-rate and multi-rate modelling. The chapter will demonstrate that the results from multi-rate modelling are closer to those from phasor-domain analysis. Moreover, the influence of DC link voltage on the real and reactive power tracing for each inverter will be demonstrated.

Keywords Autonomous · Inverter · Modelling · Multi-rate · Voltage profile

1 Introduction

Currently, modern power systems are under dynamic evolvement. To solve energy crisis, each country is open to new technologies in power sector so long they are environment-friendly, cost-effective and have no potential danger to deplete precious materials.

Large-scale microgrids play important complementary roles in modern power system. Besides the study on grid-tied operation for the microgrids, off-grid or islanded or autonomous operation is also under intensive research. The control for operating the autonomous microgrid includes f-P and V-Q method, V-P and f-Q method, and V-P control method with separate reactive power handling and at constant frequency [1–10]. Each one has its pro and cons. In comparison, both f-P, V-Q and V-P, f-Q methods need to manage both frequency variation and voltage variation while fixed-frequency or constant-frequency method only manages the

D. Zhang (✉)
School of Electrical Engineering and Telecommunication, University of New South Wales, Sydney, Australia
e-mail: daming.zhang@unsw.edu.au

F. Shahnia and J. M. Guerrero (eds.), *Microgrids and Virtual Power Plants*, Power Systems, https://doi.org/10.1007/978-981-97-6623-9_3

voltage variation in the system, and the frequency in the system takes a constant value. Therefore, the constant frequency method is easier to operate the microgrid and shows better power quality of frequency.

Autonomous microgrids could find various applications, such as power supply to remote mining sites, auxiliary power supply on trains and power supply for remote residential areas etc. Furthermore, they could also be used to supply power to existing loads, either residential ones or factory ones. In such an application, multiple solid-state transformers in parallel could be used to import planned power from large power system. This is to reduce the complexity of operating the large power systems and alleviate the necessity of their further expansion.

It is expensive to build microgrids operating at high voltage and high power. Furthermore, with un-tested methods, it is very dangerous to carry out experiment. Therefore, it is tremendously important to conduct modeling first. Matlab/Simulink is one of several platforms suitable for this purpose. To make the simulation closer to hardware implementation, usually fixed-time-step approach is taken. This is to mimic how a microcontroller with precise clocking works to control inverter and other circuits in a microgrid. Accuracy of such modelling must be improved to make it more meaningful and closer to hardware performance.

Out of available microcontrollers and FPGAs, Texas Instrument 28377 s or 28379X are good choices as they allow accurate and fast algorithms for division, multiplication and trigonometry calculation etc. Furthermore, communication between multiples of them can be established through their ADC/DAC ports to expand their functions. Their way of working to control microgrid will be mimicked by Matlab/Simulink modelling.

In this writing, both fixed-time-step single-rate and multi-rate simulations in Matlab/Simulink are conducted. In the single-rate approach, 100 kHz is chosen to be the sampling and data-processing frequency and it is also the setting for POWERGUI. The frequency of carrier triangular waveform for comparison with reference voltages, also called switching frequency here is set to be 5 kHz. Then at each of 100 kHz/5 kHz or 20 points, comparison is done between reference voltage and triangular waveform to produce gating signals. In the multi-rate approach, 100 kHz is also chosen to be the sampling and data-processing frequency. But 200 kHz is the setting for POWERGUI. 5 kHz is still chosen to be switching frequency. Then at each of 200 kHz/5 kHz or 40 points, comparison is done to produce the gating signals. The multi-rate approach is designed to simulate how two TI28377s or TI28379X are used to achieve the inverter control purpose. The processed reference voltages are sent from one microcontroller through ADC/DAC to the other. Then comparison is done in the second microcontroller to produce the gating signals. Another approach to have many more comparison points is to use EPWM comparator itself in the second microcontroller. But it comes with the compromise that the reference voltages are updated less frequently.

Besides single-rate and multi-rate modelling in Matlab/Simulink, the voltage profile at each bus is also calculated using phasor domain approach. Its comparison with the results from both single-rate and multi-rate time-domain modelling will be conducted. Moreover, the influence of DC link voltage for the inverter on accurate power tracing has been investigated. All the details are un-wrapped section by section below.

2 Formulation for Calculating Voltages at Each Bus Using Phasor Domain Approach

In the constant-frequency operated method of [9–13], each inverter is a current-controlled voltage source one. To reduce the difficulty level of start-up, emergency shut-down etc., a three-phase resistive load with large resistance is connected in parallel with each inverter. In some cases, two sets of such resistive loads could be connected with the inverter, one being before the LCL filter while the other being after it. So the equivalent circuit for such an inverter is given in the top one in Fig. 1. As such resistances are negligible, its approximate one is shown in the lower figure in Fig. 1.

Figure 2 shows the microgrid under study where transformers 1, 2 and 3 have the same ratio of 3.1125 kV/11 kV and another three transformers 4, 5 and 6 have the same ratio of 11 kV/415 V. In other applications, inverters could work at different voltage levels. Correspondingly, the ratios of transformers 1, 2 and 3 are different. Furthermore, the inverters could come with LCL filters or LCCL filters. DG1 is a combination of the grid-forming generator and reactive power compensators while either DG2 or DG3 could be a grid-supporting or grid-feeding generator. Figure 3 is a general representation of each of the distributed generators tapped to its neighbouring bus in the microgrid in Fig. 2.

The voltage profile calculation is done for a balanced system, in which each of the loads in Fig. 2 is nearly balanced and each transmission line is properly transposed and positive-sequence currents flow on the 11 kV side in the microgrid with negligible negative-sequence currents. Or it is assumed that negative-sequence

Fig. 1 Equivalence of the current source

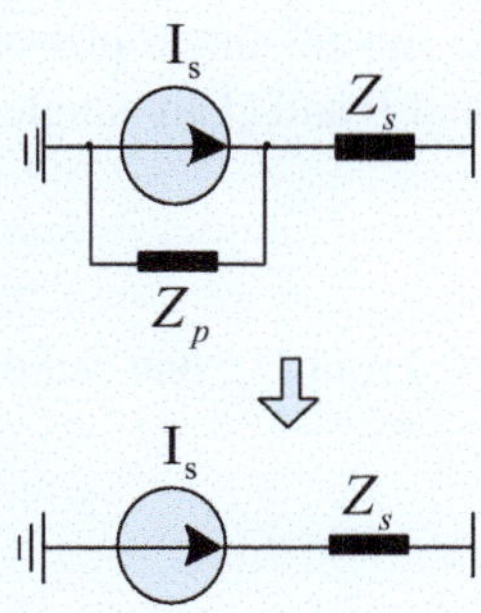

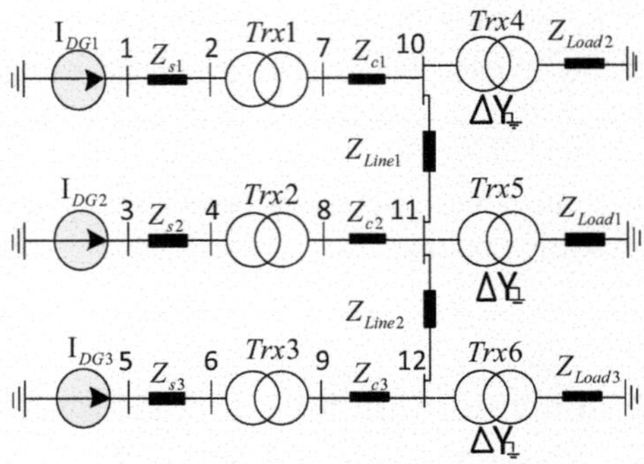

Fig. 2 Microgrid under study

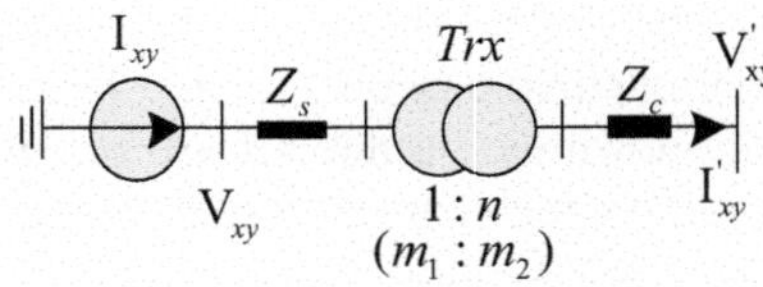

Fig. 3 A general representation of a distributed generator tapped to its neighbouring bus

current compensation is properly done to leave only the positive-sequence current seen from the high-voltage side of the transformers Trx4, Trx5 and Trx6. For such a balanced microgrid, the equivalent transform can be done to reach the circuits as shown in Fig. 4. If Z_{DG2} were negligible, one would be able to reach the equivalent circuit as shown on the right hand side of Fig. 4b. Then the method presented in [14, 15] can be used to work out the voltages at each bus. Nevertheless, for a practical microgrid system, the error produced by Z_{DG2} could be significant and its influence needs to be considered.

The formulation here takes into account the influence by each impedance between the equivalent current source and its neighbouring bus.

Below are general equations for grid-forming, grid-supporting and grid-feeding generators.

$$\mathrm{V_{xy}} \times \mathrm{I}^{*}_{\mathrm{xy}} = P + jQ \tag{1}$$

where $\mathrm{V_{xy}} = V_x + jV_y$, $\mathrm{I_{xy}} = I_x + jI_y$ are complex numbers of voltage and current in rms for single-phase system and d-q components for three-phase balanced system and P and Q are single-phase powers for three-phase system.

$$\mathrm{V_{xy}} = \left(Z_s + Z_c/n^2\right) \times \mathrm{I_{xy}} + \mathrm{V}'_{\mathrm{xy}}/n \tag{2}$$

From (2), one can derive and come to (3).

$$\mathrm{I_{xy}} = C_v \cdot \mathrm{V_{xy}} + \mathrm{I_{xy}} \tag{3}$$

where $C_v = 1/\left(Z_s + Z_c/n^2\right)$ and $\mathrm{I_{xy}} = -C_v \cdot \mathrm{V}'_{\mathrm{xy}}/n$

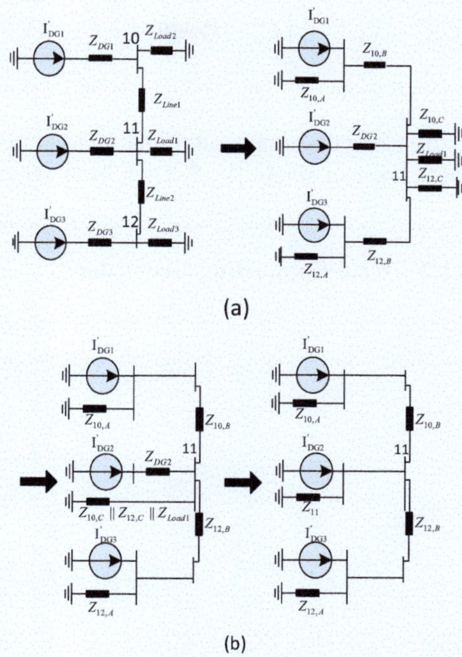

Fig. 4 Equivalent transform of the microgrid in Fig. 2

Then from (1) and (3) one has

$$C_v^* \cdot \left|\mathrm{V_{xy}}\right|^2 + \mathrm{I_{xy}^*} \cdot \mathrm{V_{xy}} = P + jQ \tag{4}$$

2.1 *Specific Formulas for Each Generator*

2.1.1 Grid-Feeding Generator

For the grid-feeding generator, P and Q are known.

By taking the conjugate of two sides of (4), one can obtain

$$C_v \cdot \left|\mathrm{V_{xy}}\right|^2 + \mathrm{I_{xy}} \cdot \mathrm{V}_{xy}^* = P - jQ \tag{5}$$

From (4) and (5), one can obtain $\left[C_v^* \cdot \left|\mathrm{V_{xy}}\right|^2 - (P + jQ)\right] \cdot \left[C_v \cdot \left|\mathrm{V_{xy}}\right|^2 - (P - jQ)\right] = \left|\mathrm{I_{xy}}\right|^2 \cdot \left|\mathrm{V_{xy}}\right|^2$.

Or

$$\left|C_v\right|^2 \cdot \left|\mathrm{V_{xy}}\right|^4 - \left[C_v^* \cdot (P - jQ) + C_v \cdot (P + jQ) + \left|\mathrm{I_{xy}}\right|^2\right] \cdot \left|\mathrm{V_{xy}}\right|^2 + P^2 + Q^2 = 0 \tag{6}$$

By using the command 'roots' in Matlab and applying proper judgement of four roots, one can obtain $\left|\mathrm{V_{xy}}\right|$. Then from (4), one can obtain $\mathrm{V_{xy}}$.

2.1.2 Grid-Supporting Generator

$$P = (K_{v0}/3) \cdot \left(K_v \cdot V_{ref} - \sqrt{V_x^2 + V_y^2}\right) \tag{7}$$

$$Q = K_{pq} \cdot P \tag{8}$$

Then

$$P + jQ = D_1 - D_2 \cdot \left|\mathrm{V_{xy}}\right| \tag{9}$$

where $D_1 = (1 + jK_{pq}) \cdot (K_{v0}/3) \cdot K_v \cdot V_{ref}$ and $D_2 = (1 + jK_{pq}) \cdot (K_{v0}/3)$.

From (4) and (9), one has

$$C_v^* \cdot \left|\mathrm{V_{xy}}\right|^2 + \mathrm{I_{xy}^*} \cdot \mathrm{V_{xy}} = D_1 - D_2 \cdot \left|\mathrm{V_{xy}}\right| \tag{10}$$

By taking the conjugate of both sides of (10), one can obtain

$$C_v \cdot \left|\mathrm{V_{xy}}\right|^2 + \mathrm{I_{xy}} \cdot \mathrm{V}_{xy}^* = D_1^* - D_2^* \cdot \left|\mathrm{V_{xy}}\right| \tag{11}$$

Then from (10) and (11), one can obtain

$$\mathrm{Coeff1} \cdot \left|\mathrm{V_{xy}}\right|^4 + \mathrm{Coeff2} \cdot \left|\mathrm{V_{xy}}\right|^3 + \mathrm{Coeff3} \cdot \left|\mathrm{V_{xy}}\right|^2 + \mathrm{Coeff4} \cdot \left|\mathrm{V_{xy}}\right| + |D_1|^2 = 0 \tag{12}$$

where Coeff1 = $\left|C_v\right|^2$, Coeff2 = $C_v \cdot D_2 + (C_v \cdot D_2)^*$, Coeff3 = $|D_2|^2 - C_v \cdot D_1 - (C_v \cdot D_1)^* - \left|\mathrm{I_{xy}}\right|^2$, Coeff4 = $-D_1 \cdot D_2^* - D_1^* \cdot D_2$

Again, by using the command 'roots' in Matlab and applying proper judgement of four roots, one can obtain $\left|\mathrm{V_{xy}}\right|$ first. Then from (10), one can obtain $\mathrm{V_{xy}}$.

2.1.3 Grid-Forming Generator

As the reactive power compensator accompanies the grid-forming generator, the overall reactive power by them together is balanced by the generation and consumption in other parts of the network.

For the microgrid system shown in Fig. 2, DG1 is the grid-forming generator. Calculation starts from DG3, then for DG2 and finally for DG1. Therefore, reactive power by other parts except a small amount by Z_{s1} and Z_{c1} can be calculated. So it is treated as known in (4).

$$P + jQ = (K_{v0}/3) \cdot \left(K_v \cdot V_{ref} - \left|\mathrm{V}_{\mathrm{xy}}\right|\right) + jQ \tag{13}$$

From (4) and (13), one has

$$C_v^* \cdot \left|\mathrm{V}_{\mathrm{xy}}\right|^2 + \mathrm{I}_{\mathrm{xy}}^* \cdot \mathrm{V}_{\mathrm{xy}} = (K_{v0}/3) \cdot \left(K_v \cdot V_{ref} - \left|\mathrm{V}_{\mathrm{xy}}\right|\right) + jQ \tag{14}$$

By taking the conjugate of both sides of (14), one can obtain

$$C_v \cdot \left|\mathrm{V}_{\mathrm{xy}}\right|^2 + \mathrm{I}_{\mathrm{xy}} \cdot \mathrm{V}_{xy}^* = (K_{v0}/3) \cdot \left(K_v \cdot V_{ref} - \left|\mathrm{V}_{\mathrm{xy}}\right|\right) - jQ \tag{15}$$

Then (14) and (15), one can obtain

$$\begin{aligned} &\mathrm{CoeffA} \cdot \left|\mathrm{V}_{\mathrm{xy}}\right|^4 + \mathrm{CoeffB} \cdot \left|\mathrm{V}_{\mathrm{xy}}\right|^3 + \mathrm{CoeffC} \cdot \left|\mathrm{V}_{\mathrm{xy}}\right|^2 + \mathrm{CoeffD} \cdot \left|\mathrm{V}_{\mathrm{xy}}\right| \\ &+X_1 \cdot X_2 = 0 \end{aligned} \tag{16}$$

where CoeffA = $\left|C_v\right|^2$, CoeffB = $\mathrm{K}_{v0} \cdot \left(C_v^* + C_v\right)$, CoeffC = $\left(C_v \cdot X_1 + C_v^* \cdot X_2\right) - \left|\mathrm{I}_{\mathrm{xy}}\right|^2 + K_{v0}^2$, CoeffD = $\mathrm{K}_{v0} \cdot (X_1 + X_2)$, $X_1 = -(K_{v0}/3) \cdot K_v \cdot V_{ref} - jQ$, $X_2 = -(K_{v0}/3) \cdot K_v \cdot V_{ref} + jQ$.

Again, by using the command 'roots' in Matlab and applying proper judgement of four roots, one can obtain $\left|\mathrm{V}_{\mathrm{xy}}\right|$ first. Then from (14), one can obtain V_{xy}.

For the microgrid system as shown in Fig. 2, the connectivity equations are

$$\mathrm{I}_{11,12} = \frac{V_{12x} + jV_{12y}}{Z_{Load3}} - \mathrm{I}'_{\mathrm{xy},3} \tag{17}$$

$$\mathrm{V}_{11} = \mathrm{I}_{11,12} \cdot Z_{Line2} + \mathrm{V}_{12} \tag{18}$$

$$\mathrm{I}_{10,11} = \frac{V_{11x} + jV_{11y}}{Z_{Load1}} + \mathrm{I}_{11,12} - \mathrm{I}'_{\mathrm{xy},2} \tag{19}$$

$$\mathrm{V}_{10} = \mathrm{I}_{10,11} \cdot Z_{Line1} + \mathrm{V}_{11} \tag{20}$$

2.2 Case a Under Study

DG1 in Fig. 2 serves as the grid-forming generator while both DG2 and DG3 serve as the grid-supporting generators.

The following steps are taken to work out the voltages at each bus,

(1) $[0.5^*V_{rated}, 1.5V_{rated}]$ is taken for the voltage at bus 12. An incremental step is assumed.
(2) A positive real value is taken for V_{12} starting from low one in the range specified in step 1.
(3) From Sect. 2.1.2, the voltage phasor at bus 5 in Fig. 2 is calculated by using Eqs. (7, 8, 9, 10, 11, 12). The current phasors $I_{xy,3}$ from DG3 and $I'_{xy,3}$ are also calculated.
(4) The current from bus 11 and 12 is calculated by using (17).
(5) Voltage at bus 11 is calculated by using (18).
(6) From Sect. 2.1.2, the voltage phasor at bus 3 in Fig. 2 is calculated by using Eqs. (7, 8, 9, 10, 11, 12). The current phasors $I_{xy,2}$ from DG2 and $I'_{xy,2}$ are also calculated.
(7) The current from bus 10 to bus 11 is calculated by using (19) while the voltage at bus 10 is calculated by using (20).
(8) The reactive power consumptions by all loads and also by Z_{Line1}, Z_{Line2}, Z_{s2}, Z_{c2}, Z_{s3}, Z_{c3} and generation by DG2 and DG3 are considered to work out the necessary reactive power from DG1. Exact reactive power from DG1 can be determined by considering the reactive power by Z_{s1} and Z_{c1} as the voltage at bus 10 has been determined. Then from Sect. 2.1.3, the voltage phasor at bus 1 is calculated. The current phasors $I_{xy,1}$ from DG1 and $I'_{xy,1}$ are also calculated.
(9) $V_{xy,1} \times I^*_{xy,1} = P_{1c} + jQ_{1c}$ is calculated. Error between the reference real power and P_{1c} or $\left|P_{1c} - P_{1ref}\right|$ is calculated and recorded, where $P_{1ref} = (K_{v0}/3) \cdot \left(K_v \cdot V_{ref} - \left|V_{xy,1}\right|\right)$.
(10) A step increment is added to the last-step voltage for bus 12; Repeat steps 3–9 until upper limit of the voltage is reached.
(11) Find the minimum value in the recorded error for each voltage value at bus 12 and the corresponding voltage at it.
(12) With the voltage at bus 12 being identified, calculate each voltage and current using steps 3–8.

Table 1 shows the parameters for each inverter where a slightly high DC side voltage is used for reducing the influence of harmonics on the inverter controller. In the current efforts, the b_n for the notch-filter based bandpass filter used in the inverter control flow is set to 4200 while the cut-off frequency for the Butterworth low-pass filter is set to 5.0 kHz [12–15]. Higher cut-off frequency for this low-pass filter is better in terms of reducing phase shift in the closed-loop transfer function of the inverter control flow for tracing the reference currents, but its injection of EMI noise is compromised.

To increase the percent load that the system can bear to turn-on and turn-off without collapse, notch-filter based bandpass filter to process the current flowing through the capacitor in the LCL filter as shown in [12–15] is a good choice. By using relatively large capacitance in the LCL filter, the 50 Hz current flowing through the capacitor can be obtained by the filter accurately. Then it participates the formation of the reference voltages. It is found that even if the capacitor's current is very high, the system still can work properly.

Table 2 shows the loads' information. One single-phase load between phase A and neutral is turned on at 2.5 s and turned off at 3.2 s. Another single-phase load between phase B and neutral is turned on at 2.8 s and turned off at 7.0 s. The modelling results for comparison with those from phasor domain approach are collected at t = 2 s and t = 8 s respectively as there are no single-phase loads at each of these two moments.

Table 1 Parameters for inverters

Parameters	Values
L_{1inv}	16.5 mH
C_{inv}	45 μF
L_{2inv}	4.5 mH
$R_{damping}$	1Ω
$f_{switching}$	5 kHz
K_p	0.275
K_i	30
K	60
K_{v0}	2000
K_v	1.15
K_{pq}	0.4843
V_{DC}	6300 V
V_{LL}(rms)	3.1125 kV

Table 2 Information on loads

Three-phase loads			Single-phase loads	
Load1	Load2	Load3	A-N	B-N
$R_{p1} = 0.27337\Omega$, $L_{p1} = 1.3053mH$	$R_{p2} = 0.27337\Omega$, $L_{p2} = 1.3053mH$	$R_{p3} = 0.20503\Omega$, $L_{p3} = 0.87017mH$	$R = 0.25\ \Omega$	$R = 0.25\ \Omega$
On: 1 s; off: 2.5 s; On: 3.5 s; off: 4.5 s	On	On	On: 2.5 s; Off: 3.2 s	On: 2.8 s; Off: 7.0 s

Tables 3 and 4 show the results for the case A under study, where the error in percent is defined as the ratio of the absolute of the difference to the value from phasor-domain calculation. On average, multi-rate modelling results in smaller difference than the single-rate modelling.

Figures 5, 6, 7, 8, 9 show the results for Case A using single-rate modelling while Figs. 10, 11, 12, 13, 14 are for Case A using multi-rate modelling. For either set of results, the power outputs from each of three DGs trace their respective reference well. Small deviation of reactive power output for DG2 and DG3 from their references is due to the single-phase loads between 2.5 and 7.0 s.

Table 3 Voltages at each bus in Fig. 2 for single-rate approach with no PV

Time		V_{Bus1} (V)	V_{Bus2} (V)	V_{Bus3} (V)	V_{Bus10} (V)	V_{Bus11} (V)	V_{Bus12} (V)
t = 2 s	Modelling	1877	1855	1853	6417	6406	6395
	Calculation	1893	1814	1720	5741	5799	5784
	Error (%)	0.85	2.26	7.73	11.8	10.46	10.56
t = 8 s	Modelling	1921	1907	1902	6634	6637	6615
	Calculation	1949	1867	1789	6061	6126	6093
	Error (%)	1.44	2.14	6.32	9.45	8.34	8.57

Table 4 Voltages at each bus in Fig. 2 for multi-rate approach with no PV

Time		V_{Bus1} (V)	V_{Bus2} (V)	V_{Bus3} (V)	V_{Bus10} (V)	V_{Bus11} (V)	V_{Bus12} (V)
t = 2 s	Modelling	1776	1756	1752	5975	5963	5949
	Calculation	1893	1814	1720	5741	5799	5784
	Error (%)	6.18	3.20	1.86	4.08	2.83	2.85
t = 8 s	Modelling	1841	1827	1821	6282	6290	6258
	Calculation	1949	1867	1789	6061	6126	6093
	Error (%)	5.54	2.14	1.79	3.65	2.68	2.71

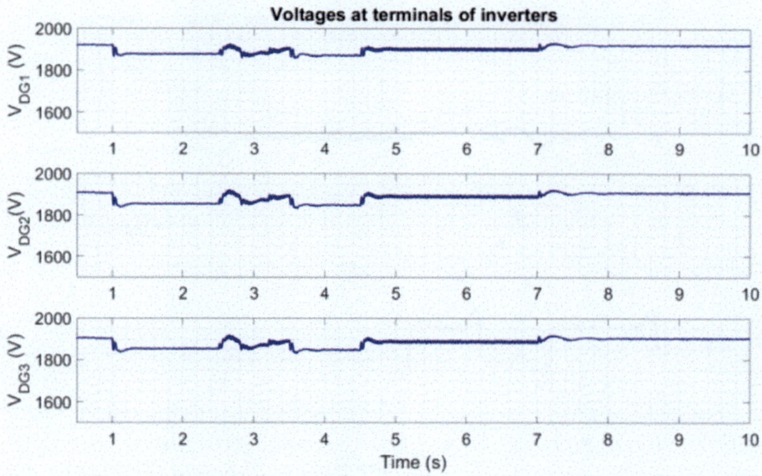

Fig. 5 Voltages at the terminals of each inverter for single-rate approach with no PV

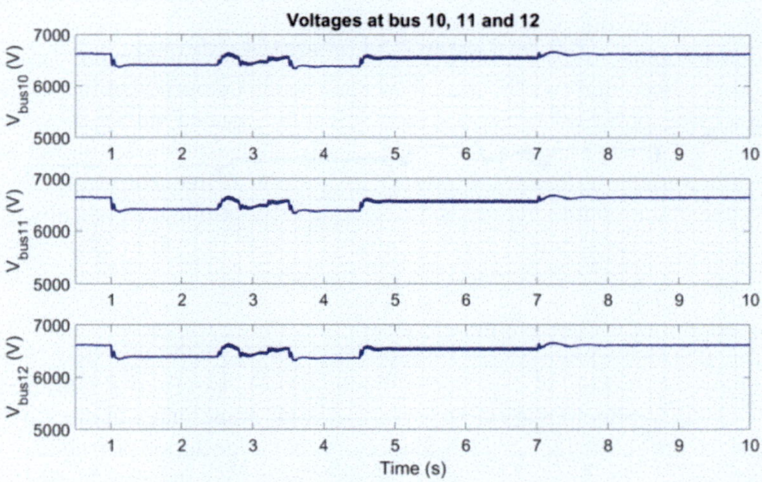

Fig. 6 Voltages at the terminals of each load referred to high voltage side for single-rate approach with no PV

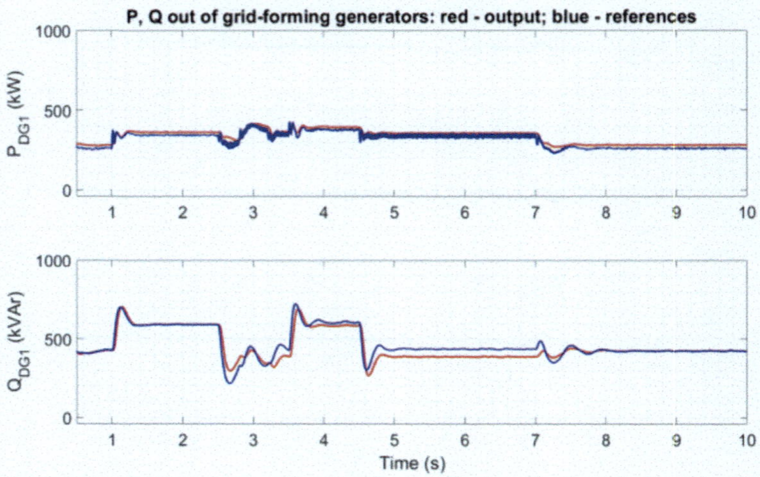

Fig. 7 Real and reactive powers from DG1 and their references for single-rate approach with no PV

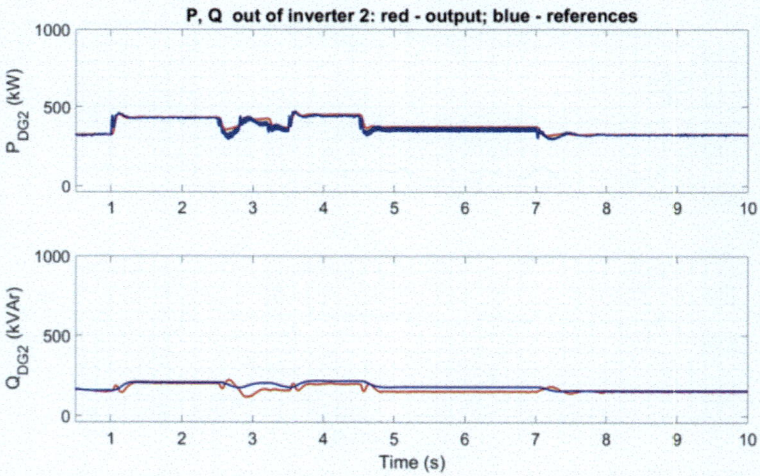

Fig. 8 Real and reactive powers from DG2 and their references for single-rate approach with no PV

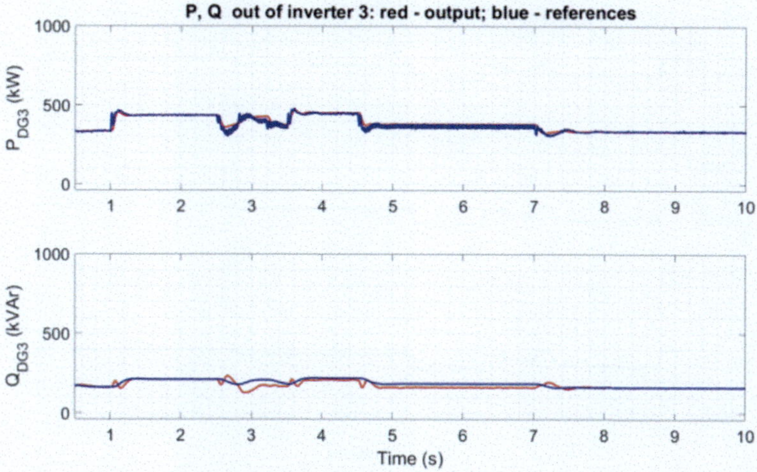

Fig. 9 Real and reactive powers from DG3 and their references for single-rate approach with no PV

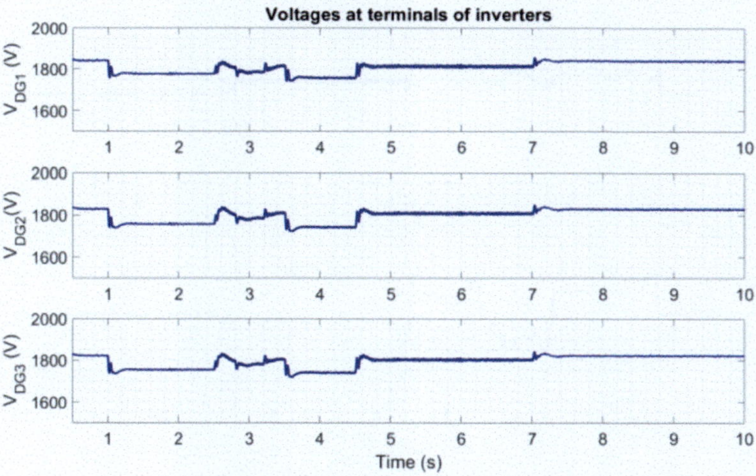

Fig. 10 Voltages at the terminals of each inverter for multi-rate approach with no PV

2.3 *Case B Under Study*

In this case, DG1 serves as the grid-forming generator while DG2 is a grid-supporting generator and DG3 is a grid-feeding generator.

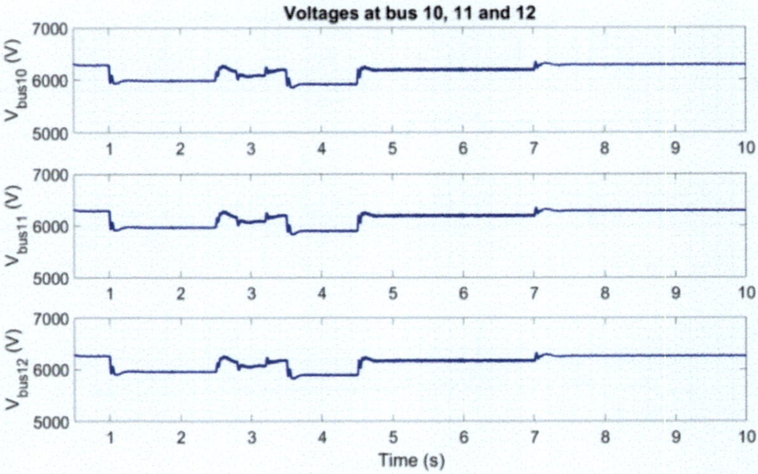

Fig. 11 Voltages at the terminals of each load for multi-rate approach with no PV

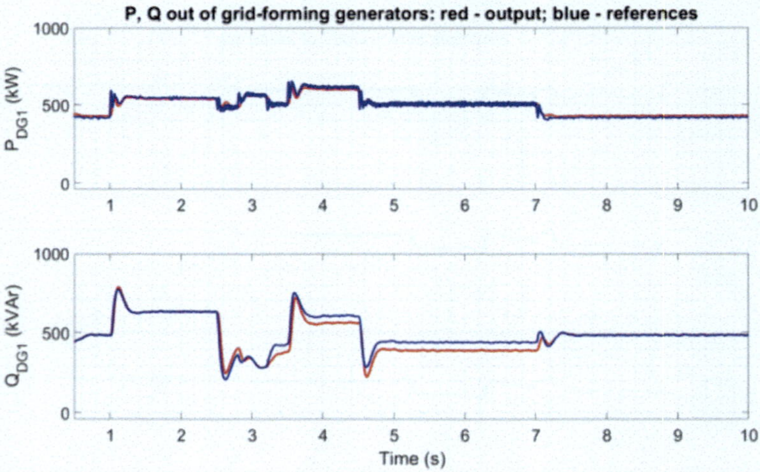

Fig. 12 Real and reactive powers from DG1 and their references for multi-rate approach with no PV

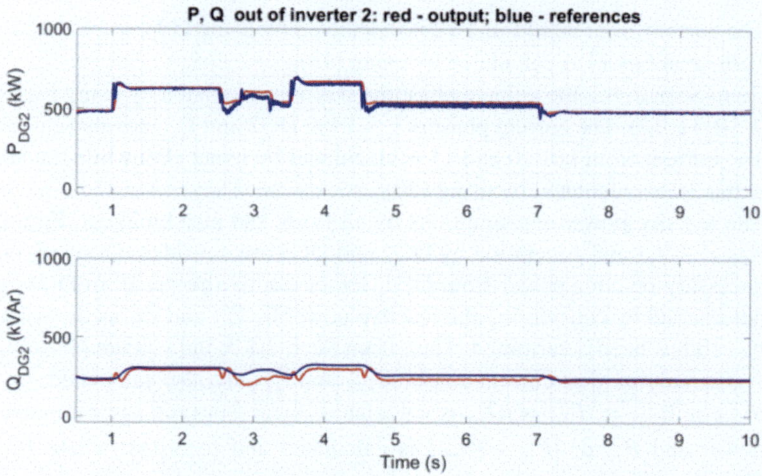

Fig. 13 Real and reactive powers from DG2 and their references for multi-rate approach with no PV

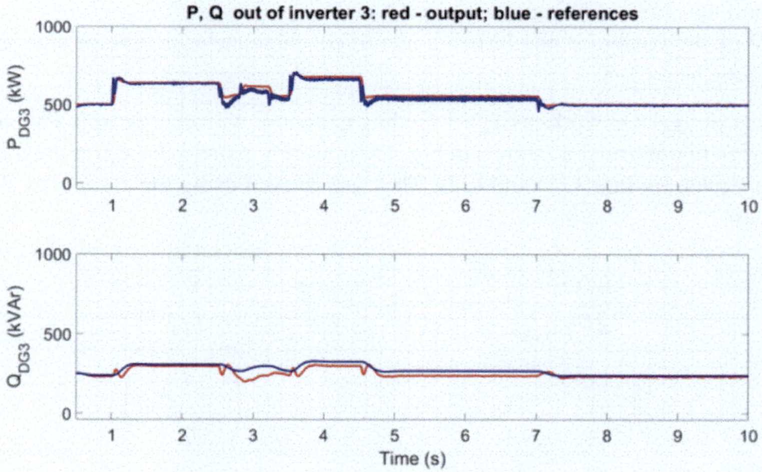

Fig. 14 Real and reactive powers from DG3 and their references for multi-rate approach with no PV

The following steps are taken to work out the voltages at each bus:

(1) [$0.5*V_{rated}$, $1.5V_{rated}$] is taken for the voltage at bus 12. An incremental step is assumed.
(2) A positive real value is taken for V_{12} starting from low one in the range specified in step 1.
(3) From Sect. 2.1.1, the voltage phasor at bus 5 in Fig. 2 is calculated by using Eqs. (5 and 6). The current phasors $I_{xy,3}$ from DG3 and $I'_{xy,3}$ are also calculated.

(4) The current from bus 11 and 12 is calculated by using (17).
(5) Voltage at bus 11 is calculated by using (18).
(6) From Sect. 2.1.2, the voltage phasor at bus 3 is calculated by using Eqs. (7, 8, 9, 10, 11, 12). The current phasors $I_{xy,2}$ from DG2 and $I'_{xy,2}$ are also calculated.
(7) The current from bus 10 and 11 is calculated by using (19) while the voltage at bus 10 is calculated by using (20).
(8) The reactive power consumptions by all loads and also by Z_{Line1}, Z_{Line2}, Z_{s2}, Z_{c2}, Z_{s3}, Z_{c3} and generation by DG2 and DG3 are considered to work out the necessary reactive power from DG1. Exact reactive power from DG1 can be determined by considering the reactive power by Z_{s1} and Z_{c1} as the voltage at bus 10 has been determined. Then from Sect. 2.1.3, the voltage phasor at bus 1 is calculated. The current phasors $I_{xy,1}$ and $I'_{xy,1}$ are also calculated.
(9) $V_{xy,1} \times I^*_{xy,1} = P_{1c} + jQ_{1c}$ is calculated. Error between the reference real power and P_{1c} or $\left|P_{1c} - P_{1ref}\right|$ is calculated and recorded, where $P_{1ref} = (K_{v0}/3) \cdot \left(K_v \cdot V_{ref} - \left|V_{xy,1}\right|\right)$.
(10) A step increment is added to the last-step voltage for bus 12; Repeat steps 3–9 until upper limit of the voltage is reached.
(11) Find the minimum value in the recorded error for each voltage value at bus 12 and the corresponding voltage at it.
(12) With the voltage at bus 12 being identified, calculate each voltage and current using steps 3–8.

In this study, DG3 works as a grid-feeding generator with energy from PV panels. The design of controller as shown in Fig. 15 for the PV panel is based on the transfer function given by Eqs. (21) and (22). They are derived using the circuits as shown in Fig. 16.

$$\left.\frac{\hat{i}_L(s)}{\hat{d}(s)}\right|_{\hat{v}_{dc}(s)=0} = \frac{s + \frac{1}{RC_{PV}}}{s^2 + \frac{1}{RC_{PV}}s + \frac{1}{L_{PV}C_{PV}}} \cdot \frac{\overline{V}_{dc}}{L_{PV}} \tag{21}$$

$$\left.\frac{\hat{v}_c(s)}{\hat{d}(s)}\right|_{\hat{v}_{dc}(s)=0} = \frac{1}{s^2 + \frac{1}{RC_{PV}}s + \frac{1}{L_{PV}C_{PV}}} \cdot \left(\frac{-\overline{V}_{dc}}{L_{PV}C_{PV}}\right) \tag{22}$$

Load information for Case B study is the same as those for Case A. Tables 5 and 6 show the results for the case under study. Both single-rate and multi-rate modelling methods produce results equally close to the values from phasor domain approach.

Figure 17 shows the irradiance level and the inductor current and its reference. By taking the controller for the PV panel above, the current flowing through the

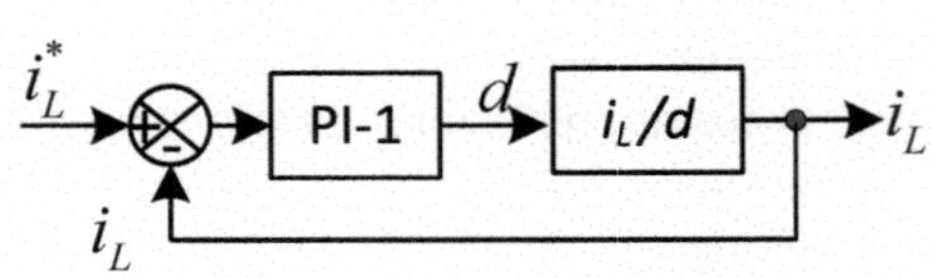

Fig. 15 MPPT control: I control

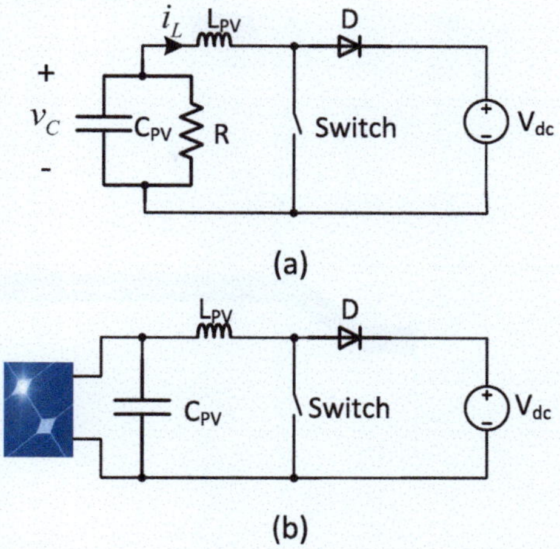

Fig. 16 **a** Modified boost converter **b** Boost converter with solar panel

Table 5 Voltages at each bus in Fig. 2 for single-rate approach with PV

Time		V_{Bus1} (V)	V_{Bus2} (V)	V_{Bus3} (V)	V_{Bus10} (V)	V_{Bus11}(V)	V_{Bus12}(V)
t = 2 s	Modelling	1744 V	1725	1724	5840	5827	5812
	Calculation	1871 V	1798	1699	5620	5694	5686
	Error (%)	6.79	4.96	4.06	3.91	2.34	2.22
t = 8 s	Modelling	1805	1793	1777	6127	6132	6090
	Calculation	1923	1846	1762	5918	6001	5975
	Error (%)	6.60	2.87	0.85	3.53	2.18	1.92

Table 6 Voltages at each bus in Fig. 2 for multi-rate approach with PV

Time		V_{Bus1} (V)	V_{Bus2} (V)	V_{Bus3} (V)	V_{Bus10} (V)	V_{Bus11} (V)	V_{Bus12} (V)
t = 2 s	Modelling	1741	1722	1715	5826	5812	5794
	Calculation	1866	1794	1694	5592	5670	5663
	Error (%)	6.70	4.01	1.24	4.18	2.50	2.31
t = 8 s	Modelling	1807	1795	1779	6132	6130	6100
	Calculation	1924	1847	1763	5923	6006	5979
	Error (%)	6.08	2.82	0.91	3.53	2.06	2.02

inductor follows its reference closely in the curtailment operation. Figure 18 shows the voltages at the PV panel terminal, output voltage of the modified boost converter and DC input for the inverter. As the PV panel terminal voltage does not change much, the control method in [13–15] works properly.

For the single-rate approach, the voltage profiles at each terminal of the DGs and the load side referred to the high voltage side of the transformers, and also the real

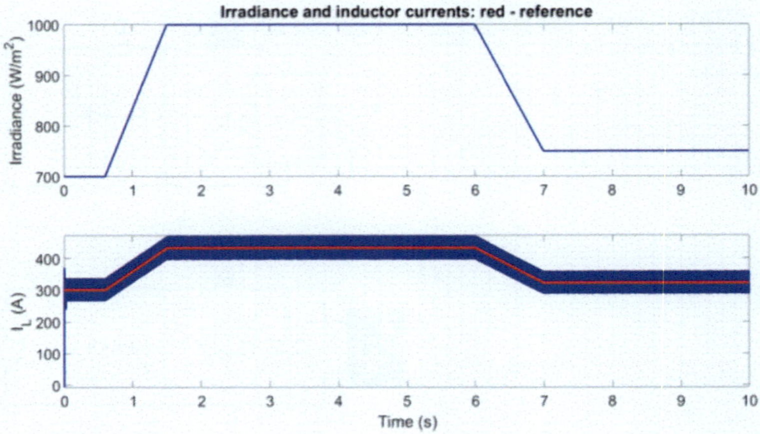

Fig. 17 Irradiance and inductor current

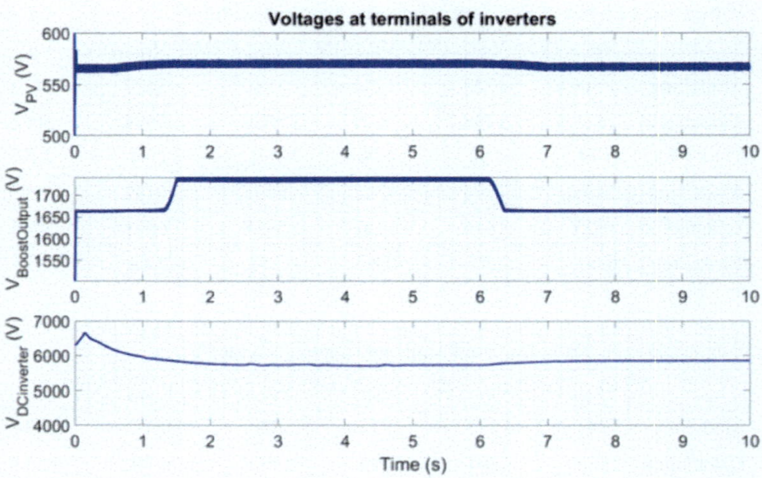

Fig. 18 Voltages for the PV panel part

and reactive powers from each DG are shown in Figs. 19, 20, 21, 22, 23. Similarly results for the multi-rate are shown in Figs. 24, 25, 26, 27, 28, 29, 30.

For this single-rate with PV, power from PV panels: at t = 2 s, P_3 = 709 kW, Q_3 = 336kVAr; t = 8 s, P_3 = 505 kW, Q_3 = 238kVAr.

For the multi-rate with PV, power from PV panels: at t = 2 s, P_3 = 680 kW, Q_3 = 325kVAr; t = 8 s, P_3 = 510 kW, Q_3 = 238kVAr.

Fig. 24Irradiance and inductor current

The significant amount of negative-sequence currents resulting from the unbalanced loads could flow through the grid-forming generator to cause malfunction to

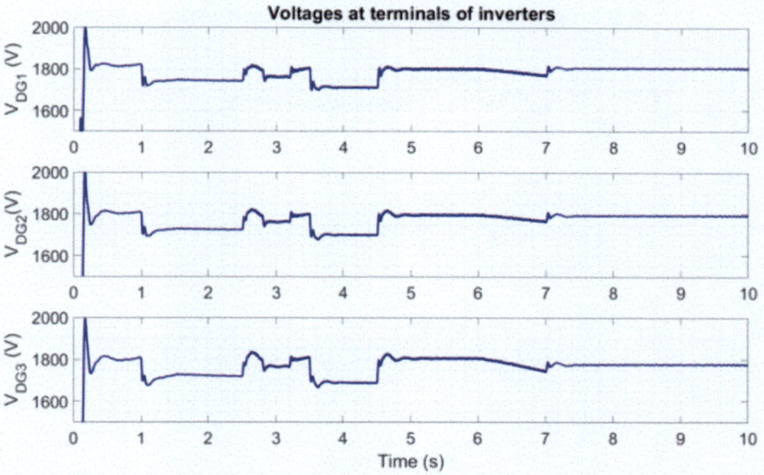

Fig. 19 Voltages at the terminals of each inverter for single-rate approach with PV

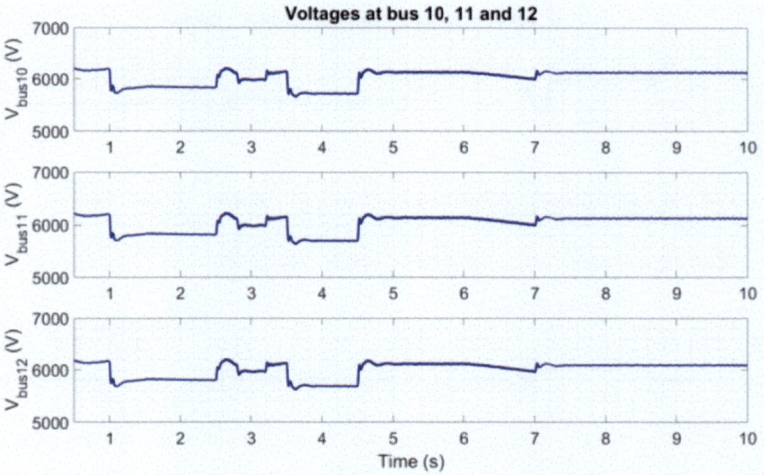

Fig. 20 Voltages at the terminals of each load

it. Therefore, it is necessary to have high enough ratio of inverters' terminal voltages to the load rated voltages. Although from the simulation results shown above, the grid-forming generators and other generators could ride through the temporary severe unbalanced load currents, caution still needs to be exerted in hardware implementation. Necessary unbalance current compensation could be implemented at loads' sides to alleviate its influence. Furthermore, besides the operating back-up grid-forming generator, one can adopt another cold-start back-up grid-forming generator to replace the agonizing grid-forming generator influenced by the negative-sequence current. That is to say, when necessary, a back-up grid-forming generator

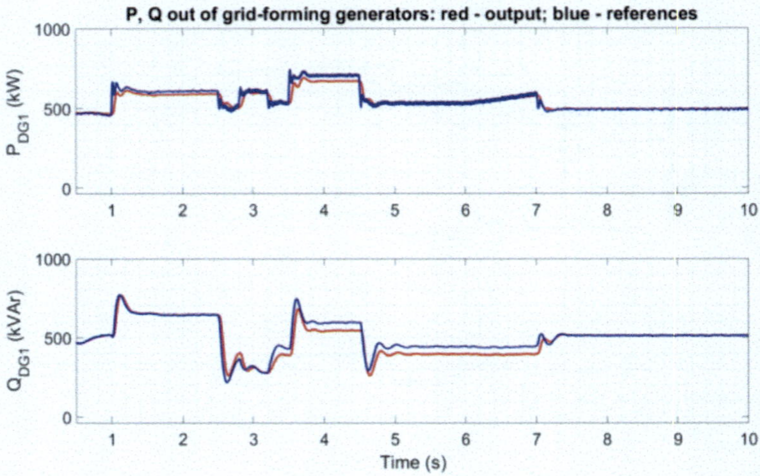

Fig. 21 Real and reactive powers from DG1 and their references for single-rate approach with PV

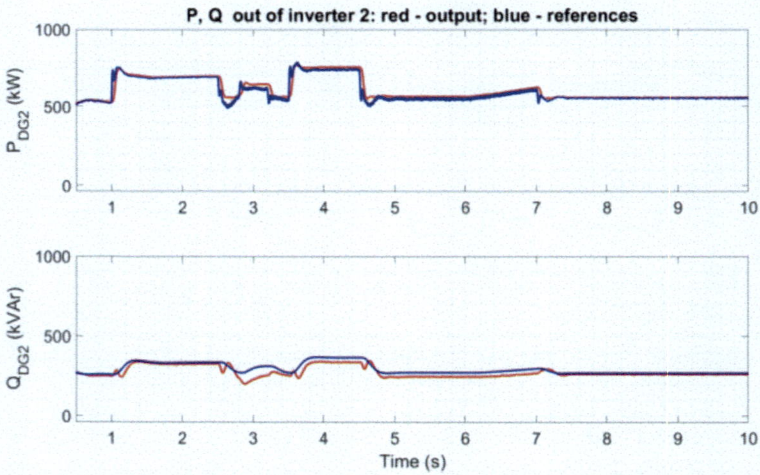

Fig. 22 Real and reactive powers from DG2 and their references for single-rate approach with PV

will start from cold state and replace the operating one to remove the influence by the negative-sequence current.

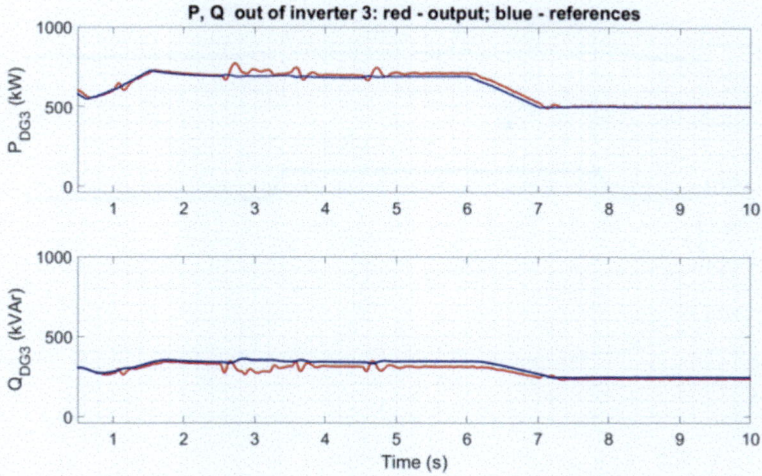

Fig. 23 Real and reactive powers from DG3 and their references for single-rate approach with PV

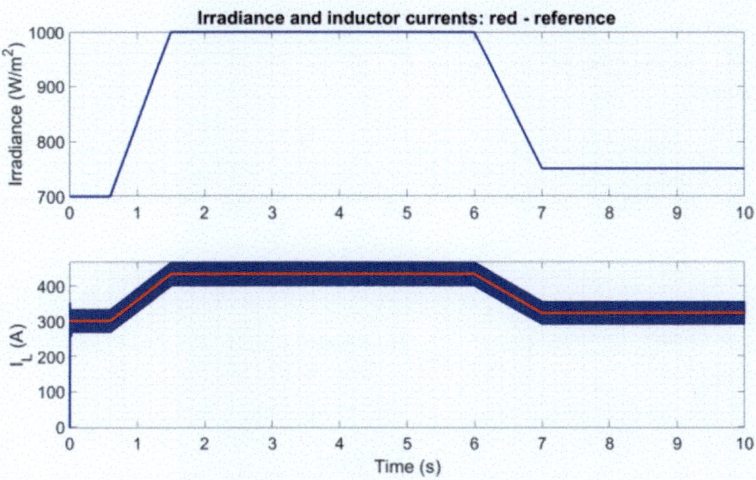

Fig. 24 Irradiance and inductor current

2.4 *Influence of Variation of DC Link Voltage on Inverter Power Outputs*

That outputs of real and reactive power accurately trace their respective references from each inverter determines whether an autonomous microgrid can work in controllable mode and power output from each inverter is manageable. In this part, the influence of variation of DC link voltage will be studied.

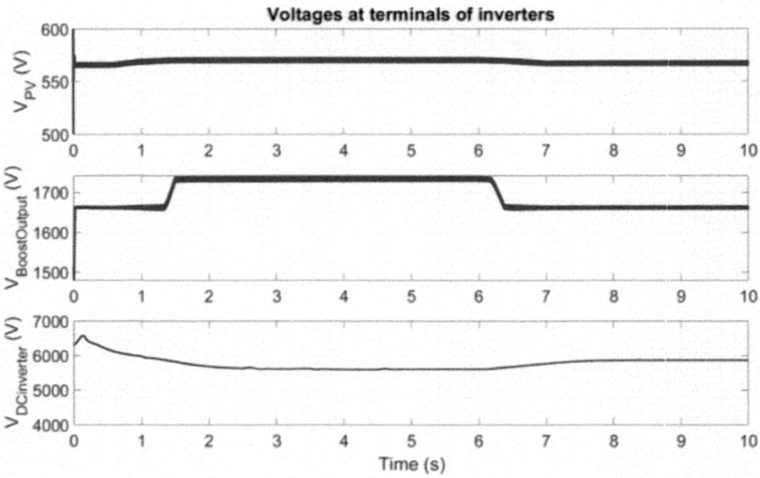

Fig. 25 Voltages for the PV panel part

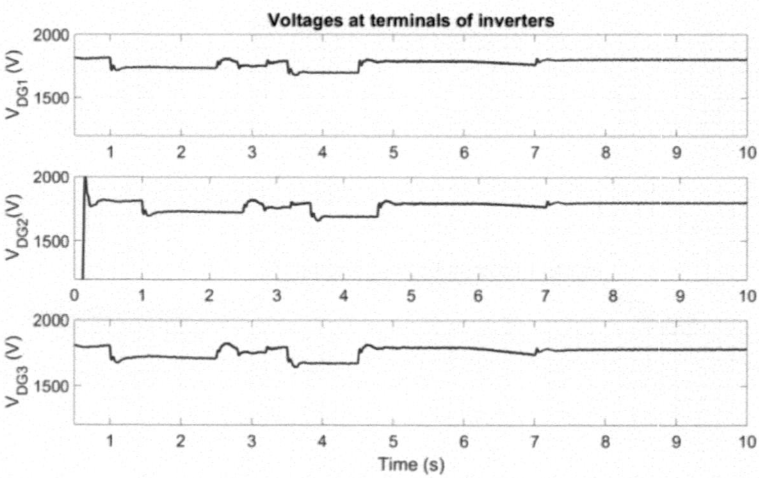

Fig. 26 Voltages at the terminals of each inverter for multi-rate approach with PV

Two cases are studied here. Figure 31 shows the variations of DC link voltage for DG1 while the voltages for the DC links of DG2 and DG3 are kept the same and equal to 5775 V for the first case. In the second case, the variations of DC link voltages for each of DG1, DG2 and DG3 are shown in Fig. 32.

Table 7 shows the output real power and reactive power for two moments, one being 2.4 s while the other being 8.0 s for both cases of DC link voltage variation as shown in Figs. 31 and 32. From it, one can see that although the DC link voltages vary, the output real powers for Cases 1 and 2 in Table 7 at each of two chosen moments are almost the same. So are the reactive powers. More comparison can be

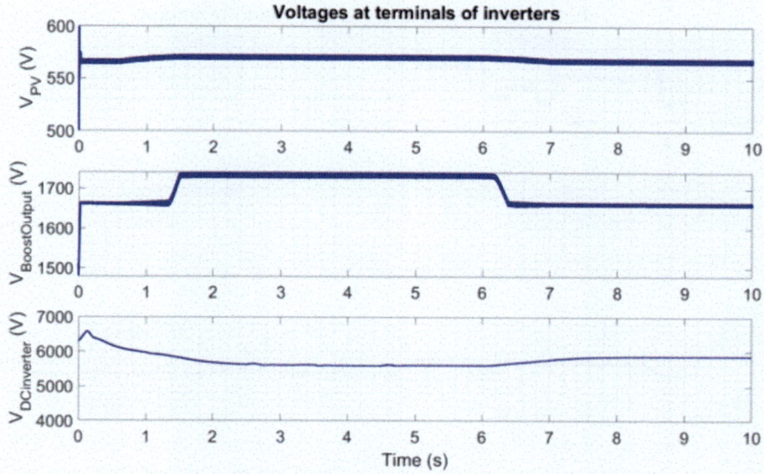

Fig. 27 Voltages at the terminals of each load for multi-rate approach with PV

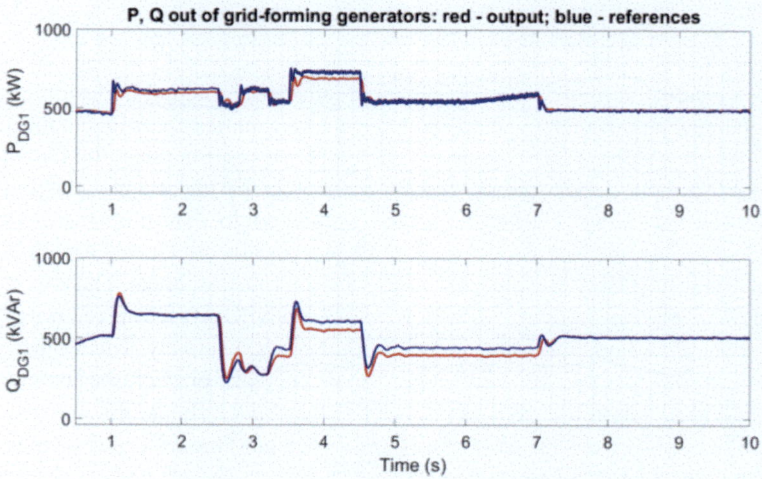

Fig. 28 Real and reactive powers from DG1 and their references for multi-rate approach with PV

done for other time moments as shown in Figs. 33 and 34. The same conclusion can be drawn for each moment when the microgrid is under steady-state with balanced loads.

Under most operating conditions, the inverters work in under-modulation mode as shown in Fig. 35 for DG1. Nevertheless, other DGs e.g. DG3 could work in over-modulation mode as shown in Fig. 36 when the loads are unbalanced. Due to the single-phase loads, the peak values of the three-phase references vary as shown in Figs. 35b and 36b. By using proper filters in the inverter control flow and also slightly

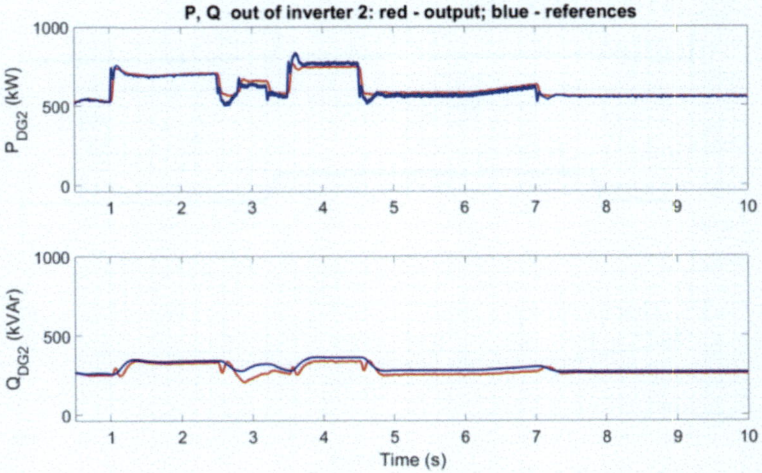

Fig. 29 Real and reactive powers from DG2 and their references for multi-rate approach with PV

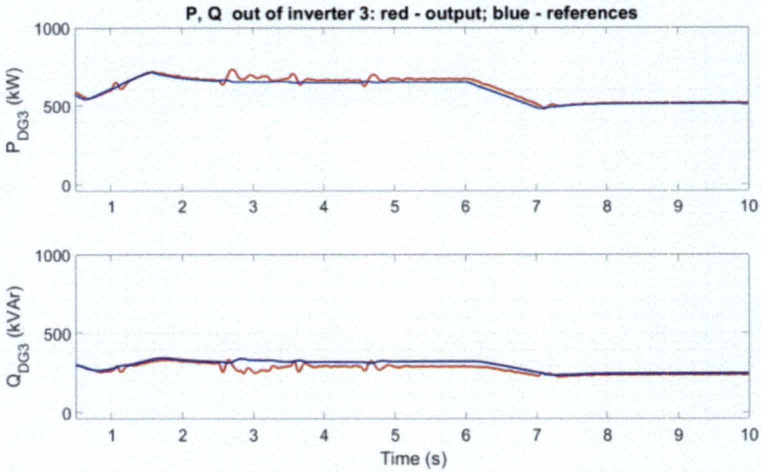

Fig. 30 Real and reactive powers from DG3 and their references for multi-rate approach with PV

high DC link voltage for each inverter, the temporary high negative-sequence currents will not be able to topple the stability of the system. Nevertheless, for a practical microgrid, it is necessary to have higher inverter terminal voltages for the same power generation in order to reduce the impact of the negative-sequence current. For the inverter with relatively low rated terminal voltages or stricter requirement on three-phase voltages' balance for powering three-phase loads such as induction motors, unbalanced current compensation at downstream load sides needs to be adopted.

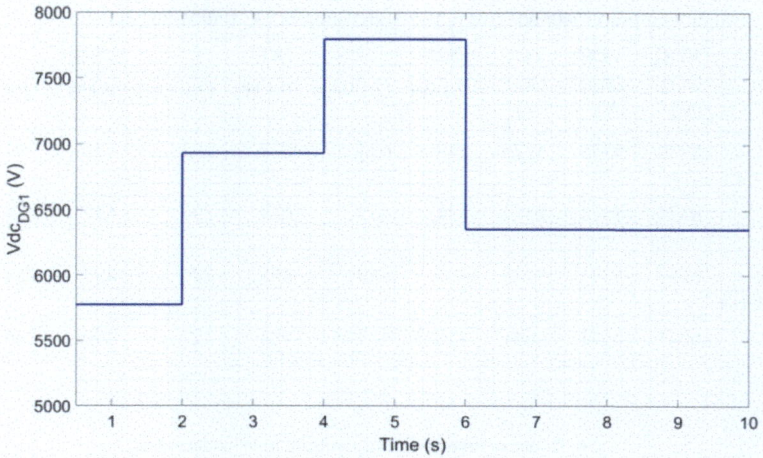

Fig. 31 Variation of DC link voltage for DG1

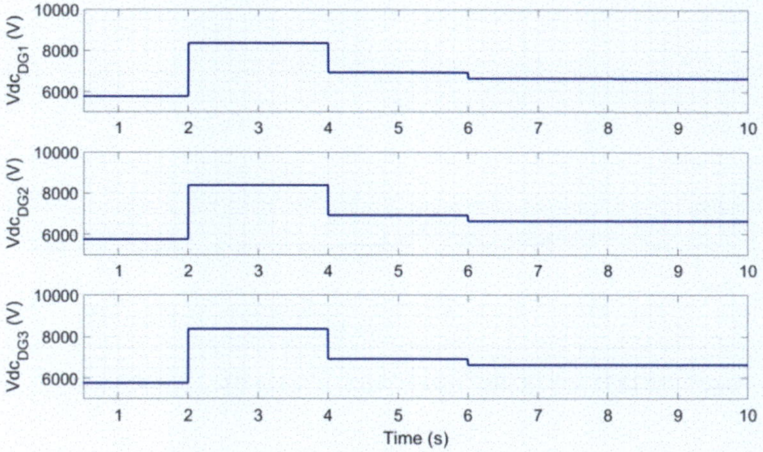

Fig. 32 Variation of DC link voltages for DG1, DG2 and DG3

2.5 Narrow Pulse Removal for Reducing Switching Losses

Figures 37 and 38 show the narrow pulse removal for the gating signal. This is to reduce switching losses. For a practical application using Texas Instrument 28377 s or 28379x, its embedded dead-time insertion serves the same purpose. Although complex space vector modulation could reach the same purpose, the adopted SPWM with narrow pulse removal is simple and yet still effective to use.

Table 7 Output real power and reactive power with different DC link voltages

		VDC_ DG1 (V)	VDC_ DG2 (V)	VDC_ DG3 (V)	DG1		DG2		DG3	
					P(kW)	Q(kVAr)	P(kW)	Q(kVAr)	P(kW)	Q(kVAr)
Case 1	t = 2.4 s	6930	5775	5775	548	634	627	292	635	297
Case 2		8374	8374	8374	546	637	629	294	637	298
Case 1	t = 8.0 s	6353	5775	5775	436	489	485	226	497	234
Case 2		6641	6641	6641	436	489	484	227	499	236

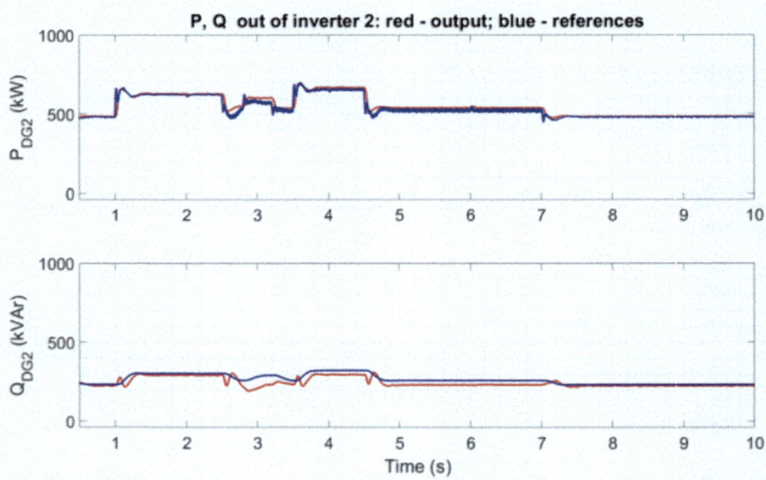

Fig. 33 Power from DG2 with DC link voltage variation in Fig. 31

2.6 Discussion

In comparison, it could be easier to build high-voltage and high-power single-phase autonomous microgrids as three-phase microgrids have the issues due to the existence of negative-sequence and zero-sequence currents. A single-phase autonomous microgrid consisting of multi-stage single-phase inverters [15, 16] could be the most cost-effective system to avoid the issues of negative and zero sequence currents in three-phase system. This is because for long-distance power transmission, distributed capacitance between lines and ground will produce zero-sequence current which could force designers to use delta-wye transformers at the inverters' side to remove it in order for the inverters to operate stably. Nevertheless, for relatively short completely transposed transmission lines between inverters and end users, such transformers could be omitted in microgrid topologies as shown in Figs. 39a and

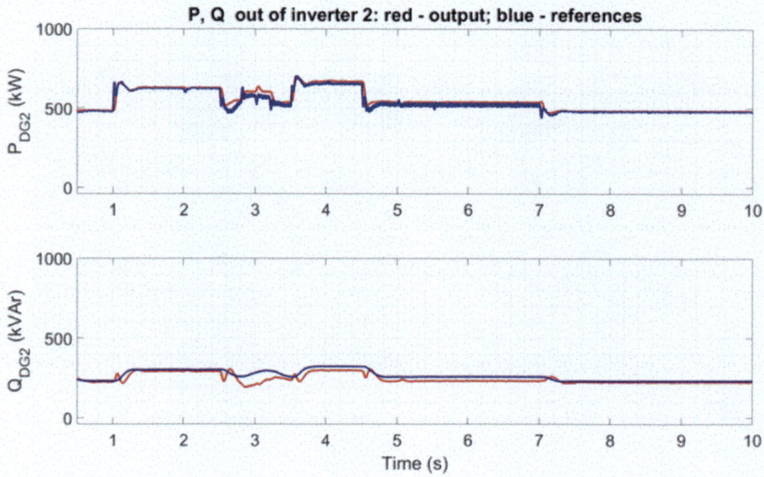

Fig. 34 Power from DG2 with DC link voltage variation in Fig. 32

b, where DG2x are an exception and they join with a 50 Hz/60 Hz step-down transformer. Grounding grids in an area with the inverter based generations are important to ensure EMI radiated emission within standard requirements and they must be designed properly. By connecting casings or metal enclosures of each equipment to the grounding grids and their proper bonding, equi-potentials are established and electromagnetic fields in the ambience are reduced to a safe level.

By using multi-stage topologies for the inverters in the microgrid operated at constant frequency [15, 16], a power level of tens even hundreds of MWs could be reached. The formation of the high power single-phase autonomous microgrid can be in such a way that its grid-forming generator and its back-up are the ones as shown in Fig. 39d and e while other grid-supporting and grid-feedings generators are formed by the circuit as shown in Fig. 39e. The reactive power compensators accompanying the grid-forming generator are composed of multiple LCL filter based inverter circuits in parallel, one of which is shown in Fig. 39e with the DC input source replaced by capacitors. The DC microgrid in Fig. 39d could be in different forms as discussed in [17–19]. A simple one could be a DC microgrid system which consists of gravity storage, fuel cell based generation, wind generation and solar generation with curtailment control, rectifier with power from conventional synchronous generation and DC loads etc. [14]. Operation of the DC microgrid needs to be in such a way that power from it to the grid-forming generator must be relatively constant within one duration. It may need to be adjusted to a new level within other durations. For small-scale microgrid systems, it could operate to have variable output from the grid-forming generator while for large microgrid systems, it could be better to use nearly constant power from it.

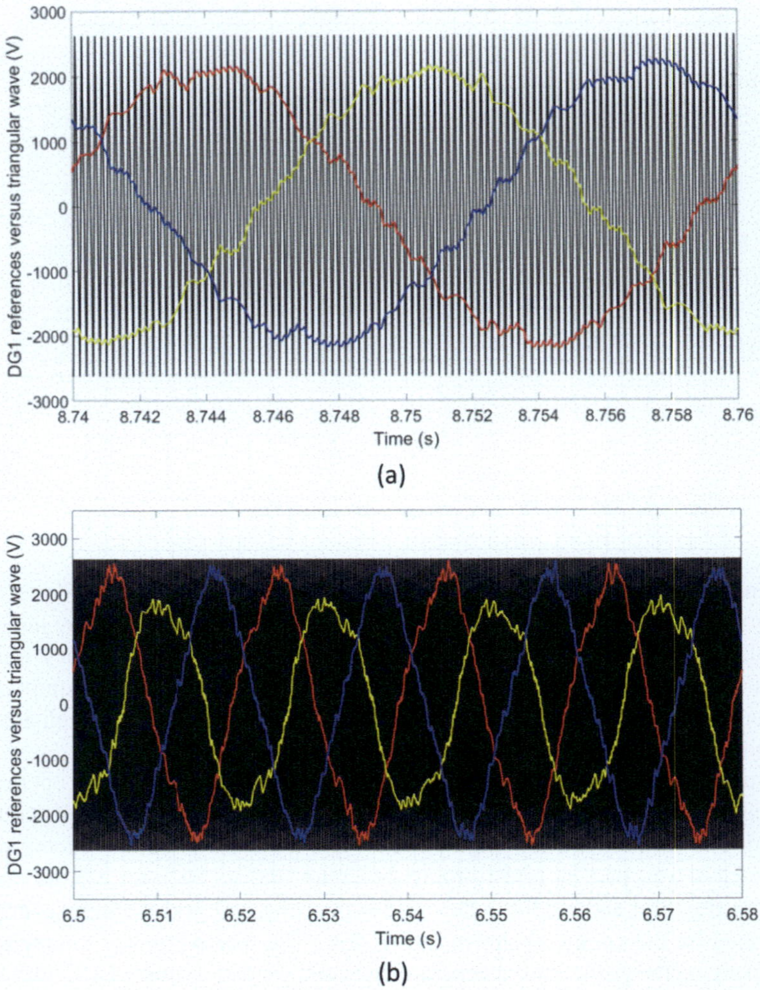

Fig. 35 Comparison of DG1 reference voltages with triangular waveform

In a similar way, a three-phase microgrid could be built. To increase power level, LCCL filter based inverters or a combination of them could also be used in the formation of microgrids [16]. For three-phase microgrids, unbalance compensation at load sides may need to be taken when the loads are severely unbalanced.

No matter which topologies the microgrid takes and whether it is a three-phase or single-phase system, the multi-rate modelling, load flow analysis methods and DC side voltage influence presented above are applicable to each of them.

As discussed earlier, autonomous microgrids could be used for power supply to remote mining sites, auxiliary power supply on trains, power supply for remote residential areas or supplying power to existing loads, either residential ones or

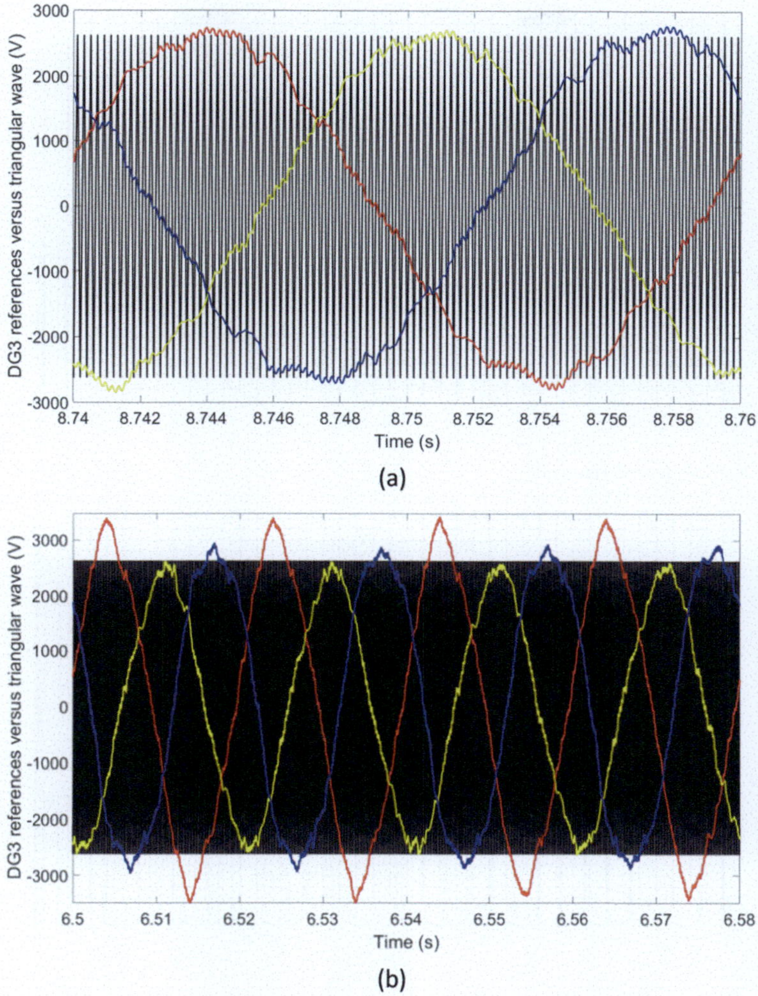

Fig. 36 Comparison of DG3 reference voltages with triangular waveform

factory ones. For the most latter application, different clusters of generation as shown in Fig. 39a could be installed at different power supply points with high voltages such as 11 kV even higher. At load sides, multiple distributed small-scale renewable generations can continue to be used or installed.

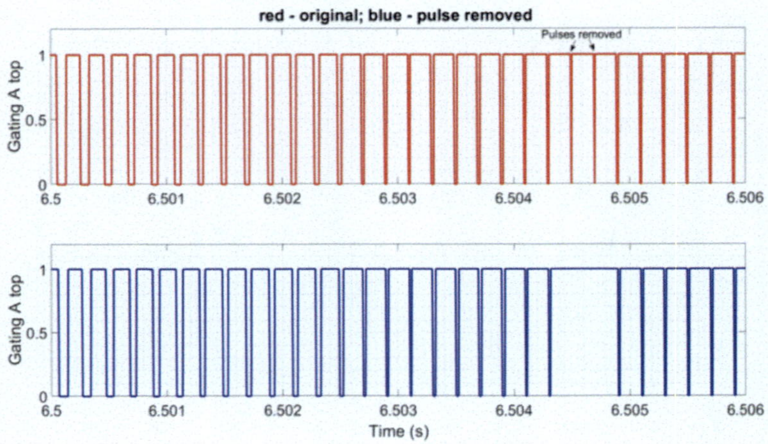

Fig. 37 Narrow pulses' removal

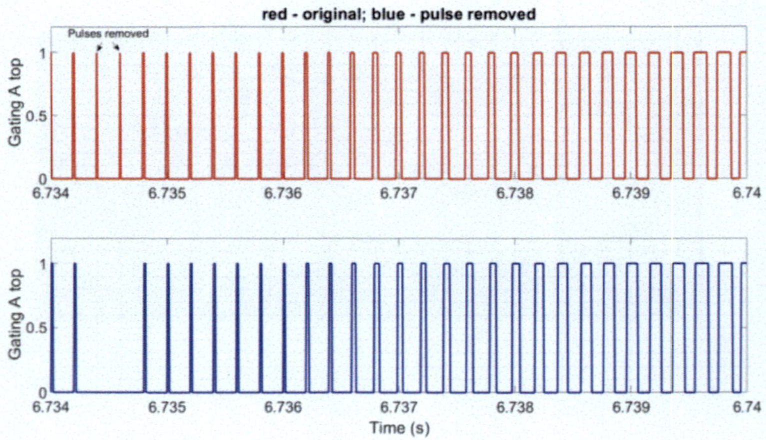

Fig. 38 Narrow pulses' removal

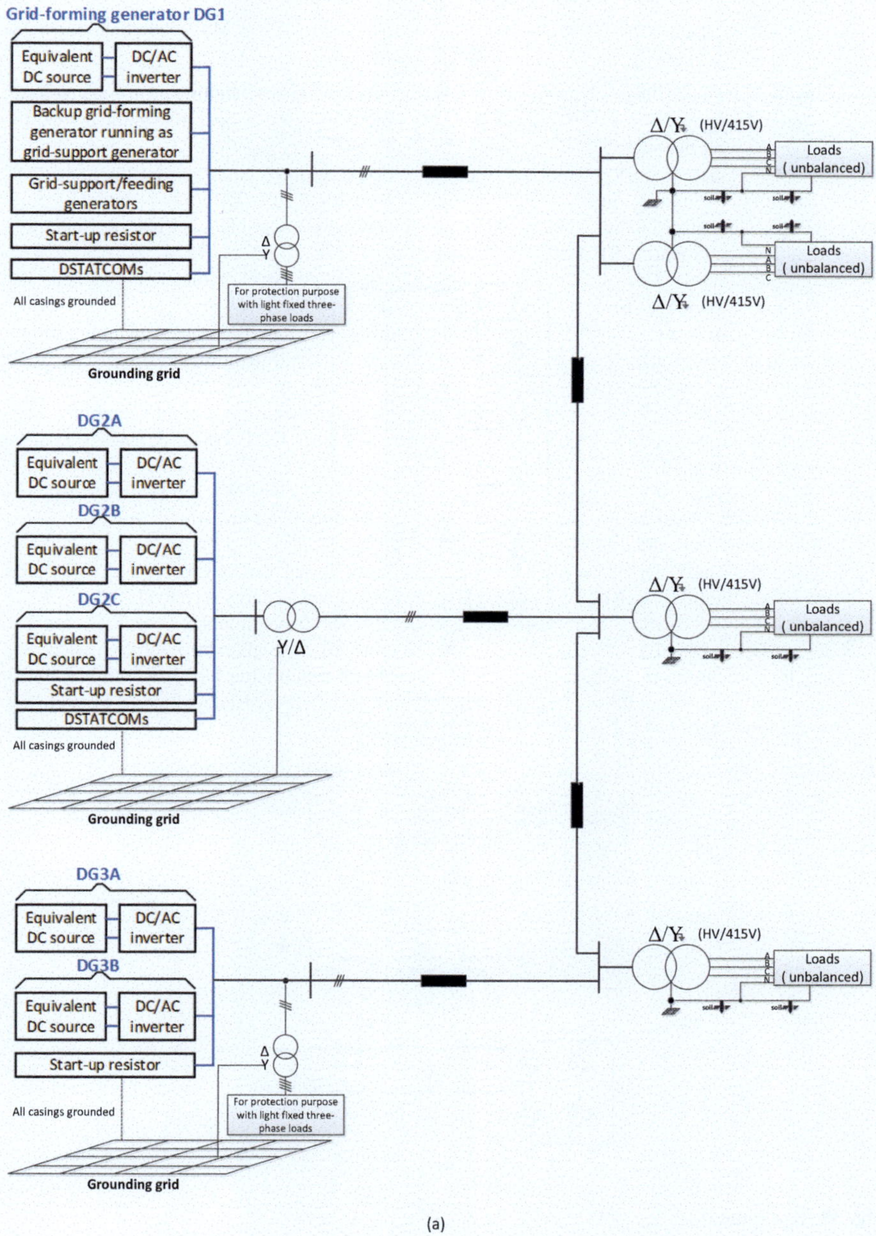

Fig. 39 Autonomous microgrid with/without 50/60 Hz transformers at inverters' sides

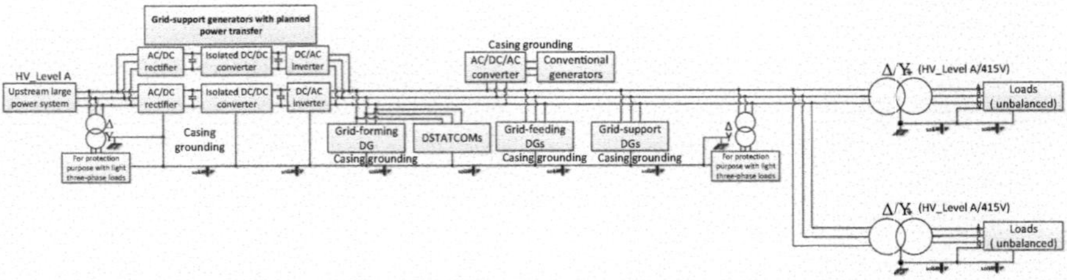

(b)

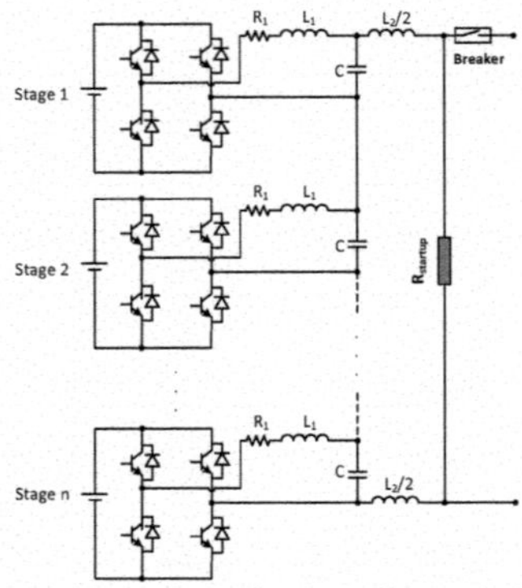

(c)

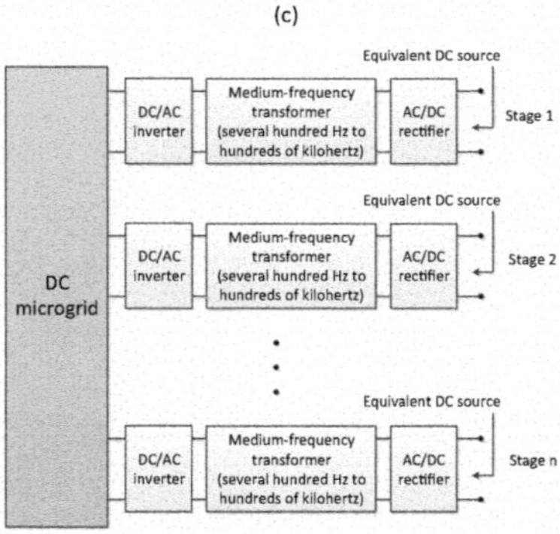

(d)

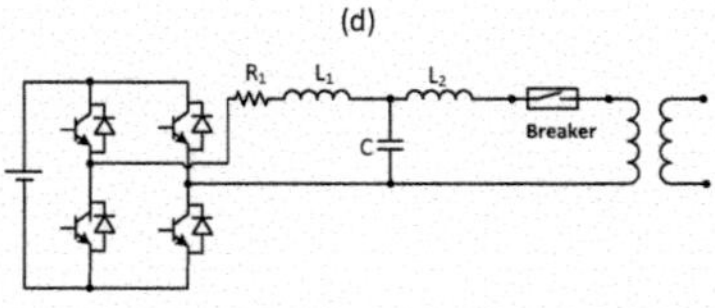

(e)

Fig. 39 (continued)

3 Conclusion

This chapter presented a fixed-time-step multi-rate modeling in Matlab/Simulink for an autonomous microgrid with an attempt to reduce the difference between modeling and hardware implementation. For a comparison purpose, single-rate modeling has also been conducted. Furthermore, phasor-domain analytical solutions of the voltages at each bus in the microgrid have also been sought and are compared with results from both single-rate and multi-rate modeling. It is found that the results from multi-rate modeling are closer to those from phasor domain analysis. This validates the effectiveness of the multi-rate modeling method, which is made closer to the method of hardware implementation of inverter control using two TI 28377 s or TI 28379X. Therefore, instead of building expensive hardware prototype, one can adopt the Matlab/Simulink based modeling to design the large-scale microgrid system and examine its steady-state and transient response. In contrast, phasor-domain analytical method can only examine the steady-state performance.

The influence of DC link voltage on the real and reactive power tracing for each inverter has also been investigated. It is found that the inverter can tolerate more than 50% higher DC link voltage without losing accurate power tracing. Nevertheless lower DC link voltage will topple the operation of the inverters straightforwardly. Furthermore, the removal of narrow pulses of gating signal has been adopted in order to reduce switching losses.

References

1. Olivares DE, Mehrizi-Sani A, Etemadi AH, Cañizares CA, Iravani R, Kazerani M et al (2014) Trends in microgrid control. IEEE Trans Smart Grid. 5:1905–1919
2. De Brabandere K, Bolsens B, Van den Keybus J, Woyte A, Driesen J, Belmans R (2007) A voltage and frequency droop control method for parallel inverters. IEEE Trans Power Electron 22:1107–1115
3. Zhong QC (2013) Robust droop controller for accurate proportional load sharing among inverters operated in parallel. IEEE Trans Ind Electron 60:1281–1290
4. Zhang Z, Huang X, Jiang J, Wu B (2010) A load-sharing control scheme for a microgrid with a fixed frequency inverter. Electr Power Syst Res 80:311–317
5. Guerrero JM, Matas J, de Vicuna LG, Castilla M, Miret J (2007) Decentralized control for parallel operation of distributed generation inverters using resistive output impedance. IEEE Trans Ind Electron 54:994–1004
6. Teodorescu R, Liserre M, Rodriguez P (2011) Grid converters for photovoltaic and wind power systems. John Wiley & Sons, Chichester
7. Coelho EAA (2002) Porfírio cabaleiro cortizo, and pedro francisco donoso garcia, "Small-signal stability for parallel-connected inverters in stand-alone AC supply systems." IEEE Trans Industry Applications 38:533–542
8. Ruben BG, João OP, Carlos AC, Ernane AAC, Alexandra MACP (2012) Differential-evolution-based optimization of the dynamic response for parallel operation of inverters with no controller interconnection. IEEE Trans Ind Electron 59:2859–2866
9. Zhang DM (2016) Operation of microgrid at constant frequency with a standby backup grid-forming generator", Proceedings of the 2016 IEEE International Conference on Power System Technology. IEEE, Wollongong, NSW, Australia. New York

10. Zhang DM, Fletcher J (2017) Design and operation of an islanded microgrid at constant frequency. In: Book Chapter in Microgrid. Rijeka: InTechOpen
11. Zhang DM (2021) Issues on autonomous AC microgrid operated at constant frequency. AUPEC, Australia
12. Zhang DM (2022) Operation of AC microgrids with PV panels' output power curtailment for minimizing the use of energy storage. J Energy Power Technol. 4:008
13. Zhang DM (2022) Constant Frequency Operated Autonomous AC Microgrid with Different MPPT Methods Applied to PV Panels and with Converting Circuit for Powering Unbalanced Loads. Asia Conference on Power and Electrical Engineering
14. Zhang DM (2023) Modeling of autonomous microgrid operated at medium-voltage level and at constant frequency and study of its voltage profile. J Energy Power Technol 5:001
15. Zhang DM, Tang X (2023) Stability analysis of autonomous microgrid operated at constant-frequency with the consideration of reactive power balance. J Energy Power Technol 5:004
16. Zhang DM (2015) Circuits of voltage source dc/ac converter with lccl or lcc filter and other modified forms, and operation of microgrid with such circuits. USA patent US10,418,916 B2
17. Kumar D, Zare F, Ghosh A (2017) DC microgrid technology: System architures, AC grid interfaces, grounding schemes, power quality, communication networks, applications, and standardizations aspects. IEEE Access 5:12230–12256
18. Dragicevic, T., Lu XN, Vasquez JC, Guerrero JM (2016) DC microgrids – Part I: A review of control strategies and stabilization techniques. IEEE Trans Power Electron 31:4876–4891
19. Dragicevic T, Lu XN, Vasquez JC, Guerrero JM (2016) DC Microgrids – Part II: A Review of Power Architectures, Applications and StandardizationIssues. IEEE Trans. Power Electronics 31(5):3528–3549

Design, Sizing, and Simulation of a DC Microgrid for Real Implementation

Mojgan Hojabri

Abstract In recent years, DC microgrids have grown in popularity because of their improved efficiency, increased reliability, and simplified control and management when compared to AC microgrids. Moreover, the number of DC appliances is increasing for residential applications. Thus, determining the best DC voltage for them is the key question. The current DC microgrid technology achieves high efficiency by utilizing sub-systems and renewable energy sources such as wind turbines, solar panels, etc. while allowing charge controllers to facilitate maximum power point tracking. In addition, energy storage systems, which help to mitigate power intermittency caused by the deployment of renewables, are an essential part of microgrid systems. Therefore, finding the ideal DC voltage level for DC home appliances is one of the goals of this chapter. Additionally, a thorough discussion of the sizing requirements for photovoltaic and storage systems for self-sufficient homes will be held. Lastly, a model for a small DC microgrid that will be installed later in a pilot region will be designed and simulated in the MATLAB/Simulink environment. The obtained simulation results show that the suggested DC microgrid is operationally feasible.

Keywords DC distribution · DC-DC converter · Buck converter · Bidirectional converter · DC appliances · DC home · Charge controller · PV sizing · Battery sizing

1 Introduction

For more than 100 years, AC has played an important role in generation, transmission, and distribution systems. The main reason for that was the higher efficiency in terms of long-distance transmission with lower power loss. Moreover, voltage levels can easily change to different levels by transformers. Therefore, most of the home and

M. Hojabri (✉)
Competence Center of Digital Energy and Electric Power, Institute of Electrical Engineering, Lucerne, Switzerland
e-mail: Mojgan.hojabri@hslu.ch; mojgan.hojabri@gmail.com

F. Shahnia and J. M. Guerrero (eds.), *Microgrids and Virtual Power Plants*, Power Systems, https://doi.org/10.1007/978-981-97-6623-9_4

office appliances were made to work with an AC system. The idea of using the DC system was back again when the power loss reduction proved for high-voltage direct current systems. However, this technology was developed by improving the rectifier to convert DC to AC. Nowadays, the increasing number of DC loads, especially in home and office applications cannot be ignored. This is because of the developing power electronics technology. Therefore, improving the system efficiency by removing the AC to DC converters in each appliance opens a new window for using DC systems for residential use cases. Moreover, no skin effect, harmonics, and the ability to store make the DC system attractive. However, the DC system is more expensive than the AC system because of insulation requirements. In addition, DC cannot be transmitted economically over long distances due to the voltage drop.

The development of microgrids, or decentralized distribution systems, is an important subject that should be considered. Simultaneously, it promotes the use of DC system-compatible renewable energy and storage technologies for residential applications. Reducing the number of converters for renewable energy systems is another factor that improves the idea of using DC systems. These all point to the benefits of employing DC microgrids for residential applications. Overall, DC microgrids lower the cost, of contemporary buildings while also improving system efficiency. Because of all these facts, in recent years, DC microgrids have grown in popularity compared to AC microgrids [1].

This chapter discusses the possibility of transitioning from an AC to a DC system for new residential loads to work as a DC microgrid. Therefore, many concepts like choosing the best voltage level, designing, and sizing requirement components will be discussed. Moreover, a designed DC microgrid will be simulated to explore results and analysis. In DC microgrids, the DC link voltage control and power management among the several sources and loads are the main concerns for the researchers. Which voltage is better for a DC microgrid with residential applications? This will be answered in Sect. 2. DC home appliances category and their new generation to make them compatible with the DC system will be discussed in Sect. 3. The sizing of DC microgrid components such as photovoltaic (PV) and battery bank will be presented in Sect. 4. Modeling, design, and simulation for a DC microgrid will be illustrated in Sect. 5. Finally, in Sect. 6, the main conclusions of the study will be listed.

2 Standard DC Voltage for Household Appliances

Voltage levels for DC microgrids in residential buildings that lay between distributed generation and loads, relying on practices and existing experience could be from 12 to 800 V DC [2]. However, standard voltage levels for DC distribution systems are introduced to reduce system complexity and losses. This section aims to find an appropriate DC voltage level that fits all household appliances based on the previous works and the recent advances in power electronics of appliances, and considering energy efficiency, user safety, and cost-saving. Therefore, voltage drop and power loss are two important factors that should be considered. In a DC system, power loss (ΔP_{DC}) and the voltage drop (ΔV_{DC}) can be calculated from the load power consumption P which is equivalent to the active power for the AC load [2] as below:

$$\Delta P_{DC} = 2R\frac{P^2}{V_{DC}^2} \quad (1)$$

$$\Delta V_{DC} = 2R\frac{P}{V_{DC}} \quad (2)$$

where R is the line resistance at 80 °C, V_{DC} is the DC voltage, and P is equivalent to the active power for the AC load.

Example 1: Calculate the voltage drop and power loss for a 10 m and 2.5 mm^2 copper wire using 24 V_{DC} and 15 A, $\rho_{cu} = 20 \times 10^{-9}$.

Resistance of the cable can be calculated as:
$\mathrm{R} = \rho\frac{\mathrm{l}}{\mathrm{A}} = 20 \times 10^{-9} \times \frac{10}{2.5 \times 10^{-6}} = 0.08\Omega, = VI = 24 \times 15 = 360W,$
Voltage drops for dual cable from (2):

$$\Delta V_{DC} = 2 \times 0.08 \times \frac{360}{24} = 2.4V, \Delta VDC\% = \frac{2.4}{24} \times 100 = 10\%$$

$$\Delta P_{DC} = 2 \times 0.08 \times \frac{360^2}{24^2} = 36W, \Delta PDC\% = \frac{36}{360} \times 100 = 10\%$$

Then, $P_{out} = 360W - 36W = 324W$.

Traditional applications of DC voltage can be found in transportation camping vehicles with 12 to 24 V DC, telecommunication with 48 V DC, and data centers within 380 to 400 V DC. Up to now, for residential applications, different DC voltage levels have been recommended or used such as 12, 24, 48, 120, 220 or 230, 311 or 326 V DC, or even higher like 750 or 1,500 V DC. Generally, DC voltages can be categorized into three main levels as below.

2.1 Low DC Voltage Level (12-48 V DC)

12 V DC system is designed for low power consumption and short distances. This voltage level is safe, and many appliances are already available for this voltage level. However, for higher power ratings, the power loss will increase according to (1). 24 V DC is within the standard of EMerge Alliance [3], which is a safe match for low power consumption. Although, 24 V DC allows higher power transmission with a longer distance than 12 V DC, power loss and voltage drop get higher in higher power consumption according to (1) and (2). 48 V DC is recommended by IEEE standard for a DC microgrid in remote and rural electrifications [4]. Therefore, this

voltage level can be used not only for telecommunication systems but also for off-grid PV systems. This voltage level doesn't need protection devices, but with the high consumption, it has a considerable power loss. Therefore, this voltage level is recommended for low-power consumption devices to achieve the most efficiency.

2.2 Medium DC Voltage Level (120-230 V DC)

120 V DC is another recommended DC voltage level. Although 120 V DC doesn't need any protection against indirect contacts, there are no 120 V DC appliances available in the market with 220 VAC or 230 VAC. 230 V DC allows the use of resistance appliances, such as incandescent lights and heating, with no changes, since it has the same RMS value as the 230 VAC. 311 and 326 V DC are equal to the peak value of the 220 VAC and 230 VAC. Most recent appliances have an input diode rectifier, which enables them to be operated under this DC level. A few modifications may be preferred by just removing that rectifier when more efficiency is required, but it is not necessary since it protects against reverse polarity.

2.3 High DC Voltage Level (380-1500 V DC)

380–400 V DC can be found within the standard of EMerge Alliance of buildings Ease compatibility with utility voltage 230 VAC. It is investigated and recommended by several researchers such as [5, 6]. It is possible to fit some AC loads directly on DC and minor modifications may be required for others [2]. This DC voltage range is more appropriate for DC or AC distribution systems in domestic applications and can be used for higher residential consumption or commercial loads. In addition, a grid-connected microgrid is possible with this voltage level. As a drawback of these voltage levels, it could be mentioned that step-down converters are required for adapting to most of the existing locally available appliances. Moreover, protection is necessary for such voltage levels.

The appropriate voltage level for high-power transmission is 1500 V DC recommended by [7]. This voltage level is introduced for industrial and commercial loads, but it can be used for aggregated buildings too. However, protection is compulsory for this level of the DC voltage step-down converters are required to adapt to most of the existing locally available appliances.

3 Next Generation of Household Appliances (DC Versus AC)

Nowadays we are using different electrical devices at home with different voltage levels and types. To replace all voltage sources from AC to DC, first, these electrical devices must be classified based on their characteristics. Home appliances can be divided into three main categories based on their characteristics:

- Resistive appliances
- Universal motor-driven appliances
- Electronics based appliances
- Switching mode power supply-based appliances

Some appliances like rice cookers, coffeemakers, electrical stoves, and old incandescent lamps are used in many houses around the world. These types of loads are resistive loads that are DC-compatible. It should be mentioned that they have to switch contacts which may produce an arch with DC voltage [2]. Meanwhile, a new generation of these appliances is electronics-based. Moreover, there are so many electronics-based devices such as LED lamps, smart TVs, laptops, and mobiles that could be mentioned here. Another type of home device is the motor-based device. A part of these loads uses universal motors like vacuum cleaners, mixers, and some electrical tools like drills and saws [2]. Universal motors are types of electric motors that can operate on either DC or AC power [8]. The old generation of air-conditioners or refrigerators used AC induction motors. However, a new generation of these devices is working with DC voltages. Therefore, they are using switching devices to change the existing AC to the required DC voltage. These devices are called switching mode power supply-based appliances.

4 Sizing DC Microgrid Components

Before sizing all the microgrid components, total power consumption should be calculated. The simple way is to check the previous year's bill to estimate the next year's consumption. Another way is to define the exact daily power consumption (E_{daily}) by considering all required equipment.

$$E_{daily} = \sum_{1}^{n} (\frac{P_{Ln}}{3600}) \times t_n \times f_n \tag{3}$$

where n, P_L, t and f indicate the number of loads, the amount of hourly power in watts, the total operating time, and the daily usage frequency of appliances, respectively. Some of the electronic devices have standby consumption. Therefore, daily consumption of standby mode ($E_{standby}$) should be defined as below:

$$E_{standby} = \sum_{1}^{n} (24 - t_n) \times P_{standby} \tag{4}$$

where $P_{standby}$ is the power of standby mode in watts. PV and battery bank sizing will be discussed in the following paragraphs.

4.1 PV System Sizing

PV power requirement can be calculated by dividing daily energy consumption as described above in kWh by the peak-sun-hour (PSH). PSH is defined as one hour in which the intensity of solar irradiance averages 1,000 watts per square meter[9]. So, it is clear that PSH depends on the location where the PV panel is going to be installed. To avoid under-sizing, PV system loss should be considered. Therefore, the peak PV power is determined as below:

$$P_{pv} = \frac{E_{daily}}{\eta \times \text{PSH}} \tag{5}$$

where PSH is the peak sun hour, and η is the PV system efficiency.

Example 2: The owner of a single-family house in Horw, Switzerland decided to be disconnected from the power grid and powered the house only by a PV system. Their previous electricity bills show that the annual energy consumption of the house is 4500 kWh. How many solar panels with 250 W power and 90% efficiency are required to supply the total load? The monthly average of the peak sun hour in Horw is shown in Table 1.

First, daily power consumption and peak sun hour per day should be calculated for this house:

Average daily energy consumption $= 4500kWh/365 = 12.3287kWh$
From Table 1, peak sun hours per day are determined as 2.93,
Then, PV power requirement: $Ppv = \frac{12.3287kW}{0.9*2.93} = 4.67kW$,
The number of panels $= \frac{(4.67 \times 1000)W}{250W} = 18.70$,
Therefore, 19 solar panels need to be installed.

Table 1 Monthly average peak sun hour in horw, switzerland [9]

Month	Jan	Feb	March	April	May	Jun	Jul	Aug	Sep	Oct	Nov	Dec
PSH	0.82	1.3	2.4	3.74	4.87	5.41	5.55	4.48	3.2	1.85	0.86	0.65

4.2 Battery Bank Sizing

Battery-based energy storage systems are used in a variety of stationary applications such as PV systems. Lead-acid and lithium-ion batteries are still commonly used in these applications. Two IEEE standards for sizing lead-acid battery banks for stationary applications: IEEE Std 485 and IEEE Std 1013. IEEE Std 485, recommended practice for sizing lead-acid batteries for stationary applications, short duration, and high discharge rate. However, IEEE Std 1013 recommended practice for sizing lead-acid batteries for stand-alone photovoltaic systems, longer duration, and lower discharge rate. This standard is used typically for residential, commercial, and remote monitoring in off-grid PV systems. According to the IEEE Std 1013, the first step for battery sizing is defining the energy consumption in ampere-hours (Ah) over the autonomy period (days without PV generation). This is called unadjusted capacity. To adjust the capacity, other factors such as battery efficiency, temperature compensation, and depth of discharge (DoD), should be considered. The recommended battery DoD varies by the type of battery and manufacturer. Typically, DoD for lead-acid batteries is 50% while for lithium-ion batteries is between 80–100% [10].

Considering all the above mentioned, the battery capacity based on ampere-hour can be calculated as:

$$Batterycapacity = \frac{E_{daily} \times t \times DoA \times K_T}{\eta_B \times DoD \times V_{DC}}(\text{Ah}) \tag{6}$$

$$DoD\% = 1 - SoC\% \tag{7}$$

where t is the battery operation hours, V_{DC} is the DC voltage or battery voltage, and SoC is the state of charge (Soc) of the battery. η_B is the battery efficiency, and K_T is the temperature compensation factor. And DoA is days of autonomy. The number of days the equipment must operate on battery power with limited solar power.

Example 3: Calculate the needed capacity of a lead-acid battery bank for a camping house that uses 600 Wh of electricity per day and has a 48 V DC system that can provide backup power for two days.

If we assume the K_T is 1 and the battery efficiency is equal to 85%, the battery bank capacity would be 58.82 Ah using (6) by considering SoC 50% for lead-acid batteries. One option would be to use four 12 V, 60 A lead-acid battery cells connected in series.

4.3 Electrical Vehicle (EV) Charging Station

According to the International Energy Agency, there will be 145 million electric vehicles worldwide by 2030 [11]. Therefore, they should be considered in the future design of a smart microgrid. Electrical vehicles (EVs) can serve as valuable energy storage devices in a smart grid, enhancing the distribution system's availability by returning power to the grid during variations in RES power generation. At this moment, there are three main types of EV charging. Based on EU standards, the first level is a single phase with 230 V and a 3.3 kW peak charging capacity. This charging category is typically belonging to home charging. Level 2, which is mostly used as public stations, are three-phase AC main of 400 V and 22 kW peak charging capacity. However, the second method is faster compared to home charging but still not fast charging. The third method is DC fast charging. These operate on a 400 VAC circuit, but before the power is sent to the car, it is converted to DC. This eliminates the need for the car's onboard charger, which transforms AC to DC and significantly speeds up the process because batteries can only charge with DC. A standard EV takes 15–45 min to become 80% full when using a fast DC charging station [2]. Therefore, a DC microgrid can be a game changer to support EVs in future smart grids. To meet the needs of various vehicles and their battery range, the charging architecture must properly account for the DC that must be delivered into a battery at a varied DC voltage level in the range of 50–1000 V DC. According to the information published by German Electrical and Electronic Manufacturers' Association [12], the voltage level for batteries in passenger vehicles is 400 V. A 400 V electric car can theoretically charge at 200 kW (400 V $\times$ 500 A) at a DC fast charging station. And with 800 V technology, has twice the charging power of 400 kW (800 V $\times$ 500 A) with the same amperage. To preserve the battery, all manufacturers are still throttling around 240–270 kW [13].

5 Modeling, Design, and Simulation for a 48 V DC Microgrid

The main components of a small DC microgrid with a PV system are solar panels, batteries, DC/DC converters, and a control system. According to the literature, 48 V DC seems to be a good choice for a small residential community in terms of efficiency, safety, and economy as discussed in previous research such as [14, 15] and recommended by IEEE standard for residential use cases [5]. Based on that, a microgrid for residential application is presented in Fig. 1. This system has two main parts: a local grid and DC-powered home (DCPH). A DCPH was designed and simulated in a MATLAB/Simulink environment and shown in Fig. 2. Assume all houses feed by their own PV and are connected to this DC bus. Therefore, a typical 4-person house in a pilot area in Switzerland is considered and used for the design of the microgrid. Figure 2 shows the simulation model that has been created in MATLAB/Simulink. This includes a solar panel, a DC-DC buck converter, a battery bank, and a control center. The control center is a combination of MPPT and battery charge controller to control the battery charging/discharging. The output of the MPPT and battery charge controller defines a signal that controls the DC-DC buck converter.

This will be explained in Sects. 5.3 and 5.4. The size of the PV panel and battery bank will be defined in Sects. 5.1 and 5.2 based on the theory explained in Sect. 4. In this local DC system, different scenarios can be considered for an energized DCPH as mentioned below.

1. Conservative
   - Always charge the battery first with the PV production
   - Only feed into the grid, when the battery is full

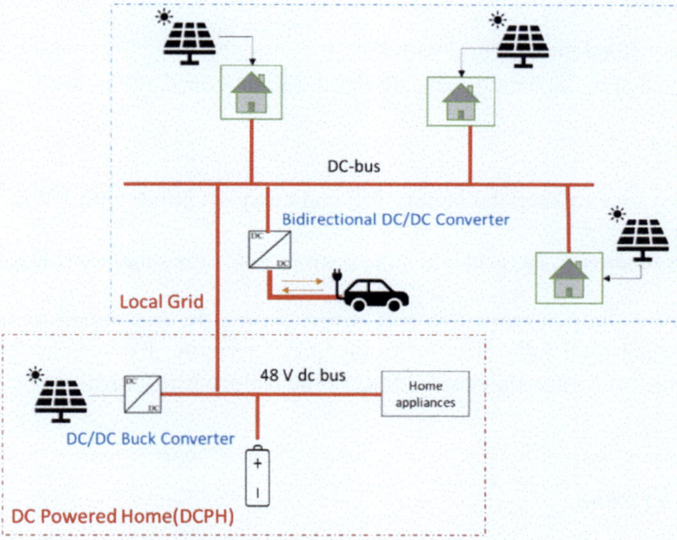

Fig. 1 Block diagram of a local DC microgrid for residential use case

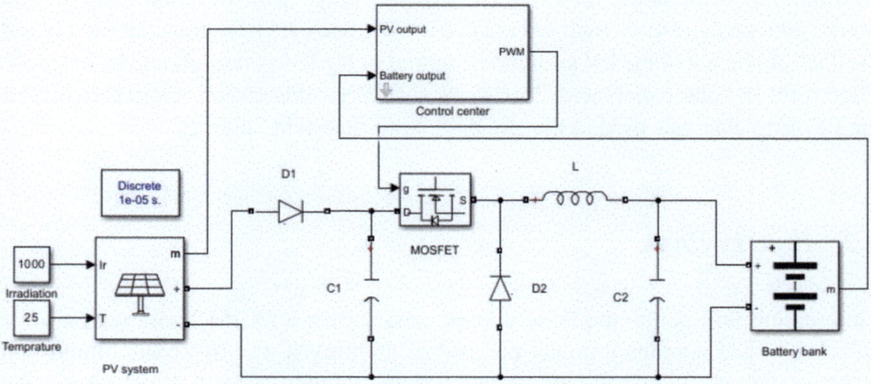

Fig. 2 A DCPH system including PV, battery, and control system

Table 2 PV array parameters

250 W solar panel	
Cell per Module (N_{cell})	60
Parallel string	2
Series-connected	9
V_{oc}, I_{sc}	36.6 V, 8.75 A
V_{mp}, I_{mp}	30.9 V, 8.1 A

2. Progressive
   - Always feed into the grid with PV
   - Only charge the battery when the grid does not need any power
3. Balanced
   - Always try to keep the battery full and charge it either with PV or from the grid
   - Start feeding to the grid when the desired SoC of the battery is reached

In this case study conservative option was chosen. In this situation, the battery will be charged until fully charged voltage (V_{max}) via the PV system. Then, start feeding to the grid when the desired SoC of the battery is reached.

5.1 PV System

Reference [1] shows an average four-person household consumes about 4500 kWh of electricity per year (including electric hot water preparation). Then, the average daily consumption would be 12.329 kWh. The annual average of the peak sun hours per day for the pilot area in Horw, Switzerland is 2.93 (see Sect. 4.1). By considering this and the 95% efficiency of the PV system, the output of the PV system should be 4.43 kW. Therefore, 18 solar panels with 250 W are needed for this reason. Characteristics of the PV array that was used in the simulation are shown in Table 2.

5.2 Battery Bank

This section will define the battery bank capacity for a DCPH home with a 48 V DC system and an annual power demand of 4500 kWh. For this case, lithium-ion batteries with 80% DoD and 90% efficiency are going to be used. If we assume the

Fig. 3 Battery bank

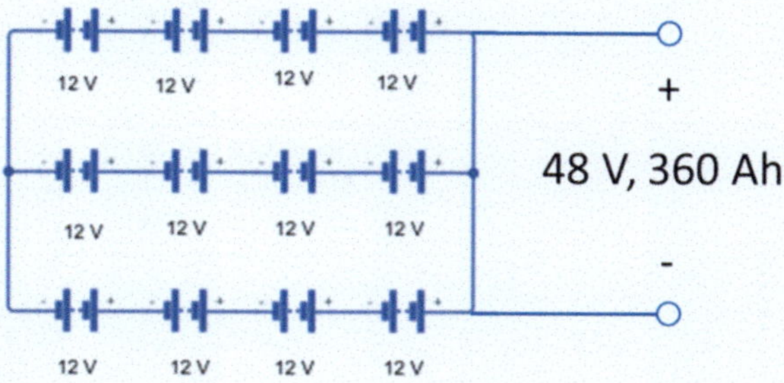

batteries are working at optimal temperature, using (6) the battery bank capacity will be around 360 Ah for one-day autonomy. One option would be to create a 48 V and 360 A battery by joining twelve 12 V and 120 Ah lithium-ion batteries in parallel and series, as seen in Fig. 3. Therefore, a 48 V DC and 360 Ah lithium-ion battery bank is used for the simulation. The fully charged voltage for this battery is 55.8714 V. It should be noted that because lead-acid batteries have a lower SoC than lithium-ion batteries [16], a larger battery bank with a higher capacity is required for the same system when using a lead-acid battery.

5.3 MPPT

MPPT is a part of the control center for DCPH. Due to its tracking efficiency and ease of use, the perturbation and observation (P&O) MPPT is frequently utilized in small and medium-sized commercial PV charge controllers and grid-connected inverters. The MPPT algorithm monitors the maximum power of the PV array and outputs its duty cycle relevant to the tracked maximum power to the battery charge controller. The flowchart of a P&O MPPT algorithm which is proposed by [17] is shown in Fig. 4. This method tracks the maximum power through a process of trial and error.

The P&O MPPT algorithm starts with reading voltage and current from the PV array. Then calculated PV power and measured voltage will be compared with their previous sample ($k-1$). The condition switch block handles the if-else conditions of the P&O method; the ΔD block lets the user adjust the duty cycle's perturbation step size; an adder with a memory block $D(K-1)$ feedback loop handles the duty cycle increment and decrement function. The $D(K)$ limit block limits the duty cycle exceeding the range between minimum and maximum duty cycle range. The output of the duty cycle is connected to the battery charge controller section.

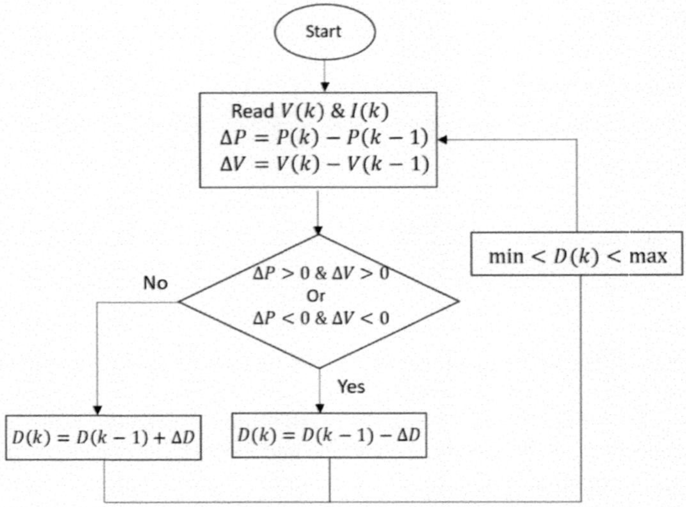

Fig. 4 MPPT flowchart

5.4 Battery Charge Controller

The battery charge controller was developed to charge a lithium-ion battery using the three-stage charging method. For this reason, the SoC and voltage of the battery are measured by the battery charge controller. Charge-controlling algorithm is shown in Fig. 5. This algorithm has three stages as described below.

Fig. 5 Battery charge controller algorithm

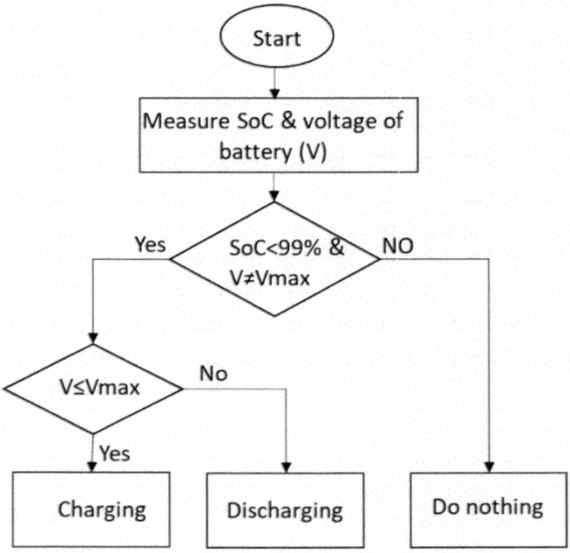

Stage 1: constant current or charging mode.

The charge controller charges the battery at the MPPT bulk charging stage when the battery SoC is less than 99% and the battery voltage is less or equal to the fully charged voltage (V_{max}). In this case duty cycle is equal to the MPPT.

Stage 2: constant voltage or discharging mode.

The charger switches to a constant voltage absorption when the battery voltage is higher than the Vmax, but the SoC is less than 99%. In this situation, the battery is going to discharge mode to meet the Vmax.

Stage 3: float charging.

When the SoC reaches 99% and the battery voltage is equal to Vmax, the charge controller switches to the float stage where the duty cycle is zero. In other words, the charge controller does nothing and continues to supply the load.

5.5 DC-DC Buck Converter

A buck converter is intended to convert the PV output voltage to 48 V since it is higher than the DC bus voltage. The buck converter converts a higher voltage from PV to a lower voltage as an input for the battery while maintaining its power delivery to charge the battery. This is obtained by reducing the input voltage and increasing the output current delivered to the battery. As shown in Fig. 2, the buck converter circuit consists of a MOSFET switching device, an inductor, a Schottky diode, and two capacitors for the input and output. During the night, the reverse blocking diode D1 is utilized to stop current from flowing backward from the battery to the PV array. The MOSFET resistance (R_{on}) is set to 0.01 Ω and the Schottky diode D2 forward voltage is set to 0.8 V. The MOSFET is switched by a pulse generator with a 1 kHz switching frequency. The output voltage of the buck converter depends upon the input voltage V_{in} and duty cycle D (8). D is the duty cycle of the PWM signal and ranges from 0 to 1 [18]. However, D_{min} can be defined by dividing the output voltage by the maximum input voltage. Maximum input voltage is considered as the output voltage of the PV panel at MPPT (V_{mp}), which is equal to 278.1 V for 1000 W/m^2 irradiance and 25 °C temperature (Fig. 6). So, the duty cycle range becomes $0.173 \leq D \leq 1$.

$$D = \frac{V_{out}}{V_{in}} \tag{8}$$

Inductance and capacitance values can be calculated from (9) and (10), where V_{in} is the input voltage from the PV array, D_{min} is the minimum value of the duty cycle, f_{sw} is the switching frequency and ΔI_L and ΔV_c are the current ripple of the inductor and capacitor voltage ripple respectively. By considering f_{sw} as 1000 Hz, ΔI_L as 1% of I_{in} and ΔV_c as 1% of V_{out}, L and C can be defined as 245.6 mH and 42.2 μF. This C value is considered for both sides (Table 3).

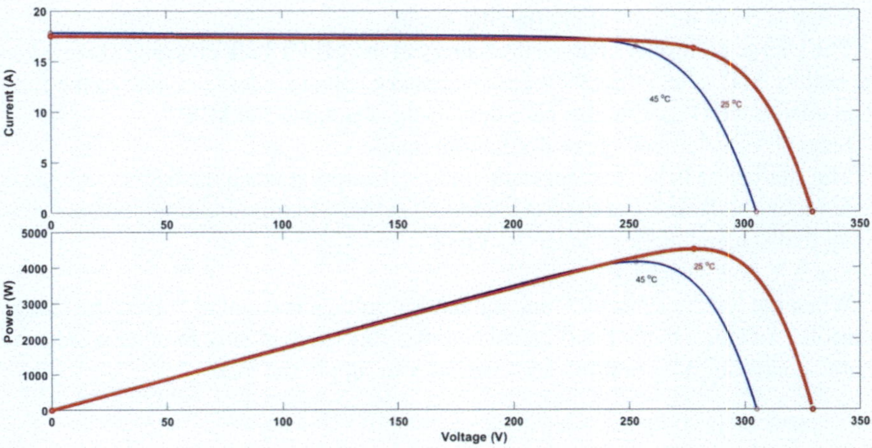

Fig. 6 PV panel output current and power

$$L = \frac{V_{in}D_{min}(1 - D_{min})}{f_{sw}\Delta I_L} \quad (9)$$

$$C = \frac{V_{in}D_{min}(1 - D_{min})}{8Lf_{sw}^2\Delta V_c} \quad (10)$$

The control center, which is shown in Fig. 2, includes maximum power and battery charge controller. The duty cycle is defined by the control center as described in Sect. 5.3. The minimum and maximum values of the $D(k)$ and ΔD are considered as 0, 0.95, and 0.0001 respectively. Based on the calculated parameters, the system, which is shown in Fig. 2 is simulated in the Simulink environment. The battery performance results, including SoC, voltage, and current, are displayed in Fig. 7. The overall efficiency of the conversion system is determined by the ratio of the output charging power delivery to the battery to the input power from the PV array and presented in Fig. 8. As is clear from this figure, the mean value of the conversion efficiency is around 95.23%. The model conversion efficiency can be further improved by utilizing a lower drain-source resistance MOSFET and a lower forward voltage diode.

5.6 Bidirectional DC-DC Converter

A bidirectional DC-DC buck-boost converter is designed to charge EVs via the DC grid. This will provide an opportunity to charge and discharge EVs and reuse them as a valuable storage system. To achieve high efficiency, this design aims for minimum current and voltage ripple. Figure 9 represents the Simulink model for a bidirectional DC-DC converter, which is connected between a 48 V DC bus and an EV battery. We assume the voltage and power of a typical EV passenger car are 400 V and

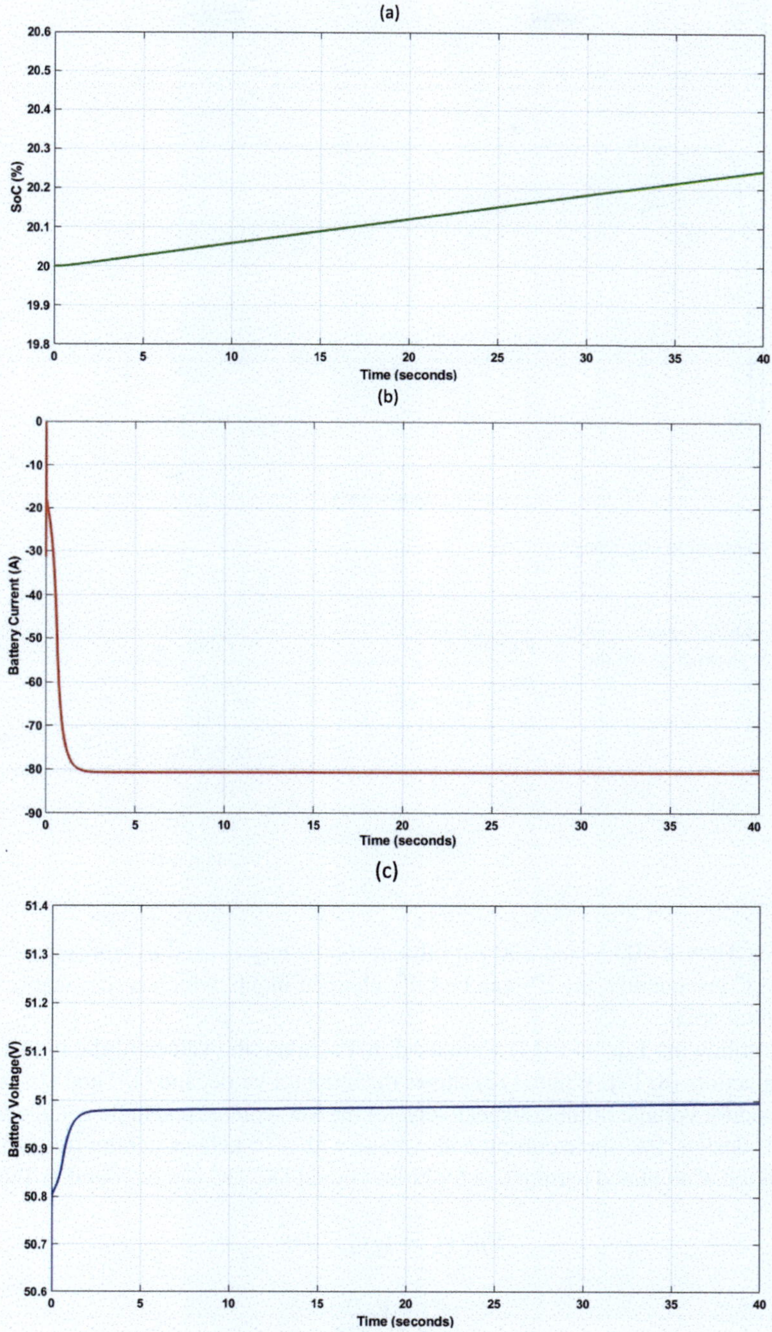

Fig. 7 Battery performance: **a** SoC **b** current **c** voltage

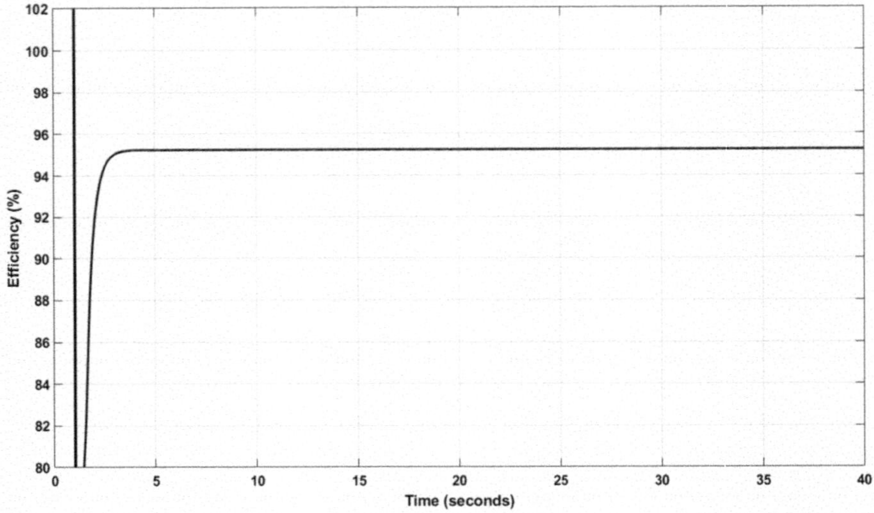

Fig. 8 Conversion efficiency

Table 3 DC/DC buck converter parameters for the simulation

Parameter	Value
Dmin	0.173
f_{sw}	1 kHz
V_{in}	278.1 V
V_{out}	48 V
C	42.2 μF
L	245.6 mH

240 kW. Thus, a 400 V and 200 Ah lithium-ion battery is used to represent the EV battery. Consequently, the EV can be fully charged in 3 h, which is reasonable for a residential area.

The following Eqs. (11–14) were used, to calculate the values of the inductance and capacitors [8, 19]. Where (12) shows the inductor value, and (13) and (14) show the capacitor voltage ripples, respectively, for the low-voltage and high-voltage sides. For this design, 1% ripple current and voltage with a 10 kHz switching frequency is considered. Calculated parameters for bidirectional DC-DC can be found in Table 4.

$$V_H/V_L = 1/(1 - D) \tag{11}$$

$$L = \frac{V_L D}{f_{sw} \Delta I_H} \tag{12}$$

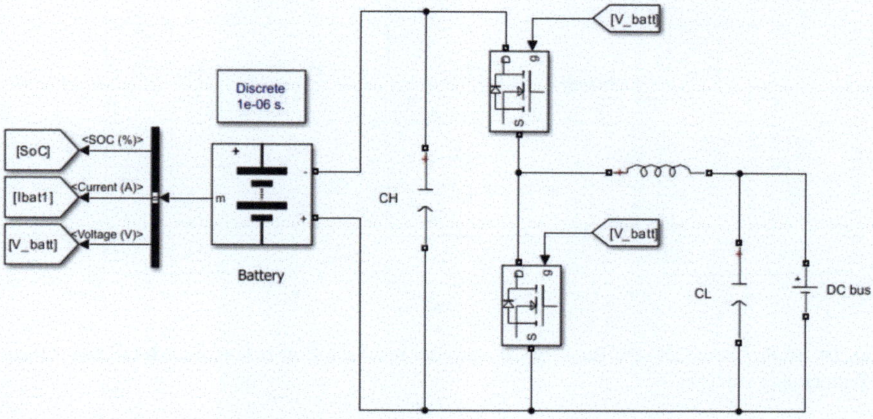

Fig. 9 Bidirectional DC-DC Buck-Boost converter for an EV charge station

Table 4 Bidirectional DC/DC converter parameters

Parameter	Value
D	0.88
f_{sw}	10 kHz
V_{in}	48 V
V_{out}	400 V
C_H	13.2mF
C_L	18.75μF
L	70.4mH

$$\Delta V_L = \frac{V_L D(1 - D)}{8 L C_L {f_{sw}}^2} \tag{13}$$

$$\Delta V_H = \frac{I_H D}{f_{sw} C_H} \tag{14}$$

Converter mode is shown in Fig. 10 with 1 denoting buck mode and 0 representing boost mode. Current moves from the EV battery to the DC bus when the converter is in the buck mode. The EV battery will discharge in this case. To charge the EV battery, the current flow direction will change while in boost mode.

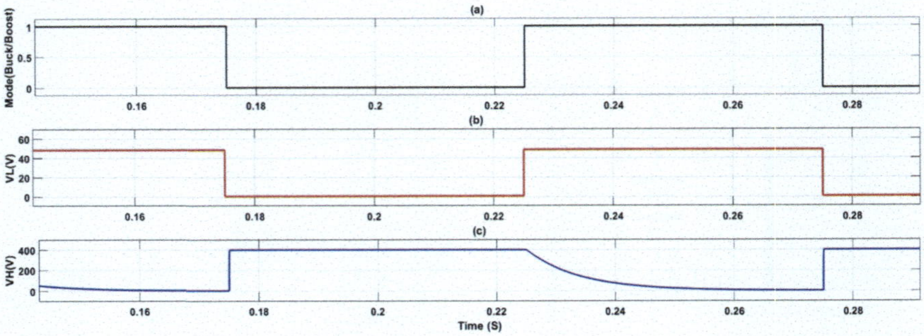

Fig. 10 Bidirectional DC-DC Buck-Boost converter results: **a** Conversion mode **b** DC bus voltage **c** Battery side voltage

6 Conclusion

This chapter discussed the modeling, design, and simulation of a testbed for a 48 V DC microgrid. Consequently, a detailed circuitry modeling of a DCPH with a 4500 kWh annual power consumption, a PV system, and a battery storage system was presented. A charge controller, a buck converter circuit, and the MPPT P&O tracking algorithm wre designed, simulated, and discussed. Using a three-stage charging method, the MPPT battery charge controller can regulate the charging of a 48 V lithium-ion battery by detecting the maximum power from the 4.43 kW PV system. The topology reaches a similar average efficiency of 95.23% as compared to the standards of numerous commercial solar PV inverters. The provided Simulink model is adaptable to any commercial MPPT charge controller with a comparable architecture. Furthermore, a detailed design and simulation of an EV charging station are performed. For this reason, a bidirectional DC/DC converter was created specifically for using the 48 V DC grid to charge and discharge EV batteries.

References

1. ZSH, TWPQ, NRT (2018) Modelling and simulation of DC microgrid. In: EEE 4th Southern Power Electronics Conference (SPEC)
2. Sabry AH, Shallal AH, Hameed HS, Ker PJ (2020) Compatibility of household appliances with DC microgrid for PV systems. Heliyon 6(12). https://doi.org/10.1016/j.heliyon.2020.e05699
3. Standards DC and Hybrid AC/DC Microgrids (2024) Emerge Alliance. Accessed: Feb. 01. [Online]. Available: www.emergealliance.org
4. IEEE Power and Energy Society (2021) IEEE Standard for DC Microgrids for Rural and Remote Electricity Access Applications. IEEE
5. Chunxue W, Mingming H, Changbin H, Zhengguo P, Jinghua Z (2019) Research on characteristics of bidirectional CLLC DC–DC transformer used in DC microgrid. J Eng 2019(18):5351–5354. https://doi.org/10.1049/joe.2019.0909
6. Arnold G, Pennell G (2020) DC Lighting and Building Microgrids opportunities and recommendations. 10.1016/j

7. Kumar D, Zare F, Ghosh A (2017) DC microgrid technology: System architectures, AC grid interfaces, grounding schemes, power quality, communication networks, applications, and standardizations aspects. IEEE Access 5:12230–12256. https://doi.org/10.1109/ACCESS.2017.2705914
8. Saleh M, Esa Y, Mhandi Y (2016) Design and Implementation of CCNY DC Microgrid Testbed. In: Portland, OR, USA: IEEE Industry Applications Society 52nd Annual Meeting
9. Hyder Z (2024) What is a peak sun hour? SolarReviews. Accessed: Feb. 01. [Online]. Available: www.solarreviews.com
10. Depth of Discharge (2024) Foxtron. Accessed: Feb. 01. [Online]. Available: www.foxtronpowersolutions.com
11. Frangoul A (2024) Sustainable Energy. CNBC. Accessed: Feb. 01. [Online]. Available: www.cnbc.com
12. Kostal L, Schmidhauser L, Kabel L, Bosch R (2013) Voltage Casses for Electric Mobility. Frankfurt. [Online]. Available: www.zvei.org
13. Welche E-Autos haben ein 800-Volt-System?, einfacheauto.de. Accessed: Feb. 01, 2024. [Online]. Available: https://einfacheauto.de/
14. Anand S, Fernandes BG (2010) Optimal voltage level for DC microgrids. In: IECON 2010 - 36th Annual Conference on IEEE Industrial Electronics Society, pp 3034–3039
15. Moussa MJ-BIlS-BS (2018) DC voltage level choice in residential remote area. In: The Ninth International Renewable Energy Congress : 2018 9th International Renewable Energy Congress (IREC), Sonia Moussa and Manel Jebali-Ben, Eds., Hammamet - Tunisia
16. Salazar D, Garcia M (2022) Estimation and comparison of SOC in batteries used in electromobility using the thevenin model and coulomb ampere counting. Energies (Basel) 15(19). https://doi.org/10.3390/en15197204
17. Tan RHG, Er CK, Solanki SG (2020) Modeling of photovoltaic MPPT lead acid battery charge controller for standalone system applications. In: E3S Web of Conferences, EDP Sciences. https://doi.org/10.1051/e3sconf/202018203005
18. Basharat S, *et al.* (2021) A duty cycle controlled ZVS buck converter with voltage doubler type auxiliary circuit. Front Energy Res 9. https://doi.org/10.3389/fenrg.2021.550115
19. Gorji SA, Sahebi HG, Ektesabi M, Rad AB (2019) Topologies and control schemes of bidirectional DC–DC power converters: An overview. IEEE Access 7:117997–118019. https://doi.org/10.1109/ACCESS.2019.2937239

Microgrid Control Assessment Using Advanced Hardware in the Loop Technologies

Sharara Rehimi, Hassan Bevrani, Chiyori T. Urabe, and Takeyoshi Kato

Abstract This chapter explores the assessment of microgrid control using advanced hardware-in-the-loop technologies. It provides an introduction to hardware-in-the-loop technologies and their applications, including virtual hardware-in-the-loop, controller hardware-in-the-loop, and power hardware-in-the-loop. The chapter highlights the significance of hardware-in-the-loop assessment for assessing microgrid control units and discusses the challenges and issues involved in hardware-in-the-loop testing. It also examines the relevant IEEE standards and grid codes for microgrid testing. Furthermore, the chapter reviews the advancements in microgrid testing with hardware-in-the-loop and presents a step-by-step implementation of a microgrid using hardware-in-the-loop technology. It covers the important components of a microgrid, the difference between switching and average components, and the integration of battery energy storage systems, photovoltaic power plants, and diesel generators. The chapter provides examples of creating a simple microgrid and a fully functional terrestrial microgrid model. It also discusses microgrid model optimization and explores new trends and topics, such as ensuring cybersecurity in microgrids and the integration of electric vehicles. Additionally, the chapter delves into the use of grid-forming converters in microgrids.

Keywords Control assessment · Hardware in the loop (HIL) · Grid-forming converters · Grid control · Energy management · System validation · Microgrid

---

S. Rehimi (✉) · C. T. Urabe · T. Kato
Institute of Materials and Systems for Sustainability, Nagoya University, Nagoya, Japan
e-mail: rehimi.sharara.z8@s.mail.nagoya-u.ac.jp

H. Bevrani
Smart/Micro Grids Research Center, Department of Electrical Engineering, University of Kurdistan, Sanandaj, Iran

F. Shahnia and J. M. Guerrero (eds.), *Microgrids and Virtual Power Plants*,
Power Systems, https://doi.org/10.1007/978-981-97-6623-9_5

1 Hardware-In-The-Loop (HIL) Technologies and Applications

HIL devices are crucial elements in modern engineering and testing practices. They enable the integration of actual hardware components with computer simulations, facilitating the evaluation of complex systems performance and functionality. HIL devices typically comprise a physical hardware unit, such as a controller or sensor, which is linked to a simulation environment. Within this environment, virtual inputs and exciters are generated and fed into the HIL device for processing. The device responds as if it were connected to the real system, enabling engineers to analyze the systems behavior under diverse operational circumstances [1]. Figure 1 shows the role of HIL devices in a microgrid test case study. I/O represents the input/output channels.

In fact, the HIL devices offer several benefits, including the ability to provide a realistic testing environment without the need for complete physical prototypes. This significantly reduces both the time and costs associated with constructing and modifying physical prototypes.

1.1 *HIL Technology*

The HIL technologies are instrumental technologies in the field of engineering testing, offering a powerful means of integrating physical hardware components

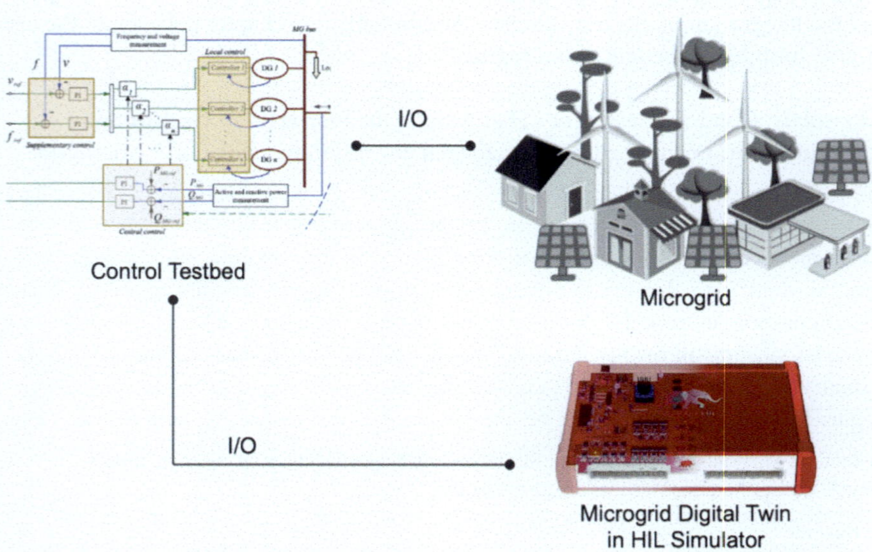

Fig. 1 The role of an HIL devices in a microgrid test case study

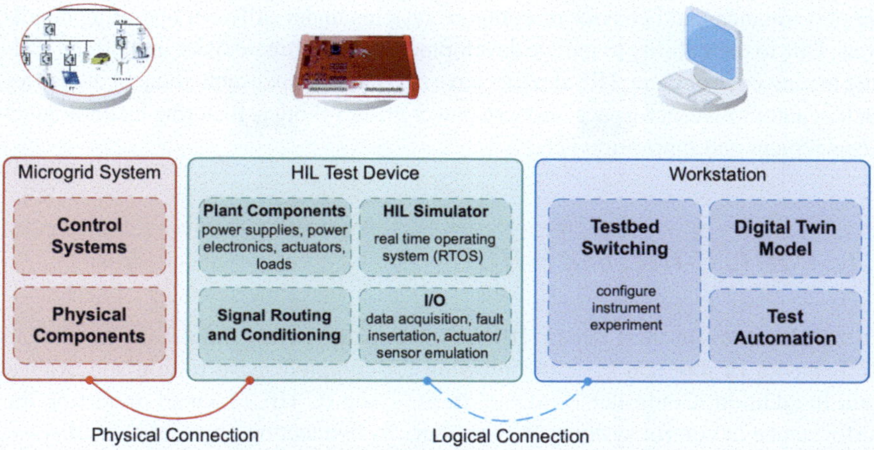

Fig. 2 Implementation of a microgrid system using the HIL technology

with computer-based simulations. These technologies facilitates the evaluation and analysis of complex systems by connecting real-world hardware elements, such as controllers and sensors, to a simulation environment. The simulation environment generates virtual inputs and stimuli, which are then fed into the HIL system for processing. This process allows engineers and researchers to examine the behavior and performance of the system under various operating conditions, enabling them to identify potential issues and optimize system design [1]. For implementation of a case study using the HIL technology, a test system, an HIL device and a computer as a workstation are needed as shown in Fig. 2.

1.2 HIL Applications

The utilization of the HIL technology provides several significant advantages in academic and industrial settings. Firstly, HIL systems enable engineers to conduct comprehensive testing and validation without the need for complete physical prototypes, thereby reducing development time and costs. By employing virtual simulations combined with real hardware components, engineers can assess the performance of systems throughout the development lifecycle, from early design stages to final validation. Secondly, HIL devices enable engineers to simulate challenging scenarios that would be difficult to replicate in real-world testing. Additionally, HIL devices ensure the reliability and robustness of systems prior to deployment. Besides, HIL devices play a vital role in contemporary engineering by providing an efficient and cost-effective approach to system testing and validation. They bridge the gap between physical hardware and computer simulations, enabling engineers to assess the performance and behavior of complex systems under different operating conditions. Due to their ability to reduce development time, mitigate risks, and offer a realistic testing environment, HIL devices have become indispensable tools in industries such as automotive, aerospace, modern power grids, robotics, fostering technological advancement and innovation [2].

1.3 V-HIL, C-HIL, and P-HIL

Virtual Hardware-in-the-Loop (V-HIL) involves using virtual simulations to emulate the behavior of physical systems, allowing for comprehensive testing and evaluation. In addition, Controller Hardware-in-the-Loop (C-HIL) focuses on testing the performance of control systems or controllers by connecting physical controllers to a simulated environment. Additionally, Power Hardware-in-the-Loop (P-HIL) is a technique used to test and validate power system components by integrating real power hardware with a simulation environment.

To provide a more detailed explanation, the V-HIL replaces the physical hardware component with a virtual model that interacts with the control system or controller being tested. The V-HIL is useful in the early stages of development or when testing complex systems that are costly to build prototype versions. It allows engineers to evaluate the control systems performance and validate its algorithms in a simulated environment without physical prototypes [1]. Table 1 presents a comparison between different implementation in a power grid.

When comparing the C-HIL and P-HIL as powerful verification methods, there are a few considerations to take into account. First, in terms of implementation, the C-HIL is generally simpler to set up compared to the P-HIL. However, the P-HIL may present interface challenges due to the integration of real power units. While both approaches have their benefits, the suitability of each method depends on the specific development context. The C-HIL is often preferred by manufacturers during the development process, as it allows for efficient testing and evaluation of control systems (controllers). On the other hand, the P-HIL offers a more realistic testing environment, enabling engineers to assess the behavior and performance of power hardware components in a simulated power system setup. Furthermore, one notable difference between the two methods is the C-HIL does not involve a real power unit, whereas the P-HIL incorporates the use of actual power hardware. This difference

Table 1 A comparison between different implementation in a power grid

Comparison	Simulation (MATLAB, PSIM and, etc.)	Demonstration	C-HIL	P-HIL
Grid Components Flexibility	**High**	Low	**High**	**High**
Converter Components Fidelity	Low	**High**	**High**	**High**

can impact the level of accuracy and fidelity in the testing process. Additionally, the C-HIL is commonly employed for evaluation purposes by utility companies [2]. The difference between the P-HIL implementation and reality is presented in Fig. 3. In this Figure, D/A and A/D represent Digital to Analog and Analog to Digital, respectively.

Furthermore, model-based systems engineering has become a valuable methodology for meeting cost, time, and quality objectives. It offers the ability to replace physical systems and prototypes with virtual models, allowing for comprehensive simulations in a secure and adaptable environment [1]. This approach saves time and reduces costs throughout the entire development process, from specification to commissioning and maintenance. Figure 4 provides a visual depiction of how model-based systems engineering can be utilized to support various stages of the development cycle. Different testing setups can be used depending on specifications, level of abstraction, application, and the device being tested, as depicted in Fig. 5. These setups include model-in-the-loop (MIL), software-in-the-loop (SIL), C-HIL, and the P-HIL.

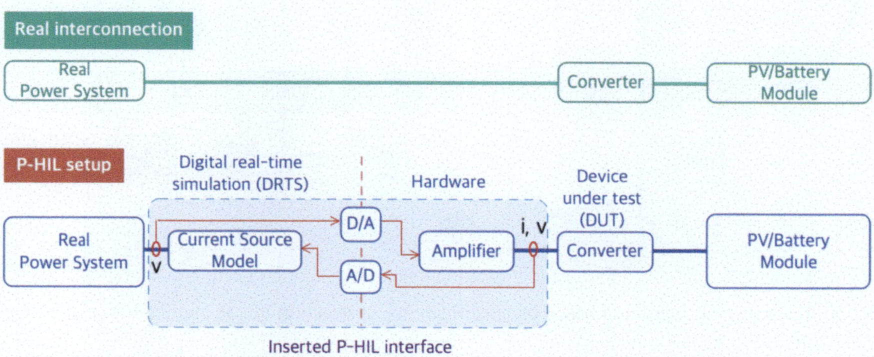

Fig. 3 Difference between the P-HIL implementation and reality

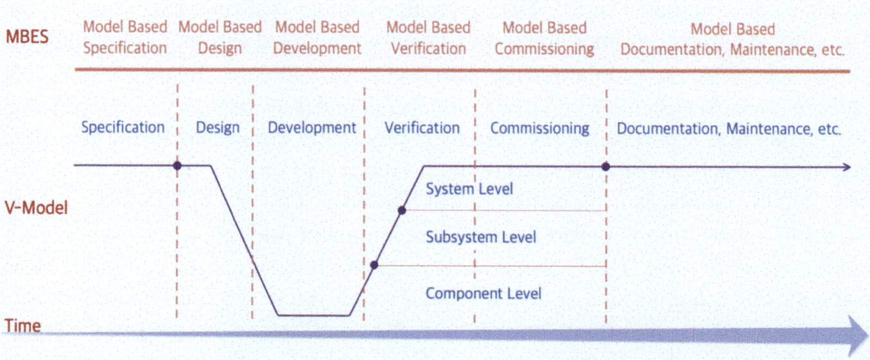

Fig. 4 Visual depiction of model-based systems engineering utilization across development cycle stages

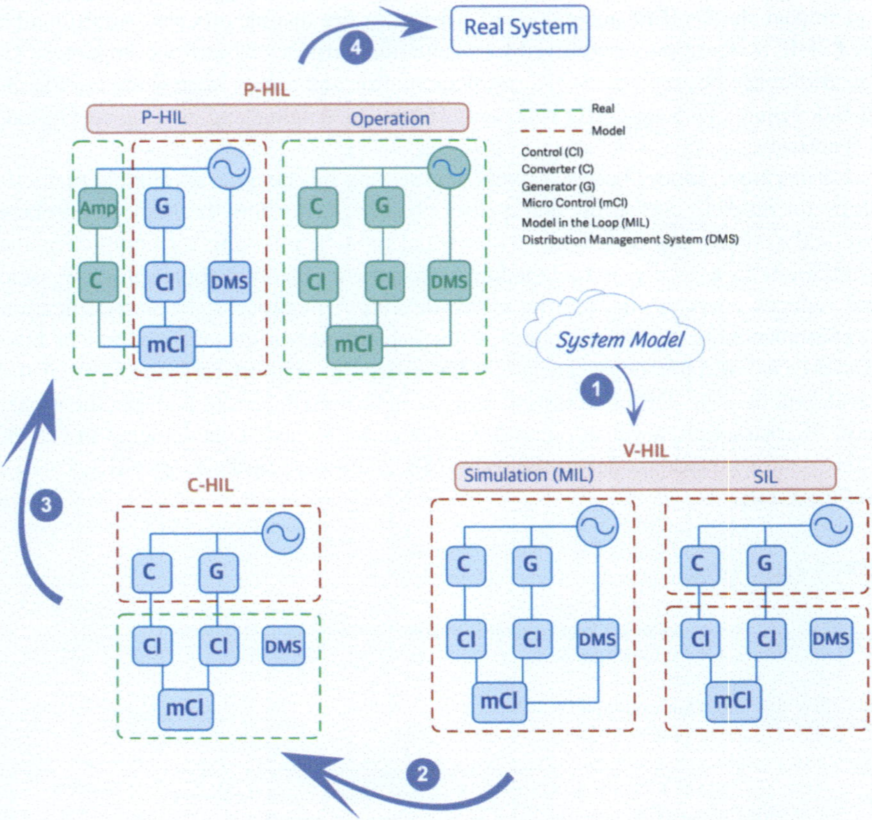

Fig. 5 Testing setups variation based on specifications, abstraction Level, application, and *device under test*

In the MIL and SIL approaches, both the control and power stages (i.e., controller and plant) are simulated in a V-HIL, typically without real-time execution. In the MIL approach, the controller and power layer are modeled together, while the SIL approach focuses on simulating the actual control software. Testing setups that combine physical systems with virtual models fall under the category of HIL testing. This means that a portion of the system is physically connected to the real-time simulation, which can be either part of the P-HIL or part of the C-HIL. In the P-HIL approach, the emphasis is on testing power hardware. Power amplifiers are used to establish a connection between the real-time simulator and the actual power hardware *device under test (DUT)*, using analog input/output signals or communication protocols. For example, analog outputs can be employed to send current and voltage references from the real-time simulation to the amplifier, while the power amplifier senses the feedback signals from the *DUT* and relays them to the simulation through the real-time simulators analog inputs. However, it is important to note that this approach typically entails higher costs, reduced flexibility, and additional safety

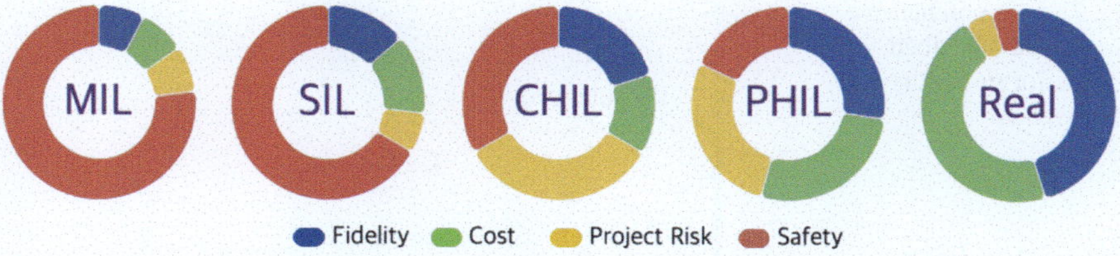

Fig. 6 Graphical comparison of different implementation test methods

precautions, despite its ability to provide highly accurate results when testing software in the presence of actual power [3]. Figure 6 shows a graphical comparison of different implementation test methods.

2 Significance of the HIL in Microgrids Control Units Assessment

Microgrids present unique challenges compared to conventional power systems due to the variety of energy sources, dynamic timescale and different operational modes, including islanded or grid-connected configurations. Additionally, the construction of remote microgrids introduces further complexities. Ensuring microgrid compliance with grid codes and achieving safe and smooth transitions is critically important. That is why testing microgrid control units is crucial to mitigate risks, safeguard equipment, and maintain project/research timelines [3].

2.1 HIL Testing to Study Control Challenges and Operation Issues

The integration and interoperability of microgrid controllers and downstream device controllers pose technical challenges in microgrid deployments. These challenges include high non-recurring engineering costs, concerns about product reliability, risks of equipment damage, cyber-physical vulnerabilities, uncharacterized behavior of control software, lack of suitable analysis tools, and absence of standardized methods to evaluate microgrid controllers compliance with industry standards.

Directly implementing a control system without adequate testing poses significant risks and dangers. Real-world testing can be risky, costly, and time-consuming. Besides, waiting for testing to be completed before receiving all necessary facilities can cause delays and financial implications. Therefore, it is essential to find alternative testing methods that minimize risks and optimize efficiency.

To address these challenges, a standardized technology platform and benchmark system for integrating and testing microgrid and device controllers are necessary.

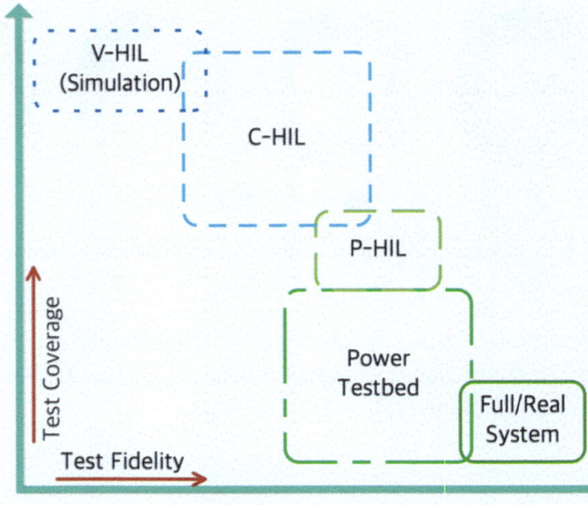

Fig. 7 A comprehensive assessment of testbed options for suitability and effectiveness

Such a platform would facilitate design, integration, interoperability evaluation, pre-commissioning testing, and compliance testing against industry standards. It would accelerate microgrid adoption by enabling realistic demonstrations, risk reduction testing, and efficient system integration and testing. Additionally, it would help substantiate microgrid operations and ensure compliance with relevant standards [3].

One approach to testing microgrid control units is the utilizing HIL testing. HIL testing platforms, have been developed to address technical integration challenges and facilitate comprehensive assessment, debugging, and compliance evaluation of microgrid and device controllers. A C-HIL-oriented approach to microgrid integration testing provides an attractive trade-off between low cost, test fidelity, and excellent test coverage. They enable the evaluation of various system conditions, guide debugging of control algorithms, demonstrate controllers capabilities, and support the revision of electrical feasibility. By utilizing the C-HIL testing platforms, engineers and project developers can reduce technical, safety, and financial risks. Furthermore, it encourages controller vendors to mature their products and prove their capabilities, leading to a higher approval rate of microgrid projects [4]. When evaluating microgrids and similar advanced power system testbeds, several key characteristics come into play for ranking purposes. These include overall cost, test fidelity, and test coverage. By considering these factors, a comprehensive assessment can be conducted to determine the suitability and effectiveness of different testbed options as shown in Fig. 7.

2.2 *IEEE Standards and Grid Codes for Microgrid Testing*

Performance standards play a crucial role in modernizing and establishing a clean grid. They offer several benefits, including facilitating the integration of renewable energy resources, enhancing cybersecurity measures, and improving overall grid reliability and resilience [5].

The rapid adoption of photovoltaics (PV) necessitates the development of interconnection standards. IEEE 1547, IEEE 1547.1, and UL 1741 are key standards undergoing revision to address this urgency. The aim is to establish interconnection and interoperability requirements that accommodate emerging distributed energy resource technologies like grid-supportive inverters and energy storage. As PV, wind, and energy storage technologies dominate new energy generation options in transmission and sub-transmission systems, there is an urgent need for performance standards. Researchers involvement in interconnection standards IEEE 2800 and IEEE 2800.2 aims to fill this gap. These standards have the potential to be as influential as IEEE 1547–2003, which laid the foundation for distributed energy resource interconnection [6].

To optimize microgrid control software algorithms, it is essential to create fully working prototypes and assess their behavior from the earliest stages. The HIL devices offer industry-standard interfaces and supports major communication protocols. This facilitates the integration of microgrid control software with existing substation communication protocols and distributed energy resources (DERs), enabling evaluation of the communication infrastructures performance. Besides, intelligent electronic devices (IEDs) using IEC 61,850 as the standardized communication protocol are gaining traction in new microgrid architectures. IEC 61,850 enables interoperability among substation devices from different vendors. The protocol facilitates the sharing of protection information and operation information for controlling, reporting, and interlocking. It relies on *Generic Object Oriented Substation Event* (GOOSE) Publish/Subscribe messaging to achieve low-latency and deterministic communication [6].

Furthermore, while the adoption of IEC 61,850 in real-world grids and infrastructures is still in progress, there is scope for further improvements and analysis. Previous works have focused on logic configuration, cyber-security vulnerabilities, and real-time simulated grid testbeds. However, comprehensive testbeds that consider physical constraints, scalability, and detailed emulation of microgrid characteristics are still limited. Hence, IEEE standards and grid codes are vital for microgrid testing and integration. They address cybersecurity concerns, distribution and transmission system integration, and interoperability with substation communication protocols. Real-time testbeds, such as those utilizing the HIL devices, contribute to the evaluation of microgrid control software and the application of communication protocols like IEC61850 [7]. Continued research and development in this field will lead to enhanced standards, improved grid integration, and optimized microgrid performance [8]. HIL testing is recommended by the IEEE 2030.8–2018 Standard for testing microgrid controllers. It allows for realistic testing of controller functionality such as islanding and dispatching [9].

2.3 Recent Advance in HIL-Based Microgrid Testing

In this subsection, recent research works on the testing of microgrid control units using HIL technology will be reviewed. These works discuss various aspects of microgrid research and development using HIL techniques. HIL simulations allow for the evaluation and validation of control strategies, power dispatch scenarios, fault detection, and cyberattack mitigation in microgrid systems. The use of HIL technology enhances the effectiveness of system simulation processes by enabling interaction with physical components. Hybrid AC/DC microgrids, which consist of both AC and DC distribution systems, require careful management of energy interaction among various components. Reference [10] focuses on validating control systems within a hybrid DC/AC microgrid using Typhoon HIL as a validation tool. The microgrid model incorporates DERs such as PV plants, diesel generators, fuel cells, wind farms, micro-turbines, battery energy storage systems, and converters. In this work, the simulation of a bidirectional converter connected between AC and DC buses is validated using both Matlab Simulink and Typhoon HIL 402-based simulations.

Furthermore, authors in [7] present real-time simulations of a hybrid microgrid, consisting of an AC microgrid, DC microgrid, and utility grid. An innovative Hill Climbing algorithm is developed to achieve even power transfer between the AC and DC microgrids, ensuring stable and reliable system operation under varying loading conditions. A generalized function is implemented to handle power flow from the utility grid to the AC and DC microgrids. The Typhoon HIL control center is utilized to implement and validate the system, with simulations performed in the Schematic Editor. Real-time simulations of the hybrid microgrid are conducted using the HIL402 emulator within a Supervisory Control and Data Acquisition (SCADA) environment. Another work introduces a multi-layer HIL methodology for the development and testing of complex microgrids [11]. A microgrid testbed is constructed, comprising real-time simulators, an industrial double active filter controller, and a solar inverter controller. The testbed incorporates various DERs, including solar inverters, diesel generators, battery storage systems, active filters, nonlinear and linear loads, and an induction machine. Experimental results of different test scenarios are presented, showcasing the flexibility and accuracy of the HIL approach in modeling of microgrid components. Besides, the presented HIL approach demonstrates significant advantages in terms of microgrid modeling flexibility. It allows for the modeling of DERs with different levels of complexity and accuracy, ranging from simple active power sources to detailed models of power stages with simulated control algorithms. The multi-layer HIL methodology proves beneficial in testing microgrids against various fault conditions, such as grid shorts, disturbances, converter faults, and large power fluctuations that are challenging to replicate in a real microgrid. The testbed itself is easily reconfigurable, enabling quick changes to the microgrid structure and the connection of external controllers or SCADA systems.

The *Smart Energy Integration Lab (SEIL)* is a microgrid testbed that utilizes the P-HIL techniques to study the grid integration of DERs. It allows for the evaluation of system dynamic performance and power dispatch scenarios in microgrids [12]. Authors in [13] focus on the emulation of a ship-board microgrid using a hybrid approach combining machine learning algorithms and electromagnetic transient based techniques. This approach enables accelerated parallel emulation and efficient resource utilization on field-programmable gate arrays (FPGAs). Furthermore, a research in [14], investigates the control of an AC microgrid utilizing PV and battery energy storage. It analyzes and implements primary and secondary control strategies in a real-time simulation software called Typhoon HIL. The power management system aims to maintain stable voltage, frequency, and power in both grid-connected and islanded modes of operation.

Additionally, Ref. [15] employs an HIL-based control and protection platform for a DC microgrid. It combines centralized control with localized parameter monitoring for fast and accurate fault detection in the microgrid. Besides, authors in [16] address optimal energy management in a microgrid incorporating renewable energy resources and battery energy storage. It uses a Lagrange multipliers-based method to obtain an analytically derived optimal economic dispatch solution, considering power flow constraints, uncertainties of renewable energy resources, microgrid operator profits, and support to the main grid. Moreover, researchers in [17] experimentally validates a primary/backup framework for preventing and mitigating the impacts of false data injection cyberattacks in a lab-scale wind/PV microgrid. It utilizes an artificial intelligence-based false data detection algorithm and a remedial action scheme (RAS) as backup mechanisms within an HIL testbed.

Reference [18] examines the control schemes and performance analysis of distributed generation sources connected in parallel in a microgrid operating in grid-connected and stand-alone modes. The stand-alone mode utilizes an inverter-based distributed generation system with a droop control scheme incorporating virtual complex impedance in the outer voltage loop. The distributed generations proportionally share the microgrid load power based on their power ratings, ensuring reliability and efficiency. The modes are automatically switched using the Phase Locked Loop (PLL) phase error $\sin(\gamma - \theta)$, which detects islanding during system disturbances and enables seamless mode transitions. The paper presents the response of the PLL phase error, islanding detection, and mode switching for various fault conditions. The microgrids performance in both modes, including islanding detection and automatic mode switching, is evaluated using an HIL platform.

In addition, authors in [19] introduce a rule-based master/slave communication-based power management system for a real-world hybrid microgrid incorporating PV and energy storage systems (ESSs). The microgrid consists of seven PV inverters, one battery inverter, and AC loads. The power management system aims to minimize electricity costs for customers and ensure a stable voltage, frequency, and power supply in both grid-connected and islanded conditions. The implementation of the microgrid, SCADA interface, and power management system is carried out using the Typhoon HIL Control Center. The power management system is developed in Python within the SCADA environment, and the microgrid is simulated using a virtual HIL device. Furthermore, in [20], researchers present a hierarchical energy management strategy for isolated microgrids using a master–slave topology, where local controllers are managed and controlled by a central controller. This approach offers several technical advantages, particularly in terms of microgrid performance and energy supply. The local controller is designed to meet specific objectives of the microgrid, such as stabilizing DC voltage and maximizing power extraction from sources. The central controller adopts a centralized approach based on fuzzy logic to preserve battery life and manage the energy balance between generation and consumption. The microgrids performance is investigated under steady-state and faulty conditions. An HIL test using the Simulink platform is conducted with an RT-LAB real-time simulator to validate the hierarchical control. The control strategy is also tested and validated using the OP1400 test bench, based on the OP4150 digital simulator. The results are compared to international standards IEEE 1547 and IEC 61,727, demonstrating excellent consistency.

Another work introduces two power flow control algorithms for a voltage source converter connected to the smart grid as a part of an ESSs, specifically designed to operate under unbalanced voltage conditions. These algorithms are enhancements of the dual vector current control algorithm (DVCC). The first algorithm, DVCC_CL, optimizes the limitation of phase currents during voltage unbalance. The second algorithm, DVCC_HB, enables high bandwidth control of active and reactive power. The choice between the two algorithms depends on the grid code requirements and constraints imposed by the specific ESS connected to the grid. The performance of both algorithms is verified using an ultralow-latency HIL emulator, which ensures accurate analysis of converter behavior under various grid conditions [21].

Furthermore, Ref. [22] presents a methodology for conducting the C-HIL testing of centralized and distributed secondary frequency control schemes in AC microgrids operating in islanded mode. It outlines the formulation of the secondary frequency control problem and explains the theory behind both centralized and distributed implementations of the control schemes. The paper also describes the testbed used for the C-HIL testing and presents the testing results, comparing the performance (measured in terms of system response time) and resilience (measured in terms of the ability to withstand the failure of a control device) of both schemes. Following, Ref. [23] uses a multitasking control algorithm for Grid-Connected Converter (GCC) in distributed generation applications. It focuses on a single-phase H-bridge voltage source converter that acts as the power electronic interface between the DG system and the grid. The control algorithm operates the GCC in current control mode to achieve the desired active power injection into the grid. Additionally, the algorithm includes capabilities for current harmonics mitigation and reactive power compensation to enhance power quality. The control algorithm optimizes the remaining capacity of the GCC for power quality improvement. It utilizes a current decomposition structure based on multiple adaptive noise cancellation filters to extract harmonic and reactive currents from the local loads. These extracted harmonics are then used to estimate compensating currents to be injected by the GCC. A single-phase PLL using adaptive noise cancellation filters is developed to synchronize the GCC at the fundamental frequency, allowing the control algorithm to adapt to grid frequency fluctuations and remain immune to grid voltage distortions. The effectiveness of the control algorithm is demonstrated through HIL testbed.

3 Step-By-Step Microgrid Implementation Using HIL

Complex systems such as microgrids require automated energy management involving multiple converter and non-converter resources. HIL devices facilitate the testing of system-level controllers, including building energy management systems, distribution management systems, and microgrid controllers. Prior to controller testing, a microgrid model must be established within a testing environment. This section presents illustrative examples that demonstrate the advantages of HIL devices and their application in the context of microgrid engineering and research [3]. The content is specifically designed to meet the needs of beginners in the field, providing them with valuable insights into the benefits of HIL devices and practical guidance on their utilization in the role of a microgrid engineer or researcher. In this chapter, the selected HIL device is Typhoon HIL, and all figures and material presented in this chapter have been prepared in the Prof. Kato Laboratory in the Nagoya University, Japan as shown in Fig. 8.

Fig. 8 Prof. Kato laboratory in the Nagoya University, Japan

3.1 An Introduction on the Most Important Components of a Microgrid

Within a modern power grid context, an HIL testbed serves several purposes, including: (i) Time domain analysis of short circuits, failure modes, and coordination studies, (ii) Testing interoperability and functionality of DERs and microgrids, (iii) Assessing the performance of microgrids under various scenarios, such as short circuits, phase losses, component failures, and voltage fluctuations, (iv) Conducting real-time sensitivity analysis of the entire grid, (v) Evaluating the hardware, firmware, software, and communication aspects of controllers across different operating conditions, including faults in both islanded and grid-connected modes. This section provides an introduction to the key components necessary for constructing a microgrid utilizing Typhoon HIL.

The implementation of a microgrid using the HIL-Typhoon Device necessitates familiarity with the essential components of a microgrid. This subsection introduces the incorporation of DERs within the Typhoon HIL environment. It covers the utilization, advantages, and disadvantages of different types of DER components. Various considerations, such as expediting modeling setup, bridging knowledge gaps, and minimizing HIL hardware utilization, may influence the selection of DER types for a project. Additionally, project scope plays a significant role in decision-making.

The Typhoon HIL examples folder offers preconfigured models and dedicated SCADA panels, facilitating the rapid construction of microgrid models. To create a new model, one can access the Schematic Editor and employ a drag-and-drop approach to add the necessary components. Prior to creating a new model, it is essential to determine the primary properties, such as the *HIL Model*, as depicted in Fig. 9. The distributed energy resources sub-library within the microgrid library contains models of DERs, including *ESSs, PV power plants, wind power plants*, and *diesel gensets*.

3.1.1 Difference Between Switching and Average Components

Notably, the Legacy sub-libraries feature *switching* and *average* components, both implemented as three-phase devices. From a controller functionality perspective, these models are indistinguishable, with similar parameterization. The main distinction lies in the power stage, where the switching model employs a detailed power electronics three-phase voltage source converter component, while the average model employs three signal-controlled voltage sources to emulate converter behavior.

The three-phase converter is controlled by an internal modulator, whereas the average model employs signal-controlled voltage sources. The Legacy (Average) components are recommended in certain scenarios, such as when integrating a battery and converter as separate components or when there is a need to conserve circuit solver resources. The choice between components primarily depends on the type of controller or *DUT*. Legacy switching components are suitable for testing converter

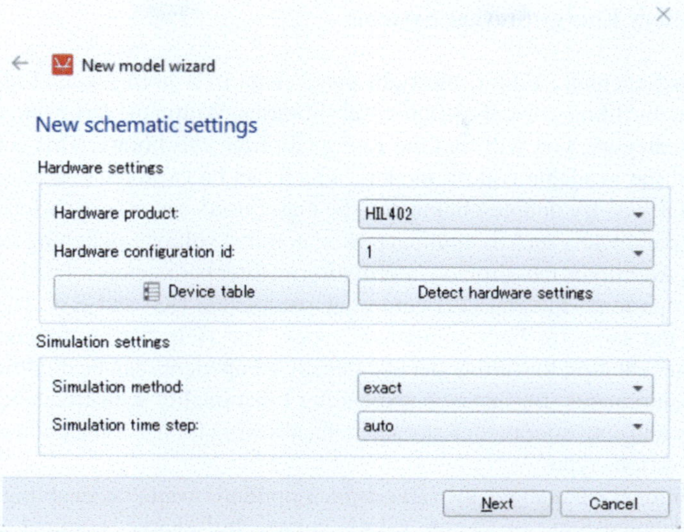

Fig. 9 Select the HIL model

controllers due to their high-fidelity power electronics models and explicitly defined inputs for gate control. In the controller design stage, Legacy switching models offer pre-implemented control subsystems for major DER applications. On the other hand, *generic* components are ideal for higher-level controls, such as microgrid modeling and testing, energy management systems, or battery management system and controle.

Microgrid controllers oversee steady-state regulation of voltage, frequency, load, and energy management. Unlike power electronics models, the slower dynamics of microgrid controllers can be captured by average and *generic* DER types. *Generic* components offer built-in functionalities commonly encountered in real-life DERs, enabling seamless integration and functional system operation. These functionalities include voltage and frequency droop, ramping, LVRT, voltage/current protection, and storm control. *Generic* components also provide self-tuning and grid support features, which are particularly valuable for individuals with limited experience in control engineering and power electronics or time constraints. *Legacy* components offer extensive parameterization options for controller gains and converter output filter values. However, their use incurs higher computational costs due to the utilization of converter and machine solvers. *Generic* components offer automatic model tuning, dedicated communication interfaces, standardized DER functions, and do not require converter or machine solver resources. Considering hardware utilization, large microgrid and power system models can strain HIL device resources.

3.1.2 Battery Energy Storage Systems

In the New Schematic Editor, the right panel features a library called *microgrid*. Clicking on this library reveals different sub-libraries. Within the *Distributed Energy Resources* category, you will find the *Energy Storage* sub-library. This sub-library contains all the available battery models, which can be explored and accessed for considered needs. By double-clicking on the mask, users can access various tabs for modifying parameters such as nominal power, nominal voltage, switching frequency, and others. Additional tabs provide information on filter parameters related to control, such as Propotional-Integral (PI) controller parameters (Kp and Ki). The parameter set is the same for both component types. The Help button directs users to online documentation for further details. Legacy components can be modified to suit specific requirements, for instance, expanding functionality with frequency droop. The battery and converter models are separate, allowing for the connection of battery models of various types.

For additional parameters, the *User defined* option is available, enabling the definition of detailed State of Charge (SOC) curves, full charge, nominal discharge current, resistance, and capacity. The signal processing tab includes features like SOC output and *Use Signal Processing look-up-table*. Enabling the SOC output property adds an extra signal processing output port to the battery model, providing the current SOC information. The execution rate property determines the execution rate of signal processing components. The *Battery* component is implemented as a controlled voltage source and a series resistance, with the voltage source controlled by a lookup table that adjusts automatically based on the battery type. The number of lookup tables is limited to 8 per device. If the *Use signal processing Look up table* property is enabled, the entire battery is modeled with a single *Signal Processing lookup table*, which is beneficial when dealing with more than 8 batteries, as it saves one lookup table resource at the expense of one *Signal Processing* resource (of which there are 16 per core).

Generic components are primarily designed to facilitate model parameterization through user-provided nominal values, removing the need for adjusting grid filter or control software details. The *Generic Battery* model comprises two main sub-components: the *Battery ESS* component, which includes a high-level control subsystem and a low-level control subsystem with the power stage, and the *Battery ESS* User Interface (*UI*) component, where inputs and outputs are defined. This component demonstrates battery inverter behaviors, considering various operation modes, limitations based on nominal parameters, and fault detection.

It can operate in simultaneous, droop, and grid following modes. The properties of the battery component allow for changes in general parameters such as nominal active power, nominal apparent power, and nominal frequency. There is also an option to include a built-in transformer. The *Battery* tab provides information on the battery, including nominal voltage and capacity, as well as the ability to set up an absolute SOC range. Grid codes offer built-in grid support functionalities like LVRT, Volt/Var, Hz/Watt, and Volt/Watt. Details about grid support functionalities can be found in the materials tab.

The Converter Extras Tab allows specification of additional parameters related to the converter, such as input filter impedance, DC link voltage margin, shunt resistance, and execution rates. The *Grid Extras* component tab is used to set up additional parameters related to the grid, including *Grid Short Circuit* power and *Grid inductive power factor*. Unlike *Legacy* components, *Generic* battery components are locked and open only towards the AC/grid side, with the battery and inverter both aggregated. Each component has its own *UI* component, which can be customized by adding a communication protocol from the available library, such as *Modbus (master/slave protocol, meaning a device operating as master will poll one or more devices operating as a slave)* or Controller Area Network (CAN). This allows for mapping the communication layer without worrying about unsupported functions in the controllers Modbus map. An example contains a dedicated example that helps test these functionalities. To find it, navigate to Example explorer, then microgrid, energy storage group, battery ESS *generic* model, and click open model. To access the information provided in this subsection, you can refer to Fig. 10. This figure presents the relevant details and visual representation of the topic being discussed.

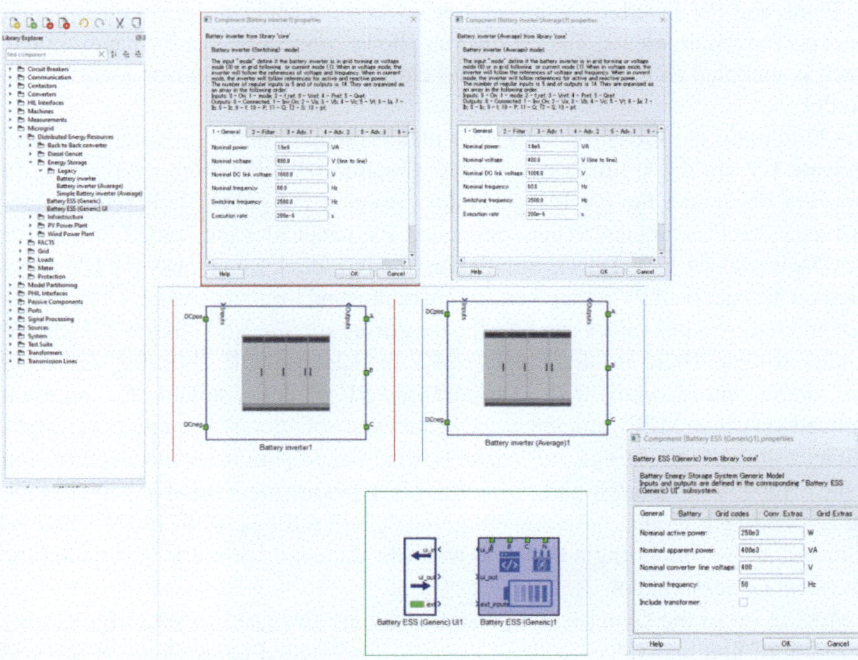

Fig. 10 Introduction to battery library

3.1.3 PV Power Plant

Under the *Distributed Energy Resources* group, there is a sub-group called PV Power Plant that contains two categories of components: *Legacy* and *Generic*. The Legacy components include neutral point clamped PV Inverter, PV Power Plant Switching and Average, and PV Inverter Average. The Legacy PV Plant (Switching) model consists of a three-phase two-level inverter connected to a current control loop. The PV plants DC side is simplified with a constant voltage source. The PV output is determined by the plant surface, and the efficiency can be adjusted using the irradiance input. The PV plant can operate in voltage or reactive power control mode. Both the Legacy PV Plant (Switching) and the Legacy PV Plant (Average) models share the same controller and have identical behavior. The Legacy PV Inverter Average component features a three-phase controlled voltage source with a current control loop. The DC link is connected to an external PV panel source. This PV inverter operates only in grid-connected mode, either in maximum power point tracking (MPPT) mode or reactive power control mode when following the grid. You can explore a dedicated example by clicking on Examples, then navigating to Microgrid, PV Plant, and PV Inverter (Average). Legacy components can be modified to suit your specific requirements. The PV inverter allows connection to a PV panel source, which is modeled as a voltage-controlled current source capacitance connected in parallel.

Additionally, the *Waveform Generator* in the schematic editor provides the ability to create I-V curves. It offers two modes: *Generate PV* and *Import I-V* curve. In *Generate PV* mode, the PV file generator creates a PV panel file based on standard data-sheet parameters. These generated PV panel files are loaded in the HIL SCADA model settings. During simulation, the HIL device generates the I-V curve based on the specified PV parameters, temperature, and irradiance values. The irradiance and temperature values can be adjusted during runtime and observed in the PV monitor widget. There are three types of PV models available: *Detailed*, *EN50530 Compatible*, and *Normalized IV*. The *Detailed PV* model is suitable for precision applications where MPPT efficiency and behavior are of interest. For microgrid applications with multiple strings of PV panels, the lumped-parameter models provided by the *EN50530 Compatible* and *Normalized IV* types are more suitable. *Detailed PV* and *EN50530 PV* models have preconfigured default settings, while the *Normalized IV* model allows importing a CSV file with raw data and normalizing it within the *Waveform Generator* tool.

Moving on to the *Generic* components, they are designed to simplify parameterization of nominal values without the need for detailed adjustments of the grid filter or controller. They are particularly useful for grid stability and system integration testing. The *Generic* PV component emulates the characteristic behavior of a grid-connected PV plant in various scenarios, including active power curtailment, reactive power control, fault detection, ramping functionalities for reference signals, and MPPT. It can operate in grid-following mode with support for active power curtailment. The component properties include tabs for specifying general parameters, PV-related parameters, and grid code functionality. The *Converter* and *Grid*

Extras tabs remain the same as in the *generic* battery example and cannot be modified. The converter and PV panel are consolidated within the *generic* component.

Note that the PV *generic* component will continue to receive updates for additional functionalities in future releases, while the *legacy* components will remain unchanged. You can explore the *PV Plant* (*Generic*) example in the example library by navigating to *Examples*, then *Microgrid*, *PV Plant*, and *PV Plant* (*Generic*). Further explanation and test cases for this example can be found in the corresponding application note, which is linked in the *Examples Explorer* page and available in the *Materials* tab. To well understanding of the information provided in this subsection, you can refer to Fig. 11. This figure presents the relevant details and visual representation of the topic being discussed.

3.1.4 Diesel Genset

Within the *Distributed Energy Resources* group, there is a subgroup called *Diesel Genset*, which consists of two categories of components: *Legacy* and *Generic*. The Legacy components represent a diesel generator modeled as a synchronous machine with speed and voltage control achieved through a governor and an exciter, respectively. Secondary control loops are used for active and reactive power control by adding an offset bias to speed and voltage references. The *diesel genset* can disconnect from and reconnect to the grid through its controlled circuit breaker, allowing it to operate in both grid-forming and grid-following modes. The component properties of the *Legacy diesel genset* provide different tabs for adjusting various parameters. You can specify basic parameters of the *genset*, as well as electrical and mechanical parameters of the synchronous machine, governor, and engine. *Woodwards DEGOV* model is employed for the governor and engine representation. An example demonstrating the functionalities of the *Legacy Diesel Genset* component can be found in the examples library under the *Microgrid* category, specifically under *Diesel Genset* and *Diesel Generator*.

One advantage of the *Legacy diesel genset* component is its suitability for research and development work related to synchronous generator control, apart from microgrid modeling. This component offers the flexibility to modify the model according to specific needs by unlinking the library and adjusting the model under the *Mask*. However, using this component for microgrid modeling has a disadvantage due to the synchronous machines utilization of the machine solver, which can pose limitations on hardware resources, especially in microgrids with multiple *diesel gensets*. To address this limitation and optimize resource usage, an alternative approach is to replace the synchronous machine in the Legacy *diesel genset* component with the *Three Phase Synchronous Machine* (*Generic*) available in the *Machines* library. This *generic* model allows for easy parameterization and utilizes a signal-processing-based solver instead of a dedicated machine solver. By modifying the *Legacy diesel genset* component accordingly, you can take advantage of the flexibility while saving on hardware resources, particularly in *multi-genset* microgrid models.

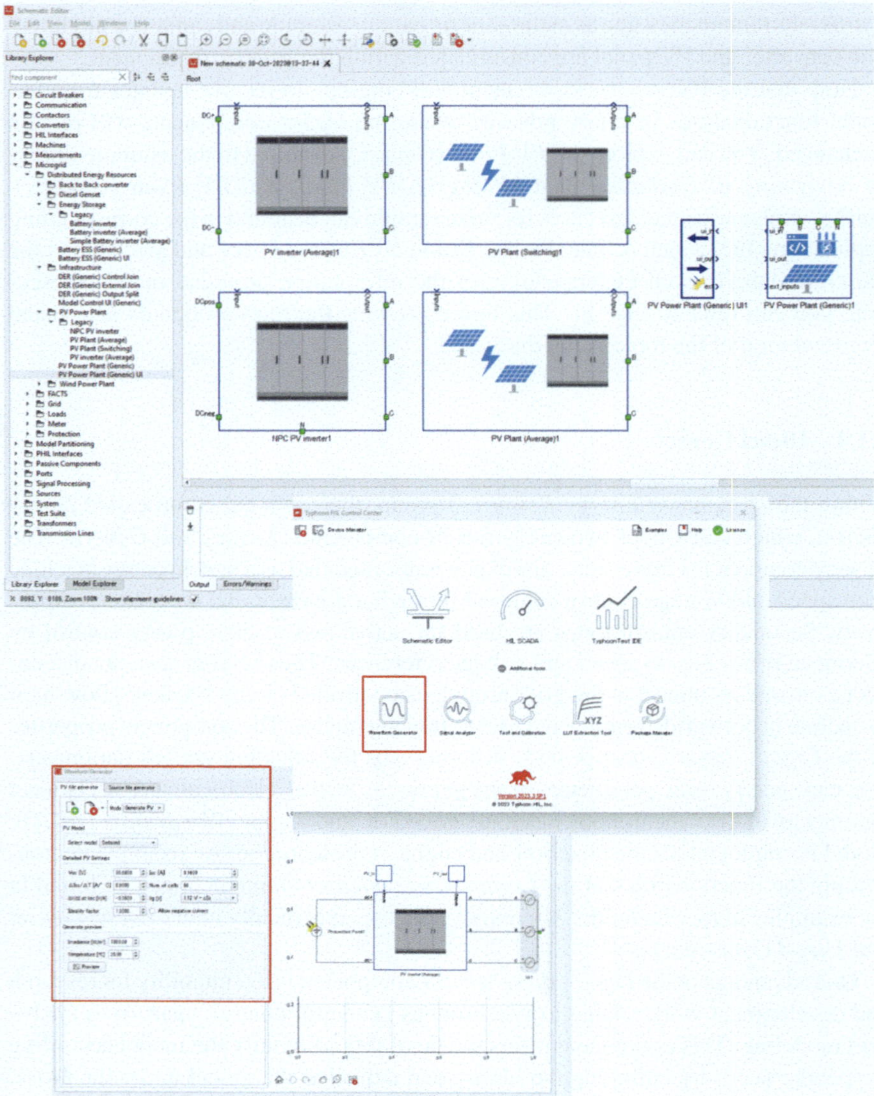

Fig. 11 Introduction to PV library

In addition to the Legacy component, a *Generic Diesel Genset* component is also available. It consists of two parts: the *Diesel Genset* (*Generic*) *UI*, which serves as an interface for inputs and outputs, and the *Diesel Genset* (*Generic*) component itself. The *UI* component facilitates the implementation of a remote terminal unit using supported protocols. The component properties of the *Diesel Genset* (*Generic*) component include tabs for specifying parameters related to nominal values, the generator system, and additional parameters. This component operates in simultaneous, droop, and grid-following modes, and its machine solver is implemented using signal processing, eliminating the need for a dedicated machine FPGA solver. The *Generic Diesel Genset* component is advantageous for system-level applications with multiple *gensets*, as it does not consume machine solver resources. The power stage of the component includes the BR model of the synchronous generator, transformer, and the miniature circuit breaker connecting the plant to the grid. The transformer can be excluded if not required. Parametrization of power stage elements can be done through the component dialog box.

To explore the functionalities of the *Generic Diesel Genset* component further, you can refer to the *Diesel Genset Generic* example in the *Microgrid* examples library. In this subsection, the entire group of *Distributed Energy Resources* components will be covered. In the upcoming sections, a demonstration will be provided on how to efficiently utilize these components to create Microgrid Models. For a comprehensive understanding of the information presented in this subsection, readers are encouraged to refer to Fig. 12. This figure provides a clear visual representation and includes all the pertinent details related to the topic being discussed.

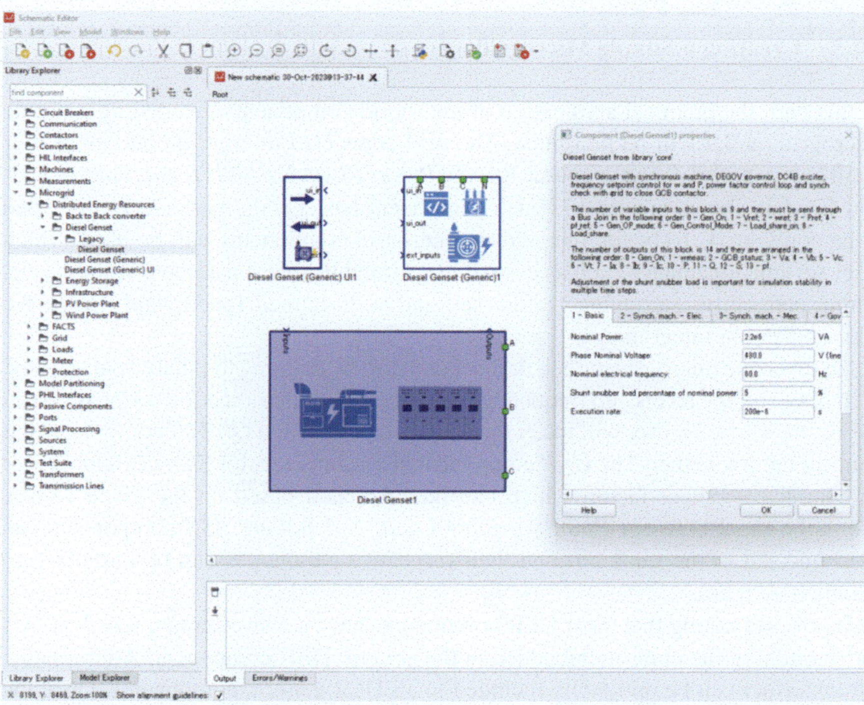

Fig. 12 Introduction to genset library

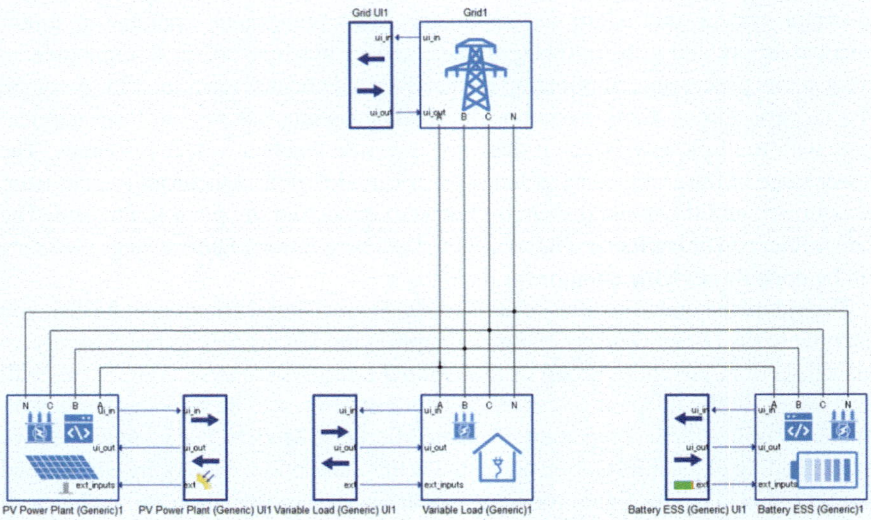

Fig. 13 Microgrid under study

3.1.5 Creating a Simple Microgrid

In this part, the practical application of microgrid components by creating a simple microgrid model will be presented. The grid-connected microgrid consists of four main components: A power source like *PV Power Plant*, *Variable Load, Battery ESS*, and *Grid* as shown in Fig. 13. Each component has specific parameters that need to be configured. To begin, the *generic* microgrid components will be used, which offer several advantages such as ease of use, straightforward parameterization, self-tuning capability, a growing feature set, and well-defined functionality maps for communication integration.

To set up the microgrid model, the user can start by defining the shared parameters, including voltage levels and frequency, using the model initialization script as shown in Fig. 14. Then, the user will insert the *Grid* component and *Grid UI* component and connect them together. The *Grid* component has parameters for *Short Circuit* power and Inductive power factor, which can be calculated based on the given values. Similarly, the *PV Power Plant, Variable Load, and Battery ESS* components can be connected to the same bus and their respective parameters can be initialized to parametrize them.

It is worth noting that most DER components have a converter tab, which allows for fine-tuning the control embedded in the *generic* DER component. Additionally, a transformer can be optionally included in all DER *generic* components. Once the model is compiled and loaded, user can create a SCADA Panel using the provided widget library. This library contains predefined widgets for the DER *generic* components, enabling quick and easy creation of the SCADA panel. The widgets can be

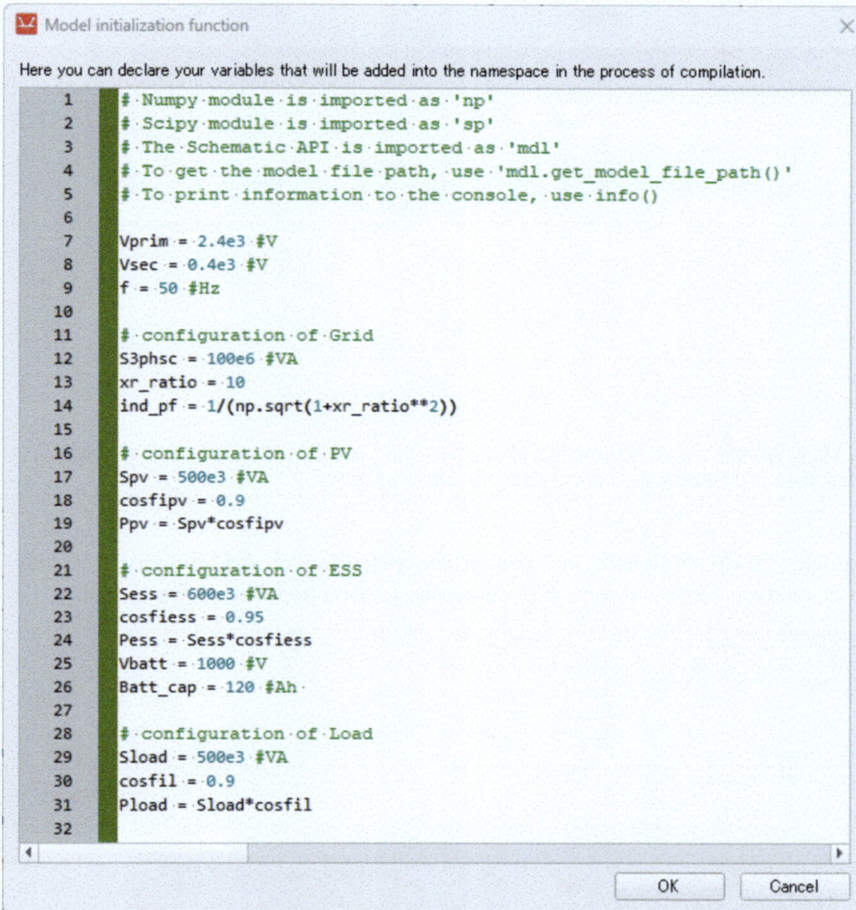

Fig. 14 Sample of Typhoon HIL script

linked to their corresponding model components, and the SCADA panel can be customized further if desired as shown in Fig. 15.

Running the simulation, user can observe the *Grid* component supplying power to the load, with active and reactive power values displayed as shown in Fig. 16. The power setpoints of the load can be modified inside the *Load* Interface sub-panel. The *Load* can be configured in different modes, and its power consumption can be influenced by voltage and frequency variations. Enabling the *PV Power Plant*, user can adjust the active power limit and observe how the active power output responds accordingly. The output power of the *PV Power Plant* is affected by irradiation and temperature values, which can be adjusted using sliders. Similarly, the *Battery ESS* component operates in different modes, such as grid following, droop, and isochronous. The droop characteristic can be defined using frequency droop offset,

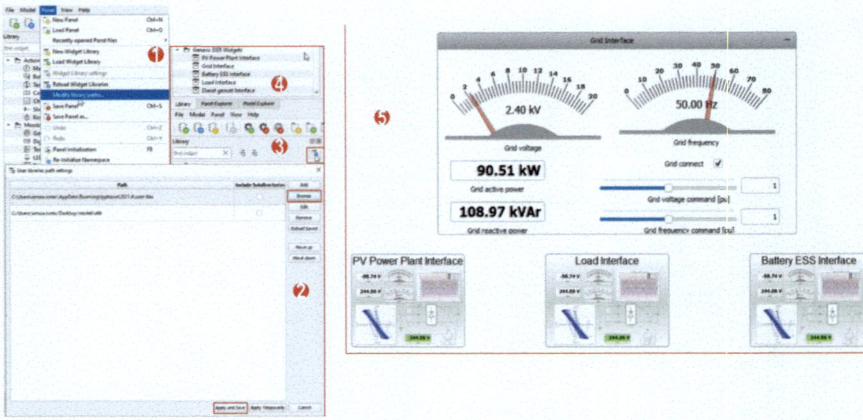

Fig. 15 Load new widget: (1) Modify library path, (2) Load the path, (3) Synch the changes, (4) Added items, (5) Graphical presentation of the added widgets

frequency droop coefficient, and voltage droop coefficient. Additionally, grid codes can be enabled for the *Battery ESS* component, providing further functionality.

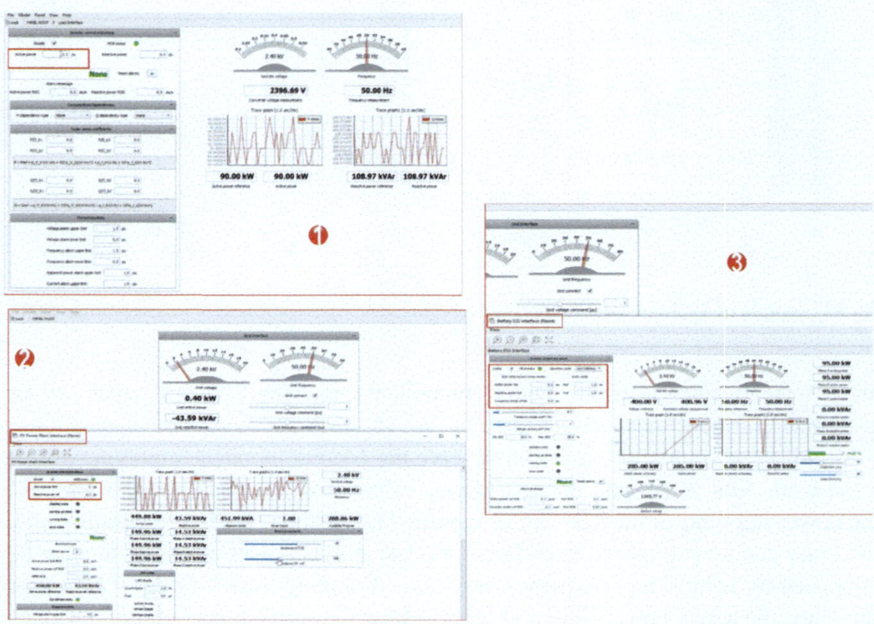

Fig. 16 (1) Overall screen after execute the model, (2) PV plant information, (3) Battery information

3.2 A Fully Functional Terrestrial Microgrid Model

In the previous subsection, a V-HIL device was utilized for simulation, but not for real time one. However, in this subsection, a terrestrial microgrid model that functions in real time and is based on DER switching components is introduced. To begin, open the model using the *Example Explorer* by navigating to *microgrid, microgrid example*, and selecting the *Plug-and-play microgrid library and testing of microgrid controller* example. Click the *Open model* button to access this example. For the sake of real time simulation, a 6-series HIL device, such as the Typhoon HIL602 + , HIL603, HIL604, or HIL606, is required. Here, the HIL606 device is used. The need for a 6-series HIL device is explained in the message console. From a resource perspective, this model utilizes four standard processing cores and a machine solver. The device configuration table illustrates how these resources are allocated in different configurations. Most configurations for the HIL606 satisfy the requirement of four signal processing components and one machine solver, except for configurations four and seven. On the other hand, a 4-series HIL device does not offer such configurations, as shown in the configuration table. The HIL606 combines the speed of the HIL404 with the capacity of the HIL604, enabling increased model complexity and the addition of more DER components. Figure 17 shows the overall structure of the selected microgrid.

This example show a microgrid controllers capability to handle a microgrid islanding event. The net power flow is observed at the point of common coupling (PCC), and system resources are commanded accordingly to prepare for and respond to islanding. In two different cases, a battery inverter and a diesel generator are utilized as the primary grid-forming DER to compare their dynamic responses during islanding.

The microgrid model consists of four DER components: an *ESS*, a *Wind Plant*, a *PV Plant*, and a *Diesel genset*. Additionally, the microgrid includes three types

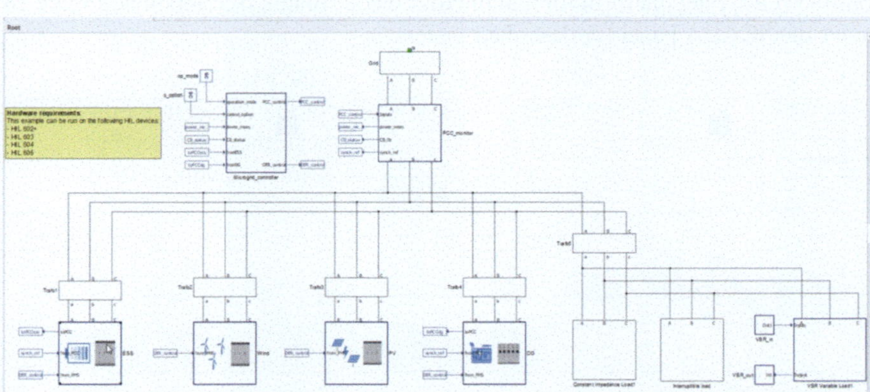

Fig. 17 Overall structure of the implemented microgrid

of loads: a constant load, an interruptible load, and a variable load. All elements of the microgrid are connected to the same PCC. Except for the interruptible load, all components are plug-and-play and sourced from the *Typhoon Microgrid Library*. Each DER has its own transformer with a ratio of 12.5 kV/480 V. The *ESS, Wind Plant*, and *PV Plant* are converter-based models. Although this example uses switched components, *generic* or *average* components can also be employed to implement the same model on 4-series devices. The microgrid control unit serves as the brain of the microgrid, which is responsible for monitoring and issuing commands to the DERs. It can command the synchronization relay to connect or disconnect the microgrid from the main grid based on the requested operating mode and grid faults. The microgrid controller also regulates power flow through the PCC by providing power references to the grid-forming DER, allowing for smooth intentional islanding. Communication protocols such as *IEC 61,850, DNP3*, or *Modus* are often used for communication between the microgrid controller and DERs in microgrids.

All DERs can be commanded through the DER control signal from the microgrid control unit, but only the *ESS* and *diesel genset* have additional inputs/outputs. As they are not based on intermittent energy sources, they can be selectively engaged by the microgrid control unit as grid-forming sources. They provide power information to the microgrid control unit and receive power references in return. The microgrid control unit is fully implemented using the signal processing toolbox, with its logic built around state machines implemented using C code components. *SM* and *SM2* are the two state machines that issue commands to the circuit breaker inside the PCC monitor subsystem and the grid-forming DERs, with the support of PI *Controllers, Signal Routing*, and *Rate Transition* components. The PCC monitor subsystem oversees power flow through the PCC, keeping the microgrid control unit informed of the main breaker status. The breaker status signal triggers the microgrid control unit to start controlling power flow by managing the DERs. The PCC monitor performs synchronization, control, and protection, including over and under voltage and frequency protection. The HIL device can be used as a control unit or rapid control prototyping platform for real microgrid control unit development. Figure 18 shows the SCADA panel for described system.

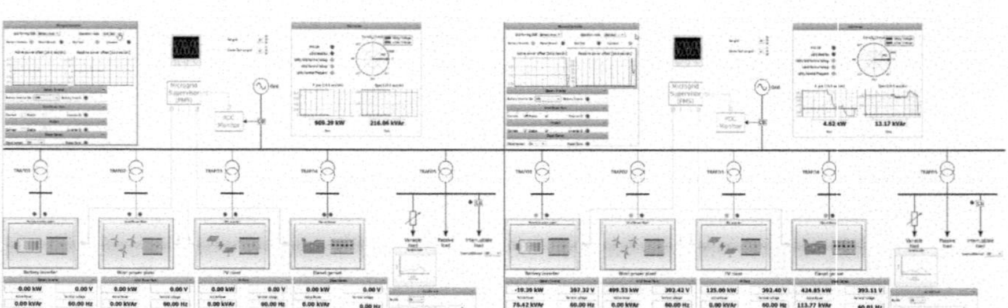

Fig. 18 SCADA panel for described system

Microgrid rapid control prototyping applications are further supported by a wide selection of connectivity options and communication protocols within the communication protocols toolbox. Connectivity and communication protocols are covered in more detail in the Communication protocols module. Typhoon HIL also provides C code export functionality, allowing the controller model to be exported from Typhoon HIL to another controller development environment. To compile and load the model, use a fourth-generation HIL606 device. Open the panel file in the model directory and run the simulation. If you are using a V-HIL device, you may notice that the simulation is slower compared to real-time simulation. This highlights the advantage of using an HIL device for real-time simulations.

3.3 Enhanced Grid Support Capabilities of Generic DER Components

In this subsection, the enhanced protection functionalities offered by the *generic Distributed Energy Resource* components, namely Volt/VAR, Hertz/WAT, and Volt/Watt, will be explored. The focus will be on examining these protective functions in the context of the *battery converter generic* component, which is located in the *examples explorer menu* of the Typhoon HIL software. The primary impetus behind the development of these new *generic* components is to provide a more streamlined and practical system integration parametrization. Moreover, these components encompass novel functionalities commonly found in real-life applications, including protection mechanisms, frequency monitoring, and phase verification. An additional advantage of the *generic* components is their independence from the machine solver of the hardware, enabling the creation of larger models featuring a greater number of generators. To access the *battery converter generic* component, navigate to the *microgrid distributed energy resources* folder in the *library explorer*. The *generic* system consists of two primary components: the *generic UI*, which facilitates communication between the generic components and the *input–output system*, and the *generic* component itself. The implementation of the *generic* component employs *signal processing* and *signal-controlled voltage sources*, with the mathematical model also relying on signal processing techniques [3].

Certain fundamental parameters can be adjusted to suit specific requirements. The volt-var, hertz-watt, and volt-watt protective functions are integrated into all inverter-based DER *generic* components. These protective functions can be *parameterized* within the grid code tab located on the mask of the *generic* component. The volt-var function primarily ensures continuous battery connectivity and regulates reactive power injection according to predefined grid codes. Similarly, the hertz-watt and volt-watt functions maintain active power injection, with the former being frequency-dependent and the latter voltage-dependent. These protective functions can be customized to meet specific requirements. To facilitate testing and analysis, the *battery converter generic* example includes its own SCADA panel. Within this

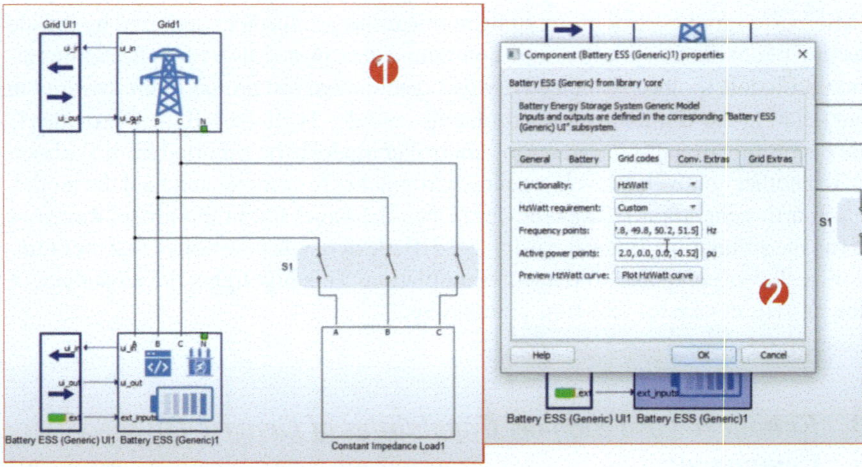

Fig. 19 Implemented microgrid: (1) microgrid model, (2) Mask options

panel, the grid interface group provides essential measurements, while the *battery ESS* interface subpanel features a converter control and status group where basic control parameters can be found. Figure 19 shows the implemented microgrid and a sample of *Mask* to find the grid codes option.

Enabling the converter requires checking the *enabled* checkbox, after which it operates in the designated operation mode. Real-time parametrization of the operation mode can be carried out through the *grid* interface panel, while the *battery* must be in grid-following mode to comply with grid codes. Automated test definitions include buttons for volt-var, hertz-watt, and volt-watt tests. To activate any of these tests, simply select the corresponding checkbox. In the grid codes panel, multiple functions can be enabled simultaneously. When volt-var, hertz-watt, and volt-watt functionalities are active, they take precedence over the LVRT deactivation and reactive power contribution settings. It is important to note that the volt-var, hertz-watt, and volt-watt protective functions exclusively operate in grid-following mode. Now the users can observe the behavior when this mode is activated. Some of the mentioned scenarios are shown in Fig. 20.

3.4 Microgrid Model Optimization

In this subsection, the microgrid model optimization will be explored. In the *Application Examples* module, authors previously examined a fully functional terrestrial microgrid model constructed using switching DER components. That example showcased the microgrid controllers ability to handle microgrid islanding events by utilizing either a diesel generator or a *Battery ESS* to form the grid. Thus, the same terrestrial microgrid model can be rebuilt using *generic DER* components. This will

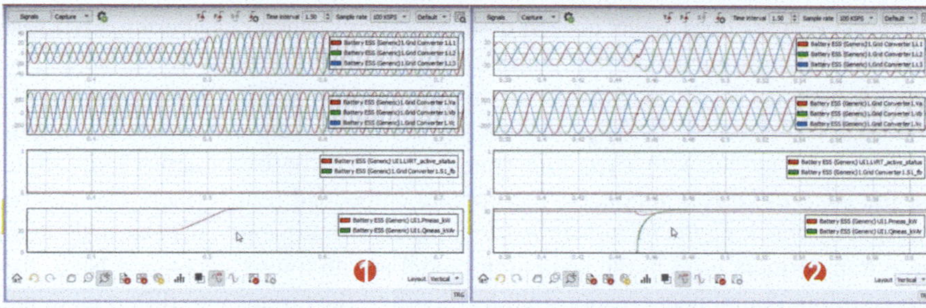

Fig. 20 Volt/VAR and Herts/Watt test, (1) Inject active power to the system, (2) Inject reactive power to the system

allow users to revisit and reimplement the model using the *generic* components. Besides, this approach allows users to optimize the models hardware utilization so that it can run on HIL 4-series devices, even though the original model was designed for 6-series devices. Additionally, the authors will maintain the microgrid controllers existing functionalities while exploring the added features of *generic* components at the system level. To access the model, open the *Example Explorer,* go to *micro-grid*, *microgrid example*, and select the *Terrestrial microgrid example using generic components*. Click the *Open model* button to proceed as shown in Fig. 21.

By employing *generic* DER components, this microgrid model can run on a 4-series device, whereas the model with switching DER components requires a 6-series HIL device. This is made possible by two factors. Firstly, the *generic* component version of the diesel generator does not require the same machine solver resources as its *legacy* counterpart. Secondly, *generic* components consume fewer resources from the standard processing cores, allowing for up to 3 DERs per core. Unlike the microgrid model with switching components, this particular model is divided into 3

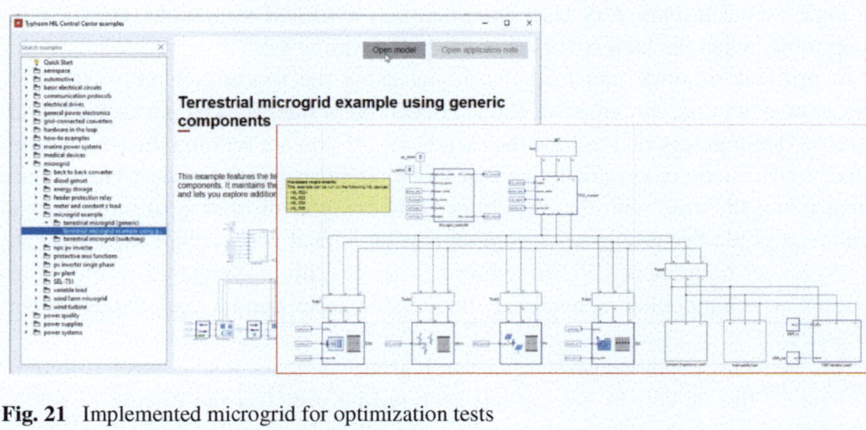

Fig. 21 Implemented microgrid for optimization tests

signal processing components instead of 4. As a result, when considering a HIL404 devices configuration table, configurations ranging from 1 to 5 are suitable. The predefined choice for this example is to run the model using HIL404 configuration 1.

Here, the differences between the microgrid model using *generic DER* components and the model with *switching* components will be examined. Firstly, note that the *generic* DER components incorporate embedded transformers that can be activated by checking the *include transformer* box. Consequently, the external three-phase transformers used in the switching model are no longer necessary. Next, the core coupling components responsible for dividing the circuit into different signal processing components will be considered. In this model, two transmission line matrix core couplings are employed and embedded in the Resistance-Inductance sections. In comparison to the ideal transformer-based core couplings used in the switching model, the transmission line matrix couplings offer the advantage of being symmetrical and not introducing any topological conflicts in the model. Additionally, transmission line matrix couplings include their own inductance and capacitance, which proves valuable in microgrid models like this one, where RL sections are used to simulate line impedances.

All functionalities of the microgrid controller remain consistent when compared to the model utilizing switching components. However, certain modifications were made to adapt the controllers interface to the *generic* DER components. You can observe these modifications by examining the DERs user interface subsystems. Notable changes in the controller interface include the normalization of reference values for voltage, frequency, active power, and reactive power since the inputs of the *generic* components user interface expect values in per unit (pu). It is essential to highlight that the *Diesel Genset* and *Battery ESS generic* components can operate in three modes by default: grid-following, simultaneous, and droop. This means that grid-forming mode can be achieved here by utilizing the voltage and frequency droop algorithms embedded in these *generic* components. This capability is not present in the switching component microgrid model. To take advantage of this functionality, the logic of the *Battery ESS* User Interface was modified to include operation in droop mode when the battery serves as the grid-forming DER.

To proceed, compile and load the model using the fourth-generation HIL404 device, as mentioned previously. The users can open the SCADA panel file found in the model directory and initiate the simulation. If you are running this simulation in the V-HIL mode, you may notice that this model can execute close to real-time, which is not the case with the switching DER component microgrid model. This is made possible because *generic* components are signal processing-based and can leverage your computers CPU to achieve faster execution compared to the more complex switching components. Here the SCADA panel can be explored as shown in Fig. 22.

The SCADA panel consists of several group widgets that control different elements in the model. In the top-left section, the user has the microgrid status indicator, which shows whether the microgrid is in grid-connected mode or islanded mode. The *Grid Status* group widget displays this information. Moving to the right, the user can see the *Microgrid Control* group widget, which allows us to select the

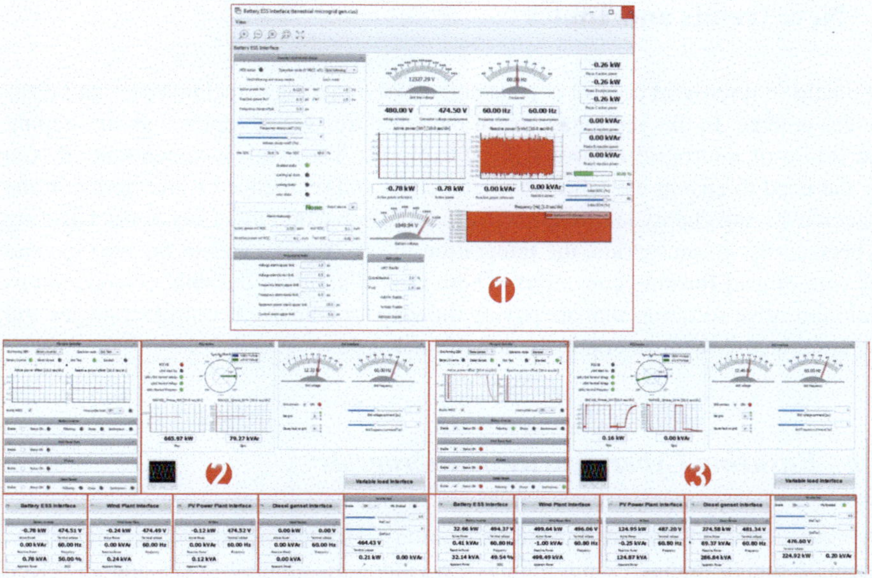

Fig. 22 SCADA panel for optimization objectives

mode of operation for the microgrid controller. The user can choose between *Grid Following*, *Droop*, and *Isochronous* modes. By selecting the appropriate mode, the behavior of the diesel generator and *Battery ESS* in response to changes in load or disturbances can be observed. Further down, there is the *Load* group widget, which allows us to control the load connected to the microgrid. The user can adjust the active and reactive power set points for the load to simulate changes in demand. Next, there is the *Diesel Generato*r group widget, which provides control over the diesel generator. The user can adjust the active power stepping and monitor the generator's active power output and fuel consumption.

Finally, there is the *Battery ESS* group widget, which allows us to control the *Battery ESS*. The user can adjust the active power stepping and monitor the SOC and active power output of the battery. To observe the behavior of the microgrid under different conditions, you can modify the load set points, switch between different control modes, and observe the responses of the diesel generator and *Battery ESS*. Pay attention to the microgrid status indicator as it switches between grid-connected and islanded modes. By exploring this model, you can gain insights into the operation of a microgrid using *generic* DER components. You can observe the advantages of utilizing *generic* components in terms of performance and hardware utilization. Additionally, you can see how the microgrid controller adapts to the different modes of operation and controls the DERs accordingly.

4 New Trends and Topics

The field of microgrid control is constantly evolving as new technologies and challenges emerge. In this section, some of the latest trends and topics that are shaping the future of microgrid control will be explored. These advancements are driven by the need to ensure the reliability, efficiency, and resilience of microgrids in the face of evolving energy landscapes. The authors delve into three key areas: ensuring cybersecurity in microgrids, the integration of electric vehicles in microgrids, and the role of grid-forming converters. These topics highlight the vital considerations and innovative solutions that are paving the way for enhanced control strategies and the continued growth and optimization of microgrids.

4.1 Ensuring Cybersecurity in Microgrids

Cybersecurity in microgrids and the application of HIL testing have gained significant attention due to the increasing reliance on communication networks and advanced control algorithms. This section explores recent advancements and trends in this field, focusing on research papers that provide insights into the challenges and solutions related to cybersecurity in microgrids and the effectiveness of HIL testing methodologies. In the field of power electronics devices and systems, the need for cybersecurity is increasing as more systems integrate into power grids with complex communication and computation systems. A real-time HIL cybersecurity testbed is demonstrated in [24] to address the lack of real-time security testbeds incorporating real cyber events. The testbed consists of a real-time power electronics system simulator, a real-time cyber system using actual network systems and a server, and penetration testing tools for generating real cyber attacks. The feasibility of the testbed is validated through the creation of various cyber attacks and their impacts on power electronics systems.

In DC microgrids, distributed control is gaining prominence due to its scalability and efficiency. However, the use of various communication network topologies and protocols exposes microgrids to unexpected cyber attacks. To address this vulnerability, a scalable cyber security testbed has been established for conducting HIL experiments and comprehensively investigating the security impact on DC microgrids. This testbed employs a Typhoon HIL 602 emulator and systematically investigates the impact of false data injection and replay attacks on the stability and performance of the microgrid [25]. Another research paper suggests a real-time HIL testing platform for analyzing the impact of false data injection attacks on microgrid control functions. The platform utilizes a real-time digital simulator to emulate the microgrid network, distributed energy resources, and their controllers. The study assesses the effects of false data injection attacks on critical microgrid control functions, such as under frequency load shedding, and uses mitigation solutions to enhance the resilience of microgrids against such attacks [26]. Furthermore, the development of an HIL testbed for studying false data injection cyberattacks on a lab-scale microgrid is presented in [27]. This testbed incorporates an OPAL-RT platform, combining software and hardware-based simulations. Experimental results demonstrate the effects of cyberattacks on the microgrids operation and verify its resilience under different attack scenarios. The setup includes data acquisition units, controllers, and various analog/digital inputs/outputs to accurately replicate real-world conditions.

Furthermore, a study evaluates the responses of residential smart PV converters to cyberattacks and assesses the performance of an intrusion detection strategy using HIL test facilities. Different types of cyberattacks are designed and tested on grid-tied smart converters, and their responses are compared with simulation results. The intrusion detection method based on power flow measurements shows promising results in detecting malicious activities and enhancing the security of smart grid devices [28]. Additionally, the growing connection of distributed energy resources, such as rooftop solar and electric vehicles, to the grid brings forth new cybersecurity threats. To effectively protect against these dynamic and diverse threats, laboratory-tested standards are necessary. These standards ensure the secure operation of interconnected technologies and communication-enabled devices.

4.2 Integration of Electric Vehicles in Microgrids

Microgrids, which incorporate distributed energy resources, are increasingly being utilized to provide more reliable power supply. However, the dynamic behavior of microgrids presents challenges in terms of their protection. The integration of renewable energy sources at a large scale and the addition of electric vehicle charging loads introduce additional protection issues due to the volatile nature of these resources. Conventional fault prediction schemes often fail to provide reliable protection in the dynamic microgrid environment. This chapter uses an adaptive scheme that accurately detects volatile changes in the microgrid under normal and fault conditions. The approach in [29] is based on a two-level support vector machine (SVM) classifier model, designed to ensure both accuracy and speed of operation. The first SVM model, identifies the mode of operation, variations in RES generation, random addition of electric vehicle charging load, and the occurrence of faults in the microgrid. The second SVM model, identifies the exact type and location of the fault. The SVM models utilize rms values of three-phase voltage and current measurements as input data. To validate the effectiveness of the scheme, simulation experiments are conducted on an IEEE 9-bus microgrid testbed using MATLAB/SIMULINK. Sensitivity analysis results demonstrate the robustness of the scheme. Additionally, real-time experiments are performed using an HIL real-time simulator, OPAL-RT, and a Raspberry Pi microcontroller. The scheme successfully achieves accurate online detection of faulted lines and fault types within a single cycle of AC in both grid-connected and islanded modes of operation.

Furthermore, examining the impact of hourly and intra-hourly battery scheduling approaches in a smart residential microgrid that incorporates an electric vehicle and a battery storage system is presented in [30]. Mathematical programming models consisting of two layers, along with a two-layer rule-based model, are employed to analyze the influence of time granularity and the unit commitment algorithm on the capacity loss of the batteries in the residential setting. The mathematical programming models include a mixed-integer programming model for optimization without considering battery degradation (wear) and a mixed-integer quadratically constrained programming model that considers battery degradation as a constraint approximated by a quadratic polynomial. To complement the theoretical analysis, an HIL experimental setup is utilized to conduct real-time testing and investigate the capacity fade of the batteries.

Vehicle-to-grid technology is an emerging concept that leverages electric vehicles as mobile power grids, enhancing their value and benefits. The integration of electric vehicles into microgrids involves several essential steps, including V2X communication with the infrastructure and bidirectional power electronic converters for power exchange with the grid. A complete structure for integrating electric vehicles into a microgrid and its practical implementation is presented in [31]. Both vehicle-to-grid and grid-to-vehicle technologies are employed, utilizing an integrated onboard bidirectional charger to facilitate power transfer between the electric vehicles and the microgrid. A mechanism is employed for utilizing surplus power from multiple electric vehicles and supplying it to the microgrid. The analysis focuses on power management, with a priority given to renewable energy (electric vehicles power) during peak hours. To validate the system and analyze its performance, a Typhoon-HIL microgrid testbed is utilized. Experimental experiments are conducted using a grid emulator to further verify the results. The practical implementation of electric vehicle integration into microgrids using vehicle-to-grid and grid-to-vehicle technologies offers valuable insights into optimizing power flow and enhancing the overall efficiency and sustainability of the microgrid system.

4.3 Grid Forming Converters in Microgrids

The environmentally friendly sources, like wind power and solar PV, use power electronic converters as interfaces to the grid, and they have a performance dramatically different from the synchronous machines used in traditional power plants. As the number of old pollutant power plants connected to the power system increases to accommodate more and more generation from renewable power sources, the operational security and resilience of the power system are negatively affected. Implementing novel control techniques on power electronic converters has become a realistic option to cope with these adverse effects. One family of control techniques known as *grid-forming* looks like a promising mechanism to overcome some of the most important effects resulting from the massive reduction in the number of synchronous generators connected to the power system [32]. Reference [1] documents the implementation of different power electronic converter control strategies in the Typhoon HIL environment. The implemented controllers belong to the family of grid-forming techniques, including droop control, virtual synchronous machine, and synchronverter. These control techniques aim to regulate the power electronic converters in a manner that emulates the behavior of synchronous generators, providing stability and resilience to the power system. To establish a fair comparison among different grid-forming converter control techniques and evaluate their behavior, a validation test has been carried out with each controller implementation connected to two isolated loads. The performance of each control technique is analyzed, highlighting their strengths and weaknesses. This analysis provides valuable insights into the effectiveness of grid-forming converter control techniques in microgrid applications, aiding in the development of more efficient and reliable energy systems.

As mentioned earlier, grid-forming converters are fundamental devices in DC microgrids for DC bus voltage regulation. Droop control is widely used for grid-forming converters control due to its Plug-and-Play feature, allowing for seamless integration of grid-forming converters. However, droop control may fail to meet some desired DC microgrid control features and performances, such as zero steady-state DC bus voltage deviation, high system inertia, and fast DC bus voltage regulation. To address these challenges and achieve the desired control performances, Ref. [33] uses a power buffer-based control framework. The power buffer, which combines a capacitor and a bidirectional DC-DC converter, acts as an interface between grid-forming converters and the DC bus. This framework enables the development of a decoupled control strategy where grid-forming converters are regulated by droop control, while advanced control strategies are applied to the power buffer. By utilizing the power buffer-based control framework, the Plug-and-Play feature of grid-forming converters is maintained while simultaneously achieving specific DC microgrid performances. This approach allows for seamless plugging in/out of grid-forming converters without introducing DC bus voltage fluctuations, ensuring high DC bus voltage regulation performance. The framework's effectiveness is demonstrated through HIL tests and real circuit experimental results, highlighting its efficacy in improving the performance of grid-forming converters in DC microgrid applications.

Smart converters, particularly grid-forming converters, offer a viable solution by providing real-time reactive power control for voltage regulation. An advanced real-time volt-var control strategy for handling voltage violations in DER-enriched distribution networks using grid-forming converters in [34]. The method focuses on controlling the voltage magnitude of affected nodes by dispatching reactive power from smart converters in real-time. To validate the real-time volt-var control strategy, a cyber-physical co-simulation is conducted between the Typhoon HIL 604 and OpenDSS platforms. The grid-forming converter is modeled in the Typhoon HIL 604, serving as the physical layer in the cyber-physical system. The OpenDSS platform simulates a CIGRE medium voltage distribution network, representing the cyber layer. A Python program facilitates the cyber-physical co-simulation between Typhoon HIL and OpenDSS, and the control algorithm is executed in real-time using a SCADA system developed for this study. The results of the real-time simulation demonstrate the effectiveness of the volt-var control strategy in handling voltage violations in DER-enriched distribution networks. By leveraging the capabilities of grid-forming converters and real-time control, this approach offers a viable solution for voltage regulation in modern distribution networks with high DER penetration.

5 Summary

This chapter presented the assessment of microgrid control through the utilization of advanced HIL technologies by introducing HIL technologies and their various applications, including the V-HIL, C-HIL, and P-HIL. The chapter emphasized the importance of HIL assessment in the context of microgrid control units and addresses the challenges and issues associated with HIL testing. It also explored the relevant IEEE standards and grid codes pertaining to microgrid testing. The chapter provided a comprehensive review of the advancements made in microgrid testing using HIL and offers a step-by-step implementation guide for constructing a microgrid using HIL. The guide covers the essential components of a microgrid, explains the distinction between switching and average components, and discusses the integration of *Battery ESSs*, PV power plants, and diesel *gensets*. Furthermore, the chapter presented examples of creating both a simple microgrid and a fully functional terrestrial microgrid model. It also delves into microgrid model optimization and explores emerging trends and topics, such as ensuring cybersecurity in microgrids and the integration of electric vehicles. Additionally, the chapter examined the utilization of grid-forming converters in microgrids, providing valuable insights into this aspect of microgrid control.

Acknowledgements The authors acknowledge the financial support of KDDI Foundation, Japan, and the received support from Smart/Micro Grids Research Center (SMGRC) at the University of Kurdistan, Sanandaj, Iran.

References

1. Caio RD, Osório AG, Sergio C (2023) Real-time simulation and hardware-in-the-loop testing using typhoon HIL. In: Amlan Chakrabarti DaP, Choudhury AK. Springer
2. Kikusato H (2023) Introduction of power hardware-in-the-loop (PHIL) simulation. Waseda University, Tokyo, Japan
3. Academy H (2022) Typhoon HIL documentation. Available from: https://www.typhoon-hil.com
4. Salcedo R et al (2019) Banshee distribution network benchmark and prototyping platform for hardware-in-the-loop integration of microgrid and device controllers. J Eng 2019(8):5365–5373
5. Narang D (2022) Grid standards and codes
6. Hemmati M, et al. (2021) Interoperability analysis of IEC61850 protocol using an emulated IED in a HIL microgrid testbed. In: 2021 IEEE International Conference on Communications, Control, and Computing Technologies for Smart Grids (SmartGridComm), pp 152–157

7. Ruchi Verma HKV, Priyanka V (2021) Hybrid microgrid: Energy management under different loading condition using typhoon HIL's real-time digital simulator. In: Materials Science and Engineering, IOP Publishing
8. Narang D (2022) Grid standards and codes. Available from: https://www.nrel.gov/grid/standards-codes.html
9. Maloney P (2019) Building a better microgrid with hardware in the loop
10. Bagudai SK, Ray O, Samantaray SR (2019) Evaluation of control strategies within hybrid DC/AC microgrids using typhoon HIL. In: 2019 8th International Conference on Power Systems (ICPS). IEEE, Jaipur, India
11. Adrien Genić PG, Danilo M, Mladen D (2016) Multi-layer hardware-in-the-loop testbed for microgrids. In: 2016 International Conference on Smart Systems and Technologies (SST). IEEE, Osijek, Croatia
12. Huerta F, Prodanovic JKGM, Matatagui P (2014) A power-HIL microgrid testbed: Smart energy integration lab (SEIL). In: 2014 IEEE Energy Conversion Congress and Exposition (ECCE). IEEE, Pittsburgh, PA, USA
13. Zhang S, Liang T, Dinavahi V (2023) Hybrid ML-EMT-Based Digital Twin for Device-Level HIL Real-Time Emulation of Ship-Board Microgrid on FPGA. IEEE J Emerg Sel Top Ind Electron 4(4):1265–1277
14. Shubham Sharma RP, Premalata J (2022) AC microgrid control using PV and battery in typhoon HIL. In: 2022 IEEE International Conference on Power Electronics, Drives and Energy Systems (PEDES)
15. Mali B, Dong CL (2023) HIL model-based control and fault detection of DC microgrid. In: 11th International Conference on Power Electronics-ECCE Asia
16. Lin FJ et al (2023) Optimal economic dispatch and power generation for microgrid using novel lagrange multipliers-based method with HIL verification. IEEE Syst J 17(3):4533–4544
17. Naderi E, Asrari A (2023) Experimental validation of a remedial action via hardware-in-the-loop system against cyberattacks targeting a lab-scale PV/wind microgrid. IEEE Trans Smart Grid 14(5):4060–4072
18. Shreeram VK, Dattatraya NG (2022) HIL implementation of an islanding detection and an automatic mode switching for droop-based microgrid. Int J Power Electron 15(1)
19. Hooman Ekhteraei Toosi AM, Amer MYMG, Andrew S (2023) Central power management system for hybrid PV/battery AC-Bus microgrid using typhoon HIL. In: 2019 IEEE 28th International Symposium on Industrial Electronics (ISIE). Vancouver, BC, Canada
20. El Zerk A, Ouassaid M (2023) Real-time fuzzy logic based energy management system for microgrid using hardware in the loop. Energies 16(5)
21. Ivanović ZR et al (2012) HIL evaluation of power flow control strategies for Energy storage connected to smart grid under unbalanced conditions. IEEE Trans Power Electron 27(11):4699–4710
22. Nigam S, et al. (2021) Controller hardware in the loop testing of microgrid secondary frequency control schemes. Electr Power Syst Res 190
23. Chilipi RR et al (2016) A multitasking control algorithm for grid-connected inverters in distributed generation applications using adaptive noise cancellation filters. IEEE Trans Energy Convers 31(2):714–727
24. Jinchun Choi DN, Bohyun A, Seerin A, Jianwu Z, Taesic K (2021) A real-time hardware-in-the-loop (HIL) cybersecurity testbed for power electronics devices and systems in cyber-physical environments. In: 2021 IEEE 12th International Symposium on Power Electronics for Distributed Generation Systems (PEDG). IEEE, Chicago, IL, USA
25. Liu M, et al. (2021) Demo abstract: A HIL emulator-based cyber security testbed for DC microgrids. In: IEEE INFOCOM 2021—IEEE Conference on Computer Communications Workshops (INFOCOM WKSHPS). pp 1–2
26. Martine Chlela GJ, Marthe K, Yves B (2016) Real-time testing platform for microgrid controllers against false data injection cybersecurity attacks. In: 2016 IEEE Power and Energy Society General Meeting (PESGM). IEEE, Boston, MA, USA

27. Asrari ENA (2021) Hardware-in-the-loop experimental validation for a lab-scale microgrid targeted by cyberattacks. In: 2021 9th International Conference on Smart Grid (icSmartGrid). IEEE, Setubal, Portugal
28. Kaewnukultorn T et al (2023) Smart PV Inverter Cyberattack Detection Using Hardware-in-the-Loop Test Facility. IEEE Access 11:90766–90779
29. RA, et al. (2023) A novel SVM based adaptive scheme for accurate fault identification in microgrid. Electr Power Syst Res 221
30. Toosi HE, Merabet A (2023) Impact of intra-hourly scheduling on state of health of battery for electric vehicle integration in smart residential microgrid. J Energy Storage 72
31. Ullah Z, Zeeshan M, Ahmed S (2023) Practical implementation of electric vehicle integration into a microgrid using V2G and G2V. In: 2023 IEEE International Conference on Smart Mobility (SM), pp 160–165
32. Hassan Bevrani TK, Toshifumi I, Kaoru I (2022) Grid connected converters: Modeling, stability and control. Elsevier
33. Jialei Su KL, Chen X (2023) Plug-and-play of grid-forming units in DC microgrids assisted with power buffers. IEEE Trans Smart Grid (Early Access)
34. Wagle R, et al. (2023) Real-time volt-var control of grid forming converters in DER-enriched distribution network. Front Energy Res 10

Stability and Control of Hybrid AC/DC Microgrids

Arash Vahidnia, Moudud Ahmed Masum, and Lasantha Meegahapola

Abstract Hybrid AC/DC microgrids are considered as viable solutions to reduce energy conversion losses in microgrids. However, hybrid AC/DC microgrids are susceptible to stability issues during high penetration of dynamic loads (e.g., induction machines). The non-linear dynamics of induction machines, result in sustained voltage/frequency oscillations following disturbances in the microgrid, which is a major challenge for stable operation of the hybrid AC/DC microgrid. This issue is even more severe when the hybrid AC/DC microgrid is operated at autonomous mode. The PEC based energy storage systems (ESSs) are used as an effective solution for power balancing in the microgrid; hence with the fast response of the PEC, microgrid voltage/frequency could be stabilised rapidly. This chapter discusses the stability and control aspects of hybrid AC/DC microgrids to address the drawbacks of these systems. A supplementary power oscillation damping controller for the ESS is demonstrated for damping the low-frequency oscillations (LFOs) in the microgrid. Furthermore, the dynamic characteristics of the loads in the microgrid can change constantly which requires re-tuning of the pod controller. This is addressed by an adaptive neuro-fuzzy inference system (ANFIS) based POD controller which has been shown to be more effective due to its ability to adjust the gain based on the frequency deviation, and handling more non-linearity in the system dynamics. Furthermore, this chapter presents a hybrid AC/DC microgrid architecture incorporating a central energy storage system with a coordinated control strategy between the central ESS and the inter-linking converter (ILC) to maintain the stable operation of the microgird.

Keywords Adaptive neuro-fuzzy inference system (ANFIS) · Energy storage system (ESS) · Inter-linking converter (ILC) · Hybrid ac/dc microgrid · Low frequency oscillations (LFOs) · Power oscillation damping (POD) controller

A. Vahidnia (✉) · M. A. Masum · L. Meegahapola
School of Engineering, RMIT University, Melbourne, Australia
e-mail: arash.vahidnia@rmit.edu.au

L. Meegahapola
e-mail: lasantha.meegahapola@rmit.edu.au

F. Shahnia and J. M. Guerrero (eds.), *Microgrids and Virtual Power Plants*, Power Systems, https://doi.org/10.1007/978-981-97-6623-9_6

1 Introduction

Microgrids are cost-effective solutions due to the integration of multiple distributed energy resources (DERs), reducing transmission and distribution costs, and higher energy efficiency; however, the major drawbacks of the renewable based DERs are their variability and intermittency. Furthermore, maintaining a reliable and stable microgrid is a big challenge in the presence of various types of DERs [1]. As conventional power grids operate in AC mode, AC microgrids require minimum modifications to be integrated to the existing power grids. Owing to technical and industrial developments in power electronic devices, a large number of DC loads and power electronic converters (PECs) are being used for different level of applications. Moreover, the DC based DERs and different types of energy storage systems (ESSs) create a new opportunity for DC microgrids. Having a limited numbers of converters along with no reactive current circulation are the main advantages of DC microgrids. The hybrid AC/DC microgrids articulate the benefits of both AC and DC microgrids, and ultimately improve the reliability, efficiency and economic operation of the microgrid [2, 3]. However, the network structure of the hybrid AC/DC microgrid is complex compared to individual AC and DC microgrids, which requires further research studies to coordinate the control of individual AC and DC sub-grids, intermittency of DERs, and reactive power compensation, etc.

The voltage source converter (VSC) based DERs do not proactively response to system frequency excursions, unless specific control schemes are designed at the DER to respond to frequency variations [4]. Hence maintaining stability of the hybrid AC/DC microgrid under autonomous mode is an enormous challenge. Many research studies on small-signal stability of hybrid AC/DC microgrids have been carried out considering static and dynamic loads [5–7]. These studies have reported that the load dynamics have significant influence on low-frequency modes. Furthermore, the droop gain of the VSC determines the power sharing and stability of the hybrid AC/DC microgrid, where a small droop gain will improve the stability margin of the microgrid. Although, a small droop gain improves the stability of the microgrid, the power sharing performance of the VSC will be deteriorated [5–7].

According to the International Energy Agency (IEA), induction motors (IMs) consume between 60 and 70% of total energy supplied to power networks [8]. The small and large IMs are generally directly connected to the low-voltage or medium-voltage distribution networks; therefore, the analysis of a hybrid AC/DC microgrid without IMs would not be practical due to the non-linear dynamics of the IM. Basically, IM electro-mechanical torque should be equal to mechanical load torque in steady-state operation. Following a transient disturbance in the microgrid, the operating point of the IM changes due to large voltage variations at the terminal of the IM. Moreover, voltage recovery following a transient disturbances is more complicated in hybrid AC/DC microgrids due to the shortage of reactive power, hence further investigations are required.

Power oscillations during the autonomous operation mode of the hybrid AC/DC microgrid pose significant control challenges. The power oscillation issue in conventional power systems is mostly related to the inertia of the large synchronous machines while in microgrids the power oscillations mainly arise from the interaction among multiple DERs. Power oscillations become more severe if IMs or dynamic loads are operating in the microgrid. An active damping controller has been proposed by researchers to damp the power oscillations in microgrids. Authors in [9] have proposed a two-degree-of-freedom (2-DOF) active damping controller for a AC microgrid with dynamic loads, while [10] proposed an integrated series compensator for an autonomous AC microgrid to damp power oscillations. DERs with decentralized control approach based virtual voltage/frequency are used in [11], whereas virtual active/reactive control strategies are proposed in [12, 13] to damp power oscillations. A dynamic droop gain in association with a virtual impedance loop is used in a DC microgrid to damp low frequency power oscillations in [14–17]. Although, the aforementioned approaches ensure improved power oscillation damping (POD) performance, they significantly affect the voltage profile and power sharing performance.

One major drawback of the hybrid AC/DC microgrid is the intermittency of renewable energy based DERs [18], hence the ESS should maintain the power balance in the microgrid. The PEC interfaced ESS can effectively minimize the power variation within the microgrid and ensure stable operation [19, 20]. The generation and load demand variations in the microgrid result in voltage/frequency variations due to the low inertia of the microgrid in comparison to the conventional power system. The PEC interfaced ESS is an effective solution for power imbalance mitigation in the microgrid, hence improve voltage/frequency oscillations [21]. Moreover, microgrid stability is significantly affected by the X/R ratio of the network, i.e., higher the X/R ratio better the microgrid stability margin compared to low X/R ratio [22, 23], hence designing a POD controller for a microgrid with a low X/R ratio is a challenging task.

Since hybrid AC/DC microgrids host various intermittent DERs, one of the key challenges is to ensure stable operation under such variable power output conditions. To minimize the effect of output power fluctuations of DERs, ESSs are usually integrated into the hybrid AC/DC microgrid [24, 25]. Common practice is to connect separate ESSs to both the AC and the DC sub-grids of a hybrid AC/DC microgrid. Complex coordinated control strategies are required to achieve accurate power management between the distributed ESSs and the inter-linking converter (ILC).

2 Electro-Mechanical Oscillation in Induction Motors

A significant percentage of the power system dynamic loads are IMs and they have a strong influence on power system stability. IM's shaft inertia and the rotor time constant substantially affect the real and reactive power recovery of the IM following a voltage dip in the power system [26]. IM is an electro-mechanical device, and electromagnetic field in the air-gap produces electro-magnetic force between the rotor and the stator. In steady-state condition, both the electrical torque and mechanical torques are equal. During transient disturbances, electro-mechanical torque imbalance causes rotor speed oscillations. With the increase of inertia of the IM, damping of electro-mechanical oscillation decreases. Another important parameter of IM is the rotor time constant ($T_{dom} = (L_m + L_r)/R_r$), which depend on mutual inductance (L_m), rotor inductance (L_r) and rotor resistance (R_r), and increases with the machine power rating. Also it adversely affects the damping of electro-mechanical rotor oscillations [27].

Sub-synchronous (or torsional) resonance (SSR) is an important problem in power system, which was first occurred in south Nevada in 1970 [28]. SSR is a condition where turbine-generator exchange energy at one or more frequency bellow system frequency following a transient disturbance from a equilibrium condition [29]. Transient interactions may give rise to voltage and frequency oscillations. When SSR occurs, sub-synchronous frequency current components flow into the stator of the IM. The stator flux rotates at a speed lower than the synchronous speed. The new stator current frequency is determined by the superposition of the SSR frequency (ΨSSR) components and system frequency (ΨSys) components. As a result, stator magnetic flux oscillates in between ($\Psi Sys \pm \Psi SSR$) and electro-magnetic torque (Te) of IM, hence speed starts to oscillate. In conventional power systems, low-frequency electro-mechanical oscillation frequency range varies between 0.1 and 2.0 Hz [30] while SSR oscillation frequency range varies between 10 and 50 Hz [31].

Generally, industrial machines are three-phase IMs which are connected either as direct-on-line (DOL) or using a variable frequency drive (VFD) system. Approximately 80% of the total industrial machine loads are operated as DOL starting mode [32]. In DOL start motors, the stator of the IM is directly connected to the network, hence any voltage/frequency deviations at the motor terminal makes motor vulnerable to instability [33, 34]. With the development of power electronics technology, VFD plays an important role in IM control, and VFD based IMs will constitute a significant share in near future. Constant torque, constant power, and variable torque are the three basic topologies used in VFDs. Among these three basic topologies, constant torque is widely used in industrial applications, such as elevators, feeders, conveyors, hoists, mixers, and bottling machines etc. The equation of maximum torque is given by:

$$T_{\max} = \frac{K(\frac{v}{f})^2}{\frac{R_s}{f} \pm \sqrt{\frac{R_s}{f} + 4\pi^2(L_s + L_r)^2}} \tag{1}$$

Here, K is a machine constant, and v, f are voltage and frequency respectively. As R_s/f is much less than $2\pi(L_s + L_r)$, the above equation can be simplified as;

$$T_{\max} = \frac{K(\frac{v}{f})^2}{2\pi(L_s + L_r)} \tag{2}$$

Constant maximum torque is achieved by maintaining a constant voltage/frequency (v/f) ratio. Transient interactions may give rise to voltage and frequency oscillations which causes electro-mechanical rotor oscillations, and hence rotor speed oscillations. One significant characteristic of the torsional oscillation is that it can propagate through power electronic interface of the VFD. As torsional oscillation frequency is less than the system frequency, it can easily propagates through the low-pass filter, which requires additional damping controller with existing VFD. Authors in [34] proposed an additional controller with existing VFD to damp torsional oscillations, while in [35] authors have employed a damping control unit at the IM to mitigate torsional oscillations during transient conditions. Therefore, low-frequency oscillations associated with IMs would significantly influence on microgrid stability.

The electro-mechanical oscillations of increasing amplitude as a result of insufficient damping will lead to unstable conditions in the microgrid. Under steady-state operating conditions, total generation and load demand remain equal, and speed of rotating machines remains constant. The equilibrium between the generation and load demand will be lost during transient disturbances, hence voltage and power oscillations are inevitable. When a power system undergoes a transient perturbation, the variation in the electrical torque of the synchronous machine can be divided into the following two components;

$$\Delta T_{ele} = T_{sys}\Delta\delta + T_{damp}\Delta\omega \tag{3}$$

$T_{damp}\Delta\omega$, is the damping torque component, which is in phase with the rotor speed deviation ($\Delta\omega$), while T_{damp} is the damping torque coefficient. $T_{sys}\Delta\delta$, is the synchronizing torque component, which is in phase with the rotor angle ($\Delta\delta$), while T_{syn} is the synchronizing torque coefficient. Inadequate synchronizing torque causes a progressive increase of rotor angle, whereas damping torque deficiency leads to oscillatory instability [36, 37]. It is clearly noticed that damping torque is associated with speed deviation.

The oscillation dynamics of the synchronous generator rotor is represented by the swing equation. This standard concept is used to develop a POD controller for the hybrid AC/DC microgrid to damp oscillations. The swing equation can be written as;

$$M\frac{d^2\delta}{dt^2} = p_{mec} - p_{ele} \tag{4}$$

where M, P_{mec}, and P_{ele} are inertia constant, mechanical power input to prime-mover and electrical power output of the synchronous generator. Consider a DER connected to the microgrid as shown in Fig. 1. The reference voltage of the microgrid is $E_{mg}\angle 0$, and R_1, R_2, X_1, and X_2 are line resistance and reactance respectively. If the DER voltage angle $E_{dg}\angle\delta$ leads the microgrid voltage by an angle δ, which keep changing as the DER undergoes small power oscillations following a perturbation in the microgrid. The current equation can be written as;

$$I = \frac{E_{dg}\angle\delta - E_{mg}\angle 0^\circ}{R_1 + R_2 + j(X_1 + X_2)} \tag{5}$$

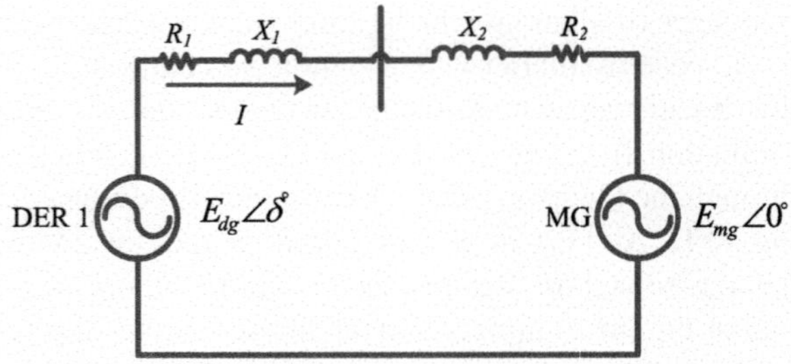

Fig. 1 A simplified circuit of a DER connected to the microgrid

where, $Z_T = \sqrt{(R_1 + R_2)^2 + (X_1 + X_2)^2}\angle \tan^{-1}(\frac{X_1+X_2}{R_1+R_2})$.
The complex power is determined as;

$$S = E_{mg}\angle 0°.I^* = E_{mg}\angle 0°.\left(\frac{E_{dg}\angle\delta - E_{mg}\angle 0°}{Z_T}\right)^* \tag{6}$$

$$S = \frac{E_{dg}.E_{mg}.\cos\delta - E_{mg}^2}{Z_T} - j\left(\frac{E_{dg}.E_{mg}.\sin\delta}{Z_T}\right) \tag{7}$$

From the above equation, the total generated power can be written as;

$$P_{ele} = Re\{S\} = \left(\frac{E_{dg}.E_{mg}.\cos\delta - E_{mg}^2}{Z_T}\right) \tag{8}$$

Now, considering the small change of angle ($\delta = \delta_0 + \Delta\delta$) around the operating points;

$$P_{ele}(\delta_0 + \Delta\delta) = \frac{E_{dg}.E_{mg}.\cos(\delta_0 + \Delta\delta)}{Z_T} - \frac{E_{mg}^2}{Z_T} \tag{9}$$

$$P_{ele} + \Delta p_{ele} = \frac{E_{dg}.E_{mg}}{Z_T}.\left[\cos\delta_0.\cos\Delta\delta - \sin\delta_0.\sin\Delta\delta\right] - \frac{E_{mg}^2}{Z_T} \tag{10}$$

$$= \frac{E_{dg}.E_{mg}}{Z_T}.\left[\cos\delta_0.1 - \sin\delta_0.\Delta\delta\right] - \frac{E_{mg}^2}{Z_T} \tag{11}$$

Now, linearizing the power equation for small change of angle around the operating points as;

$$\begin{gathered}\Delta p_{ele} + \frac{E_{dg}.E_{mg}}{Z_T}\sin\delta_0\Delta\delta = 0 \\ \Delta p_{ele} + P_{sys}.\Delta\delta = 0\end{gathered} \tag{12}$$

where, $p_{sys} = \frac{E_{dg}.E_{mg}}{Z_T}.\sin\delta_o$ is the sysnchronzing torque coefficient. The linearized form of equation for small change of operating points to represent small-scale power oscillations is gives as;

$$M\frac{d^2\Delta\delta}{dt^2} = \Delta p_{mec} - \Delta p_{ele} \tag{13}$$

The electro-magnetic torque of an IM can be sub-divided into synchronizing torque and damping torque. Dynamic stability is significantly affected by the damping torque component, which is the prime interest of this research study. The damping torque is induced by the interaction between the rotor side-band current with the field produced by the fundamental stator current, and the stator side band current with the field produced by the slip frequency rotor current. The upper side-band frequency current produces a positive damping torque, and lower side band frequency current produces a negative damping torque. The damping torque produced by the side-band current is the fundamental characteristics of unstable operation of electro-mechanical devices. The stator side-band current produce a damping torque which would be positive or negative depending on the machine inertia, relative magnitude of supply frequency and the oscillation frequency. The rotor side-band current produces damping torque which would be positive or negative depending on the oscillation frequency and the slip [38]. The damping torque is proportional to the speed deviation and this is the basis of the POD controller design, which will be described here. The damping torque is given by [36];

$$\Delta T_{damp} = \Delta p_{damp} = D\frac{d\delta}{dt} = D\Delta\omega \tag{14}$$

Here, damping coefficient is represented by D. If damping torque is added without any change in the mechanical power input (i.e., $\Delta T_{mec} = 0$), then the swing equation becomes;

$$M\frac{d^2\Delta\delta}{dt^2} = -P_{sys}\Delta\delta - D\Delta\omega \tag{15}$$

δ_0 is initial position of the rotor in a synchronous reference frame at t = 0;

$$\begin{aligned} \delta &= \delta_0 + \omega_r t - \omega_0 t \\ \frac{d\delta}{dt} &= \omega_r - \omega_0 \end{aligned} \tag{16}$$

After linearising the above equation;

$$\begin{aligned} \frac{d\Delta\delta}{dt} &= \Delta\omega_r \\ \frac{1}{\omega_0}\frac{d\Delta\delta}{dt} = \frac{\Delta\omega_r}{\omega_0} &= \Delta\omega \end{aligned} \tag{17}$$

Now, substituting the value of $\Delta\omega$ into the Eq. (15) it becomes;

$$M\frac{d^2\Delta\delta}{dt^2} = -P_{sys}\Delta\delta - D\frac{1}{\omega_0}\frac{d\Delta\delta}{dt} \tag{18}$$

In Eq. (15), the term $\frac{E_{dg}.E_{mg}.\cos\delta - E_{mg}^2}{Z_T}$ associated to $\Delta\delta$ is the synchronizing torque coefficient (P_{sys}).

$$M\frac{d^2\Delta\delta}{dt^2} + P_{sys}\Delta\delta + D\frac{1}{\omega_0}\frac{d\Delta\delta}{dt} = 0 \tag{19}$$

The characteristic equation of the above equation can be written as;

$$MS^2 + \frac{D}{\omega_0}S + P_{sys} = 0 \tag{20}$$

The roots of the above characteristics equation are;
$S_1, S_2 = -\frac{D}{2M\omega_0} \pm j\frac{1}{2}\sqrt{\frac{D^2 - 4MP_{sys}\omega_0^2}{M^2\omega_0^2}}$
From Eq. (20), it is clear that for a positive damping torque, the roots of the characteristics equation have a negative real part. Thus, for the positive damping torque the system is in stable condition.

The ESS controller compares the voltage and frequency deviations between the measured and reference values and is used to determine the necessary active/reactive power adjustments to maintain stable voltage/frequency of the hybrid AC/DC microgrid. Any disturbance, such as a sudden load change in the microgrid, would result in voltage and frequency fluctuations, and ultimately will lead to power oscillations around the steady-state operating point. These power oscillations result in rotor speed oscillations at the IM. To damp the rotor oscillations, the voltage and frequency mismatch should be reduced as much as possible. Therefore, a supplementary POD controller is developed for the ESS to provide a stabilizing signal proportional to the frequency deviation (since the frequency deviation is proportional to the speed deviation) to assist mitigate rotor speed oscillations.

3 POD Controller for Energy Storage Systems

The ESS is utilized to compensate the intermittent behaviour of distributed renewable generators in the microgrid. Which also provides reactive power support to the AC sub-grid of the hybrid AC/DC microgrid, and regulates voltage/frequency deviations in the microgrid. An ESS along with the control architecture is illustrated in Fig. 2(a), which comprises an inner current control loop, an outer voltage control loop, and a supplementary POD controller. The inner current control loop and outer voltage/frequency control loops of the ESS determine the power output of the ESS.

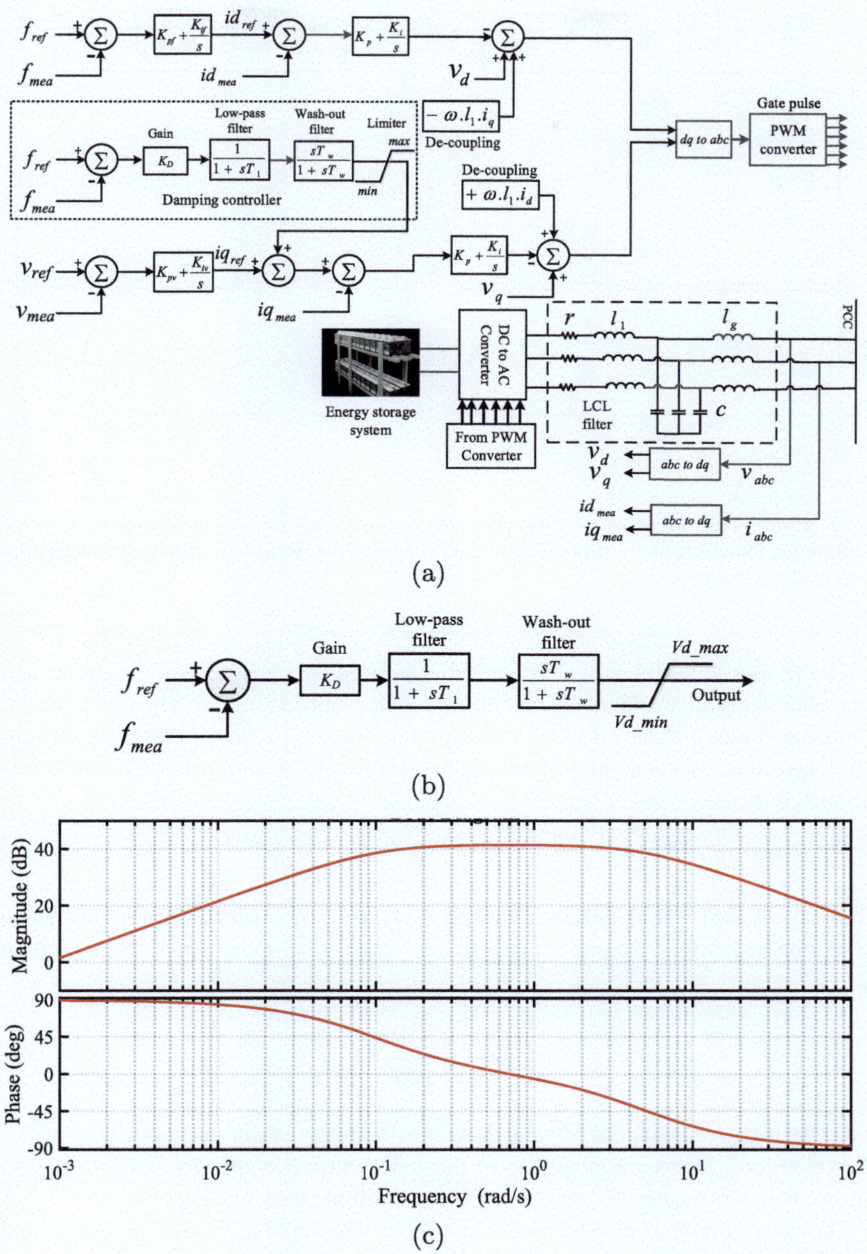

Fig. 2 **a** The ESS controller with POD controller, **b** POD controller, **c** Frequency response characteristics of the POD controller

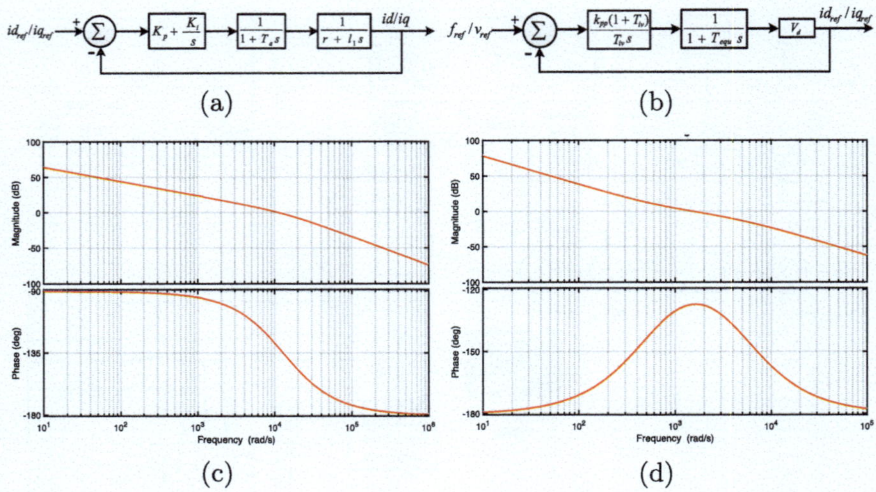

Fig. 3 **a** Block diagram of inner current controller, **b** Block diagram of outer voltage controller, **c** Frequency response characteristics of inner current controller, **d** Frequency response characteristics of outer voltage controller

The proportional integral (PI) controller is used to generate id/iq current, and controller gains are evaluated using the approach proposed in [39]. The inner current controller block diagram is shown in Fig. 3a. Here, T_a, r, and l represent switching delay, line resistance and inductance respectively. The proportional gain kp is calculated as;

$$k_p = \frac{\omega l}{v_{dc}} \tag{21}$$

where, v_{dc} is the DC-link voltage of the ESS, ω is the crossover frequency, which can be written in terms of the phase margin (ϕm) and the sampling time (T), according to;

$$\omega = \frac{\frac{\pi}{2} - \phi m}{1.5T} \tag{22}$$

In this section, ϕm is fixed at 50^0 and the controller integral time constant, $T_i = 10/\omega$. The open-loop frequency response of the inner current controller is illustrated in Fig. 3c, which indicates 46^0 phase-margin and infinite gain margin.

The outer voltage controller block diagram is illustrated in Fig. 3b. The switching delay (T_{equ}) is considered twice the switching delay of the inner current controller i.e. $T_{equ} = 2 * T_a$. Based on the criteria that the inner loop controller should be faster than outer loop controller, the time constant of the outer loop voltage controller is fixed to $T_{iv} = 10 * T_i$. The proportional gain (k_{pv}) is selected to ensure 50^0 phase margin. The open-loop frequency response of the outer voltage/frequency controller is illustrated in Fig. 3(d), which indicates 53.16^0 phase-margin and infinite gain margin.

The supplementary POD controller is implemented at the ESS, as it can respond faster in comparison to the other sources in the microgrid. The POD controller provides an additional stabilizing input to the ESS controller to damp the low frequency oscillations (LFOs). If a network disturbance occurs, such as a load change, an outage of DERs, or a short-circuit fault, it would give rise to voltage/frequency fluctuations in the microgrid, and ultimately result in rotor speed oscillations at induction machine loads. The POD controller is designed at the ESS to generate a supplementary control signal proportional to frequency oscillations, hence damp the rotor speed oscillations. Figure 10b illustrates the control architecture of the POD controller, which comprises a gain block, a low-pass filter, and a wash-out filter followed by a limiter. The input signal of the POD controller is the frequency deviation $(f_{ref} - f_{mea})$ at the point of common coupling (PCC) of the ESS, and it provides additional damping power component proportional to the frequency oscillation at the AC sub-grid. The POD controller design process is based on the procedure outlined in [40, 41]. A transient disturbance can causes a large speed deviation at the induction machine. More speed oscillations means poor damping performance which requires additional damping torque/power. The transfer function of the POD controller is given by;

$$H(s) = k_D.\left(\frac{1}{(1 + T_1 s)}\right).\left(\frac{T_w s}{(1 + T_w s)}\right) \tag{23}$$

Here, k_D, T_1, and T_w are gain, low-pass filter time constant and wash-out filter time constant respectively. The logical approach of selecting the gain of the POD controller is based on speed deviation i.e. more the speed deviation more the damping torque. The gain of the POD controller is fixed at 120. The time constant (T_1) of the low-pass filter is chosen as 0.2 s to pass the LFO unaltered to the POD controller. The wash-out filter time constant is fixed to 10 s. After substituting all these values the transfer function of the POD controller can be written as;

$$H(s) = 120.\left(\frac{1}{(1 + 0.2\,s)}\right).\left(\frac{10\,s}{(1 + 10\,s)}\right) \tag{24}$$

The frequency response of the POD controller is depicted in Fig. 2c which indicates a 90.49^0 phase margin.

4 Study on the Microgrid Test System

The hybrid AC/DC microgrid model illustrated in Fig. 4 is developed in MATLAB Simulink which is inspired by the microgrid architectures presented in [42, 43]. The total installed capacity of the AC sub-grid of the hybrid AC/DC microgrid is 1000 kW, and the AC sub-grid bus voltage and frequency are 400 V and 50 Hz respectively. Due to the intermittent behaviour of the DERs, a 500 kVA battery storage system is connected to the AC sub-grid of the hybrid AC/DC microgrid.

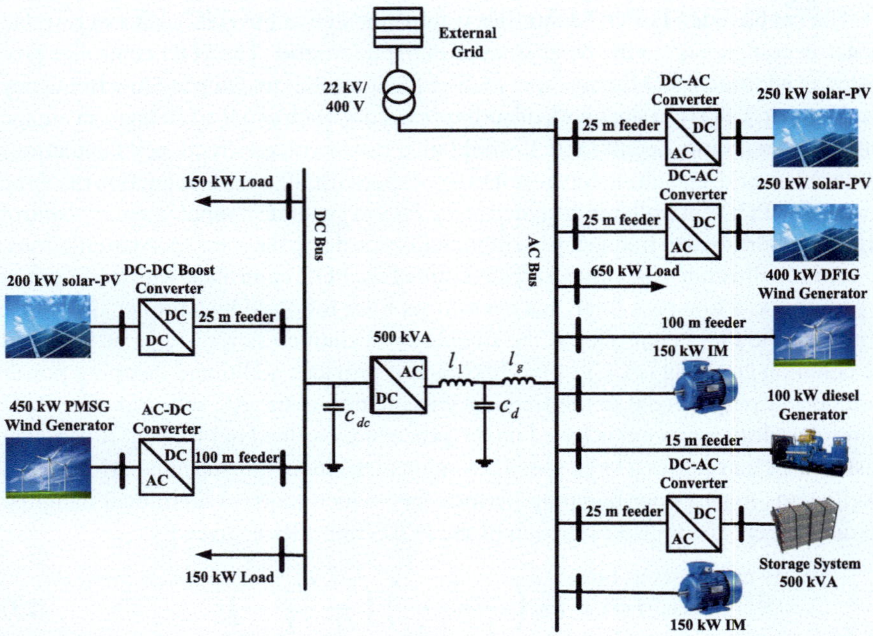

Fig. 4 Architecture of the test system

Because of high energy density, a lithium-ion battery bank has been used as the ESS, and the charging *(I < 0)* and discharging *(I > 0)* characteristics are determined by the following equations [44];

$$V_t = V_0 - rI - \left(\frac{Q.K}{IT - Q.0.1}\right) - I^* - \left(\frac{K.Q.IT}{Q - IT}\right) + A.\left(\exp^{-(B.IT)}\right) \tag{25}$$

Here, subscript K denotes the peukert constant, Q is the battery ampere-hour (Ah) capacity, IT extracted AH capacity, and A and B denote exponential voltage and time constant respectively. The initial state-of-charge (SoC) of the battery bank is considered as 90%. The open-circuit voltage and the ampere-hour (Ah) capacity of the battery bank are 650 V and 850 Ah respectively. The controllers shown in Figs. 2 and 3 are implemented at the ESS.

A 400 kW doubly-fed induction generator (DFIG) based wind generator is equipped with back to back PECs, and the DFIG is represented by the two-mass model. The wind speed at the DFIG is fixed at rated wind speed (e.g. 12 m/s). As the total solar-PV capacity is 500 kW, two parallel solar-PV systems each rated at

250 kW are used to improve the overall reliability of solar-PV generation in the AC sub-grid [45]. Each 250 kW solar-PV system is equipped with a maximum power point tracking (MPPT) system, and is connected to the AC sub-grid through a 300 kVA voltage source converter. The solar-PV array is consisted of 88 parallel strings with 7 modules per each string. The open-circuit voltage and short-circuit current at MPPT for each module are 72.9 V and 5.69 A respectively. A 100 kW diesel generator is directly connected to the PCC of the AC sub-grid of the AC/DC hybrid microgrid. Approximately, 80% of residential and industrial motor are DOL start type [34], hence two 150 kW DOL start IMs are used as the dynamic load in the hybrid AC/DC microgrid, which amounts to 37.50% of the total load at the AC sub-grid.

A 500 kVA interlinking converter (ILC) connects both the AC sub-grid and the DC sub-grid. The LCL filter and the DC link filter parameters are designed according to [46]. The PCC voltage of the AC sub-grid and the DC sub-grid are 400 V and 650 V respectively. The X/R ratio of the AC sub-grid is 0.4. The AC sub-grid is connected to the distribution feeder through a 22 kV/400 V distribution transformer. It must be noted that the voltage and frequency are measured at the AC busbar of the AC sub-grid, and the frequency is measured using a phase-locked loop (PLL).

The total installed capacity of the DC sub-grid of the hybrid AC/DC microgrid is 650 kW, and the DC sub-grid bus voltage is maintained at 650 V. A 450 kW direct-drive permanent magnet synchronous generator (PMSG) is connected to the DC bus through an AC/DC rectifier, and rectifier ensures constant DC voltage under varying wind conditions. The wind speed at the PMSG is fixed at rated wind speed (e.g. 12 m/s). The active power of the PMSG is regulated by the pitch-angle controller when the wind speed exceeds rated wind speed of the PMSG. As the PMSG is operating at the rated power, pitch-angle is maintained at zero degrees in this simulation [47]. A 200 kW solar-PV system is equipped with a MPPT system, and is connected to the DC sub-grid of the hybrid AC/DC microgrid through a 250 kVA DC-DC boost converter.

5 Simulation Scenarios and Results

The effectiveness of this POD controller of the ESS is verified through non-linear time-domain simulations considering different penetration levels of dynamic loads (i.e., IMs) and disturbances in the hybrid AC/DC microgrid. The simulation studies and analysis have been carried out on various scenarios and the select following cases are discussed here;

- 37.50% dynamic load (300 kW step load increase at the AC sub-grid)
- 18.75% dynamic load (200 kW step-load increase at the DC sub-grid)
- 37.50% dynamic load (A line-to-ground fault at the AC sub-grid of the PCC).

5.1 37.50% Dynamic Load (300 kW Step Load Increase at the AC Sub-Grid)

In this scenario another DOL start 150 kW IM was added to the AC sub-grid, and it is operated at full-load. Now, the AC sub-grid comprises a 500 kW constant power load and a 300 kW IM load, while the DC sub-grid comprises a 300 kW constant impedance load. The performance of this POD controller by simulating the same load disturbance as the previous scenario (i.e. connecting and disconnecting a 300 kW load). The IM rotor speed, and the AC sub-grid PCC voltage during the load disturbance are illustrated in Fig. 5a and b respectively, whereas Fig. 5c and d illustrate the DC voltage of the DC sub-grid, and the amount of power injected to the AC sub-grid from DC sub-grid through the ILC respectively.

As similar to the previous scenario, the sudden load disturbance has caused voltage and frequency oscillations in the AC sub-grid, DC voltage oscillations in the DC sub-grid, and IM rotor speed oscillations. The IM rotor speed indicates more oscillations without the POD controller compared to the scenario with this POD controller. The AC and the DC sub-grid voltage dip at the moment of load addition and removal are quickly recovered with the support from the POD controller. Furthermore, the ILC power transfer is also affected by the sudden load addition, which is effectively reduced by the POD controller.

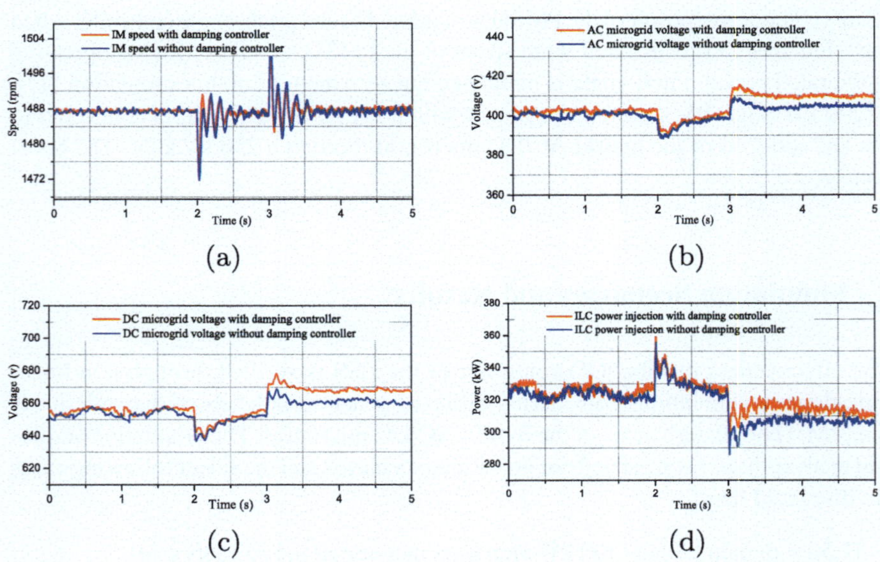

Fig. 5 **a** Rotor speed characteristics of the IM during the sudden load change at the AC sub-grid with 37.50% IM loading, **b** The PCC voltage of the AC sub-grid during sudden load change at the AC sub-grid with 37.50% IM loading, **c** The PCC voltage of the DC sub-grid during sudden load change at the AC sub-grid with 37.50% IM loading, **d** The ILC power transfer to the AC sub-grid during sudden load change at the AC sub-grid with 37.50% IM loading

The POD controller has attenuated the rotor speed oscillations by 11.4% compared to rotor speed oscillations without the POD controller. The settling time of the system with and without POD controller are 2.55 s and 2.73 s respectively, and the POD controller has reduced the settling time approximately by 0.18 s. Hence the POD controller can effectively damp the rotor speed oscillations event at high penetrations of dynamic loads.

5.2 18.75% *Dynamic Load (200 kW Step Load Increase at the DC Sub-Grid)*

In this scenario the total generation capacity of the AC sub-grid is maintained at 1000 kW and it is operated at 80% loading condition, which comprises a 650 kW constant power load and a 150 kW IM load. The POD controller performance was investigated by adding a 200 kW load at t = 2 s in to the DC sub-grid, and subsequently it was removed at t = 3 s. The IM rotor speed and the AC sub-grid PCC voltage are illustrated in Fig. 6a and b, whereas Fig. 6c and d illustrate the DC sub-grid voltage at the PCC and the amount of power injected to the AC sub-grid from the DC sub-grid through the ILC respectively.

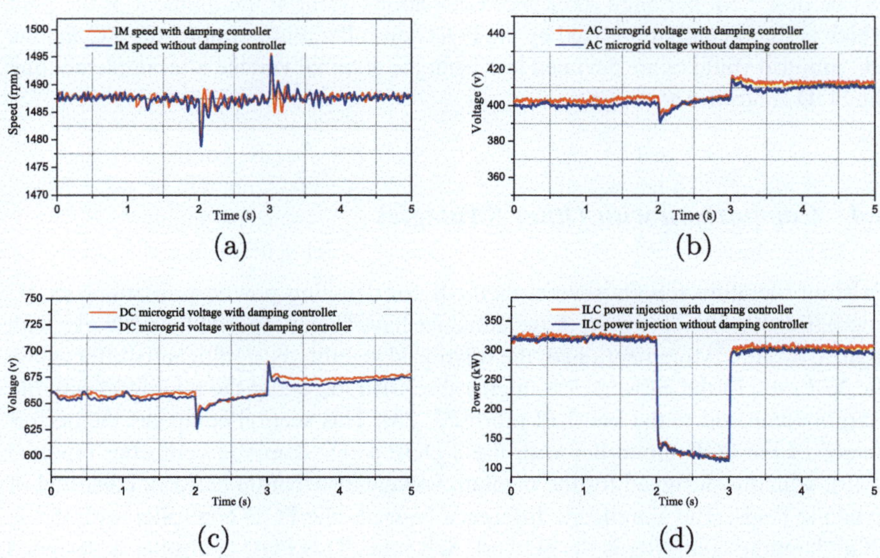

Fig. 6 **a** The IM rotor speed during sudden load change at the DC sub-grid with 18.75% IM loading, **b** The PCC voltage of the AC sub-grid during sudden load change at the DC sub-grid with 18.75% IM loading, **c** The PCC voltage of the DC sub-grid during sudden load change at the DC sub-grid with 18.75% IM loading, **d** The ILC power transfer to the AC sub-grid during sudden load change at the DC sub-grid with 18.75% IM loading

When the 200 kW constant impedance load is connected to the DC sub-grid, the DC sub-grid has to supply this additional load, and hence the ILC has reduced power transfer to the AC sub-grid by the same amount. As expected the load disturbance has caused voltage and frequency oscillations, and has also caused IM rotor speed oscillations. The IM rotor speed shows more oscillations without the POD controller, and the oscillations have significantly dampen after applying the POD controller. The AC and the DC sub-grid voltage dips at the moment of load disturbance have quickly recovered after implementing the POD controller. Furthermore, the ILC power transfer is also affected by the sudden load addition, which has also significantly improved by the POD controller.

5.3 37.5% Dynamic Load (A Line-to-Ground Fault at the AC Sub-Grid PCC)

A line-to-ground fault is applied at the PCC of the AC sub-grid of the hybrid AC/DC microgrid at t = 2 s and the fault has been cleared after 150 ms. The IM rotor speed and the AC sub-grid voltage are illustrated in Fig. 7a and b respectively, whereas Fig. 7c and d illustrate the DC sub-grid voltage and the power injected to the AC sub-grid from the DC sub-grid through the ILC respectively. According to Fig. 7, it can be observed that large rotor speed deviations occur at the moment of the fault which is significantly improved by the POD controller. Furthermore, the AC and the DC sub-grid voltages are deviated less from the nominal voltage after implementing the POD controller at the ESS.

5.4 Comparison with Other Strategies

Existing literature reports several methods for damping power oscillations in AC microgrids [9, 48]. However, those strategies have been proposed for networks with high X/R ratio. Aforementioned, the microgrid stability is significantly affected by the X/R ratio of the network, i.e., higher the X/R ratio better the microgrid stability margin compared to the low X/R ratio [22, 23]. This section compares the performance of the POD controller with the 2-DOF active damping controller reported in [9], which is designed for the medium voltage distribution network consisted of dynamic loads. The significant difference between the POD controller with the 2-DOF damping controller is the network X/R ratio. This POD controller is designed for networks with low X/R ratio (less than 1), i.e., for low voltage resistive distribution networks, whereas the 2-DOF damping controller is designed for a network with X/R ratio of 4.42 (i.e. for medium voltage inductive distribution networks).

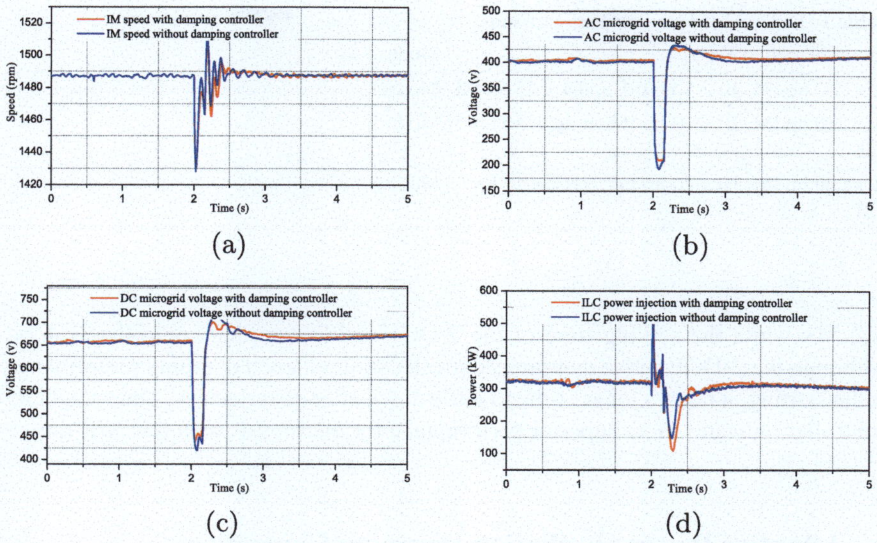

Fig. 7 **a** Rotor speed characteristics of the IM at fault condition, **b** The PCC voltage of the AC sub-grid at fault condition, **c** The PCC voltage of the DC sub-grid at fault condition, **d** The ILC power injection to the AC sub-grid at fault condition

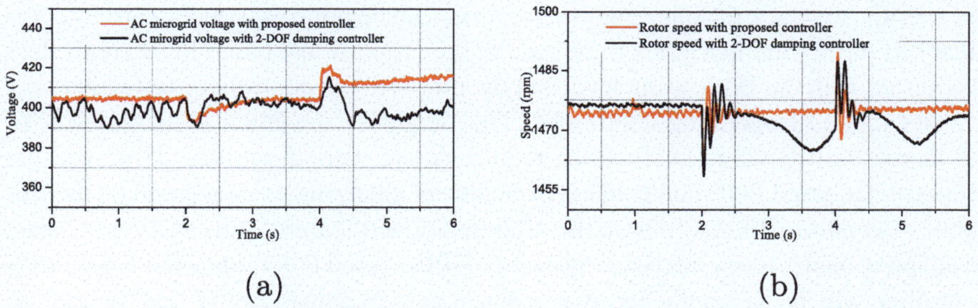

Fig. 8 **a** PCC voltage of the AC sub-grid during a sudden load change at the AC sub-grid, **b** Rotor speed characteristics of the IM during a sudden load change at the AC sub-grid

Figure 8a illustrates the the AC sub-grid voltage during sudden addition of the 300 kW load at the AC sub-grid with POD controller and 2-DOF damping controller, whereas Fig. 8b illustrates the rotor speed of the IM during sudden addition of the 300 kW load at the AC sub-grid with POD controller and 2-DOF damping controller. It can be seen from the Fig. 8b that the load disturbance causes voltage fluctuations in the AC sub-grid. The voltage recovery after disturbances is fast in case of POD controller and large for the 2-DOF damping controller. Same as before, the POD controller performance was investigated by adding a 300 kW load at t = 2 s into the AC sub-grid, and subsequently it was removed at t = 4 s. After implementing the 2-DOF active damping controller, the rotor speed oscillation is attenuated by 68.42% while it is 88.24% with the POD controller. The settling time of the POD controller

Table 1 Different IM loading configurations

Combination of IM's	Mode
1 × 200 hp IM, 40 × 5-hp IM, and 4 × 25 hp IM	Mode-1
1 × 200 hp IM, 20 × 10-hp IM, and 2 × 50 hp IM	Mode-2
1 × 200 hp IM, 10 × 20-hp IM, and 7 × 15 hp IM	Mode-3

is 2.54 s while the settling time of the 2-DOF active damping controller is 4.2 s. Although the 2-DOF active damping controller shows a good damping performance at the starting, it takes a longer time to settle down after disturbances. Thus, the POD controller indicates more superior performance for microgrids with low X/R ratio.

6 Adaptive Power Oscillation Damping Controller

In previous sections, it is demonstrated that supplementary power oscillation damping (POD) controller implemented in the ESS is capable to damp the oscillations in the microgrid. However, the damping performance can be improved under various combinations of IMs in microgrids if the gain of the proportional based POD controllers is adjusted with the IM loading level and the proportional controller is effective over a wider range of operating conditions. The frequency deviation is increased with the increase of dynamic load, which requires more damping torque. However, the proportional based POD controller is incapable of changing its gain based on the frequency deviation. To incorporate the aforementioned features, this section presents an adaptive neuro-fuzzy inference system (ANFIS) based POD controller for the ESS controller, which can handle highly non-linear system dynamics [49–51]. In order to analyse the impact and performance of the ANFIS based POD, the microgrid in Fig. 4 is modified as illustrated in Fig. 9 with detailed description provided in [52, 53].

According to the investigation carried out in the Ontario hydro system found that the total IM load configuration consists of 28.90% small residential IMs, 18.1376% medium commercial IMs, and 52% large industrial IMs [54]. Therefore, in this study, the test system IM loading configuration consists of IMs with different power ratings, which mimic the practical IM loading configurations as listed down in Table 1.

For the 300 kW step load addition scenario, the rotor speed with no controller and non-adaptive controller (NA-POD)—based on [52]—is illustrated in Fig. 10a, where a single 200-hp IM is used as a dynamic load. The voltage and frequency oscillations in the AC sub-grid are caused by the load disturbances, which subsequently caused IM rotor speed oscillations. It can be seen from Fig. 10a that IM's rotor have significant oscillations, and improvement is achieved by implementing a NA-POD controller. Later on the dynamic load is increased by adding another 200-hp IM to the AC sub-grid and the rotor speed oscillations were observed while the setting of the NA-POD

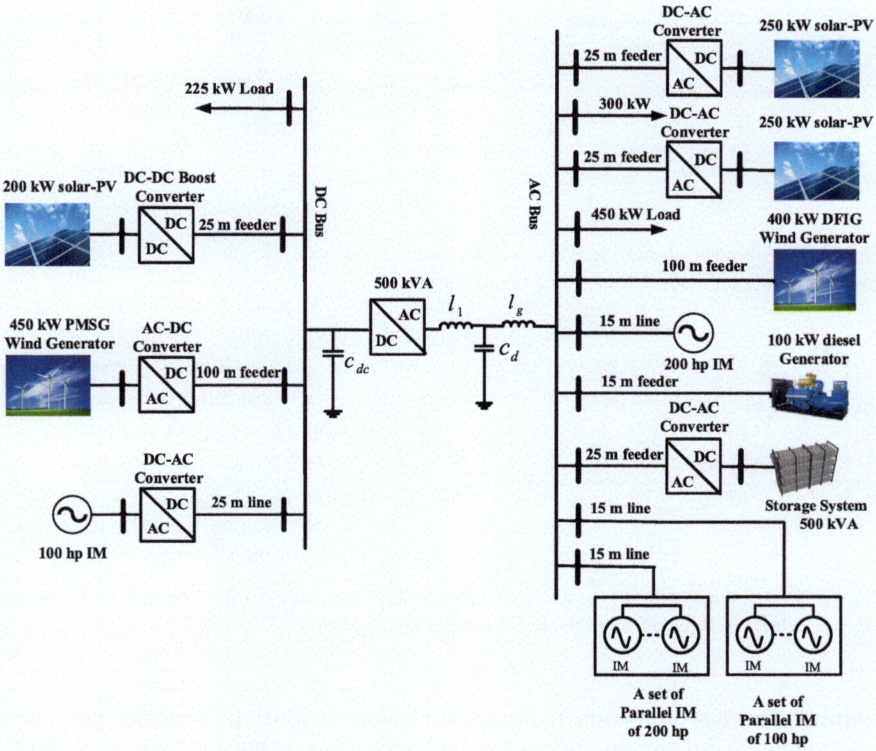

Fig. 9 The hybrid AC/DC microgrid model with various IMs

controller remained unchanged. The IM rotor speed oscillations with single 200-hp IM and two 200-hp IM are illustrated in Fig. 10b which shows insufficient damping performance of the NA-POD controller for the two 200-hp IM loading conditions. Basically, the NA-POD controller provides a supplementary control signal to the ESS controller proportional to speed/frequency deviation. The gain of the NA-POD controller is fixed and can not be changed with the loading condition of the system. Therefore, under high IM loading level more speed/frequency deviations can be expected, and hence more damping (D) torque is required to damp speed oscillations as D is proportional to the speed/frequency (ω) deviation ($D \propto \omega$) [36].

Now, the microgrid IM loading configuration is changed to mode-1 (i.e., one 200 hp IM, forty 5 hp IM, and four 25 hp IM) and the performance of the NA-POD is evaluated by applying the aforementioned load disturbances which is depicted in Fig. 10c. With the same setting of the POD controller, it provides little improvement in damping performance compared with no control case. The power spectrum density (PSD) analysis of the AC sub-grid system frequency with 40 parallel 5-hp IMs and single 200-hp IM is illustrated in Fig. 11a. It can be inferred that the system frequency (Fig. 11a) has one dominant oscillation frequency around 5 Hz in the case of single 200-hp IM, whereas the system frequency with 40 parallel 5-hp IMs has

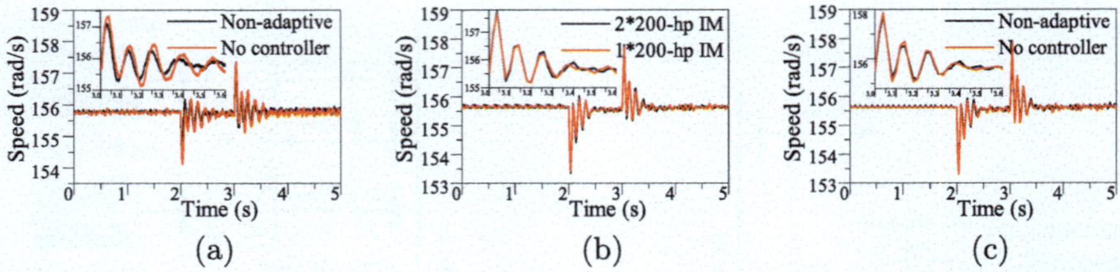

Fig. 10 IM rotor speed characteristics; **a** Rotor speed, **b** Rotor speed with one and two 200-hp IM, **c** Rotor speed with mode-1 IM loading configuration

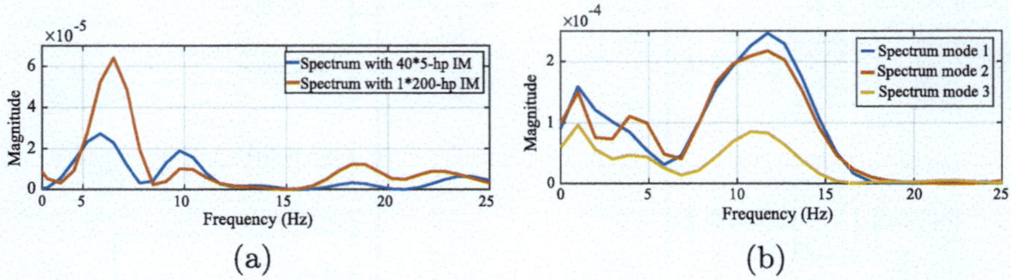

Fig. 11 Power spectral density of the system frequency; **a** spectrum of forty 5-hp IM and 1–200-hp IM, **b** spectrum of three different mode IM loading configuration

multiple dominant oscillation frequencies of around 5 and 10 Hz. Similarly, the power spectrum of the AC sub-grid frequency with three distinct IM modes (shown in Table I) is presented in Fig. 11b. It can be seen from Fig. 11b that each IM loading mode has three dominant frequencies which shows the increased non-linearity of the hybrid AC/DC microgrid.

7 Energy Storage System with Adaptive POD Controller

In order the improve the performance of the POD controller, ANFIS based adaptive controller [51, 55] is added to the ESS asillustrated in Fig. 12. The control scheme is designed as a supplementary controller of the outer voltage control loop of the ESS controller. This supplementary POD controller provides additional control signal proportional to the AC sub-grid frequency deviation.

The general architecture of the ANFIS is depicted in Fig. 13 which consists of five layers, i.e., an input layer, three hidden layers, and an output layer; and each layer or node are connected through directional links. Three basic steps should be followed to design ANFIS; (1) learning paradigm, which means learning process based on the system information; (2) learning algorithm, which refers to a rule that is used to adjust the synapses and parameters of the ANFIS; and (3) assess the network capacity and samples required for training to train the network [56]. The deviation of the frequency (Δf) from the reference point and the prediction of future deviation,

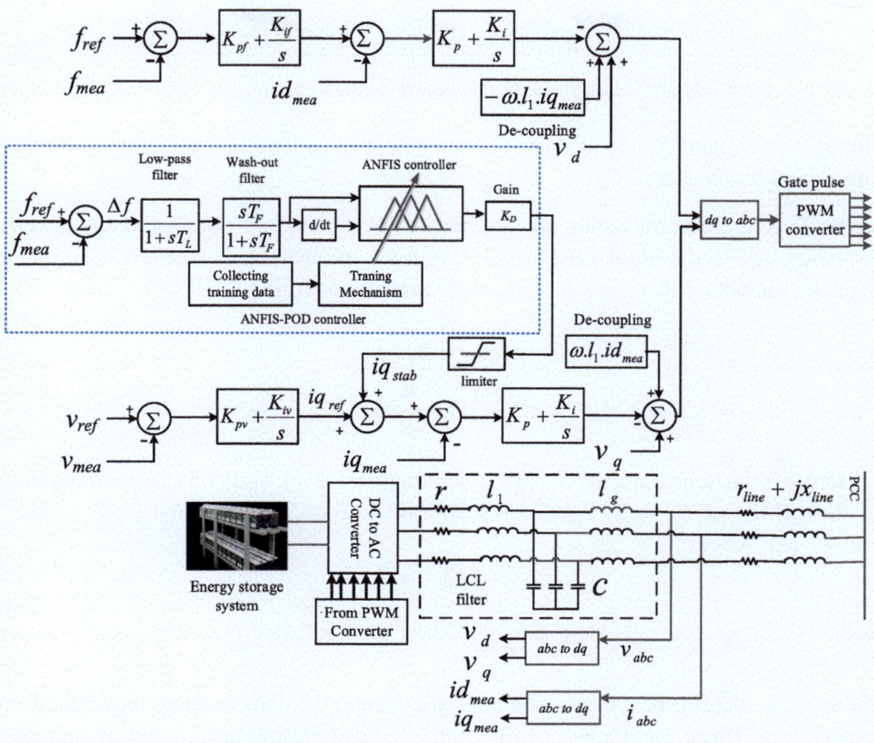

Fig. 12 The ESS controller with an adaptive POD controller

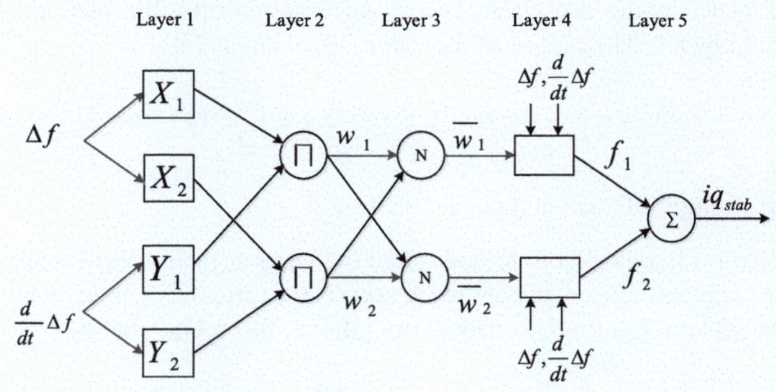

Fig. 13 ANFIS control architecture [51]

i.e. derivation of frequency deviation ($\frac{d}{dt}\Delta f$) from the reference set point are given as input to the ANFIS based adaptive POD controller. The basic rules used in the Takagi-Sugeno type ANFIS is given bellow;

$$\begin{aligned}
&\text{if}(\Delta f = X_1)\&(\frac{d}{dt}\Delta f = Y_1), \text{then } f_1 = p_1\Delta f + q_1\frac{d}{dt}\Delta f + r_1 \\
&\text{if}(\Delta f = X_2)\&(\frac{d}{dt}\Delta f = Y_2), \text{then } f_2 = p_2\Delta f + q_2\frac{d}{dt}\Delta f + r_2
\end{aligned} \tag{26}$$

where, X_1, X_2 and Y_1, Y_2 are fuzzy sets, while p_1,p_2, q_1, q_2, r_1, and r_2 are the consequent parameters.

Layer 1: The input variables are applied to the layer 1 to obtain the fuzzy sets corresponding to the input variables. Every node in layer 1 is a square node which gives the membership value of each input variables as follows;

$$\begin{aligned}
O_i &= \mu X_i(\Delta f) \\
O_i &= \mu Y_i(\frac{d}{dt}\Delta f)
\end{aligned} \tag{27}$$

Here, i is the node number, X_i(or Y_i) is a linguistic level, and O_i is the membership function (MF) of X_i or Y_i. In this research, Gaussian MF is chosen (μX_i) and is defined as follows;

$$\begin{aligned}
\mu X_i(\Delta f) &= e^{\frac{-1}{2}(\frac{x-c}{\sigma})^2} \\
\mu Y_i(\frac{d}{dt}\Delta f) &= e^{\frac{-1}{2}(\frac{x-c}{\sigma})^2}
\end{aligned} \tag{28}$$

where σ, c determines the width and the center of the membership function respectively. These parameters of the membership function are known as premise parameters.

Layer 2: Layer 2 nodes are circle nodes and they multiply the incoming signals coming from layer 1. The output of the layer 2 is given as follows;

$$O_{2,i} = W_i = \mu X_i(\Delta f) \times \mu Y_i(\frac{d}{dt}\Delta f) \tag{29}$$

Here, firing strength of each node is denoted by W_i.

Layer 3: Layer 3 nodes are circle nodes and the number of fuzzy rules are equal to the number of layers. The output of the ith node is the ratio of ith node firing strength to the weighted sum of all nodes firing strengths, as formulated bellow;

$$O_{3,i} = \bar{W}_i = \frac{W_i}{\Sigma(W_1 + W_2)} \tag{30}$$

Layer 4: Layer 4 nodes are circle nodes and output node function can be formulated as follows;

$$\begin{aligned}
O_{4,i} &= f_1 = \bar{W}_1(p_1\Delta f + q_1\frac{d}{dt}\Delta f + r_1) \\
O_{4,i} &= f_2 = \bar{W}_2(p_2\Delta f + q_2\frac{d}{dt}\Delta f + r_2)
\end{aligned} \tag{31}$$

Layer 5: Layer 5 provides the ANFIS final output by summing all the incoming signals from previous layer. The output signal (iq_{stab}) can be formulated as follows;

$$O_{5,i} = iq_{stab} = \Sigma(f_1 + f_2) \tag{32}$$

The ANFIS controller has to set the values of the premise and the consequent parameters by the forward pass and the backward pass respectively. In the forward pass, consequent parameters are evaluated while premise parameters kept constant, whereas in the backward pass, premise parameters are evaluated while consequent parameters are kept constant. Now, consider that the training data set has N number of input data, the error function for the nth $(1 \leq n \leq N)$ training data set can be formulated as follows;

$$E_n = \sum_{p=1}^{m} (T_{p,n} - O_{p,n}^{m})^2 \tag{33}$$

where, $O_{p,n}$, $T_{p,n}$, and E_n are the pth component of the nth output vector, reference output vector, and error respectively. Therefore, the total measured error is:

$$E = \sum_{n=1}^{N} E_n \tag{34}$$

The rate of error at node (m, i) from (33) can be formulated as;

$$\frac{\delta E_n}{\delta O_{i,n}^{m}} = -2(T_{i,n} - O_{i,n}^{m}) \tag{35}$$

If σ is a parameter of the ANFIS, then rate of change of error with respect to σ is;

$$\begin{gathered} \frac{\delta E_n}{\delta \sigma} = \sum_{O \varepsilon S} \frac{\delta E_n}{\delta O} \frac{\delta O}{\delta \sigma} \\ \frac{\delta E}{\delta \sigma} = \sum_{n=1}^{N} \frac{\delta E_n}{\delta \sigma} \end{gathered} \tag{36}$$

Here, S denotes the nodes whose output depends on σ. Also, $\frac{\delta E}{\delta \sigma}$ denotes the rate of change of the total measured error with respect to σ. Therefore, updated generic parameter, σ can be written as;

$$\begin{gathered} \Delta\sigma = -\beta \frac{\delta E}{\delta \sigma} \\ \beta = -\frac{J}{\sqrt{\sum_{\sigma} (\frac{\delta E}{\delta \sigma})^2}} \end{gathered} \tag{37}$$

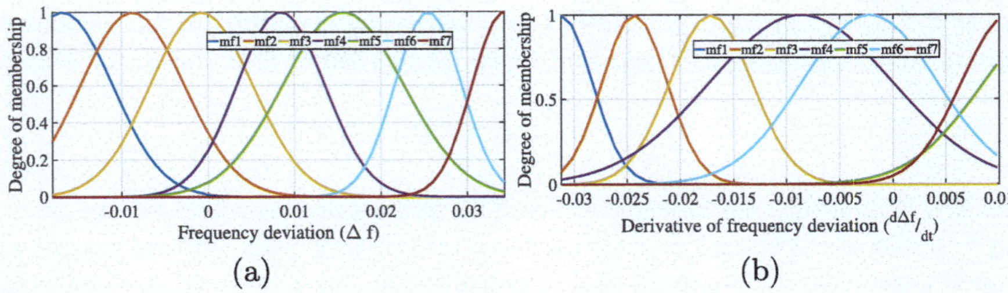

Fig. 14 MF after training; **a** frequency deviation **b** derivative of frequency deviation

where β is the learning rate and J is the step size which is varied to enhance the rate of convergence. Off-line learning is used to update σ based on (36), and the detail description of the training procedures is presented in the subsequent section.

7.1 ANFIS Based POD Controller Training

To train the ANFIS based POD controller, the training data sets were collected from the non-adaptive POD controller implemented at the ESS controller. The test system was simulated for different values of the POD controller gains (K_D), and 5000 training and testing data points were obtained to design the ANFIS controller. Mode 1 IM loading configuration is connected to the AC sub-grid of the hybrid microgrid as the dynamic loads. Gaussian MF is used here, and MF parameters are tuned by 70% of the total collected data (i.e., training data set). The initial FIS structure created by the grid partitioning technique, and back propagation algorithm is used for training. Before starting the training, the number of training epochs is selected as 100 and the model over-fitting was closely observed. The rest of the data set (30% of the collected data), including the testing data set (15% data), checking data set (15% data) is used to validate the trained FIS structure [57]. The MF of the frequency deviation after training is shown in Fig. 14a, while Fig. 14b shows the MF of derivative of frequency deviation after training. After training, the Δf and the $\frac{d\Delta f}{dt}$ MFs are varied between –0.018 to 0.0343 and –0.0312 to 0.0099 respectively.

8 Simulation Results with Adaptive POD Controller

The ANFIS based POD controller is trained with mode 1 IM loading configuration (given in Table I) and tested with untrained mode 2 and 3 IMs loading configuration under disturbances to demonstrate the robustness of this approach. Non-linear dynamic simulations have been performed to evaluate the effectiveness of the ANFIS based adaptive POD controller of the ESS considering known and unknown set of

IM loading configurations and disturbances in the hybrid AC/DC microgrid and the following select scenarios are presented below;

- Training mode (26.72% load disturbances at the AC sub-grid)
- Test mode-2 (26.72% load disturbances at the AC sub-grid)
- Test mode-1 (40.03% load disturbances at the DC sub-grid)
- Test mode-1 (outage of a solar-PV at the AC sub-grid).

8.1 Training Mode (26.72% Load Disturbances at the AC Sub-Grid)

Three types of IM configurations (see Table I) are considered in this study and the ANFIS based adaptive POD controller was trained for mode-1 IM loading configuration (known as the training mode). The total generation capacity of the AC sub-grid is 1 MW, and it is operated at 82.3% loading condition (300 kW CPL, 150 kW constant impedance load, and 373 kW IM), while the DC sub-grid installed capacity is 650 kW and it is operated at 46.09% loading (225 kW as a constant impedance load and a 100 hp IM). The disturbances are created by adding a 300 kW load at $t = 2.0$ s in the AC sub-grid, and subsequently, it is removed at $t = 3.0$ s, and the performance of the ANFIS based adaptive POD controller is recorded. The IM rotor speed and the AC sub-grid PCC voltage are illustrated in Fig. 15a and b respectively, whereas Fig. 15 (c) and (d) depict the AC sub-grid system frequency and the ESS power output, respectively. The voltage and frequency oscillations in the AC sub-grid and the DC voltage oscillations in the DC sub-grid are caused by the load disturbances. The voltage and frequency oscillations in the AC sub-grid will cause rotor speed oscillations due to following reasons, (1) IM power consumption (both active and reactive power) is influenced by the stator voltage and rotor speed, (2) IM rotor speed oscillations are caused by frequency variations, and (3) The change of rotor speed will affect power consumption. Furthermore, the above issues are even more severe for a system with high penetration of dynamic loads with a combination of small, medium, and large IMs. As highlighted in Sect. III the presence of multiple operating IMs introduces more non-linearity into the systems dynamics.

When an additional 300 kW load is connected to the AC sub-grid, the ESS supplies the required load power to the system based on the deviation of system voltage and frequency from the reference set point. It can be seen from the ESS output that with the same gain for both controllers, it injects more active power and ensures better response. More active power injection from the ESS to the AC sub-grid ensures better voltage profile during load disturbances which can be noticed from Fig. 15b. With the increase of IM loading level, more damping torque is required to damp out speed oscillations. The significant difference of this controller with non-adaptive POD controller is that the controller will change the gain based on the frequency deviation while the non-adaptive controller has the same gain.

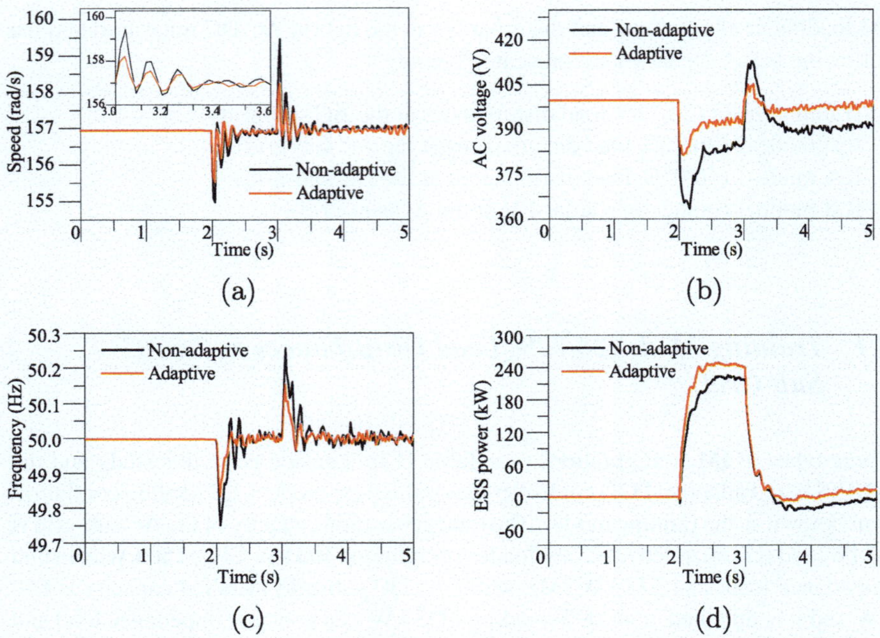

Fig. 15 Hybrid AC/DC microgrid with training mode IM load under 26.72% load disturbances at the AC sub-grid, **a** IM rotor speed, **b** AC sub-grid PCC voltage, **c** AC sub-grid frequency, **d** ESS power injection to the AC sub-grid

The rotor speed oscillation attenuation factor between the first and the forth peaks is utilized to identify the improvement of the damping performance of the ANFIS based adaptive POD controller and the non-adaptive POD controller. The controller has attenuated the speed oscillations by 88.71% while it was 80.40% with a non-adaptive POD controller. Therefore, the ANFIS based adaptive POD controller provides better damping attenuation factor compared to the non-adaptive POD controller. Moreover, the maximum frequency deviation during the 300 kW step load addition into the AC sub-grid is 49.84 Hz and 49.75 Hz for the ANFIS-POD controller and non-adaptive POD controller respectively. Furthermore, the controller improved AC sub-grid voltage deviation during the 300 kW step load addition into the AC sub-grid by 15 V compared to the non-adaptive POD controller. The performance comparison of the controller in terms of rotor speed attenuation, maximum voltage, and frequency deviation during load disturbance is summarized in Table 2.

Table 2 Performance comparison

	NA-POD controller	ANFIS-POD controller
Attenuation of speed oscillation	80.84%	88.71%
Maximum frequency deviation	49.75 Hz	49.84 Hz
Maximum voltage deviation	364 V	379 V

8.2 *Test Mode-2 (26.72% Load Disturbances at the AC Sub-Grid)*

The robustness of the ANFIS-POD controller with another set of IM loading configuration (mode 3) known as test mode 2 is provided in this sub-section. Same as the previous scenario, previously trained ANFIS-POD controller with different untrained IM configurations are connected to the AC sub-grid while IM loading level will remain the same. An additional 300 kW load is connected at $t = 2.0$ s and subsequently, removed at $t = 3$ s. The effect of the load disturbance on IM rotor speed and AC sub-grid voltage are illustrated in Fig. 16a and b respectively. The effect of load disturbance on DC sub-grid voltage and ILC power injection from the DC sub-grid to the AC sub-grid through ILC are also illustrated in Fig. 16c and d respectively.

The voltage and frequency oscillations are caused by the load disturbance which has given rise to AC sub-grid voltage oscillations, DC sub-grid voltage oscillations, and IM rotor speed oscillations. Figure 16a shows that for an un-trained set of IMs connected to the AC sub-grid, the controller has damp out IM rotor speed oscillations by 10% compared to the non-adaptive controller. It is noteworthy that the voltage profile of both the AC and the DC sub-grid is better compared to the non-adaptive controller.

8.3 *Test Mode-1 (40.03% Load Disturbances at the DC Sub-Grid)*

In this scenario, an un-trained set of IM loading configuration (mode 2) known as the test mode 1 is connected to the AC sub-grid. The effectiveness of the ANFIS-POD control scheme is evaluated through a connection (at $t = 2.0$ s) and disconnection (at $t = 3.0$ s) of a 200 kW constant impedance load at the DC sub-grid of the hybrid AC/DC microgrid. Figure 17a and b show the response of the IM rotor and the PCC voltage of the AC sub-grid for this case. Figure 17c and d depict the comparison of the DC sub-grid voltage and ILC power injection between the non-adaptive and adaptive controller. An energy imbalance at the DC sub-grid is transferred through the ILC into the AC sub-grid and subsequently leads to DC bus voltage oscillations, AC sub-grid voltage oscillations, and rotor speed oscillations. Moreover, different number

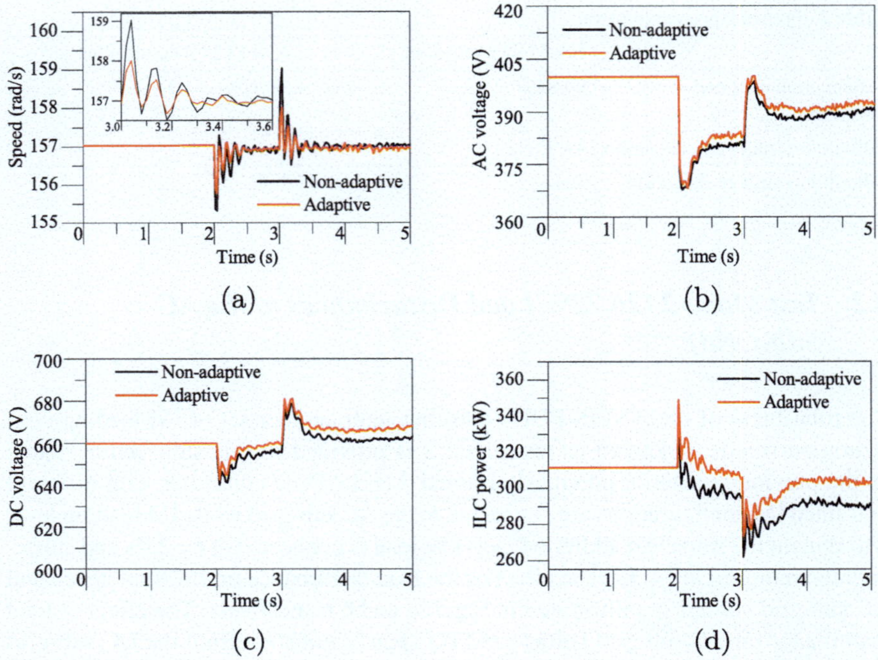

Fig. 16 Hybrid AC/DC microgrid with mode 3 IM load under 26.72% load disturbances at the AC sub-grid (Test mode-2), **a** IM rotor speed, **b** AC sub-grid PCC voltage, **c** DC sub-grid PCC voltage, (d) The injected power to the AC sub-grid through ILC

of parallel operating IMs with various power ratings have changed the dynamics of the entire hybrid AC/DC microgrid by introducing more non-linearity. Figure 17a shows that the ESS with adaptive POD controller has resulted in lower IM rotor speed oscillations and small AC sub-grid voltage dip during load perturbations by injecting required amount of active power.

8.4 Test Mode-1 (Outage of a 250 kW Solar-PV at the AC Sub-Grid)

In this scenario, an un-trained set of IM loading configuration (mode 2) known as the test mode 1 is connected to the AC sub-grid. The effectiveness of the ANFIS-POD control scheme is evaluated through an outage (at $t = 2.5$ s) and re-connection (at $t = 3.5$ s) of a 250 kW solar-PV at the AC sub-grid of the hybrid AC/DC microgrid. Figure 18a and b show the response of the IM rotor speed and the frequency of the AC sub-grid for this case. When the 250 kW solar-PV is disconnected from the AC sub-grid at $t = 2.5$ s, it has caused a generation and load-demand imbalance at the microgrid (from $t = 2.5$ s to 3.5 s). This imbalance was sensed by the ESS and it supplied the

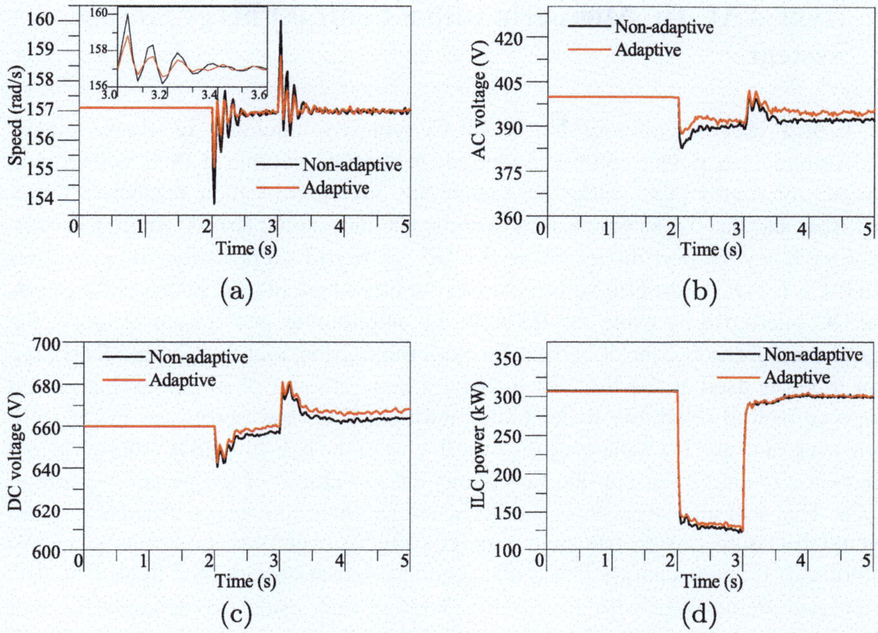

Fig. 17 Hybrid AC/DC microgrid with mode 2 IM load under 40.03% load disturbances at the DC sub-grid (Test mode-1), **a** IM rotor speed, **b** AC sub-grid PCC voltage, **c** DC sub-grid PCC voltage, **d** The injected power to the AC sub-grid through ILC

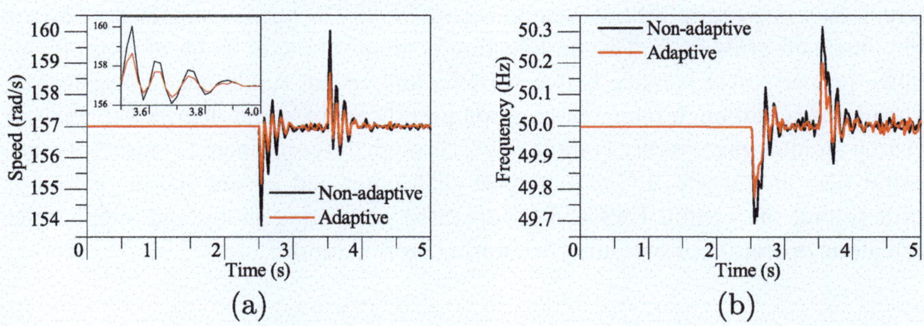

Fig. 18 Hybrid AC/DC microgrid with mode-2 IM load under outage of a solar-PV at the AC sub-grid (Test mode-1), **a** IM rotor speed, **b** AC sub-grid frequency

additional load-demand. The supplementary ANFIS-POD controller of the ESS has effectively dampen the rotor speed oscillations and frequency response. Therefore, the supplementary ANFIS-POD controller is effectively damping oscillations even for untrained scenarios.

9 Hybrid AC/DC Microgrid with a Central Energy Storage System

In hybrid AC/DC microgrids, the ILC typically operates in three modes: (a) Autonomous power control mode, where the DC microgrid PCC voltage and the AC microgrid PCC voltage/frequency are governed by their respective DERs or ESSs, and the ILCs operate in power dispatching mode, (b) AC microgrid voltage/frequency support mode, where the DC microgrid supports the AC microgrid via ILCs, (c) DC microgrid voltage support mode, where the AC microgrid supports the DC microgrid by using the ILCs. In the autonomous power control mode, the conventional droop control is used for both microgrids, and the power (PQ) control mode is used at the ILC. To manage a large amount of power exchange and improve overall reliability under faults, multiple ILCs were proposed in [58, 59]. However, multiple ILCs increase the overall cost of the hybrid AC/DC microgrid and create a circulating current due to the unequal reactances of the parallel operating ILCs. This will also over-stress the ILCs, reduce their conversion efficiencies, and eventually could cause permanent damage [60]. To overcome the issues related to multiple ILCs and separate ESSs, this section discusses a modified hybrid AC/DC microgrid architecture with a central ESS rather than using two separate ESSs at two individual sub-grids. This modified architecture eliminates the necessity of multiple ILCs and will ensure economic benefits to the users [61]. The schematic of the modified hybrid AC/DC microgrid architecture diagram is shown in Fig. 19. The key difference between this architecture and the traditional architecture is that there is only one central ESS, where the central ESS DC bus is connected with two individual sub-grids PCC. This architecture is quite simple in its basic form; it can realize proper power balance between DERs and ensure stable operation between two sub-grids, although interactions among multiple ILCs and distributed ESSs in existing architecture are very complicated. Through the coordinated control scheme, central ESS and single ILC can support each other and ensure stable operation. Furthermore, the central ESS can be operated in floating mode and support the standalone operation of two sub-grids during an ILC failure.

10 Co-ordination Control Strategy with the ILC and the Central ESS

The hybrid AC/DC microgrid considered here is operating in the autonomous operation mode. The associated coordinated control strategy between the ILC and the central ESS are described in the subsequent sub-sections, which will support the power management and control modes of the ILC.

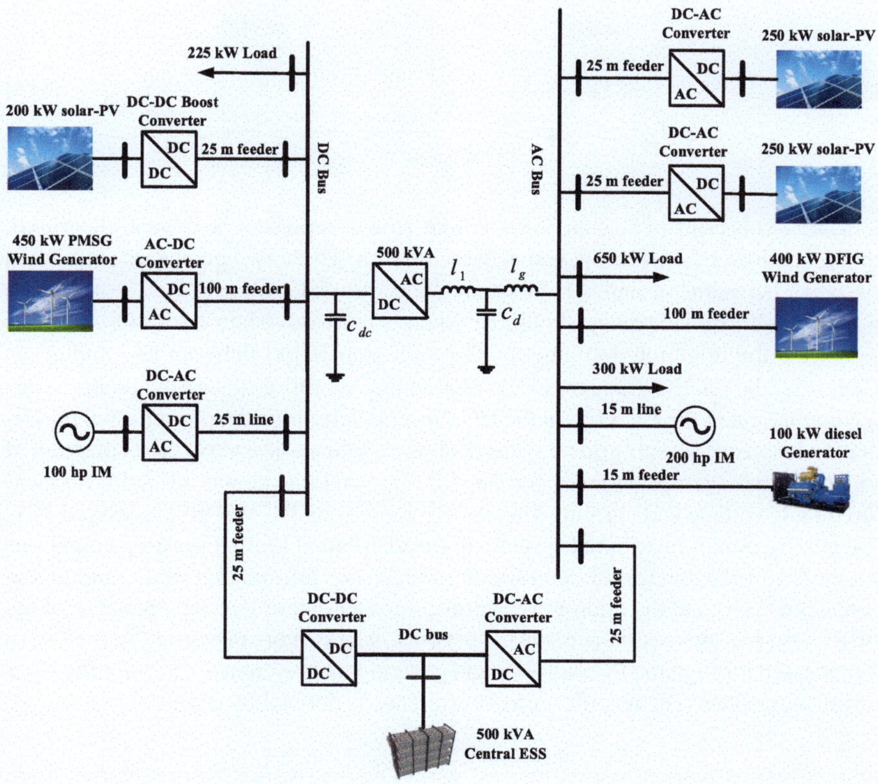

Fig. 19 AC/DC microgrid system with a centralized ESS

10.1 Control of the ILC

The circulating current issue which arises due to the parallel operation of multiple ILCs is resolved here by using a single ILC. The main reason of using the multiple ILCs is to transfer a large amount of power between two sub-grids and improve the overall system reliability in an event of failure of one ILC. Along with the central ESS, the proportional power-sharing between the DC and the AC sub-grids is properly managed by the single ILC. The AC sub-grid frequency and the DC sub-grid PCC voltage are considered as the global parameters, and it is assumed that these parameters are identical throughout the hybrid AC/DC microgrid operation. Important control parameters such as voltage and frequency are measured at the PCC of both sub-grids, and the phase-locked loop is utilized to measure frequency [62]. AC sub-grid frequency and the DC sub-grid PCC voltage are converted to the per-unit ($p.u.$) quantity by using (38) [58];

$$I_{pu} = \frac{f_{ref} - \frac{1}{2}(f_{upper} + f_{lower})}{\frac{1}{2}(f_{upper} - f_{lower})}$$
$$I_{dc,pu} = \frac{V_{dc,ref} - \frac{1}{2}(V_{dc,upper} + V_{dc,lower})}{\frac{1}{2}(V_{dc,upper} - V_{dc,lower})} \quad (38)$$

where, subscript ref, mea, $upper$, and $lower$ represent reference, measured, upper, and lower limits of the respective variables. These two quantities are now in the range between -1 and $+1$. In steady-state, $p.u.$ DC sub-grid PCC voltage and $p.u.$ AC sub-grid frequency should be equal. The steady-state error $(I_{pu} - I_{dc,pu})$ represents the direction and amount of power transferred between two sub-grids. Therefore, the relative frequency deviation in the AC sub-grid, and the relative voltage fluctuation in the DC PCC in the DC sub-grid determine the proportional power-sharing between two sub-grids. Figure 20 shows a schematic control block diagram of the power management strategy for the ILC. V_{abc} and I_{abc} denote AC sub-grid measured PCC voltage and current, whereas V_{dc} represents DC sub-grid measured PCC voltage. In a smart distributed system, all the distributed DERs operating conditions are stored in the microgrid central controller using information and communication technology, and the microgrid central controller determines the operation of the DERs based on the requirements of the network. The steady-state error $(I_{pu} - I_{dc,pu})$ is processed through the PI controller and generates the reference, $I_{d,ref}$ for the inner current controller. The reactive current reference is derived by [63];

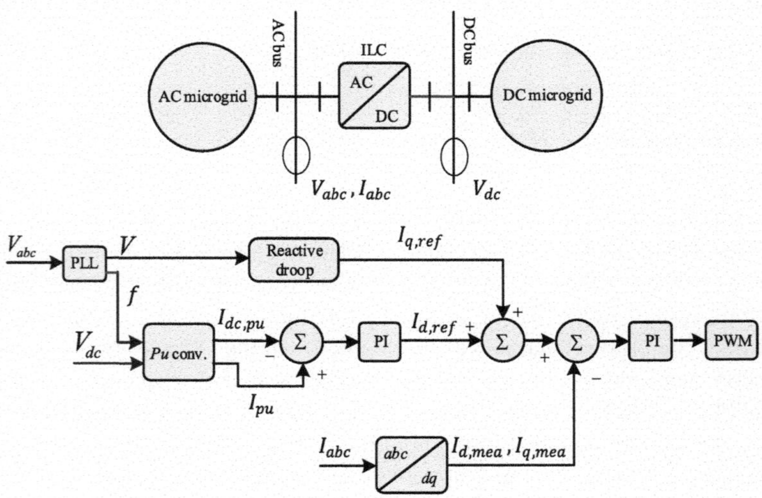

Fig. 20 The ILC control schemes block diagram

$$
\begin{aligned}
i_{q,ref} &= -\frac{2}{3}\frac{Q_{ref}}{V} \\
Q_{ref} &= \frac{V - V_{\max}}{n_Q}
\end{aligned}
\tag{39}
$$

For unity power factor operation, $i_{q,ref}$ is set to zero. The d-axis current error ($I_{d,ref} - I_{d,mea}$) and the q-axis current error ($I_{q,ref} - I_{q,mea}$) are processed through the inner PI controller and generates the reference voltage signals for the ILC.

10.2 Control of the Central ESS

A central ESS is deployed here to reduce the control complexity, and hence to improve the system stability. Figure 21 shows a schematic flowchart of the control logic of the central ESS to support AC and DC sub-grids in autonomous operation modes. The central ESS continuously monitors the DC sub-grid PCC voltage and the AC sub-grid PCC voltage/frequency to support both sub-grids. The central ESS and the ILC work together to maintain a stable operation. The central ESS charging or discharging mode is determined by the SoC level ($SoC_{\min} < SoC < SoC_{\max}$) of the ESS. Therefore, based on the SoC level, the central ESS can be operated either in: (1) charging mode, (2) discharging mode, or (3) remain idle.

10.2.1 Control of Central ESS to Support the DC Sub-Grid

A bi-directional DC/DC buck-boost converter topology is utilized to connect the central ESS to the DC sub-grid. This is aimed to maintain stable PCC voltage at the DC sub-grid. For the charging scenario, the DC/DC converter operates in the buck mode and vice-versa. Figure 22 shows a schematic control block diagram of the central ESS to support the DC sub-grid voltage. It consists of two control loops: (1) an outer voltage control loop and (2) an inner current control loop. The capacitance (C) and inductance (L) of the DC-DC buck-boost converter can be obtained as [64];

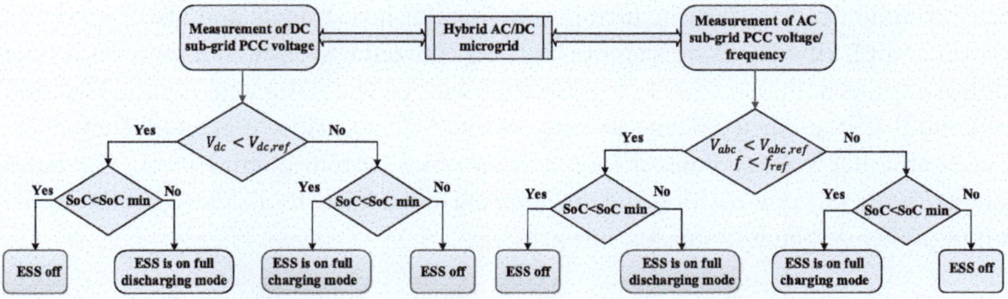

Fig. 21 Flow chart of the control logic of the central ESS

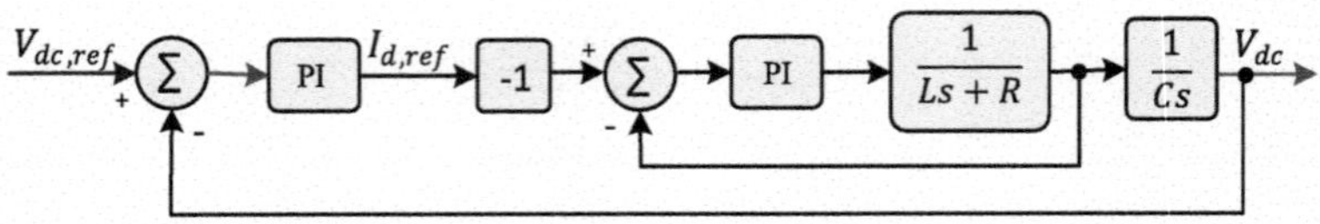

Fig. 22 The central ESS DC sub-grid support control scheme.

$$C = \frac{D}{2 \times R \times f}$$
$$L = \frac{R}{2 \times f} D(1 - D) \tag{40}$$

where, D, f, and R represent duty cycle, switching frequency, and load resistance, respectively. In this design, the switching frequency and the duty cycle are maintained at 5 kHz and 50%, respectively. The voltage error signal between the reference DC bus voltage ($V_{dc,ref}$) and the measured PCC DC voltage (V_{dc}) is processed through the outer PI controller and generates the reference current ($I_{d,ref}$) for the inner current controller. The current error signal between the reference current and the measured load current has processed through the inner PI controller and subsequently generates a voltage signal for the ESS converter.

10.2.2 Control of Central ESS to Support the AC Sub-Grid

Rather than using a central ESS, when distributed ESSs are employed at the AC and DC sub-grids, it is required that the each ESS unit should be charged or discharged simultaneously. Considering equal power sharing, in the discharging mode, the ESS with lower SoC dispatch the same amount of power as the ESS with higher SoC. Therefore, ESS with lower SoC will run out of stored energy. Similarly, when the ESSs are in the charging mode, ESS with the higher and lower SoC absorb the same amount of power. Therefore, ESS with higher SoC charge first and cannot store a large amount of generated power, and hence reduce the overall efficiency. To solve this issue, a combination of centralized control, distributed control, and adaptive droop control strategies are proposed in [19, 65, 66]. This control approach requires many communication links incurring more capital investments and also introduces control complexity. By using a single large centralized ESS, which requires a simple control approach, the previously mentioned issues can be mitigated. Figure 23 shows the central ESS control scheme to support the AC sub-grid voltage and frequency. It uses an outer voltage control loop and an inner current control loop. The outer voltage/frequency control loop and the inner current control loop determine the ESS's output power dispatch to the AC sub-grid.

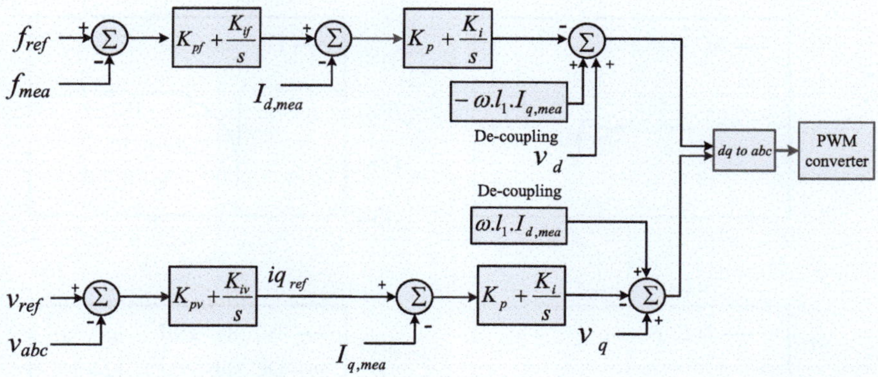

Fig. 23 The central ESS AC sub-grid support control scheme.

11 Simulation Scenarios with Central ESS

The hybrid AC/DC microgrid with central ESS is simulated in the Matlab/SIMULINK environment considering various operating and load switching scenarios to validate and analyze the performance of the coordinated control strategy of this hybrid AC/DC microgrid architecture. The following select scenarios are presented below;

11.1 Case 1: DC Sub-Grid is Overloaded and is Supported by the ILC and the ESS

This simulation scenario emulates a practical scenario, where the DC sub-grid is supported by the AC sub-grid and the central ESS. Meanwhile, load switching events have occurred at both sub-grids. The following load switching events are simulated in the the hybrid AC/DC microgrid, (1) a 200 kW load plugged on and off from the DC sub-grid at $t = 2$ s and $t = 3$ s respectively, and (2) a 300 kW load plugged on and off from the AC sub-grid at t= 4 s and $t = 5$ s respectively. The DC sub-grid generation capacity is 650 kW while the total 800 kW load is connected, making the DC sub-grid to be overloaded. Therefore, the AC sub-grid is supporting the DC sub-grid by supplying approximately 150 kW of active power via the ILC. Figure 24a and b show the AC and the DC sub-grids PCC voltage at the two successive 200 kW and 300 kW step load perturbations at the DC and the AC sub-grids respectively. Figure 24c and d show the ILC power injection from the AC to the DC sub-grid and the central ESS output power dispatch due to two successive load perturbations respectively. A sudden addition of a 200 kW load (at $t = 2$ s) at the DC sub-grid causes power shortage at the DC sub-grid. Therefore, the central ESS attached to the DC sub-grid PCC also detects the power shortage at the DC sub-grid by measuring the DC sub-grid PCC

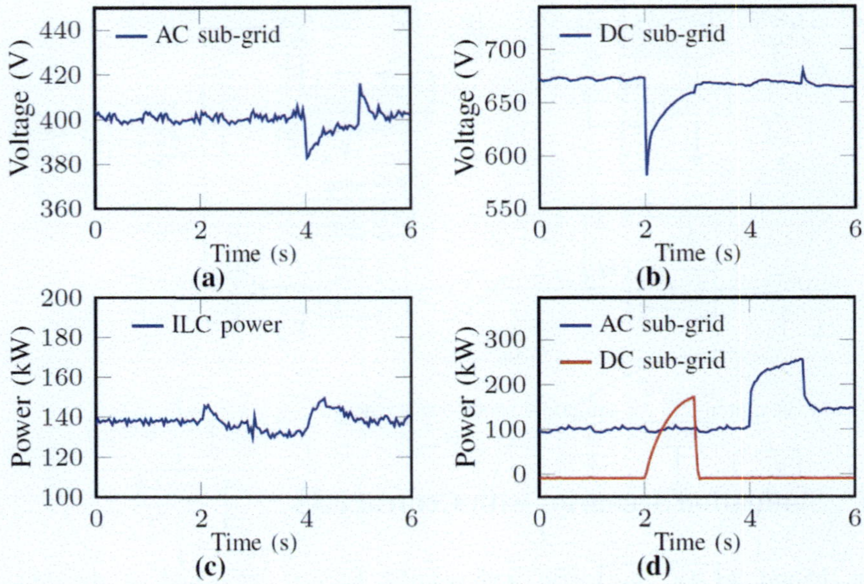

Fig. 24 Hybrid AC/DC microgrid parameters for case 1, **a** AC PCC voltage, **b** DC PCC voltage, **c** ILC power, **d** central ESS power

voltage and supplies the required power to restore the DC sub-grid PCC voltage. Meanwhile, AC sub-grid PCC voltage remains stable and the ILC power injection from AC sub-grid to DC sub-grid stays the same.

In addition to the DC bus load disturbances, a 300 kW load is suddenly plugged on the AC sub-grid at $t = 4$ s, and subsequently plugged off at $t = 5$ s. The central ESS attached to the AC sub-grid PCC detects the power shortage at the AC sub-grid by measuring the AC sub-grid PCC voltage/frequency and supplies the required power to restore the AC sub-grid PCC voltage/frequency to its reference value. The sudden load addition causes a deviation of AC sub-grid PCC voltage/frequency, and the ILC power dispatch is staying at the same value. The central ESS connected to the AC sub-grid momentarily supplies the additional power to the AC sub-grid to restore the AC sub-grid PCC voltage/frequency. Meanwhile, the ILC power dispatch from the AC to the DC sub-grid remains the same as prior to the load disturbance. In this way, the central ESS and the ILC are coordinating with each other and maintain hybrid AC/DC microgrid stable operation.

11.2 *Case VI: The ILC Power Flows from the DC to the AC Sub-Grid and a Single Line to Ground Fault Has Occurred at the AC Sub-Grid*

Different types of faults frequently occurred in real power networks. This sub-section tests the fault tolerance capability of the control architecture. To analyse this, a 200 kW load is connected (at $t = 2$ s) and disconnected (at $t = 4$) in DC sub-grid, together with a single line to ground fault (that occurs at $t = 3.0$ s and is cleared after 100 ms) at the AC sub-grid PCC. The AC sub-grid is running at overloading condition, and the DC sub-grid supports it by transferring 250 kW power through the ILC. Figure 25a–d show the AC sub-grid voltage, DC sub-grid voltage, ILC power transfer from the DC to the AC sub-grid, and the central ESS output power dispatch under these operating conditions. The AC sub-grid PCC voltage is significantly dropped due to the fault, and the central ESS connected to the AC sub-grid momentarily supplied the required amount of power to recover the voltage. The ILC power transfer and the DC sub-grid PCC voltage are also affected by the fault, but recovers without leading to an unstable condition. The results reveal that the control architecture can provide stable operation of the hybrid AC/DC microgrid under the fault at the AC sub-grid PCC.

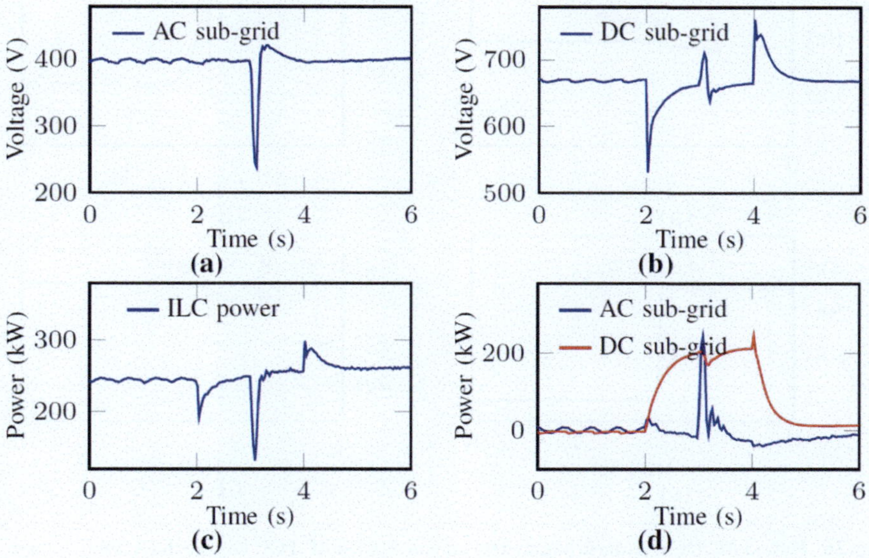

Fig. 25 Hybrid AC/DC microgrid parameters for case2, **a** AC PCC voltage, **b** DC PCC voltage, **c** ILC power, (d) ESS power

11.3 Case 3: Standalone Mode-Isolation of the AC Sub-Grid from the DC Sub-Grid

Two sub-grids of the hybrid AC/DC microgrid may need to operate in standalone mode due to the fault in ILC. This scenario analyses the ability of the central ESS to manage the power balance between two isolated sub-grids without the ILC. In this case, the AC sub-grid is disconnected from the AC sub-grid at t = 2.5 s. Before the standalone mode of operation, the DC sub-grid is transferring 250 kW via the ILC. The PCC voltage of two sub-grids, ILC power, and central ESS power are shown in Fig. 26a–d. When two sub-grids are operating at standalone mode, the excess power of the DC sub-grid is stored into the central ESS via the DC-DC converter, while the central ESS connected to the AC sub-grid is now supplying the power, which was supplied by the ILC prior to the isolation. Therefore, this control approach can ensure stable operation during the standalone mode. The installation of ESS requires more capital investment than the ILC, other than that, it will ensure hybrid AC/DC microgrid stable operation over a wide range of operating conditions, ILC failure, and disturbances.

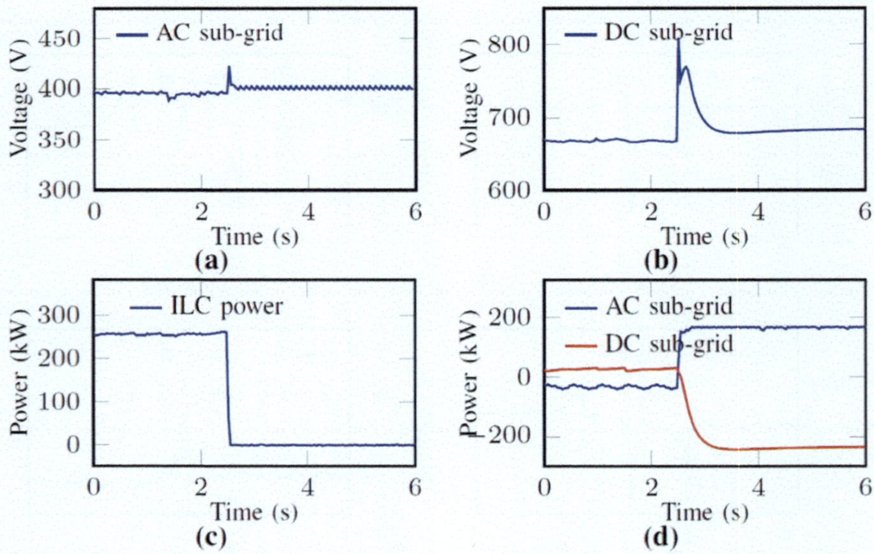

Fig. 26 Hybrid AC/DC microgrid parameters for case VIII, **a** AC PCC voltage, **b** DC PCC voltage, **c** ILC power, **d** ESS power

12 Conclusions

In this chapter the stability and control aspects of hybrid AC/DC microgrid are presented and dicussed. This study has revealed that dynamic loads, such as IMs cause voltage and frequency oscillations at the hybrid AC/DC microgrid due to electromechanical oscillations at the IM rotor. Also, the voltage and frequency variations are severe in-comparison to the conventional power system due to lack of inertia in the microgrid. The PEC interfaced ESS controllers can rapidly minimise the generation and load demand imbalance, and hence ensure stable operation of the hybrid AC/DC microgrid over a wide range of operating conditions. Therefore, a POD controller was implemented at the ESS, and it can provide additional damping power in proportional to the frequency oscillations in the AC sub-grid. The presented studies have demonstrated that the POD controller can effectively dampe the LFOs in the hybrid AC/DC microgrid, therefore ensure microgrid stability. Considering that conventional POD controllers can't adapt to the changes in the microgrid load composition, an ANFIS based POD controller is demonstrated to mitigate the oscillations in the microgrid with various combinations of IMs.

This chapter also presented a hybrid AC/DC microgrid structure with a central ESS rather than having two separate ESSs at each AC and the DC sub-grids. A coordinated control strategy between the ILC and the central ESS is demonstrated for the hybrid AC/DC microgrid architecture. The central ESS can facilitate the power transfer between the AC and DC sub-girds and eliminates the use of multiple ILCs when a large amount of power transfer is required. Furthermore, by implementing this new hybrid AC/DC microgrid structure and the coordinated control strategy, control complexity has reduced, and the system stability has further improved. The hybrid AC/DC microgrid architecture can also provide additional benefits by eliminating disadvantages of hybrid AC/DC microgrids with multiple ILCs, such as unequal SoC between multiple ESSs, circulating current issues, and over reliant on ILC for power balancing in conventional hybrid AC/DC microgrids.

References

1. Hatziargyriou N, Asano H, Iravani R, Marnay C (2007) Microgrids. IEEE Power Energy Mag 5(4):78–94
2. Justo JJ, Mwasilu F, Lee J, Jung J-W (2013) Ac-microgrids versus dc-microgrids with distributed energy resources: a review. Renew Sustain Energy Rev 24:387–405
3. Asmus P (2010) Microgrids, virtual power plants and our distributed energy future. Electr J 23(10):72–82
4. Tan Y, Muttaqi KM, Ciufo P, Meegahapola L (2017) Enhanced frequency response strategy for a PMSG-based wind energy conversion system using ultracapacitor in remote area power supply systems. IEEE Trans Ind Appl 53(1):549–558
5. Ahmed M, Vahidnia A, Meegahapola L, Datta M (2017) Small signal stability analysis of a hybrid ac, dc microgrid with static and dynamic loads. In: Universities power engineering conference (AUPEC), 2017 Australasian. IEEE, pp. 1–6

6. Pogaku N, Prodanovic M, Green TC (2007) Modeling, analysis and testing of autonomous operation of an inverter-based microgrid. IEEE Trans Power Electron 22(2):613–625
7. Dheer DK, Soni N, Doolla S (2016) Improvement of small signal stability margin and transient response in inverter-dominated microgrids. Sustain Energy Grids Netw 5:135–147
8. Administration UEI et al (2011) Annual energy outlook 2011: with projections to 2035. Government Printing Office
9. Kahrobaeian A, Mohamed YA-RI (2014) Analysis and mitigation of low-frequency instabilities in autonomous medium-voltage converter-based microgrids with dynamic loads. IEEE Trans Ind Electron 61(4):1643–1658
10. Saghafi H, Karshenas H, Bakhshai A (2015) Application of integrated series compensator in damping power oscillations in standalone microgrids. Can J Electr Comput Eng 38(1):2–9
11. Li Y, Li YW (2011) Power management of inverter interfaced autonomous microgrid based on virtual frequency-voltage frame. IEEE Trans Smart Grid 2(1):30–40
12. Majumder R, Ledwich G, Ghosh A, Chakrabarti S, Zare F (2010) Droop control of converter-interfaced microsources in rural distributed generation. IEEE Trans Power Deliv 25(4):2768–2778
13. Yao W, Chen M, Matas J, Guerrero JM, Qian Z-M (2011) Design and analysis of the droop control method for parallel inverters considering the impact of the complex impedance on the power sharing. IEEE Trans Ind Electron 58(2):576–588
14. Hamzeh M, Ghafouri M, Karimi H, Sheshyekani K, Guerrero JM (2016) Power oscillations damping in dc microgrids. IEEE Trans Energy Convers 31(3):970–980
15. Rashidirad N, Hamzeh M, Sheshyekani K, Afjei E (2017) An effective method for low-frequency oscillations damping in multibus dc microgrids. IEEE J Emerg Sel Top Circuits Syst 7(3):403–412
16. Mahdavyfakhr M, Rashidirad N, Hamzeh M, Sheshyekani K, Afjei E (2017) Stability improvement of dc grids involving a large number of parallel solar power optimizers: an active damping approach. Appl Energy 203:364–372
17. Rashidirad N, Hamzeh M, Sheshyekani K, Afjei E (2017) High-frequency oscillations and their leading causes in dc microgrids. IEEE Trans Energy Convers 32(4):1479–1491
18. Hua K, Vahidnia A, Mishra Y, Ledwich G (2016) Efficient probabilistic contingency analysis through a stability measure considering wind perturbation. IET Gen Trans Dist 10(4):897–905
19. Palizban O, Kauhaniemi K (2015) Distributed cooperative control of battery energy storage system in ac microgrid applications. J Energy Storage 3:43–51
20. Xu Y, Zhang W, Hug G, Kar S, Li Z (2015) Cooperative control of distributed energy storage systems in a microgrid. IEEE Trans Smart Grid 6(1):238–248
21. Kim J-Y, Jeon J-H, Kim S-K, Cho C, Park JH, Kim H-M, Nam K-Y (2010) Cooperative control strategy of energy storage system and microsources for stabilizing the microgrid during islanded operation. IEEE Trans Power Electron 25(12):3037–3048
22. Vorobev P, Huang P-H, Al Hosani M, Kirtley JL, Turitsyn K (2018) High-fidelity model order reduction for microgrids stability assessment. IEEE Trans Power Syst 33(1):874–887
23. Tabatabaee S, Karshenas HR, Bakhshai A, Jain P (2011) Investigation of droop characteristics and x/r ratio on small-signal stability of autonomous microgrid. In: Power electronics, drive systems and technologies conference (PEDSTC), 2011 2nd. IEEE, pp 223–228
24. He L, Li Y, Shuai Z, Guerrero JM, Cao Y, Wen M, Wang W, Shi J (2018) A flexible power control strategy for hybrid ac/dc zones of shipboard power system with distributed energy storages. IEEE Trans Ind Infor 14(12):5496–5508
25. Nejabatkhah F, Li YW (2014) Overview of power management strategies of hybrid ac/dc microgrid. IEEE Trans Power Electron 30(12):7072–7089
26. Popović D, Hiskens I, Hill D (1998) Stability analysis of induction motor networks. Int J Electr Power Energy Syst 20(7):475–487
27. Roy N, Pota H, Mahmud M, Hossain M (2014) Voltage control of emerging distribution systems with induction motor loads using robust LQG approach. Int Trans Electr Energy Syst 24(7):927–943

28. Hall M, Hodges D (1976) Experience with 500 kv subsynchronous resonance and resulting turbine generator shaft damage at mohave generating station, vol 76. IEEE Publication, pp 22–29
29. Group ISRW et al (1985) Terms, definitions and symbols for subsynchronous oscillations. IEEE Trans Power Appar Syst 104(6):1326–1334
30. Padiyar K (2012) Analysis of subsynchronous resonance in power systems. Springer Science & Business Media
31. Klein M, Rogers G, Kundur P (1991) A fundamental study of inter-area oscillations in power systems. IEEE Trans Power Syst 6(3):914–921
32. Dattaray P, Chakravorty D, Wall P, Yu J, Terzija V (2018) A novel control strategy for subsynchronous resonance mitigation using 11 kv VFD-based auxiliary power plant loads. IEEE Trans Power Deliv 33(2):728–740
33. Dattaray P, Wall P, Terzija V, Mohapatra P, Yu J (2017) Impact of location and composition of dynamic load on the severity of ssr in meshed power systems. In: PowerTech, 2017 IEEE Manchester. IEEE, pp 1–6
34. Dattaray P, Wall P, Terzija V, Chakravorty D, Mohapatra P, Yu J (2016) Impact of load dynamics on torsional interactions. In: Power systems computation conference (PSCC). IEEE, pp 1–7
35. Purushothaman S, De Leon F (2011) Eliminating subsynchronous oscillations with an induction machine damping unit (IMDU). IEEE Trans Power Syst 26(1):225–232
36. Kundur P, Balu NJ, Lauby MG (1994) *Power system stability and control*, vol 7. McGraw-hill New York
37. Saadat H (1999) Power system analysis. McGraw-Hill
38. Russell A, Pickup I (1982) An analysis of the induction motor. part 2: damping and synchronising torque production. In: IEE proceedings B-electric power applications, vol 129, no 5. IET, pp 238–242
39. Holmes D, Lipo T, McGrath B, Kong W (2009) Optimized design of stationary frame three phase ac current regulators. IEEE Trans Power Electron 24(11):2417–2426
40. Hadjikypris M (2015) Supervisory control scheme for facts and HVDC based damping of inter-area power oscillations in hybrid ac-dc power systems," PhD dissertation, The University of Manchester
41. Yee SK (2005) Coordinated tuning of power system damping controllers for robust stabilization of the system. University of Manchester
42. Liu X, Wang P, Loh PC (2011) A hybrid ac/dc microgrid and its coordination control. IEEE Trans Smart Grid 2(2):278–286
43. Meegahapola L, Nutkani IU, McGrath B, Holmes DG (2017) Fault ride-through capability of hybrid ac/dc microgrids during ac and dc network faults. In: 2017 IEEE energy conversion congress and exposition (ECCE), Oct 2017, pp 44–51
44. Saw L, Somasundaram K, Ye Y, Tay A (2014) Electro-thermal analysis of lithium iron phosphate battery for electric vehicles. J Power Sources 249:231–238
45. Meegahapola L, Robinson D (2016) Smart power systems and renewable energy system integration. Springer, Cham. Ch. Dynamic modelling, simulation and control of a commercial building microgrid
46. Reznik A, Simões MG, Al-Durra A, Muyeen S (2014) *lcl* filter design and performance analysis for grid-interconnected systems. IEEE Trans Ind Appl 50(2):1225–1232
47. Haque ME, Negnevitsky M, Muttaqi KM (2008) A novel control strategy for a variable speed wind turbine with a permanent magnet synchronous generator. In: Industry applications society annual meeting (2008) IAS'08. IEEE, pp 1–8
48. Sun Y, Hou X, Yang J, Han H, Su M, Guerrero JM (2017) New perspectives on droop control in ac microgrid. IEEE Trans Ind Electron 64(7):5741–5745
49. Espi JM, Castello J, García-Gil R, Garcera G, Figueres E (2011) An adaptive robust predictive current control for three-phase grid-connected inverters. IEEE Trans Ind Electron 58(8):3537–3546
50. Li H, Shi K, McLaren P (2005) Neural-network-based sensorless maximum wind energy capture with compensated power coefficient. IEEE Trans Ind Appl 41(6):1548–1556

51. Jang J-S (1993) ANFIS: adaptive-network-based fuzzy inference system. IEEE Trans Syst Man Cyb 23(3):665–685
52. Ahmed M, Meegahapola L, Vahidnia A, Datta M (2019) Analysis and mitigation of low-frequency oscillations in hybrid ac/dc microgrids with dynamic loads. IET Gen Trans Dist 13(9):1477–1488
53. Ahmed M, Vahidnia A, Datta M, Meegahapola L (2020) An adaptive power oscillation damping controller for a hybrid ac/dc microgrid. IEEE Access 8:69 482–69 495
54. Price WW, Wirgau KA, Murdoch A, Mitsche JV, Vaahedi E, El-Kady M (1988) Load modeling for power flow and transient stability computer studies. IEEE Trans Power Syst 3(1):180–187
55. Jang J-SR, Sun C-T, Mizutani E (1997) Neuro-fuzzy and soft computing-a computational approach to learning and machine intelligence [book review]. IEEE Trans Autom Cont 42(10):1482–1484
56. Jain AK, Mao J, Mohiuddin KM (1996) Artificial neural networks: a tutorial. Computer 29(3):31–44
57. Taher A et al (Feb 2010) Adaptive neuro-fuzzy systems. In: Fuzzy systems. Intech
58. Loh PC, Li D, Chai YK, Blaabjerg F (2012) Autonomous operation of hybrid microgrid with ac and dc subgrids. IEEE Trans Power Electron 28(5):2214–2223
59. Peyghami S, Mokhtari H, Blaabjerg F (2017) Autonomous operation of a hybrid ac/dc microgrid with multiple interlinking converters. IEEE Trans Smart Grid 9(6):6480–6488
60. Xiao H, Luo A, Shuai Z, Jin G, Huang Y (2015) An improved control method for multiple bidirectional power converters in hybrid ac/dc microgrid. IEEE Trans Smart Grid 7(1):340–347
61. Ahmed M, Meegahapola L, Datta M, Vahidnia A (2021) A novel hybrid ac/dc microgrid architecture with a central energy storage system. IEEE Trans Power Deliv 37(3):2060–2070
62. Zhou JZ, Ding H, Fan S, Zhang Y, Gole AM (2014) Impact of short-circuit ratio and phase-locked-loop parameters on the small-signal behavior of a VSC-HVDC converter. IEEE Trans Power Deliv 29(5):2287–2296
63. Loh PC, Li D, Chai YK, Blaabjerg F (2013) Autonomous control of interlinking converter with energy storage in hybrid ac-dc microgrid. IEEE Trans Ind Appl 49(3):1374–1382
64. Rashid MH (2017) Power electronics. Butterworth-Heinemann
65. Lu X, Sun K, Guerrero JM, Vasquez JC, Huang L (2013) State-of-charge balance using adaptive droop control for distributed energy storage systems in dc microgrid applications. IEEE Trans Ind Electron 61(6):2804–2815
66. Semënov D, Mirzaeva G, Townsend C, Goodwin G (2017) A battery storage control scheme for ac microgrids. In: Proceedings of the IEEE 20th international conference on electrical machines and systems (ICEMS), Sydney, NSW, Australia, Oct 2017, pp 1–6

Quantifying Transient Dynamics for Microgrid's Inverter-Based Resources

Touria El Mezyani, Farzad Ferdowsi, Mohammad Seyedi, and Chris S. Edrington

Abstract Power converters are becoming increasingly important in power systems due to the high penetration of inverter-based resources. Converters are used to implement control strategies and to interface different subsystems (e.g., distributed energy resources, loads, batteries, etc.) that produce, store, or consume electrical energy. They are suitable for microgrids because they provide fast operation and the ability to implement control algorithms at the primary and secondary levels. This chapter studies the complexity associated with the emergence of new complex dynamics in power systems that arise with the high penetration of inverter-based resources. Complexity manifests as bifurcation, oscillation, instability, and chaos, which traditional Models or data-based prediction, such as machine learning, cannot predict. It is mainly due to power electronics nonlinearity, coupling between these components, different feedbacks such as control, and human and environmental feedbacks. Complexity quantification is crucial to modern microgrids with high penetration of inverter-based resources as it allows them to indicate the system's vulnerability, instability, and hidden problems. In this chapter, two complexity metrics are investigated and computed using the permutation entropy approach and ordinal pattern technique, and approximate entropy. Various case studies are conducted on simulated data from noise-coupled buck converters, two parallel-connected buck converters, and solid-state transformers integrated into a microgrid.

Keywords Power converters · Behavioral complexity · Shannon entropy · Approximate entropy · Ordinal patterns · IBRs

T. E. Mezyani (✉)
School of Engineering, University of North Florida, Jacksonville, USA
e-mail: t.elmezyani@unf.edu

F. Ferdowsi · M. Seyedi
College of Engineering, University of Louisiana at Lafayette, Lafayette, USA
e-mail: ferdowsi@louisiana.edu

C. S. Edrington
College of Engineering, Clemson University, Clemson, USA
e-mail: cedring@clemson.edu

F. Shahnia and J. M. Guerrero (eds.), *Microgrids and Virtual Power Plants*, Power Systems, https://doi.org/10.1007/978-981-97-6623-9_7

1 Introduction

The growing number of smaller distributed energy resources (DERs) and power converters, namely inverter-based resources (IBRs), will impact the overall system's behavior. There have been growing concerns as the number of power electronic devices used in power systems, particularly inverters, has increased [1]. These concerns are related to new emergent IBR behaviors, referred to as behavioral complexity [2–4], manifesting as oscillations, bifurcations, instability, and chaos. If these new emergent states are not detected and identified correctly, the system will be pushed far from the equilibrium state to the critical points [5, 6], known as self-organized criticality [6].

Due to the increasing penetration of IBRs into modern microgrids, power systems will exhibit more of these emergent, unpredictable dynamics, which may hinder achieving adequate situation awareness needed to make the right decisions and respond effectively to events. Situation awareness relies on accurate prediction and estimation of the system state variables, performed via physics-based models or data-based prediction, such as machine learning. However, both prediction approaches rely on correlations of the state variables, periodicity, or patterns. Therefore, both approaches may fail to predict the system states when they exhibit emergent dynamics that are very complex and cannot replicated by the physics-based model or machine learning tools. In addition, complexity analysis of the power system reveals the operating characteristics of the system, which indicate transient changes in system status. As a result, the variation in complexity can provide information about the system's behavior and its dynamical changes.

Complexity and entropy provide diverse viewpoints on the system's evolution, which examine the interactions between interconnected components and may be able to detect system instability, vulnerability, and hidden problems early on, enabling proactive solutions and a more dependable grid. Several mechanisms have been investigated in nonlinear systems by which a process can generate complex behavior, such as deterministic chaos.

Therefore, the nonlinearity and interconnectivity of IBRs and feedback mechanisms make the systems function close to their operating limits (i.e., near critical points) [6–9]. Hence, it is important to track the emergent behaviors when they occur to enable the system to take appropriate mitigation solutions. Several metrics, such as Shannon entropy [10], Kolmogorov-Sinai entropy [11], etc., were the subject of intensive studies regarding complexity. However, these entropy metrics have been mainly considered to evaluate the randomness or the information uncertainty. In contrast to entropy metrics, the complexity metric should take values between two extreme conditions: complete knowledge about the system dynamic (determinism) and total ignorance (randomness). In the first case of complete knowledge about the system, the exact behavior of the system can be predicted by a simple analytical model, while in the second case of total ignorance, a simple statistical description of the system can also be easily used for prediction. However, between these extreme cases of knowledge and ignorance about the system, describing it using either deterministic or stochastic tools is challenging as it encompasses a mixture of deterministic and random behaviors.

Measuring a system's behavioral complexity by observing time series was studied in various research fields, such as physics, biology, social sciences, and economics [12–16]. The complexity seen in the time series can indicate that the system is shifting to a quasi-state in the form of oscillation and bifurcation. This chapter reviews two behavioral complexity metrics using ordinal pattern and permutation entropy (OPPEn) [17, 18] and approximate entropy (ApEn) [19] for quantifying the complexity in the large spectrum of a possible power system behaviors. Both APPE and ApEn are suitable for identifying behavioral complexity in real-time, where OPPE identifies hidden patterns and unobserved states in time series measurements, and ApEn identifies irregularities. Both techniques are beneficial as they don't require prior knowledge of the system and are easy to implement for real-time operation. In addition, they are robust to the noises and easily adaptable to different systems' dynamics [17, 18].

2 Entropy and Complexity Measures

Nonlinearity and feedback mechanisms, such as controllers, human, and environmental feedbacks, are the main drivers of IBRs' structural complexity. This complexity is mainly seen as a cohabitation of order (regularities) and disorder (randomness) in the time series data. In general, IBRs, particularly power electronics, can exhibit various dynamics ranging from regular periodic (order) to chaotic behaviors (disorder). In between these behaviors there exist a plethora of periodic dynamics for example such as saddle node bifurcations, Hopf bifurcations, umbilic bifurcations, period doubling bifurcations, reconnection bifurcations, tangent bifurcation, and cusp bifurcations [21, 22].

For example, a population growth evolution can exhibit various behaviors depending on a feedback parameter. The evolution of this system, called the logistic map, is given by the following model:

$$X_{n+1} = kX_n(1 - X_n) \tag{1}$$

With k is the feedback parameter that may take values on the interval [0.4], by varying the value of the parameter k ($k \in [14]$), the system in (1) exhibits different types of dynamics as shown in Fig. 1a and. in bifurcation diagram Fig. 1b. In a given time window, the time series of the variable Xn show regularities and symmetries, which correspond to a deterministic behavior (called order), and randomness, which corresponds to a random behavior (called disorder). Each of these behaviors exhibits a certain degree of complexity and instability.

In [15], the authors have defined the structural complexity of a process as a degree of organization presented in the system's behavior as patterns, regularity,

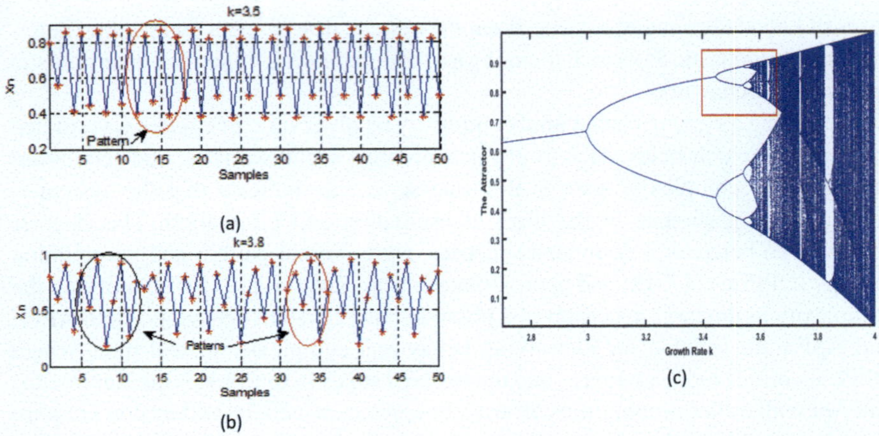

Fig. 1 Example nonlinear complex systems—**a** Logistic map showing period doubling for k = 3.5, **b** and bifurcation for k = 3.8, **c** changes of dynamics for k ∈ [04]

and symmetry. Several researchers in computational mechanics, computer programming, and signal and image processing have been interested in the pattern's finding and identification. The pattern evaluation looks at the system dynamics' repetitions and relationships over time and space. IBR systems may exhibit wide behavior ranges, including periodic, bifurcation, and chaotic dynamics. Structural complexity measures should be evaluated in these different dynamics, which tells us how complex these dynamics are.

Researchers have widely investigated several metrics for the disorder [10, 11], which are commonly expressed using entropy such as Bozman-Gibbs-Shannon entropy [10]:

$$H[X_i] = -\sum_{i=1}^{N} p_i \log_2 p_i \tag{2}$$

With p_i is the probability of the state X_i. The normalized entropy can be expressed as follows:

$$H_N = H/H_{\max}$$

With $H_{\max}$ is the maximum entropy achieved for the equiprobable distribution that is $p_i = 1/N$.

Thus, one we can define the complexity of the IBR behavior that reflects the disorder and the degree of unpredictability of the IBR response [14], is defined as:

$$\Gamma_{\alpha\beta}(Y) = \left(\frac{H(Y)}{H_{\max}(Y)}\right)^{\alpha}\left(1 - \frac{H(Y)}{H_{\max}(Y)}\right)^{\beta} \tag{3}$$

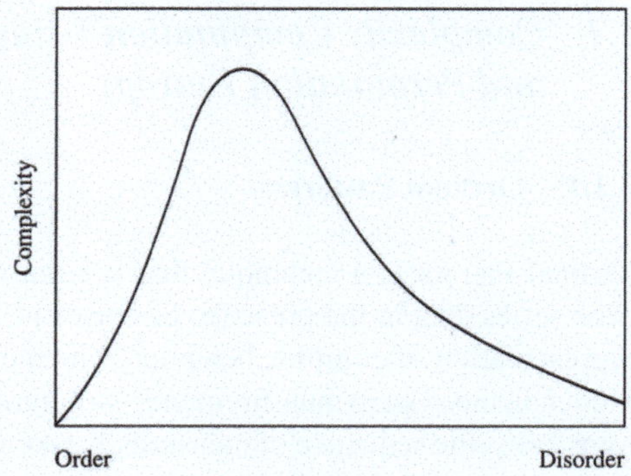

Fig. 2 $\Gamma_{\alpha\beta}$ versus entropy evolution [14]

With,

- $\alpha > 0$ and $\beta > 0$ are constant parameters that can be evaluated from the experiments.
- $Y = [y_1, \ldots, y_n]$: is the time series measurements that correspond to IBR's current, voltage, etc..
- $\psi = (1 - \frac{H}{H_{\max}})$: is the distance between normalized Shannon entropy and its maximum realization which is equal to 1.

$\Gamma_{\alpha\beta}$ is the function of disorder (Fig. 2) that $\Gamma_{\alpha\beta}$ takes values between two main entropy conditions: minimal at the steady state and maximal in the case of total ignorance where the system response is a random In contrast, the entropy, as shown in Fig. 2, is maximal when the states are of equal probability, that is the case of stochastic systems such is the case of renewable energy resources. Thus, it is trivial that structural complexity will be minimal in these two boundaries of maximum and minimum entropy because we can predict the system behavior either based on deterministic model or probabilistic function respectively. Therefore, in between the two specials values of maximum and minimum entropy, there exists a wide range of possible degrees of correlation in the process that can be captured by a structural complexity metric $\Gamma_{\alpha\beta}$.

3 Quantification of Complexity

The computation of complexity from the time series measurements is not a straightforward process. Several techniques have been used in [17–20] to calculate the entropy and complexity. However, most of these procedures are not suitable for real-time operation. This section examines two real-time complexity quantification approaches of OPPEn and ApEn [12, 13].

3.1 Complexity Computation Using Ordinal Pattern and Permutation Entropy

3.1.1 Ordinal Pattern

Ordinal Pattern is a technique that is suitable for capturing hidden patterns of the time series data in the presence of noises [4, 15]. It uses symbolic dynamics for the representation of complex behavior, which enables the identification of the patterns hidden in the system measurements' structure [12]. The symbolic dynamic technique transforms the sequence of measures Y_t taken on time windows [t-D, t] of length D to a sequence of symbols S_t. Given a one dimension time series $(Y_{t-n}, ...Y_{t-2\tau}, Y_{t-1}, Y_t)$ with n the length of time series, the set of symbols S_t are built in a given length $D \in IN$ and a lag $\tau \in IN$ such as:

$$S_t \rightarrow \left(Y_{t-D\tau}, Y_{t-(D-1)\tau}, ...Y_{t-2\tau}, Y_{t-\tau}, Y_t\right).$$

The objective of transforming the time series, which reflects the system dynamic, to a symbolic dynamic is to extract patterns and evaluate their probability. In Bandt and Pompe [17], the authors suggested considering D and τ such that $3 \leq D \leq 7$ and $\tau = 1$. However, it has been proven that the analysis with lagged data points, i.e., $\tau \geq 2$, can be useful for reaching a better comprehension of the underlying dynamics.

Different coding techniques are feasible and can be directly applied to the data sets. This chapter applies ordinal patterns-based technique, which consists of comparing the neighboring values in data series. This method aims to transform the data sequence of the sequence S_t into an ordinal pattern represented by a symbol $\pi^i_D withi \in [1, D!]$. The design of ordinal patterns π^i_D is realized by the transformation of data series to symbols that captures the changes of data series. Different symbolic representations of ordinal patterns can be proposed as sequence of rank numbers as shown in Fig. 3a. For example, we consider a symbol composed of numbers 1, 2 and 3, with 1 denotes the smallest, 2 the intermediate, and 3threethe largest of the values. For each sequence of series of measurement $\{Y_{t-2}, Y_{t-1}, Y_t\}$ taken over a time window of length two with $Y_{t-2} < Y_{t-1} < Y_t$, one can associate the pattern <3,2,1> of dimension D = 3, as depicted in Fig. 3b. The number of possible ordinal patterns is equal to D! = $(3 \times 2 \times 1)$.

3.1.2 Permutation Entropy

In the case of two elements being equal, by convention we take the first element as the smaller element. This methodology allows pattern recognition through topology of the symbolic sequences. Numerical computation of complexity can be computed from the probability related to these symbols. By counting the number of times each ordinal pattern π^i_D appears in the symbolic series divided by the total number of ordinal patterns $3 \times 2 \times 1$, one can compute a probability distribution function. For

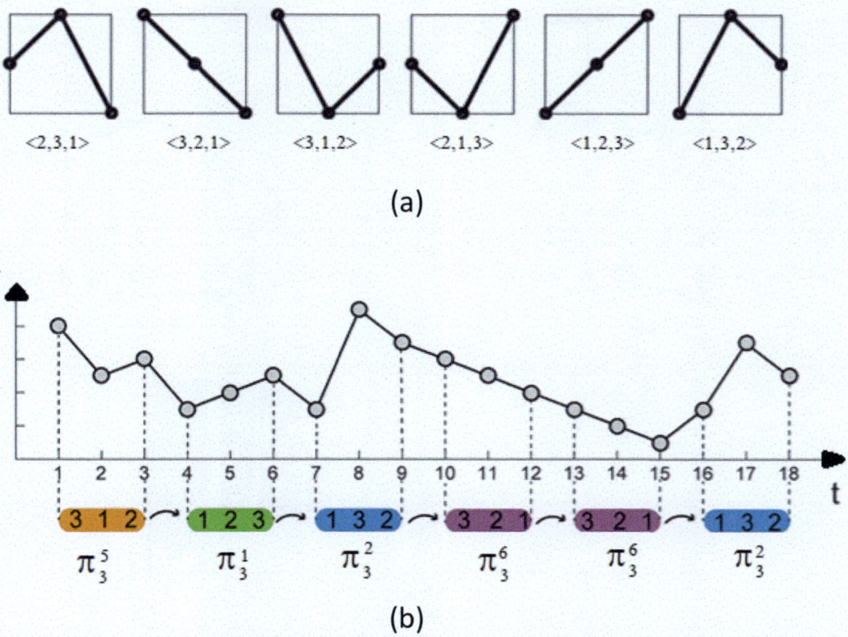

Fig. 3 **a** Symbolic illustration of ordinal patterns of order $D = 3$ *with no overlap*, **b** An example of transformation from a time series measurements to ordinal patterns sequence

each pattern π_D^i (i = 1, ..., D!) we determine its probability of the realization as follow:

$$p_i(\pi_D^i) = \frac{card\left(\pi_D^i\right)}{L}, with i \in [1, D!] \tag{4}$$

where *card* represents the cardinality of π_D^i. For a time series of length M, the total number of patterns generated is $L = M - (D - 1)\tau$.

From (4) the entropy, called permutation entropy, is given by:

$$H(\pi_D) = -\sum_{i=1}^{D!} p_i\left(\pi_D^i\right) \log p_i\left(\pi_D^i\right) \tag{5}$$

And PE per symbol is given as follows:

$$h_D = \frac{H(\pi_D)}{D - 1} \tag{6}$$

Using the entropy formulation in (5), $\Gamma_{\alpha\beta}$ can be computed using the complexity metric in (3). The simplest case of computing maximum entropy is to consider all the patterns π_D^i (i = 1,…,D!) with equal probability. However, knowing the system

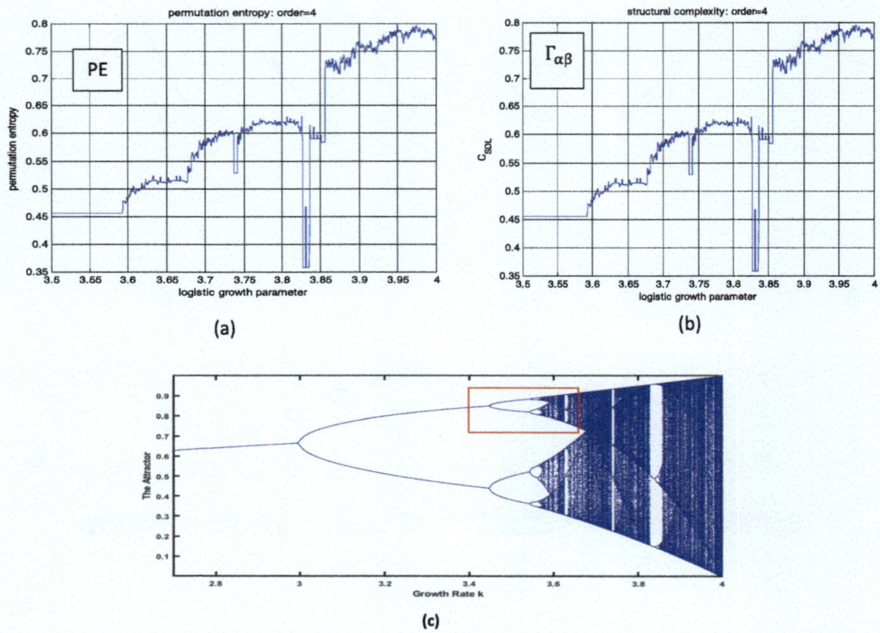

Fig. 4 **a** Permutation entropy, **b** Behavioral complexity $\Gamma_{\alpha\beta}$, **c** Bifurcation diagram of Logistic map

configuration, limitations and control specifications allows to identify the set of forbidden symbols (unreached state), which are patterns that do not exist in the data series. Thus, the maximum entropy accounts only for all possible allowed symbols.

To illustrate the OPPEn approach, we consider Logistic map equation in (1). The calculation is performed in real-time using sliding window of length T = 10. Figure 4 shows OPPEn (in Fig. 4a and complexity $\Gamma_{\alpha\beta}$ (in Fig. 4b evolution of logistic map with the feedback parameter k vary from [3.5 4]. $\Gamma_{\alpha\beta}$ and PE shows that they have a very similar evolution. For different value of k, as depicted in Fig. 4(b) that $\Gamma_{\alpha\beta}$ decreases when k is low increases when k is high (e.g., close to 4).

3.1.3 Robustness Against Noise in Deterministic Time Series

There are two basic approaches to represent the noise effect on the system trajectory.

- Dynamical noise is due to errors in determining the initial state and propagates with the dynamics. Let's consider a dynamical system described in (7),

$$\frac{dx(t)}{dt} = f(x(t), u(t), \theta_i(t))x(t_0) = x_0 + \eta_0 \tag{7}$$

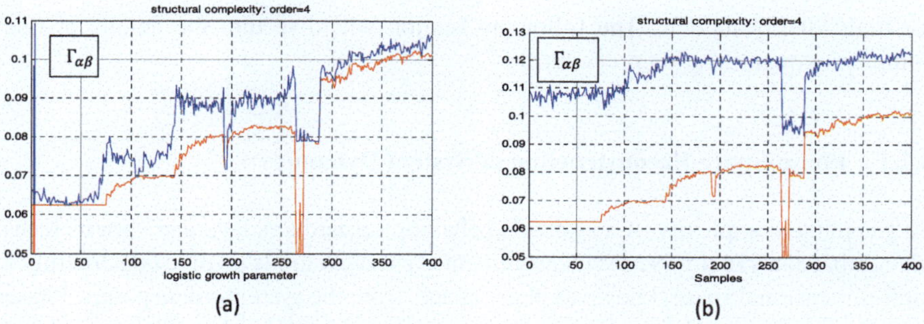

Fig. 5 **a** $\Gamma_{\alpha\beta}$ (lower line) and $\Gamma_{\alpha\beta}$ (upper line) with Gaussian observational noise with a standard deviations $s = 0.00025$. **b** $\Gamma_{\alpha\beta L}$ (lower line) and $\Gamma_{\alpha\beta}$ (upper line) with additive Gaussian noise at standard deviations $s = 0{:}05$

where, $x(t) \in IR^n$ is the state variables,$u(t) \in IR^{\varsigma}$ is the input vector,θ is the dynamical noises, and x_0 is the initial condition,

- Observational (or additive) noise adds a random fluctuation to the true value of measurements Y:

$$Y(t) = g(x(t), u(t), \theta) + \eta(t) \quad (8)$$

with $Y(t) \in IR^r$ is the measurement vector of dimension r ($Y(t) = [Y_1, Y_2, ..., Y_r]^t$ with r denoting the total number of sensors), and η denotes the noise vector that affect the constraints in (8).

Practically, the measurements are affected by dynamical and observational noises, which may lead to false symbols and patterns. However, permutation entropies have a practically important invariance property against the noises. It has been proven in [13] that: (i) Forbidden patterns are unconditionally robust against dynamical noise, (ii) Visible patterns are conditionally robust against dynamical noise, (iii) Visible and forbidden patterns are conditionally robust against observational noise. To evaluate the effect of noise on $\Gamma_{\alpha\beta}$, Gaussian noise was added to the deterministic time series obtained from logistic map. For small level of noises, one can notice that noises cause only a small increase of $\Gamma_{\alpha\beta}$ (in Fig. 5a). In practice, the data series is usually filtered prior to their process. Even though the data is filtered, a small level of noise cannot be removed from the data series. However, the remaining noises in data series do not appear to cause the problem in the computation of $\Gamma_{\alpha\beta}$.

3.2 *Complexity Measure Using Approximate Entropy*

ApEn is another technique suitable to quantify complexity in real-time from observable measurements. It measures allows to identify irregularities of time series data obtained from direct measurements such as currents and voltages, avoiding the need for high fidelity models. The following section will dive into the details of this complexity quantification procedure.

3.2.1 Phase-Space Reconstruction of System Dynamics

To gauge the complexity of a system's behavior, researchers leverage various techniques that analyze time series measurements. These methods involve constructing a high-dimensional space, known as phase space, from the system's data points. Phase space is a mathematical concept built using the system's key dynamical variables. A sequence of measurements is transformed into a series of vectors with a specific dimensionality (m). These vectors incorporate a time lag, ensuring they capture not only the current state but also the influence of past states on the system's behavior. Considering the time lag τ, each vector is defined as follows:

$$\hat{x}_n = \langle u_n, u_{n+\tau}, u_{n+2\tau}, \ldots, u_{n+(m-1)\tau} \rangle \tag{9}$$

For a two-dimensional phase-space, the vectors are formed as follows:

$$\begin{cases} \hat{x}_1 = < u_1, u_{1+\tau} > \\ \hat{x}_2 = < u_2, u_{2+\tau} > \\ \vdots \\ \hat{x}_{N-\tau} = < u_{N-\tau}, u_N > \end{cases} \tag{10}$$

For a three-dimensional phase-space, the vectors are reformed as follows:

$$\begin{cases} \hat{x}_1 = < u_1, u_{1+\tau}, u_{1+2\tau} > \\ \hat{x}_2 = < u_2, u_{2+\tau}, u_{2+2\tau} > \\ \vdots \\ \hat{x}_{N-2\tau} = < u_{N-2\tau}, u_{N-\tau}, u_N > \end{cases} \tag{11}$$

To illustrate the concept further, Fig. 6 depicts the reconstruction of phase space for two cycles of a normalized sinusoidal signal. The sampling time used is 2 (ms), resulting in 17 data points extracted from these two cycles. As shown in the Fig. 6, this data sequence (denoted as $\langle u_1, u_2, \ldots, u_{17} \rangle$) is transformed into a series of vectors for embedding in the phase space.

3.2.2 Real-Time Estimation of Embedded Phase-Space Parameters

Researchers have explored various methods for selecting optimal parameters in phase-space reconstruction, with a particular focus on the embedding dimension. This study leverages the false nearest neighbor technique to identify the most appropriate

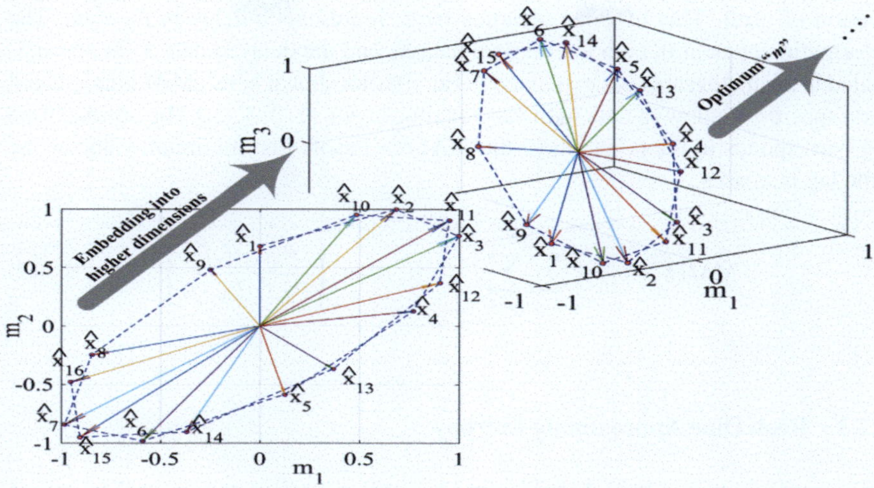

Fig. 6 Construction of the phase-space for a normalized sinusoidal time series in "*m*" dimension

embedding dimension. The challenge lies in ensuring that data points that appear close in a low-dimensional space are indeed true neighbors in the higher-dimensional phase space. The false nearest neighbor method addresses this by detecting "false neighbors" based on how distances between neighboring points change as the embedding dimension increases. This approach prioritizes finding the minimum embedding dimension needed for an accurate representation, as higher dimensions become computationally expensive for complexity analysis. The computation of the optimum number of dimensions is as follows:

$$\hat{x}_n{}^{(i)} = \langle u_n{}^{(i)}, u_{n+\tau}{}^{(i)}, u_{n+2\tau}{}^{(i)}, \ldots, u_{n+(m-1)\tau}{}^{i} \rangle \tag{12}$$

$$D = \sqrt{\begin{array}{c}\left(u_n{}^{(i)} - u_n\right)^2 + \left(u_{n+\tau}{}^{(i)} - u_{n+\tau}\right)^2 + \ldots \\ +\left(u_{n+(m-1)\tau}{}^{(i)} - u_{n+(m-1)\tau}\right)^2\end{array}} \tag{13}$$

$$D^{new} = \sqrt{D^2 + \left(u_{n+(m)\tau}{}^{(i)} - u_{n+(m)\tau}\right)^2} \tag{14}$$

$$D^{new} - D \leq T \tag{15}$$

$$\sqrt{\frac{D_{new}^2 - D^2}{D^2}} \leq T' \tag{16}$$

To identify the most suitable time lag τ for phase-space reconstruction, this research employs the Mutual Average Information (MAI) technique. MAI quantifies the information shared between the original time series data and a time-shifted version of itself. This time lag, denoted by τ, represents a delay in the data. The information content in both the original data X_1 and the delayed data X_2 is approximated using histograms. As explained in Eq. (9) (not shown here), MAI is calculated for a specific window of data with varying time lags τ $(1, 2, 3, \ldots)$.The optimal time lag corresponds to the point where the MAI reaches its first minimum value as the time lag increases.

$$MAI(X_1, X_2) = \sum_{i=1}^{N_{X_1}} \sum_{j=1}^{N_{X_2}} P_{X_1X_2}(i,j)\log_2\left\{\frac{P_{X_1X_2}(i,j)}{P_{X_1}(i)P_{X_2}(j)}\right\} \tag{17}$$

3.2.3 Real-Time Approximate Entropy

This section explores how ApEn is used to compute complexity in real-time time series when they occur. In contrast with OPPEn that allow to identify patterns in time series measurements, ApEn is a technique that enable to identify irregularity in the time series.

measures the dynamic variation within the data by analysing how close data points remain in higher dimensional spaces. Before calculating ApEn, several parameters need to be chosen, including the embedding dimension (m), time lag τ, and a similarity threshold (r) that defines how close data points (vectors) should be considered similar in the phase space. This research focuses on a real-time implementation, where a window of data is used with a size determined by the data acquisition rate. The similarity threshold (r) is typically set as a percentage of the standard deviation of the data. The method then calculates the number of points within a specific distance (r) of each point in the window. Finally, ApEn is computed based on the ratio of similar points to the total number of data points, considering both the current dimension and a higher dimension.

$$T_i^m(r) = \left.\frac{\left(N_{x_j}\middle| d\left(\underset{i}{\widehat{x}}, \underset{j}{\widehat{x}}\right) \leq r\right)}{ws - (m-1)\tau}\right|_{\substack{1\leq \forall i,j \leq ws-(m-1) \\ j\neq i \\ d\left(\underset{i}{\widehat{x}},\underset{j}{\widehat{x}}\right)=\max(|\Delta(u_i,u_j)|)}} \tag{18}$$

$$Q^m(r, ws, \tau) = \frac{1}{ws - (m-1)\tau} \sum_{i=1}^{ws-(m-1)\tau} \ln T_i^m(r) \tag{19}$$

$$ApEn(m, r, \tau, ws) = Q^m(r, ws, \tau) - Q^{m+1}(r, ws, \tau) \tag{20}$$

Figure 7 illustrates the regular entropy and ApEn measures of Logistic map in (1) with an additive noises. It shows that ApEn. The Entropy and ApEn measures were

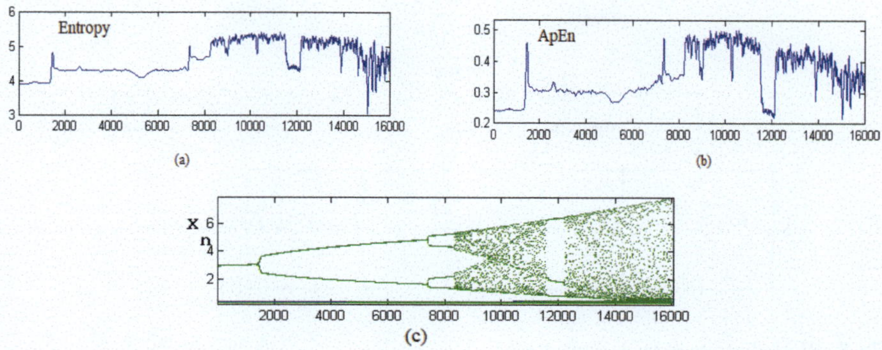

Fig. 7 Real-time entropy and ApEn measures of Logistic map with Gaussian observational noise with a standard deviations $s = 0.00025$. **a** Regular entropy, **b** ApEn, **c** Bifurcation diagram of logistic map

performed on 1600 data with variation of feedback parameter $k \in [1,4]$. Similarly, to OPPEn, the ApEn showed robustness to the added nose than regular entropy as shown in Figure (a) and Figure (b).

4 Use Cases

To examine the evolution of behavioral complexity, this chapter studies three power converters integrated in microgrids: (i) a single buck converter with a control feedback, (ii) two coupled buck converters with controllers, and (iii) a three node microgrid equipped with three Solid State Transformers (SSTs). The Buck converter is considered in simulation to illustrate the dynamic that power electronics systems can exhibit due control feedback and perturbations. Two cascade buck converters were investigated to illustrate the effect of interaction between these power electronic devices. Lastly SSTs were investigated to show the effect of load fluctuation on the system behavior. These use cases serve as a suitable representation of the broader category of Power Electronic-based Components and particularly IBRs within the microgrid environment.

4.1 Complexity Computation Using OPPEn Applied to Power Electronics Systems

4.1.1 Case of One Power Converter

OPPEn is used to analyze the structural complexity evolution in a DC-DC buck converter in continuous conduction mode controlled by a pulse-width modulator

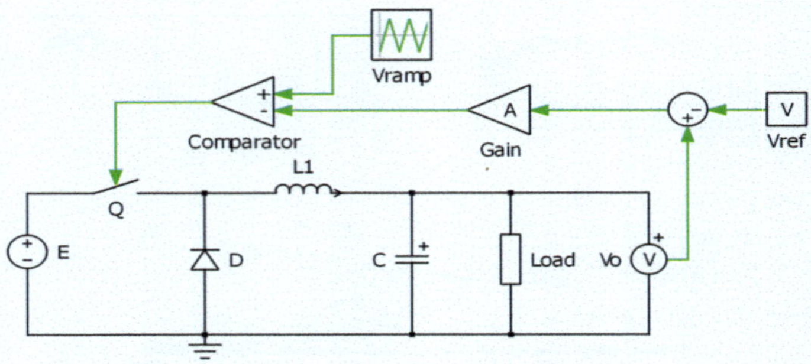

Fig. 8 Block diagram of buck converter

connected to a PV source, as depicted in Fig. 8. The model uses negative feedback α with voltage-mode control to automatically correct the duty cycle of the buck converter. In voltage-mode control, the output voltage is compared with a reference voltage to generate a control signal which drives the pulse-width modulator by using some typical feedback compensation setup. The duty ratio is continuously controlled by the feedback circuit in order to maintain the desired output voltage with good accuracy, regardless of disturbances in the input voltage or load current.

The output voltage $v_0(t)$ is measured and compared with a reference voltage V_{ref} that is coupled to a spurious signal. This forces the output voltage to follow the reference voltage regardless of any disturbances or component variations. The spurious signal is introduced to illustrate input noise, or poor construction of the circuit, within the feedback control system. If the frequency of this spurious signal is different than the switching frequency, bifurcations occur in the converter. The system parameters are presented in Table 1. The bifurcation parameter used is γ, which is defined as γ = vs/Vref, where vs is the amplitude of the noise signal, and V_{ref} is the reference voltage.

The behavior being analyzed is the occurrence of intermittent bifurcations, and chaotic dynamics in the output of a buck converter. Figure 9a–d (in upper panels) show the dynamic evolution of the current output from the buck converter for different feedback disturbances γ value, with $\gamma = 0.0019$, $\gamma = 0.0044$, $\gamma = 0.005$, and $\gamma = 0.007$ in (a), (b), (c) and (d), respectively. When the value of the spurious signal is relatively low, the phenomena of intermittent sub-harmonics occur, shown as period-4 sub-harmonics (in Fig. 9a, b). As the amplitude increases, the buck

Table 1 Buck parameter

V_{in}	24 V
V_{ref}	11 V
L	20 mH
C	47 μF
R	22 Ω
f	2500 Hz
f_s	2501 Hz

converter is driven into intermittent chaos, shown (in Fig. 9c, d). The recurrence of the bifurcation events is dictated by the difference between the noise signal frequency and the power supply's switching frequency, i.e. larger noise signal frequencies lead to more frequent bifurcations.

The numerical computation of behavioral complexity evolution in the buck converter with different value of feedback γ is computed by using a sliding window of thousand the time series measurement. Figure 9a–d (in lower panels) show the evolution of $\Gamma_{\alpha\beta}$ computed from the different current responses with$\alpha = \beta = 1$. Our computation of $\Gamma_{\alpha\beta}$ reveals its ability to detect these bifurcation structures, as well as

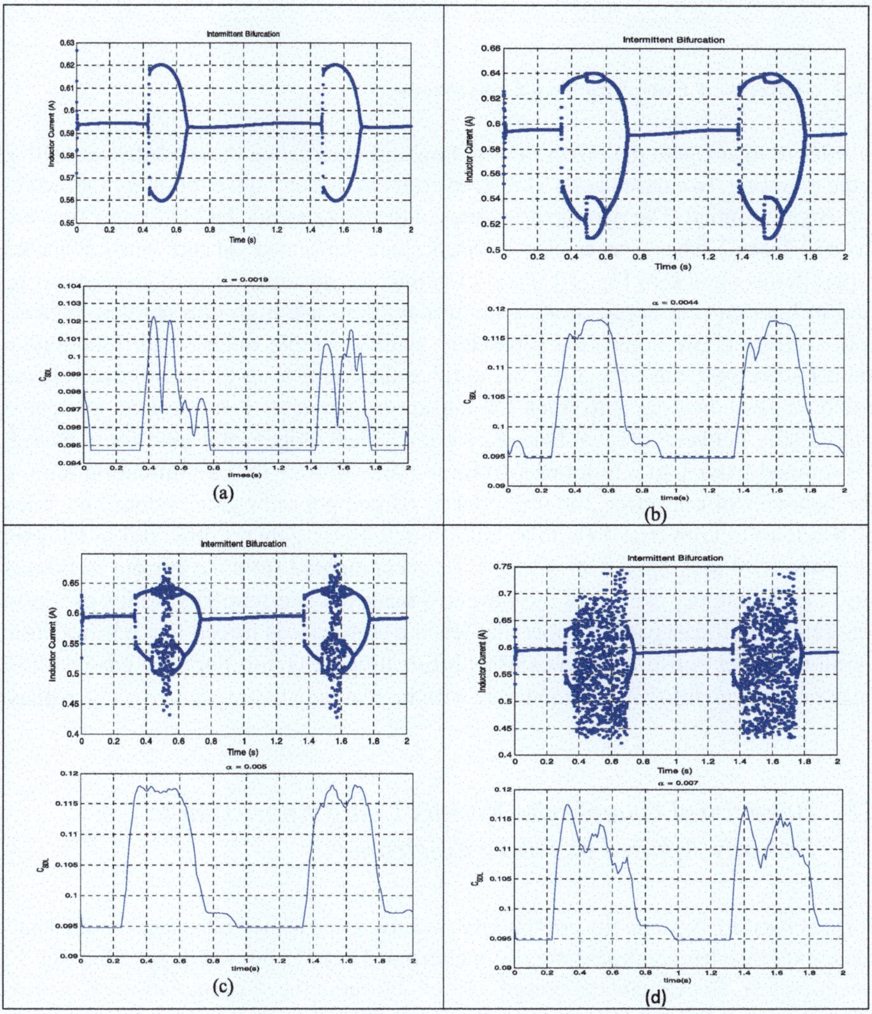

Fig. 9 $\Gamma_{\alpha\beta}$ evolution (in lower panel), and current evolution (in upper panel) for, **a** $\gamma = 0.0019$, **b** $\gamma = 0.0044$, and **c** $\gamma = 0.005$, **d** $\gamma = 0.007$

other changes in the system dynamics. The computed $\Gamma_{\alpha\beta}$ responds as expected, that is higher when the structure of the data is chaotic and lower when the structure of data becomes more organized. When the converter current is perfectly periodic, $\Gamma_{\alpha\beta}$ is minimal. If, as a result of the bifurcation, chaotic dynamics emerge, then the $\Gamma_{\alpha\beta}$ registers an increase, spiking to 0.12 and above. Notice in Fig. 8a–d (lower panels), $\Gamma_{\alpha\beta}$ increases slightly before and after each spike. This increase in the value of structural complexity can be explained by the fact that the moving windows can get two different sorts of data coming from two different structures. The entropy computed from mix data is higher due to the higher amount of disorder that can be seen in this data series compared to other data series taken from other windows, thereby the complexity is lower.

4.1.2 Effects of Coupling Two Converters

In order to investigate the effect of coupling and emergence of new bifurcation due to the coupling, two cascade DC-DC converters in master /slave topology (shown in Fig. 10) are studied. The parameters values of the new cascade DC-DC converters are given in Table 2. These two buck converters share the same load and were connected in parallel as shown in Fig. 10. Parallel connected systems allow converters to be used in high current applications without having to use high power rated devices. Both converters are controlled separately using a simple pulse width modulation control strategy. Converter 2 has an additional current loop in the control scheme which causes converter 2 to track the current in converter 1 so the load current is split equally between the two buck converters. The compensator gain for converter 1 is ramped from 1 to 7.5 in order to force the converter into a bifurcation region. The behavior of converter 2 is observed as its compensator gain is constant. Then the behavior of converter 1 (with a stable compensator gain) is observed as the gain for converter 2 is ramped from 1 to 7.5. $\Gamma_{\alpha\beta}$ is computed from the current responses (Fig. 11a, b in upper panels) of the slave and master converters. The results of $\Gamma_{\alpha\beta}$ in Fig. 11a, b (in lower panels) show that when a converter is forced into a bifurcation region the other converter also exhibits bifurcation behavior. For each case both of the converters go into bifurcation mode with increases in their respective $\Gamma_{\alpha\beta}$ values.

4.2 Behavioral Complexity Results Using Approximate Entropy Applied to Power Electronics

ApEn is used to evaluate the complexity evolution in a SST. Subsequently, the analysis is extended to encompass the entire microgrid with 3 nodes as depicted in Fig. 12 (lower panel). To evaluate the complexity evolution in the system, various scenarios will be explored in the following sections. Building upon previous research, this

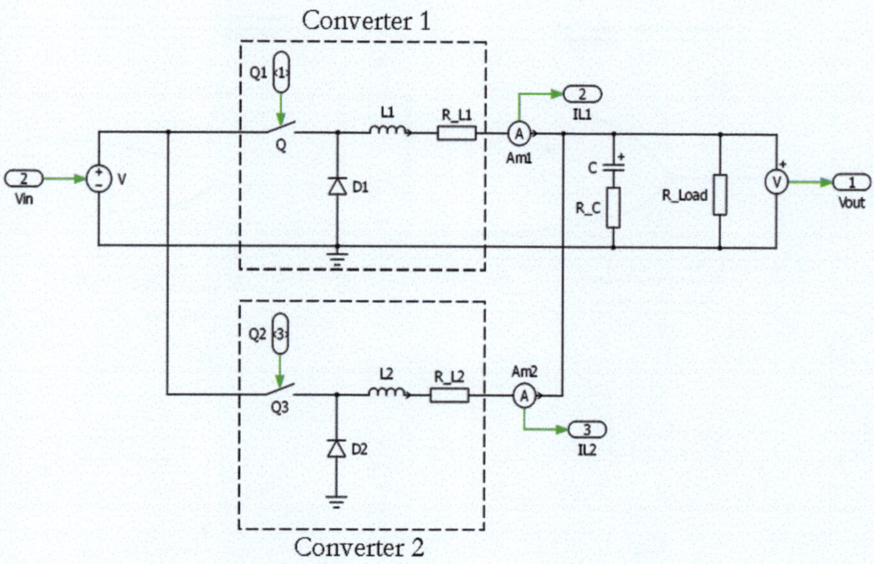

Fig. 10 Parallel buck converters scheme

Table 2 Component values

Output Voltage υ	20 V
Inpu Voltage E	40 V
Offset Voltage V_{offset}	3 V
Inductance L_1,, ESR R L1	1.5 mH, 0.05 Ω
Inductance L_2, ESR R_L2	3.0 mH, 0.2 Ω
Capacita ce C, ESR R_C	4.7 μF, 0.01 Ω
Load Resistance R	10 Ω
Switching Peri id T	40 μs

study utilizes a 1 MW ring main microgrid equipped with three single-phase SSTs as the test system [23].

In following SSTs is used as an alternative to conventional 60Hz transformers within the distribution system. Compared to traditional transformers, SSTs offer significant size and weight reduction, ultimately enhancing system performance. The dynamic behavior of the SST is influenced by a combination of parameters related to the power source, load characteristics, and control strategies. Figure 12 (upper panel) illustrates the schematic model of a single-phase SST alongside the complexity processing units.

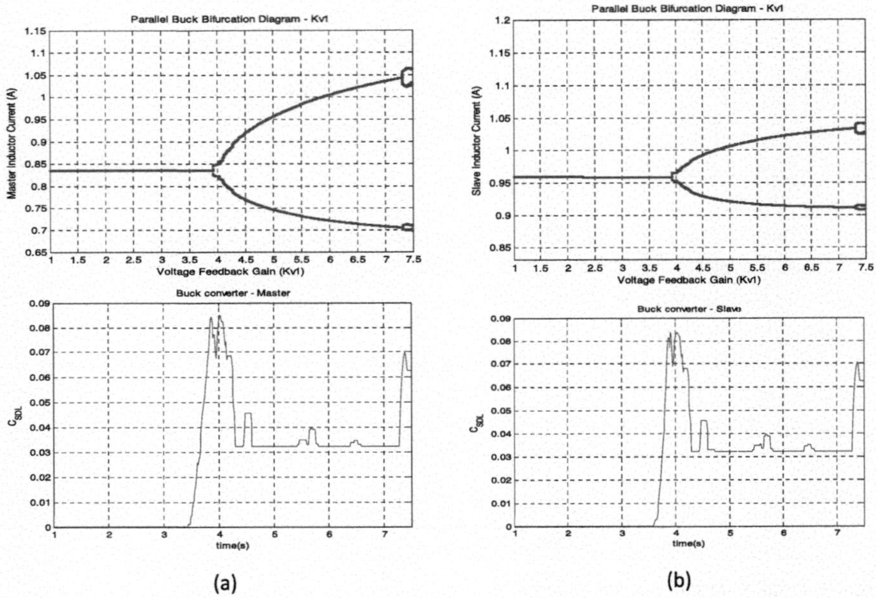

Fig.11 **a** master inductor currents I_{L1} (upper panel) and $\Gamma_{\alpha\beta}$ evolution (lower panel), **b** slave inductor current I_{L2} (upper panel) and $\Gamma_{\alpha\beta}$ evolution (lower panel)

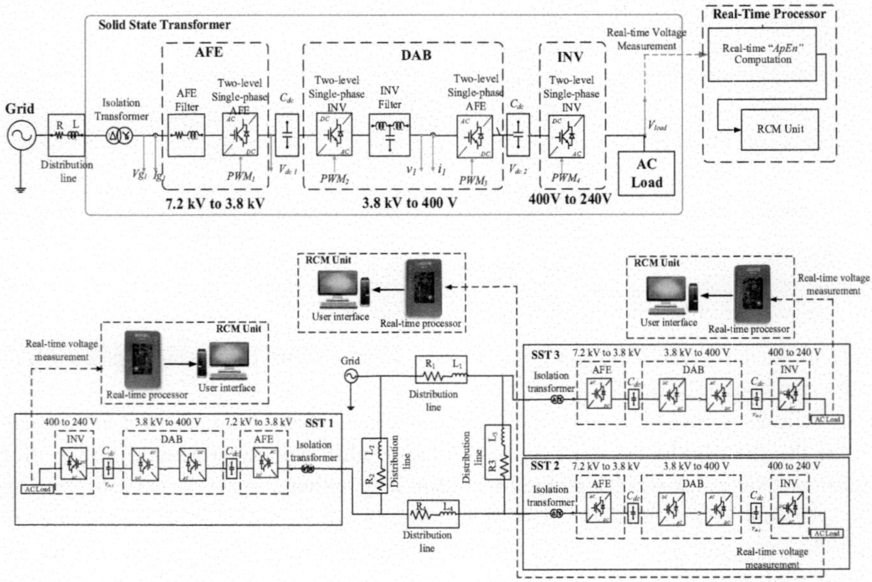

Fig. 12 Schematic of the experimental setup for the complexity processing unit with multiple SSTs

4.2.1 Load Variation for a Single SST

The initial scenario assesses the ApEn computation under load switching conditions. Each single-phase output of the three SSTs has a nominal power rating of 5 kVA, corresponding to a 3-wire single-phase topology. In this scenario, a constant power RLC load is connected to an SST at 1 s and disconnected at 2 s. The resistive component of the load is set to 5 kW, while both the inductive and capacitive components are set to 1.5 kVAR each. To detect subtle dynamic changes within the system, an amplifier is employed to magnify the detected signal. This amplifier reads the rms value of a parameter (voltage or current) and compares it to a reference value. Any deviation between the measured value and the reference is then amplified by a factor of "kv." The amplified RMS voltage (V_{A-RMS}) is defined as follows:

$$V_{A-RMS} = \left|k_v\left(V_{measure} - V_{ref}\right)\right| \tag{21}$$

Equation (21) defines the amplified RMS voltage (V_{A-RMS}), where "k_v" represents the gain factor, "$V_{measure}$" is the measured voltage, and "V_{ref}" is the reference voltage. In this experiment, a gain factor of 5 (determined through trial and error in omitted simulations) was chosen for the amplifier. Figure 13a displays both the instantaneous voltage and the amplified RMS voltage (V_{A-RMS}). While the instantaneous voltage waveform shows no significant change during load connection and disconnection (1 s and 2 s, respectively), the V_{A-RMS} profile exhibits two distinct shifts. To confirm the RCM method's ability to detect load switching, the V_{A-RMS} data is embedded into a phase space with a sampling time of 2 ms. For a system with non-complex behaviour, the phase space would depict a ring-like shape for the instantaneous voltage and a single point (one attractor) for the V_{A-RMS}. However, Fig. 13b reveals a more complex scenario for the V_{A-RMS} data, as the smaller three-dimensional phase space indicates multiple attractors. Before calculating complexity, the embedding dimension (m) and filtering level (r) need to be determined in real-time. Figure 13c illustrates this real-time parameter determination process. The filtering level remains relatively constant, suggesting that new attractors introduced during high load operation (twice).

the nominal load) are close together. However, as shown in Fig. 13b, these attractors require embedding in a higher dimensional space. Finally, Fig. 13d depicts the real-time computation of Approximate Entropy.

In addition, ApEn was tested on the system that undergo fault. Figure 14 shows the simulation results of the calculate for the case of fault. and the resulting complexity measure. The detected load switching events at both 1 and 2 s. This demonstrates a key advantage of the method, i.e., its ability to identify subtle dynamic changes that conventional protection devices might miss.

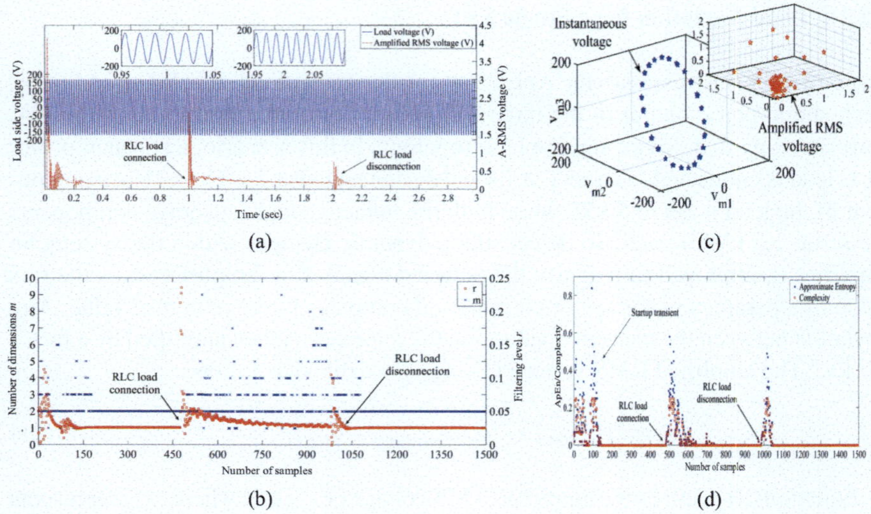

Fig. 13 Complexity quantification in case of load variation, **a** Load voltage during the load switching, **b** Real-time determination of *m* and *r* for the load switching scenario, **c** Phase-space reconstruction for the load switching scenario, **d** Real-time computation of *ApEn* and complexity for the load switching scenario

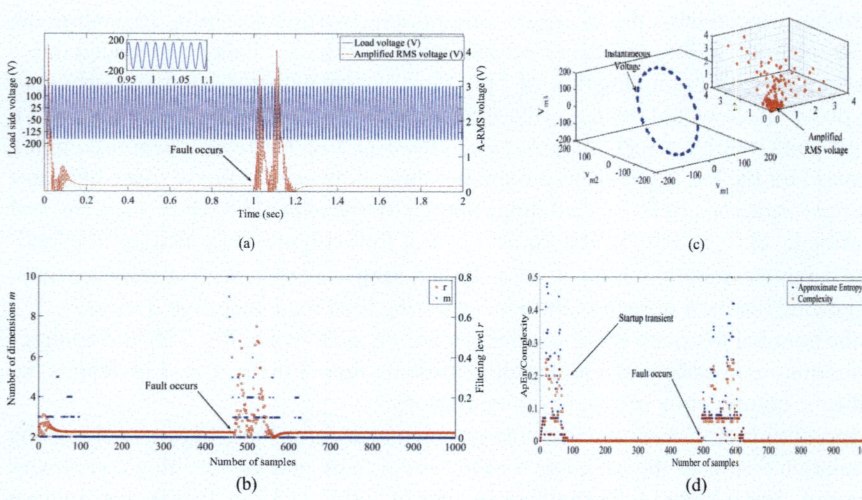

Fig. 14 Complexity quantification in case of fault, **a** Load voltage during the fault, **b** Real-time determination of *m* and *r* for the fault scenario, **c** Phase-space reconstruction for the fault scenario, **d** Real-time computation of *ApEn* and complexity for the fault scenario

5 Conclusion

This chapter discussed two complexity quantification methodologies of OPPEn and ApEn. OPPEn uses symbolic dynamic to provide an analytic approach to calculate the structural complexity for several dynamical behaviors from only system measurements, without the need of prior knowledge about the configuration or model of the systems. The ApEn-based real-time complexity measurement metric offers a novel approach to monitoring power system behavior. Unlike traditional methods, it relies solely on real-world measurements, avoiding model-dependent simplifications. This study on behavioral complexity will help power engineers analyze and evaluate the emergent phenomena that arise when adding more IBRs to the system. For this purpose, behavioral complexity was investigated, and a metric to measure that complexity was evaluated. Real-time complexity calculation allows to effectively identifies anomalies in time series measurement due to instability, faults, load variation, and others. It proved that is promising tool that will enhance situation awareness in microgrids.

References

1. Gu Y, Green TC (2022) Power system stability with a high penetration of inverter-based resources. Proc IEEE
2. El-Mezyani T, Wilson R, Sattler M, Srivastava SK, Edrington CS, Cartes DA (2012) Complexity quantication to enhance the shipboard power systems design and modelling. IET Electr Syst Transp
3. Ferdowsi F, Vahedi H, Abianeha A, Edrington C, El Mezyani T (Jan 2019) A data driven real-time stability metric for SST-based microgrids. Sci Direct Elsevier: Int J Electr Power Energy Syst, 134
4. Ferdowsi F, Vahedi H, Edrington CS, El-Mezyani T (2018) Dynamic behavioral observation in power systems utilizing real-time complexity computation. IEEE Trans Smart Grid 9(6):6008–6017
5. Dobson I, Carreras BA, Lynch VE, Newman DE (2007) Complex systems analysis of series of blackouts: cascading failure, critical points, and self-organization. Chaos 17:026103
6. Wei MS, Xue AC, Zhang XM (2008) On power system blackout modeling and analysis on self-organized criticality, vol 51. Springer, pp 209–219
7. Mitleton-Kelly E (2000) 'Complexity: partial support for BPR'. Systems engineering for business process change, Springer, England, pp 24–37
8. Tse CK (2004) Complex behavior of switching power converters. CRC Press, Boca Raton, FL
9. Wilson R, Sattler M, El-Mezyani T, Srivastava S, Edrington C, Cartes D (2010) Issues of complexity in power systems; an introductory study. Cybern Inform (JSCI)
10. Shannon C (1948) A mathematical theory of communication. Bell Syst Tech J 1948(27):397–423
11. Keller K, Sinn M (2010) Kolmogorov-Sinai entropy from the ordinal viewpoint. Phys D, 997–1000
12. Heylighen F (1996) The growth of structural and functional complexity during evolution. In: Heylighen F, Aerts D (eds)
13. Crutchfield J, Young K (1989) Inferring statistical complexity. Phys Rev Lett 63:105–109
14. Shiner JS, Davison M, Landsberg PT (1999) Simple measure for complexity. Phys Rev 59:1459–2146

15. Feldman D, McTague CS, Crutchfield JP (2008) The organization of intrinsic computation: complexity-entropy diagrams and the diversity of natural information processing. Chaos 18
16. Martin MT, Plastino A, Rosso OA (2006) Generalized statistical complexity measures: geometrical and analytical properties. Phys A 369:439–462
17. Bandt C, Pompe B (2002) Permutation entropy: a natural complexity, measure for time series. Phys Rev Lett 88(17)
18. Amigó J (2010) Permutation complexity in dynamical systems, 1st ed. Springer Series in Synergetics
19. Pincus SM (1991) Approximate entropy as a measure of system complexity. Proc Natl Acad Sci USA 88(6):2297–2301. https://doi.org/10.1073/pnas.88.6.2297. PMID: 11607165; PMCID: PMC51218
20. Li W (1991) On the relationship between complexity and entropy for markov chains and regular languages. Complex Syst 5:381–399
21. Hale J, Kocak H (1991) Dynamics and bifurcation, vol 3. ISBN: 978-1-4612-8765-0
22. Costa M et al (2003) Multiscale entropy analysis of human gait dynamics. Phys A: Stat Mech Its Appl 330(1–2):53–60
23. Zhao T et al (2009) An average model of solid state transformer for dynamic system simulation. In: IEEE power & energy society general meeting. IEEE

Machine Learning and Internet-of-Things Solutions for Microgrid Resilient Operation

Chun-Lien Su, Seyed Hossein Rouhani, Mahmoud Elsisi, Zulfiqar Ali, Hoang Le Quang Nhat, and Muhammad Sadiq

Abstract Microgrids inherently have lower inertia in comparison with the power system due to the high penetration of renewable generation sources. On the other hand, the lower spring reserve, high dynamic loads, and inherent volatility of renewable generation sources have demanded real-time control of them. The development of intelligent electrical devices, communication infrastructure, and Internet-of-Things (IoT) technology has enhanced real-time microgrid operation. In contrast, they have exposed the smart microgrid to cyber malicious activities. Cyber attackers try to disrupt the microgrid operation by manipulating the sensors, control center, communication infrastructure, and intelligent electrical devices, leading to decreasing power quality, cascading instability, and, finally, blackout. The effect of a cyber-attack in a standalone microgrid will be more distractive. Given the advantages inherent in these technologies, this chapter conducts an extensive examination of their integration into the management, control, and upkeep of both onshore and offshore microgrids. It provides a detailed account of the communication infrastructure within microgrids and identifies potential vulnerabilities. The chapter then elucidates methods for detecting cyber-attacks, placing a particular emphasis on machine learning, IoT, and Digital Twins. Subsequently, the chapter delves into the convergence of these technologies with maritime microgrids. It expounds upon how these advancements render maritime microgrids susceptible to cyber threats and emphasizes the instrumental role of machine learning in fortifying their cyber resilience. Furthermore, the chapter explores the application of IoT and Digital Twins, combined with machine learning, to enhance the maintenance and operational resilience of maritime microgrids. Finally, the chapter addresses the technical challenges faced by smart seaports and offers resilient operational solutions to mitigate potential disruptions.

Keywords Low inertia · Machine learning · IoT and digital twins · Cyber attack

C.-L. Su (✉) · S. H. Rouhani · M. Elsisi · Z. Ali · H. Le Quang Nhat · M. Sadiq
Department of Electrical Engineering, National Kaohsiung University of Science and Technology, Kaohsiung City, Taiwan
e-mail: cls@nkust.edu.tw

F. Shahnia and J. M. Guerrero (eds.), *Microgrids and Virtual Power Plants*,
Power Systems, https://doi.org/10.1007/978-981-97-6623-9_8

1 Introduction

In the power grid future, sensors and transducers will have a significant role in continuously monitoring energy levels in reaction to changes in demand. Intelligent sensing system integration gives up new opportunities for power measurement and real-time data analysis, facilitating quick decision-making. There is an increasing need to switch to sustainable energy generation from renewable sources as the demand for electricity keeps rising. We are currently seeing a significant transformation in the public power system. The bidirectional flow of energy caused by the presence of distributed power plants serves as an example of this transformation. This two-way energy exchange takes place within microgrids, which are regional groups of power grid nodes that distribute electricity according to regional demand. As a result, energy flow dynamics, including its direction, must change dynamically. The idea states that energy should be transferred from microgrids with surplus to those with deficits, as shown in Fig. 1. However, a number of factors, including intermittent energy output from renewable sources, significantly affect this general principle. Furthermore, further restrictions on energy routing are imposed by the qualities of voltage and current signals, such as their characteristics. The creation of cutting-edge technology that enables the power grid to effectively regulate the constantly shifting, bidirectional energy flow is required by this new environment. These systems must also guarantee equipment interoperability between newer and older models. The current situation where multiple dispersed production facilities provide energy to end customers and provide excess energy to the overall electric network. However, there is a risk of blackouts and service inefficiencies due to the disruptions that might occur during the production of electricity from renewable sources. The smart power grid must be able to prevent supply problems in advance. These requirements emphasize the value of cutting-edge sensor systems. For these sensor technologies to provide a thorough picture of the state of the power grid at each individual node, measurements must be quickly acquired, and results must be quickly analyzed. Power meters, which precisely measure various power parameters, are an example of such sensing devices. These measurements can include not only power consumption but also additional information about power quality, depending on their intended use. These sensor-driven technologies will be essential in maintaining the stability and effectiveness of the power grid as it develops into a more sophisticated and convoluted entity [1].

Distributed Energy Resources (DERs) are used by a sophisticated microgrid to generate electricity and reduce reliance on the main utility grid, which lowers prices for users [2]. The loads and DERs held by various owners are connected through community resilience microgrids. Through distribution lines controlled by a local power utility provider, these microgrids ensure a consistent and sustainable electrical supply to the entire neighborhood, especially during outages [3]. In the wake of large weather-related disasters, community resilience microgrids excel at maximizing electrical, economic, and societal goals while serving important loads like

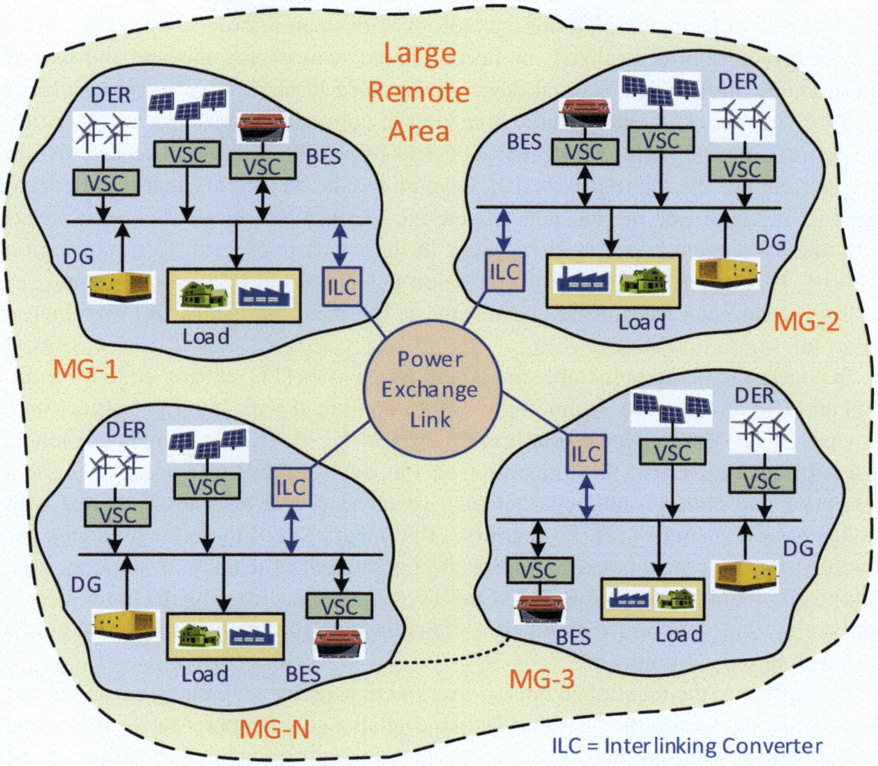

Fig. 1 Community resilience microgrids network interconnection through a common power exchange link

pharmacies, gas stations, and hospitals [4]. As a result, demand for intelligent microgrids is growing since they can provide many electricity consumers during extended outages of the primary power system. Community resilience microgrids, which stand for the next generation of electrical grid models, offer unique opportunities for operation and management inside the local distribution network. They reduce the need for local customers to share loads, promote greater integration of renewable sources, and reduce carbon emissions. The reliability, sustainability, and resilience of smart microgrids can be further improved by incorporating data produced by deployed Internet-of-Things (IoT) devices together with other elements, including DERs, controllers, and loads [5]. Additionally, quick data distribution via the communication infrastructure improves the recognition of major problems experienced by smart microgrids, such as islanding and grid-connected events. Additionally, as communication lags for transmitted data decrease, the capacity to quickly react to urgent and important circumstances greatly improves. The digitalization of energy systems, as illustrated by smart grids, is important for increasing the use of renewable energy and for boosting resilience against conventional weather and technological failures, but it also creates new openings for cyberattacks [6].

On the other hand, merging this technology with smart microgrid operation under a centralized, semi-centralized, or decentralized control scheme and the use of autonomous intelligent electrical devices has made it susceptible to cyber-attacks [7, 8]. Cyber attackers, through the hijacking of communications and manipulating the performance of intelligent electrical devices by remote or physical access, disrupt the operation of the microgrid [9, 10]. Given the reduced stability margin in microgrids owing to lower inertia, it is imperative to swiftly neutralize any malicious activities to prevent a potential blackout in the smart microgrid. Communication security, hardening access control, intrusion detection and prevention systems, as well as continuous monitoring and testing, is the most vital steps in lower inertia smart microgrid operation.

In contrast to being vulnerable to malicious activities [11], cutting-edge Information and Communication Technologies (ICT) combined with Intelligent Electronic Devices (IEDs) have enhanced maritime microgrids' resilient operation. For instance, in a shipboard microgrid, this integration encompasses propulsion systems, electrical networks, and battery units, guaranteeing seamless power allocation aligned with navigational dynamics [12]. Furthermore, the integration of these technologies has emerged novel approaches to optimize the operational efficiency of ships through equipment maintenance strategies. Employing machine learning technologies to analyze system data and maintenance plans ensures the most optimal conditions for machines and equipment.

In addition to the mentioned operational and management challenges, the seaport microgrids are facing the Variable Demand challenges. Seaports have fluctuating power demands due to their diverse activities, including loading/unloading operations, maintenance, and vessel docking. To address these challenges, the integration of digitized, IoT-driven applications with an intelligent energy management system effectively supervises the allocation of electricity. This is achieved through the implementation of advanced control techniques facilitated by robust communication networks [13].

This chapter has been structured as follows: Sect. 2 investigates Microgrid Infrastructures and Vulnerabilities. In Sect. 3, Cyber-Attack Detection and Prevention Solutions are provided. Section 4 studies the Machine Learning and IoT Solutions for cybersecurity in smart microgrids. Maritime Microgrids are studied in Sect. 5.

2 Microgrid Infrastructures and Vulnerabilities

2.1 Microgrid Infrastructures

Consider a network of Fig. 1 where autonomous microgrids, consisting of DERs, distributed generators, loads, and one interlinking converter are coupled to a power exchange link, as shown in Fig. 2. The DERs are connected to the microgrid through power electronic interfaces which are controlled in different operating modes to regulate the power and frequency of each microgrid. The DERs are assumed to be operating in the $P - f$ and $Q - V$ droop control mode [14, 15]. Communication between IEDs and the microgrid control center can be implemented both through wired and wireless networks within each autonomous microgrid, between neighboring microgrids, and between the microgrid and the utility power system. Both have advantages and disadvantages, and their application differs based on the microgrid location and legacy components. Ethernet, Fiber Optics, and Power Line Communication are the most common forms of wire-based communication in microgrids. These wire-based communications, based on their transmission rate, ability for application over long distances, construction, and cost consideration, are used in microgrid operation. For example, Ethernet has a reasonable transmission rate (up to 100 Mbps) and is used for short distances. In contrast, Optical Fiber also has a high transmission rate and is used for long distances. The most distinguishing features of new wireless technology are easy implementation, lower cost, and adaptation to challenging geographical conditions. However, problems in severe weather conditions, lack of penetration through obstacles (a problem in urban areas), packet loss in long-distance, and uncertain time delay are their practical limitations in low inertia microgrids. The most commonly used wireless technologies in microgrids include Wi-Fi, Bluetooth, ZigBee, and mobile networks (3G/4G/5G, GPRS) [16–18].

In order to transmit information within microgrids, either standalone or grid-connected, communications infrastructure is necessary. Moreover, for transmitting measured data, implementing additional functionalities like routers and merging units are required. For reliable operation, a protocol is used to exchange data between the

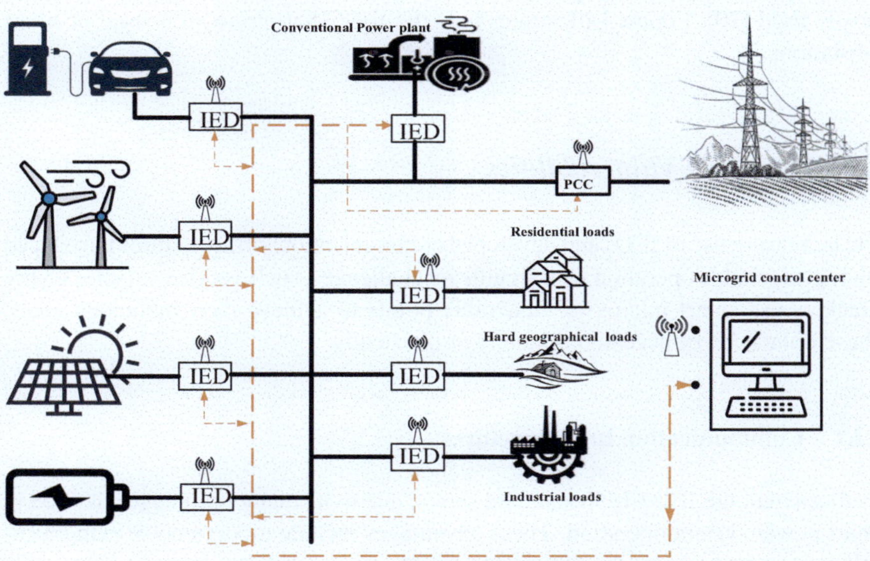

Fig. 2 Conceptual description of microgrids

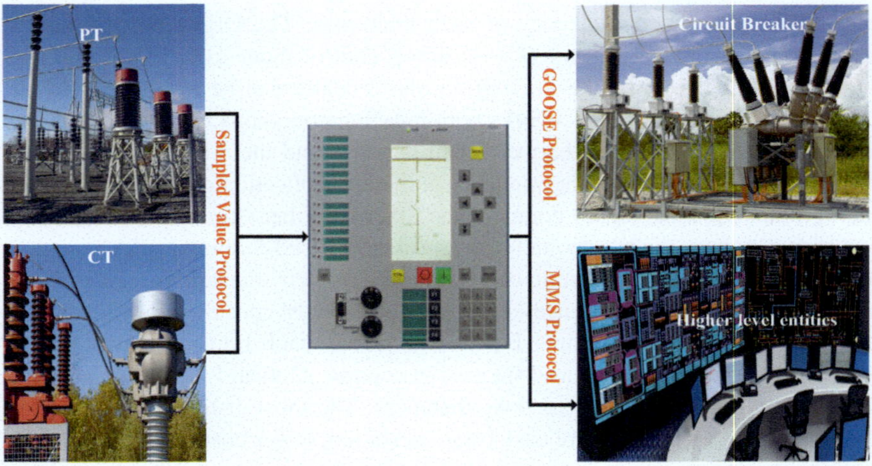

Fig. 3 Schematic of the IEC61850 protocol's locations in Substations

components. Protocol is a set of rules that define how data is exchanged between components. Some of the communication protocols in the power system are Modbus, OPC/OPC UA, DNP3, TASE.2/ICCP, and IEC series. DNP3 and IEC60870-5-104 are designed for usage in supervisory control, and data acquisition (SCADA), and IEC61850, the newest standard, has been developed for communication networks in substations [19]. Nowadays, the IEC standards, due to their interoperability with the legacy power system, easy implementation, lower costs, and more reliability, are mostly used [20]. Figure 3 illustrates how the IEC61850 protocol is used in smart substations.

2.2 *Microgrid Vulnerabilities*

The increasing use of IEDs and developed communication infrastructures have faced microgrid reliable operation with significant challenges, such as cyber-attacks. Cyber attackers use smart microgrid vulnerable points to achieve their malicious goals. These points can be accessed remotely or physically.

2.2.1 Communication Infrastructures

As discussed, the IEC 61850 standard series, due to its many advantages, is used in smart power system operation. These advantages include modern networking technologies, enhanced information management, interoperability between the power system equipment, simplified system integration, and reduced implementation costs [21]. In contrast, their main drawback is the lack of cyber security mechanisms,

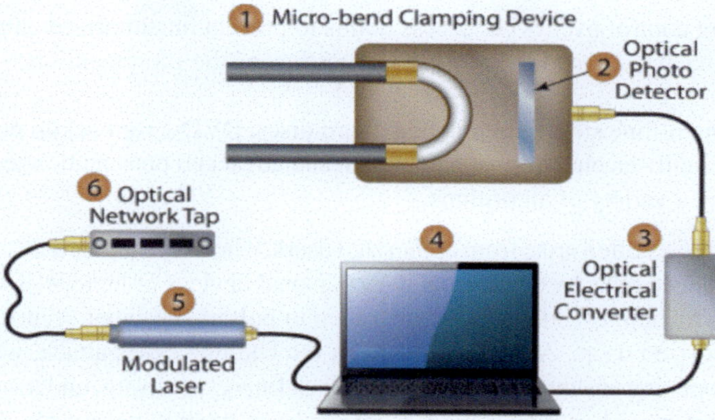

Fig. 4 Tap fiber optic lines without breaking the fiber [23]

which make the power system vulnerable to cyber-attacks [22]. Therefore, attackers can hijack the communication between the IEDs or between the IEDs and the control center. For example, they can tap fiber optic lines without breaking the fiber through simple hardware, then they can launch malicious attacks such as false data injection, creating delays, fake traffic and, etc., as shown in Fig. 4 [23].

2.2.2 Intelligent Electrical Devices

To attack the IEDs, attackers try installing malware, Trojan horses, or worms by hijacking the communication or connecting their portable storage devices or personal laptops. Stuxnet and Black energy are examples of such malicious activities [24, 25]. These malware are designed to damage the IEDs performance on a special condition, make them unavailable, or send false information. They focus on IEDs, disabling sensors, disrupting getaway functions, circuit breaker statuses, backup storage devices, and managing various intelligent devices. Consequently, false commands, misleading statuses, and inaccurate measurements circulate throughout the network, leading to the malfunctioning of crucial system components.

To ensure protection against malware, implementing a range of strategies is essential [26–28]:

- Employing dedicated communication protocols within the smart microgrid instead of relying solely on the Internet.
- Dividing the smart microgrid into distinct segments for enhanced security.
- Installing firewalls in each segment to thwart unauthorized external access.
- Employing antimalware software to preemptively identify and combat virus infections.
- Implementing password expiration policies and employing two-step authentication for software installations and patch updates.
- Identifying potential points of attack and implementing physical safeguards.

– Exercising control over cyber access points to prevent unauthorized entry.
– Conducting regular audits and rigorous testing of the aforementioned security measures.
– Employing offline storage media (e.g., hard discs, DVDs) for backup purposes.
– Utilizing multi-faceted intrusion detection and advanced presentation techniques employing a variety of algorithms.

Internal threats often arise from authorized users, who may inadvertently or intentionally introduce risks by connecting their personal devices or laptops. To mitigate unconscious malicious actions, it is imperative to bolster the cyber security awareness of authorized users. A prime example is the Ukraine cyber-attack, which was initiated by opening an email containing the Black Energy malware, underscoring the critical importance of user education. Conversely, to thwart consciously malicious activities, implementing robust security measures is essential. This includes hardening access and adopting two-step authentication, wherein the second step involves authentication by security personnel. Furthermore, establishing password lifetimes and enforcing regular changes (thus preventing access by former staff) is crucial. Additionally, continuous refinement and enhancement of detection and prevention methods are paramount in fortifying against deliberate internal threats.

3 Cyber-Attack Detection and Prevention Solutions

In general, cyber-attack detection methods in microgrids are classified into two categories: model-based and model-free based, also known as the data-driven based methods, as shown in Fig. 5. These methods are used for anomaly detection based on the availability of the system model and its accuracy [29–32].

Model-based methods use the information extracted from the transmitted measurements. Observer-based, parity-space, and parameter estimation-based methods are examples of model-based intrusion detection methods. These methods

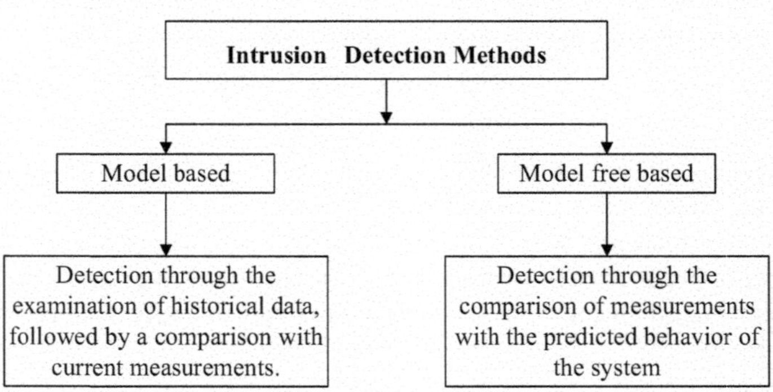

Fig. 5 General cyber-attack detection methods

estimate the system's behavior and compare it with the actual measurements for intrusion detection.

Model-free-based methods don't need a system model for anomaly detection. The idea behind these methods is cyber-attack detection based on the historical system performance data, Waveform Analysis. These methods utilize Artificial Intelligence (AI) and Machine Learning (ML) to analyze historical. Integration of these methods with the legacy power system brings various advantages, such as intrusion detection, access control management, and anomaly detection. These methods discover patterns and insights from data, subsequently generating their own algorithms for cyber-attack detection. ML-based methods, including neural network-based deep learning, play a significant role in cyber-attack detection.

4 Machine Learning and IoT Solutions

4.1 IoT

IoT is one of the technologies that combine a variety of already-existing technologies, including sensors, the Internet, software, real-time operating systems, embedded systems, etc. IoT primarily serves as a conduit to connect physical objects to the virtual world by way of the Internet. Mark Weiser first proposed the idea of connecting the physical and virtual dimensions in the early 1990s. The IoT paradigm enables remote control of physical things using developing Internet-based technologies [33]. IoT has gained popularity since it was formally launched by the International Telecommunication Union at the 2005 World Summit on the Information Society in Tunisia. The International Telecommunication Union released an exhaustive report on the subject, delving into IoT's concepts and its global ramifications [34].

IoT acts as a vast networking platform that makes it possible to communicate with and manage many different systems online. It entails the coordination of sensors, computers, and actuators over an internet connection. Data streams from sensors and other intelligent devices that track changes in the environment converge on a common platform. Important data is processed, calculated, and evaluated here. The refined data can be saved for later use or immediately used to carry out tasks like controlling or actuating switches on other devices connected to the same networked platform. The main target of this idea is to increase efficiency by reducing the requirement for human oversight and management of digital systems. These systems run instantly, require less human effort, increase human independence, maximise resource use, save time, and encourage transparency. Figure 6 provides a graphic illustration of various IoT ideas.

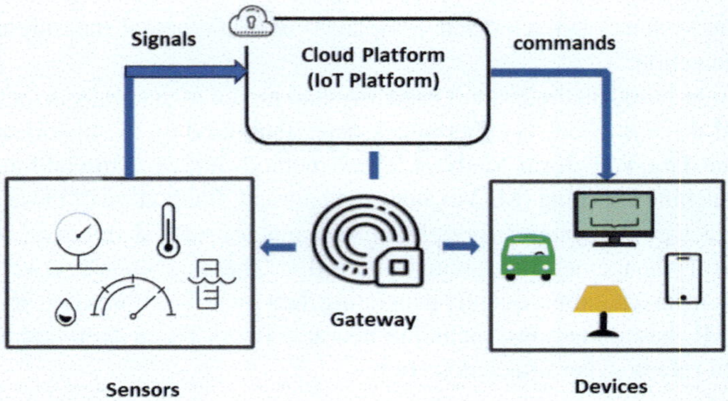

Fig. 6 Related concepts of IoT

4.2 IoT Applications

The need for an adaptable and intelligent electrical grid that can meet the various demands of diverse customers has increased as metropolitan regions and their populations expand. The National Institute of Standards and Technology presented a design for the potential electrical grid in 2007. This design intends to retain cost effectiveness while ensuring the scalable, secure, interoperable, and controllable operation of electricity grids [35]. Figure 7 provides an example of the IoT framework created for a smart grid [36].

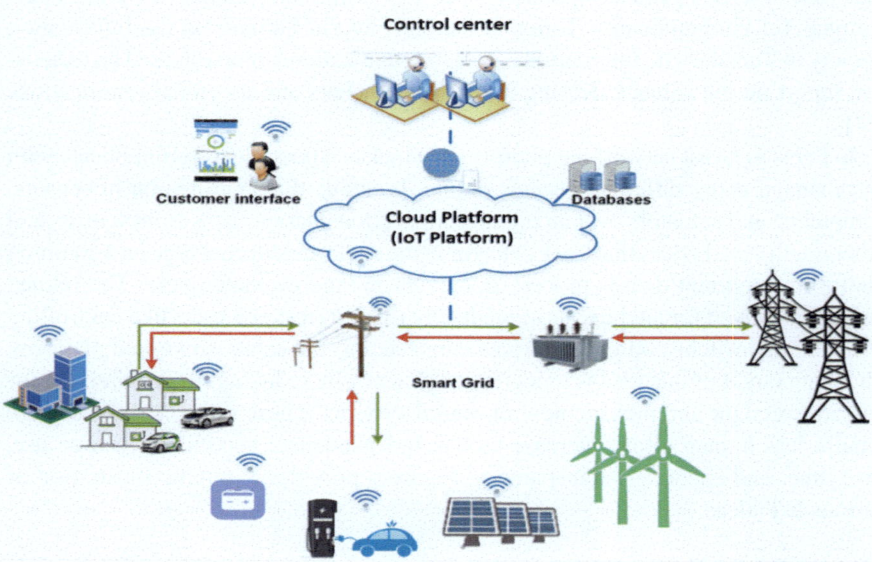

Fig. 7 The IoT structure for smart grid

The IoT provides dispersed computing power and bidirectional connectivity, making it a potential solution to the problems that come with converting traditional energy networks to more modern smart grids [37, 38]. The integration of distributed renewable energy resources on a large scale within a smart grid framework, the creation of real-time data communication between consumers and service providers for tariff and energy consumption information, the infrastructure creation for the collection and relay of grid parameter statistics for analysis, and the implementation of actions based on these analyses are just a few essential services [39]. The smart energy grid produces a significant amount of data and information that must be transported, processed, and stored in order to enable informed decision-making [40]. The IoT emerges as a suitable option with significant potential for integration into the smart grid system in this context given its multiple advantages across numerous industries. The IoT can help with the smooth transition from the conventional power grid to a smart energy system that is more efficient, in addition to improving accuracy and efficiency through its intelligent and proactive capabilities [41].

The transition from a traditional power grid to a smart grid can be facilitated by advanced metering infrastructure supported by smart metering technologies by introducing intelligent information-processing capabilities into the exchange of electricity between service providers and consumers [42]. The IoT holds tremendous promise for regulating and optimizing energy use thanks to the integration of sensor and actuation technologies into advanced metering infrastructure. Energy use, voltage, current, and phase measurements are all part of the considerable data that this integrated system collects from different grid segments. This data can be gathered, communicated, and intelligently evaluated using state-of-the-art IoT technology, leading to improved energy grid management [43]. The management of power generation infrastructure, SCADA for transmission and distribution operations, advanced metering infrastructure, as well as carbon footprint and environmental monitoring, are just a few of the areas where IoT technologies can have a significant impact on smart energy grid systems. By enabling decentralised monitoring and management of distributed energy resources, advanced cloud, and edge computing technologies can reduce the cybersecurity risks associated with conventional centralized SCADA systems [44]. By seamlessly integrating with other smart entities like appliances, homes, buildings, and cities to access and control many devices over the Internet, an IoT-enabled smart grid may operate and manage the electrical grid more effectively. However, this integration calls for sophisticated computing powers and resource management systems. Implementing an IoT-enabled smart grid has its own set of difficulties despite the improvement in the monitoring and operation of the energy system.

4.3 IoT Challenges

The Internet of Things faces various difficulties that prevent mainstream adoption, much like many modern technologies do. These issues cut across both the technological and moral spectrums. IoT faces challenges in terms of technology, including interoperability, scalability, security, privacy issues, and data management. The likelihood of interoperability issues is growing as more and more IoT systems are deployed. Due to the proliferation of proprietary systems that are incompatible with one another, it can be challenging to integrate many systems into a cohesive whole. Harmonization across hardware, protocols, and other aspects is necessary to achieve compatibility [45]. It is crucial to develop standards in order to overcome these challenges. To improve interoperability, such standards must include technical requirements, such as hardware and software requirements for all devices. Notably, attempts have been made to define standards for fundamental IoT technologies like RFID and WSNs by a number of standards bodies, including International Telecommunication Union, ISO, IEEE, and IEC [46].

Scalability, a key component of IoT, presents a number of extra difficulties, especially those pertaining to security and privacy. The majority of IoT systems are used on a big scale, which raises issues regarding security and privacy. IoT combines numerous technologies that have varying degrees of security flaws. Every component of the IoT ecosystem needs to be resilient to external interruptions and interference and protect against vulnerabilities. Because a sizable portion of IoT devices have weak processing skills, putting effective security measures in place is a difficult undertaking. Such restrictions put the ecosystem at risk of weaknesses. An important component of IoT is scalability, which presents several new difficulties, particularly those pertaining to security and privacy. Concerns about security and privacy are brought to the fore because most IoT systems are used on a big scale. IoT combines numerous technologies, which introduce security flaws at multiple levels. Every component of the IoT ecosystem needs to be protected against weaknesses and resilient against disruptions and influence from outside sources. The deployment of strong security measures is a challenging issue because a significant proportion of IoT devices have insufficient processing capabilities. The ecosystem is potentially vulnerable due to these constraints [47].

Furthermore, privacy issues are made worse by the transmission of data across a vast network by gadgets that may observe daily activities. Any device with Internet connectivity is vulnerable to a variety of assaults, including those involving identity theft, unauthorized data access, and others. Users have legal access to other people's devices, such as IP cameras, thanks to websites like Shodan [48]. IoT devices need to keep the highest level of privacy protection feasible in a time when it is of the utmost importance to preserve one's private.

4.4 Machine Learning Concept and Applications

A branch of AL called ML has a lot of potential for improving the management and operation of microgrids. ML offers a variety of paths for development and is divided into four categories depending on learning techniques: supervised, unsupervised, semi-supervised, and reinforcement learning. In supervised learning, continuous value target features are classified or predicted using training algorithms on labelled datasets [49]. Regression and classification are two learning activities that are pertinent to microgrid control. Classification in microgrids extends to power quality, voltage stability assessment, fault detection and classification, and disturbances due to load variations and transient situations [50]. Predicting PV generation, power demand, and market pricing are all forecasting-related issues [51].

Unsupervised learning is particularly effective in analyzing and grouping unlabelled datasets, finding patterns devoid of human input, and forecasting unknown values. It streamlines and speeds up learning. Clustering reveals data patterns and entity separation. Unsupervised learning's microgrid applications include network topology detection, consumer/prosumer segmentation, and load profile clustering [52]. Labelled and unlabelled data have both recently been used for ML model training. Models may now learn from both labelled and unlabelled data and make predictions for upcoming points using semi-supervised learning. A noteworthy semi-supervised learning approach is the Generative Adversarial Network architecture [53]. To construct time-series power generation profiles for DERs [54] and plan energy storage for solar PV microgrids [55], semi-supervised learning like Generative Adversarial Network for microgrids can generate trained data from noise and close the gap between trained and real data.

An environment, agent, reward, and action are the components of reinforcement learning, which is unique. It requires an agent to actively engage with and shape the environment while earning incentives for their efforts [49]. Maximizing cumulative rewards through ongoing interactions is the aim. Reinforcement learning improves the management and operation of decision-driven microgrids. The potential of reinforcement learning is utilized in areas including optimal energy management, autonomous electricity market participation, and multi-microgrid interaction [56]. A schematic overview of AI methods that can be used in microgrid control is shown in Fig. 8.

4.5 Machine Learning and IoT Case Study

The authors' group has developed an IoT architecture for cyberattacks and data loss detection based on ML techniques using low-cost systems embedded in the smart meters, as shown in Fig. 9 [57]. To classify data from smart meters, an ML model based on decision trees was developed. This decision tree model was then linked with the smart system to allow for real-time meter output validation through IoT. The

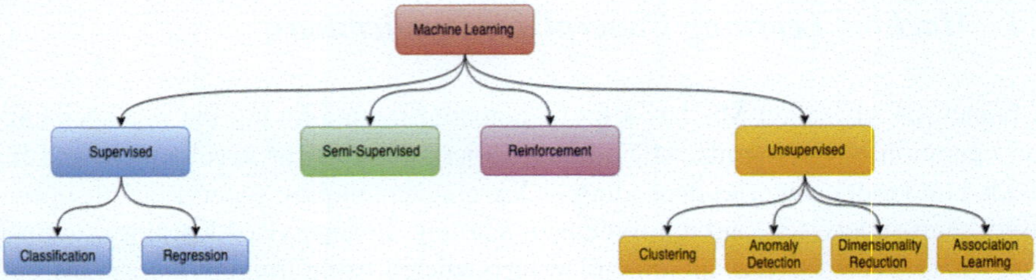

Fig. 8 AI techniques that can be implemented in microgrid control

smart meter's real-time dataset, which was gathered under a variety of operational conditions, served as training and test data for the decision tree model. Fake data was added to this dataset to represent the effect of cyberattacks. The data from a real smart meter was given a label of 1, whereas data that was faked was given a label of 2. The decision tree model's training and testing were made easier by the labelling. Furthermore, during online operation, the validation of data loss due to fluctuations in internet speed was verified. The following pseudo-code (Algorithm 1) provides a comprehensive depiction of the complete process encompassing data acquisition, validation, and visualization.

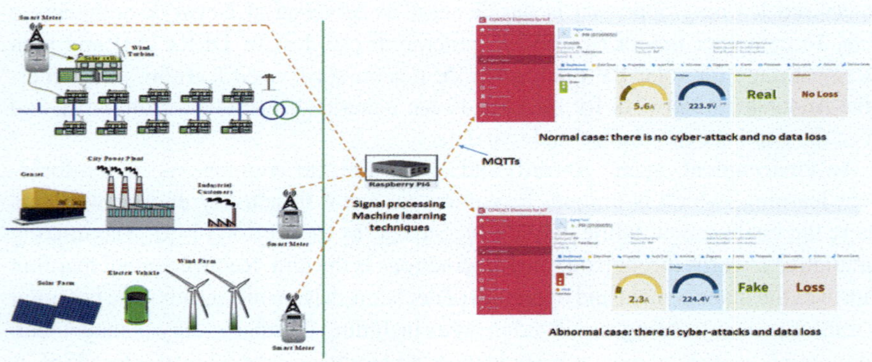

Fig. 9 Smart meters monitoring for cyber-attacks and data loss detection using ML

Algorithm 1 Data collection, validation, and visualization pseudo-code

```
Send signals from the smart meter
Enter the signals to the ML model
Classify the signals by the ML model
  if the output of the ML model == 1
      Publish that the signal is 'Real'
  else
      Publish that the signal is 'Fake'
  end if
Store the receiving time of the signal
Determine the change of time
  if the change of time ≤ sample rate of the smart meter
      Publish 'No loss data'
  else
      Publish 'There is loss data'
  end if
Publish the reading of the smart meter
```

The IoT platform's dashboard will provide decisive findings with regard to data validation and possible data loss brought on by problems with internet access. The output of the IoT system as a result of this test is shown in Fig. 10 on the dashboard. The data is accurate and has never been lost, as shown. The traffic light indicator also shows that the operating state is "green," which denotes system stability and the absence of any detected events or alarms. This provides evidence that the decision tree model operates correctly and effectively.

For IoT systems, the Internet's speed is a major barrier. As a result, the intelligent system must consider data loss as well as the network's required internet speed. In this case, the capabilities of the Internet and the data type are jointly evaluated. The dashboard of the IoT platform is shown in Fig. 11 and shows faked data along with instances of data loss. This is a severe abnormality that denotes both fabricated data and insufficient network speed for the smart system. Additionally, the system is considered unstable, resulting in abnormal occurrences and potential alerts when the operational state is shown by a red hue in the traffic light indicator.

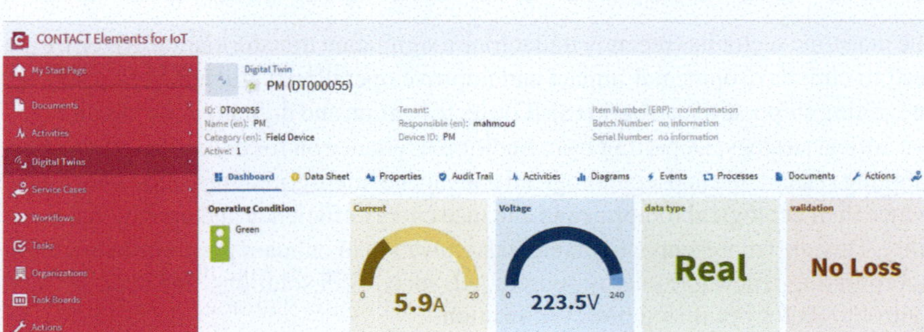

Fig. 10 The dashboard of the IoT platform results in the case of real data and no data loss

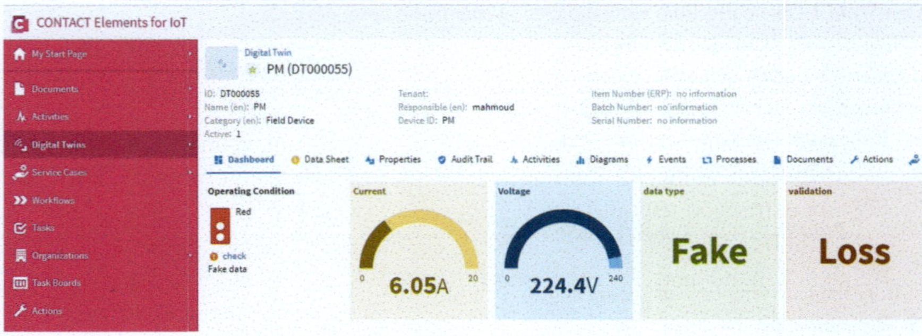

Fig. 11 The dashboard of the IoT platform results in a case of testing the fake data and internet capability

5 Cyber Resiliency of Maritime Microgrids

Maritime sector microgrids supply power to maritime facilities, such as ports, ships, offshore platforms, and coastal installations, which can operate localized and independently or can be connected to the main grid in some places. The maritime microgrid seamlessly integrates a diverse array of energy sources, encompassing traditional options like diesel generators, as well as renewable sources such as solar panels, wind turbines, and potentially even wave or tidal energy. Additionally, it incorporates advanced energy storage systems for enhanced reliability and efficiency. The incorporation of cutting-edge ICT in conjunction with IEDs has significantly enhanced the operational management of intricate maritime sector microgrids. This progress is particularly evident in demanding geographical settings such as offshore energy farms. However, it has concurrently exposed these systems to potential cyber threats, underscoring the critical importance of robust cybersecurity measures.

5.1 Shipboard Microgrid

The maritime sector has recently undergone a significant transformation driven by the need to curb environmental impact and improve operational efficiency. In response, integrating shipboard microgrid (SMG) electrification and digitalization has emerged as a critical strategy. Departing from traditional reliance on fossil fuel-based systems, adopting sustainable and technologically advanced microgrid setups allows vessels to reduce their ecological footprint and enhance energy efficiency and operational flexibility. This shift represents a fundamental renovation of onboard power infrastructure, incorporating renewable energy sources, advanced ESS solutions, and sophisticated control tools for optimized power utilization.

The growing momentum behind the electrification and digitization of SMGs stems from the desire to overcome the limitations posed by conventional power systems.

The maritime industry faces various challenges, including stringent environmental standards, fluctuating fuel costs, and the need for increased operational reliability. By incorporating cutting-edge technologies within SMGs, such as intelligent power management systems and the integration of advanced ICT, vessels can fine-tune energy consumption, reduce dependence on fuel, and enhance overall operational safety.

In Fig. 12, a simplified representation of the AC electrical power infrastructure in a semi-submersible deck cargo and heavy lift carrier vessel is illustrated. This intricate system comprises propellers, thrusters (which consume approximately 90% of the ship's power), diesel generators, windlasses, mooring winches, ballasting pumps, and hotel loads. Integrating ESSs, such as batteries and fuel cells, is pivotal in significantly enhancing the vessel's energy efficiency and fault protection. Overseeing power generation and distribution within the SMGs, the power management systems ensures optimal energy utilization. It manages the activation, synchronization, and operational speed of generators, which can be facilitated remotely through autonomous, ICT-connected, and IoT-monitored smart power management systems. These systems execute critical functions to improve security and fuel efficiency, encompassing voltage/frequency control, load sharing, priority generator selection, blackout recovery, load shedding, and intelligent energy storage control. Nevertheless, the growing digital integration of this setup simultaneously exposes the system to potential cyber threats directed at generators, propeller drives, or other sensitive equipment faults, leading to significant blackout risks [58, 59].

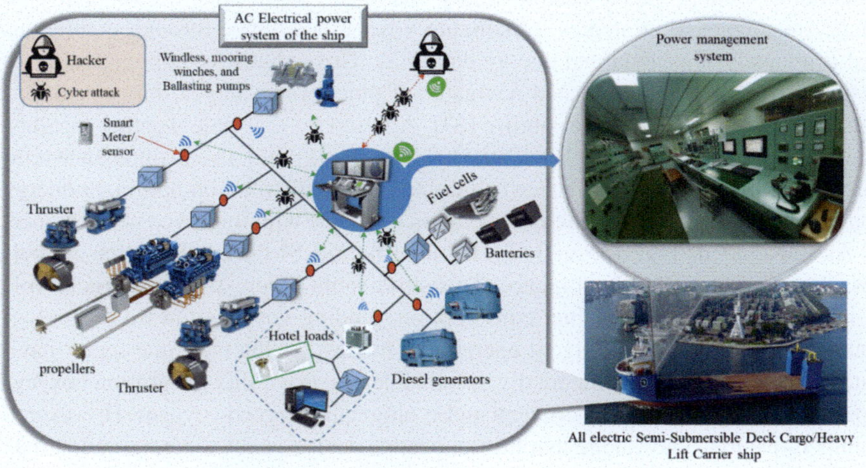

Fig. 12 AC Electrical power system of a SMG

5.1.1 Utilizing IoT for Optimized Voyage Planning and Compact Fuel Feeding

In the realm of SMGs within the maritime sector, operational expenses, notably fuel costs comprising 50–60% of the total operating expenses, wield considerable influence [60]. This context has prompted the adoption of an innovative approach combining IoT and ICT networks, leading to a transformative shift in voyage planning within SMGs. This shift leverages real-time monitoring, predictive analytics, and intelligent power management systems, driving a significant evolution in navigation that optimizes routes and fuel consumption in real-time [61].

This evolution appears vessels beyond conventional limitations, positioning them as interconnected units. SMGs proactively tackle challenges by incorporating real-time data, weather forecasts, navigational risks, and energy demands, facilitating precise route adjustments while accommodating energy generation and consumption patterns. This integration encompasses propulsion systems, electrical networks, and battery units, ensuring seamless power allocation aligned with navigational dynamics [12].

Consequently, fuel efficiency is a pivotal attribute achieved through meticulous energy management during operations. By harnessing sensors, data analytics, and adept predictive modelling, SMGs effectively curtail fuel consumption and greenhouse gas emissions, contributing to sustainable SMG operations.

Empowered by IoT capabilities, ESSs cooperate with voyage planners to efficiently allocate stored energy to various system components, unlocking unexplored potential for both propulsion and auxiliary functions. SMGs engage in sophisticated energy management strategies, prolonging voyage endurance through continuous real-time analysis of load and energy data [62]. This exhaustive approach aims to design voyage plans that exceed navigational objectives, safeguarding operations while enhancing economic feasibility [63]. Advised by real-time weather insights, intelligent routing circumvents potential hazards, while predictive maintenance reinforces equipment reliability. The synchronized power distribution optimizes energy consumption, ensuring the efficient operation of SMGs with improved performance.

As depicted in Fig. 13, an IoT-integrated hybrid SMG configuration propels vessels toward enriched intelligence. This sustainable progress plays a crucial role in mitigating the impact of fuel consumption, guiding voyages toward technological sophistication and intelligent operational modes. As the maritime sector navigates towards more environmentally sustainable practices, this integration conveys a pivotal milestone, ensuring efficient and ecologically aligned voyages. However, it is imperative to acknowledge that this innovative device-enabled configuration may be susceptible to attacks targeting power equipment, potentially leading to a system blackout.

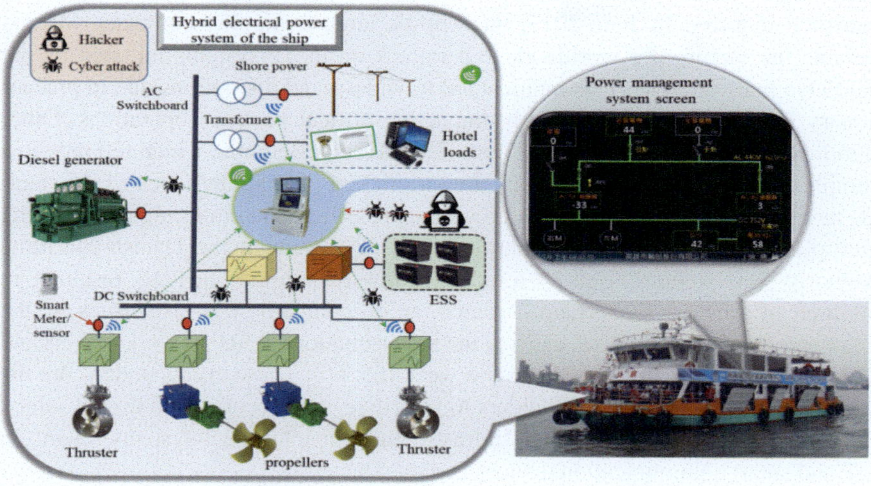

Fig. 13 Hybrid electrical power system of SMGs

5.1.2 Ensuring Cybersecurity for SMGs

The integration of ICT and IEDs with SMGs has rendered SMGs more susceptible to cyber threats, emphasizing the urgent need for cyber resilience to guarantee secure and efficient vessel operations in the face of evolving risks. Moreover, the resilience of microgrid systems, facilitated by fifth-generation communication systems, is directly impacted by cyber-physical attacks and uncertainties. To fortify these systems against potential breaches, the enactment of resilience-based control techniques and the consideration of a thorough cyber-attack model are imperative. Several control mechanisms shield the system from cyber threats and uncertainties, evaluating their resilience through various performance measures and validating them within a real-time software synchronization environment. Furthermore, the seamless integration of SMGs with external networks and shore-based systems exposes them to potential malware infiltration, jeopardizing the safety and security of the vessels and their onboard systems. Building redundancy and contingency measures into the SMG infrastructure can intensify its resilience against cyberattacks. Accessible redundant systems and backup power sources should be readily available to ensure uninterrupted power supply and prevent disruptions to the vessel's operations. In addition, developing secure data backup and recovery strategies can mitigate the risks associated with data loss or corruption stemming from cyber incidents. By implementing comprehensive cybersecurity strategies and fostering a culture of cyber vigilance among maritime personnel, the cyber resiliency of SMGs can be effectively reinforced, securing the safe and reliable operation of vessels within the marine industry.

The escalating presence of cyber threats and potential pathways for malicious digital activities has emerged as a significant concern for global leaders in the maritime industry. In June 2017, the world's largest shipping company, Mærsk, became the victim of a worldwide cyberattack on its IT systems initiated by the NotPetya malware [64]. This incident led to widespread disruptions due to precautionary IT shutdowns, significantly impacting critical business operations. Other significant occurrences include the port of Antwerp in Belgium, which fell prey to a complex cyberattack orchestrated by drug traffickers, and South Korea's GPS satellite navigation system, disrupted by coordinated electronic jamming signals affecting multiple ships and aircraft. Furthermore, the IT system of the Danish Maritime Authority was compromised by a hidden virus transmitted via PDF, resulting in the theft of sensitive data. Saudi Aramco, an oil company, was also struck by the Shamoon cyberwarfare virus, causing the incapacitation of 30,000 workstations and a complete halt in operations for over a week [65, 66]. These events underscore the critical need for industry stakeholders to prioritize cybersecurity and digital safety. Establishing a secure and resilient infrastructure for intelligent systems reliant on IoT and smart communication devices is paramount.

Cybersecurity embodies a contemporary concept that assesses the risk associated with potential scenarios or occurrences such as data corruption, subversion, and harm to IT assets, which can disrupt operational services and lead to unfavorable outcomes [11, 67]. The International Maritime Organization defines effective cyber risk management as identifying, analysing, evaluating, and reporting cyber threats and establishing acceptable tolerance levels, preventive strategies, and mitigation measures. Within the maritime sector, two primary cyber-attack categories prevail: untargeted and targeted attacks [68, 69].

Untargeted attacks seek to exploit vulnerabilities across various companies or vessels within the maritime ecosystem, whereas targeted attacks are explicitly directed at entities within the maritime domain.

The following measures are important to ensure cyber resiliency in SMGs:

- Advanced Cybersecurity Measures: Implementing robust cybersecurity protocols and encryption techniques is vital to safeguard SMGs' critical data and communication channels. Enhanced authentication mechanisms and regular security assessments are imperative to identify and mitigate potential vulnerabilities.
- Integrated Cyber-Physical Security Infrastructure: A comprehensive cyber-physical security framework is essential to protect SMGs from cyber and physical attacks. This includes deploying intrusion detection systems, firewalls, secure communication protocols, and physical security measures to prevent unauthorized access to critical shipboard infrastructure.
- Continuous Monitoring and Incident Response: Real-time monitoring of SMG operations is crucial to detect anomalous activities and potential cyber intrusions. Implementing proactive incident response strategies and disaster recovery plans can significantly minimize the impact of cyberattacks, ensuring the uninterrupted functionality of SMGs and onboard systems.

– Training and Awareness Programs: Educating shipboard personnel about cybersecurity best practices and raising awareness about potential cyber threats play a pivotal role in fortifying the cyber resiliency of SMGs. Training programs should emphasize the importance of adhering to security protocols and promptly reporting suspicious activities to the designated authorities.

The development of autonomous maritime ecosystems, as depicted in Fig. 14, leverages technological advances to unlock extensive opportunities for independent, smart SMG technologies. This progress enhances operational visibility, communication, cost-effectiveness, increased productivity, reduced greenhouse gas emissions, optimization, and the satisfaction of customer expectations [14]. Utilizing technological advances-enabled remote ship connectivity to onshore centres facilitates various tasks such as navigation optimization, power system control, maintenance, data analysis, and predictive analytics. However, these advancements also bring forth unforeseen cybersecurity challenges. The growing complexity of cyber threats within critical maritime ecosystems poses significant risks to data integrity [69, 70].

The vulnerability of the maritime ecosystem to various cyberattacks is illustrated in Fig. 15, derived from a cyber survey conducted by IHS Fair play in collaboration with the Baltic and International Maritime Council (BIMCO) in 2016 and 2018. Over this time, a broader spectrum of distinct cyberattacks has emerged. Notably, guidelines on shipboard cybersecurity, acknowledged by multiple organizations and entities (CLIA, ICS, INTERCARGO, INTERTANKO, IUMI, BIMCO), have highlighted various types of attacks, as depicted in Fig. 15. These attacks are categorized

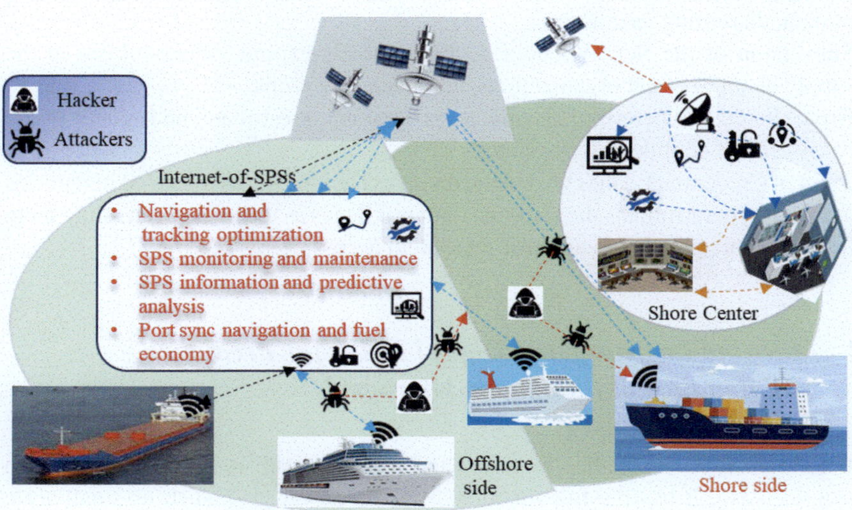

Fig. 14 Concept of cyber security in autonomous maritime ecosystems

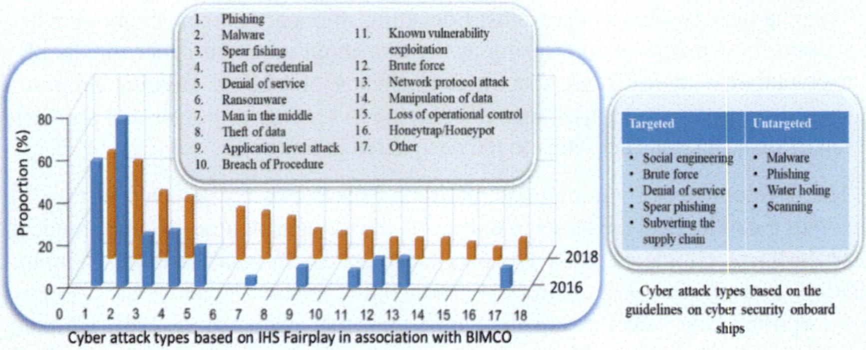

Fig. 15 Maritime vulnerability

as untargeted and targeted attacks [11], with each attack in Fig. 15. concisely defined within the context of the maritime SMG domain.

- Phishing: Cybercriminals disseminate emails or fake websites across various sections of the maritime ecosystem to deceive users, gain access to systems, or extract sensitive data.
- Spear-Phishing: Similar to phishing, this attack targets a specific ship or company.
- Malware: This term refers to malicious software utilizing harmful elements such as spyware, trojan horses, worms, and viruses to infiltrate various targets, pilfer confidential information, or disrupt systems.
- Denial of Service: Attackers inundate the targeted network with a large volume of data, disrupting user access.
- Subversion of the Supply Chain: Cybercriminals exploit vulnerabilities in the supply chain to target organizations, companies, or ships.
- Brute Force: Cybercriminals create programs to estimate and guess weak passwords of confidential systems.
- Theft of Credentials: This cybercrime involves hacking the user's proof of identity, granting cybercriminals the same account privileges as the owners, enabling them to manipulate passwords and restrict system access.

5.2 *Machine Learning and Digital Twins for Predictive Maintenance of Shipboard Microgrids*

Although an SMG is smaller in scale than a power grid, it is still considered a microgrid. Also, it has the full functionality of a power grid, such as small-scale generation, transmission, and distribution, where equipment is installed as key nodes to ensure the power quality of the grid. SMG has some different challenges to deploy from the general microgrids, such as the high density of power equipment and harsh operating environment. The SMG has some issues, making it more complex than other microgrid, such as environmental factors changing during ship's trip on the sea, such as temperature, storm, lightning, sea wave, etc. [71, 72]. Besides, electrical noises due to the large device integration density in a narrow space bring some concerns. The difference is that the SMG operating is related to water, where repairing or replacing equipment when damaged is difficult. More energy is required to make the bigger cargo ships and faster yachts. As discussed, ML and IoT play an important role in recent power management strategies. In addition, AI applications in power systems have attracted great interest in ship power electrification.

To reduce fuel use and emissions, diesel propulsion is converted into electric motors, replacing generators with greener sources of electricity. This also poses a requirement for electrical energy storage systems (ESS), thereby evolving to increase energy storage density and system performance.

From the first electric-powered passenger ships, Queen Elizabeth II (1987) to all-electric ships such as the passenger ferry Ampere (2015), a major limitation in the development of electric ships is the traveling distance because the battery capacity is smaller than the demand of large ships. Hybrid system-based ESS can fill this gap and prove to be more suitable for system designs while waiting for a breakthrough from battery technology or renewable energy. Despite being converted to electric propulsion like Queen Elizabeth II, it still uses a diesel power generation system, so emissions are still maintained; all-electric ships are usually short-travel vessels such as ferries or small cruises. In addition, the advantages of dc grid systems and medium-voltage direct current architecture also make them the first choice for building an SMG now and in the future. Along with the trend of an SMG becoming increasingly complex by integration, the role of converters becomes increasingly important to ensure power quality and reliability for the ship system. Power converters are used to convert the electrical energy from ac to dc (vice versa) and change the voltage levels for the required demands. Therefore, the number of converters has recently increased in SMG and placed a requirement for the reliability of the shipboard grid, which is very important, especially for ships on duty. The more power electronic devices are in, the more noise and fault occur. However, some modern inverters and converters have built-in fault detectors using fuzzy algorithms, and fault diagnosis has not been performed in advance. Therefore, the development of fault management strategies is necessary for an independent and highly reliable power system such as SMG. Therefore, the application of ML techniques is a potential direction to expand the fault management strategy, which approaches the self-correcting capabilities of shipboard power equipment (SPE). Furthermore, based on computational technology development, data analysis from ML techniques will yield more specific and accurate diagnostic results when applied. Recent studies have paid much attention to applying ML techniques to electrical systems analysis [73], and some of these methods perform very well. For example, the Naive Bayes method can classify virtually up to 97% of faults in SMG [74], or the decision tree methods can predict false alarms at 44% probability [75]. The artificial neural network algorithm has been used in many studies to predict energy demands and detect faults in SMG. However, studies of fault management on specific power equipment of SMG, especially for naval ships, are very primitive. Although SPE from different manufacturers can provide different

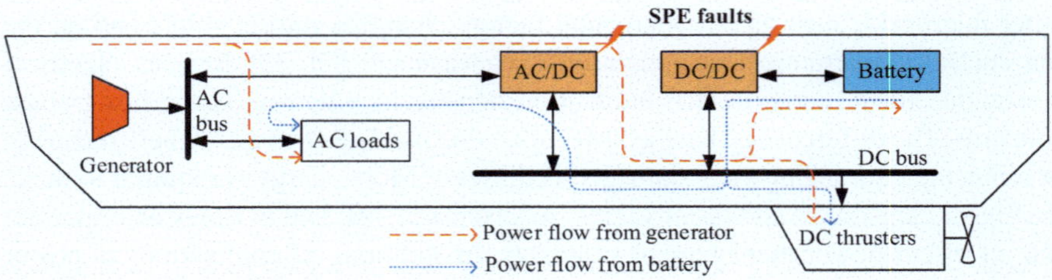

Fig. 16 An illustrator of shipboard power equipment (SPE) faults, which can interrupt the power flow of a hybrid ship

error diagnosis information, ML will train existing datasets into diagnostic models without depending on the equipment brand. Therefore, this technique promises to be more flexible in fault management. Figure 16 depicts a common model structure of a power converter-based ship power system, where SPE faults can occur anytime and critically affect the ship's performance.

The hybrid ships are integrated into an extending power source, stored in the battery packages, which can supply energy for ship electric propellers and other shipboard activities. In this case, the main reason for integrating other energy sources, such as batteries, is to reduce dependence on generators or diesel engines and towards zero emissions. For a hybrid ship with various power sources, using medium-voltage direct current and converters gives many advantages, such as reducing installation volume, wiring, and power loss, and more flexibility to implement new integration or ideas when required. However, many types of faults come from both hardware and software, making SMG interrupt. The hardware can be a short circuit or open circuit fault when any device in the grid gets physical damage; from the other, they are concerned about device incompatibility, e.g., the difference in standards, controlling methods, operating at a high frequency of power electronics components causing noise in the grid. In any case, the system will also be disrupted in need of repair and restoration. Thus, fault management strategies are beneficial to ensure a smooth operation.

In addition, a new direction is opened to optimize the operating capacity of ships by equipment maintenance strategies. There is an association between fault management and equipment maintenance: good maintenance strategies reduce damages caused by the fault, and a combination of maintenance strategies and fault management strategies helps to increase the reliability of the power system. Maintenance plans are compulsory to maintain equipment's health; manufacturers also give users methods and schedules for maintaining their equipment and components. However, the determination of the health of the equipment depends on the operation; the manufacturer's recommendations become too general and no longer seem optimal enough for the life of the device. Maintenance engineers are the ones who know best what needs to be done with the equipment in the system they have built. Applying ML technologies to analyze system data and maintenance plans ensures the most optimal conditions for machines and equipment. Three main strategies to ensure machines operate well or prevent damage from failure that can be applied for SPE are shown in Table 1.

Reactive maintenance occurs immediately after the fault is discovered by repairing or replacing; for example, open or short circuit faults occur on power cables due to unpredictable factors such as rats, fires, natural disasters, etc., equipment that needs to be replaced. The preventive maintenance method is currently implemented for almost all ships; this is the planned maintenance for preventive purposes. It means this method doesn't care about equipment status and can take more time than necessary. To optimize maintenance activities the predictive maintenance strategy is a higher level of maintenance. Figure 17 gives a more precise illustration for the comparison of the mentioned strategies.

Table 1 Three main Maintenance strategies

Maintenance strategies	Describe	Applied to
Reactive maintenance	The maintenance action is implemented after faults The faults cannot be managed Spend too much time handling faults and bring them back to work	Simple systems, easy to maintain Require an operator to supervise machines' health
Preventive maintenance	The maintenance action is implemented before faults No fault occurs, but it may increase unnecessary maintenance actions	Important system or equipment
Predictive maintenance	It is the same as preventive maintenance, but it optimizes operation time and reduces the number of maintenance actions. However, prediction accuracy is the key to evaluating the efficiency of maintaining	Suitable for automatic systems and requires optimal operation

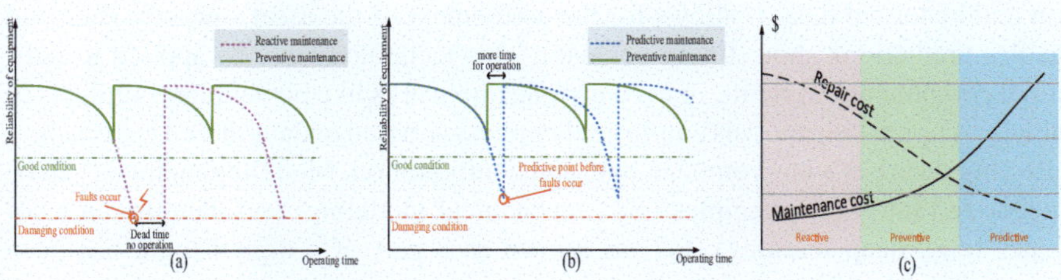

Fig. 17 Comparisons of maintenance strategies. **a** Reactive maintenance versus preventive maintenance: the preventive maintenance activity continuously ensures device reliability is always good. **b** Predictive maintenance versus preventive maintenance: the equipment in both cases can work continuously, and predictive maintenance tries to extend the working time of the equipment. **c** Cost of three types of maintenance

As in Fig. 17, there is a dead operation time for handling faults in reactive maintenance, including isolating faults and reconfiguring the system. Unexpected faults make the system waste time for recovery and increase the cost. Thus, pre-fault strategies of maintenance are worth the objects of protecting power converters. It allows power converters to work for the longest duration and prevent unexpected dead time; the pre-fault strategies include two types of maintenance: preventive and predictive. Preventive maintenance ensures that the device condition is always in a good state through planned maintenance activities; the other can increase the time-in-use of the device by predicting a safe point for maintenance before faults occur. In the context of preventive maintenance, the approach taken is one based on manufacturer recommendations and accumulated experience. This regimen is consistently implemented in a timely manner, ensuring that the equipment remains in optimal condition. The method based on prediction seems to be effective in saving time and money for maintenance activities but also meets some problems with the reliability of prediction. However, prediction accuracy is generally improved by supporting the latest mathematic algorithms, such as deep learning or ML, which depends on the amount of data for analysis activities. Moreover, this also increases the density of installed sensors on each device, making predictive maintenance more expensive than the others, but this is worth considering for the stability and confidence it brings. A model of predictive maintenance starts with a set of historical data. This dataset is the real data of failures in whole operation conditions, including normal stage, run-to-failure stage, and faulty stage. Data is classified and preprocessed to prepare for the predictive model. Prediction is an endless process with two main tasks: training and validating. Predictive maintenance can be deployed by using current SCADA and sharing the collected data for running the model without installing additional measuring equipment.

5.2.1 Deploy Predictive Maintenance-Based Digital Twin and ML for SMG

A digital twin is a real-time model of a physical asset, capturing its current condition and historical data. It allows for the assessment of the asset's present state and enables predictions about future behavior. This technology can be applied to individual components, systems, or even complex networks like power plants or fleets of vehicles. Digital twin models employ either physics-based or data-driven approaches, reflecting the asset's environment, age, and configuration. Modeling methods can be categorized into first principles (e.g., mechanical modeling) and data-driven techniques (e.g., deep learning). IoT, as known as a part of a digital twin, has been upgraded to AI of Things recently. A digital twin model will include the IoT asset's required components, behaviors, and dynamics. The IoT application drives what you need to model as part of a digital twin. The proliferation of the IoT and AI of Things has significantly contributed to the growth of online services. Among these, cloud servers stand out as a critical component, facilitating bidirectional data flow and creating new possibilities. This surge in technological capabilities has sparked an innovative concept: transforming predictive maintenance models into cloud-operated software services. Each node in this network is equipped with integrated or attached sensors capable of monitoring the device's status and relaying this data to the central server. Companies specializing in predictive maintenance services utilize this data for in-depth analysis, providing insights into potential failures or estimating the Remaining Useful Life (RUL) of equipment. This concept, aptly termed as Predictive Maintenance as a Service (PaaS), offers a glimpse into a streamlined and efficient future solution.

As mentioned above, data is the material for computers to visualize the object under consideration. For an electrical system in general, three common types of data can be measured: voltage, current, and frequency. The change in these quantities provides a lot of information about the system. Evolving situations can be learned by ML and deep learning models. This process is depicted in Fig. 18, where the SMG data is trained to become the feature components of the DT. The two-way arrow in the schema shows the connection of the DT and affects physical devices. The purpose of this connection is to predict a possible future failure of the SMG, which could be an overload, short circuit, or deterioration of the device's health index, such as RUL.

Figures 19 and 20 exemplify a DT implementation for an SMG. The complexity of SMG with many power sources at different voltage levels causes system instability, thereby easily causing serious errors. Although devices have the prescribed life and maintenance intervals under ideal conditions, as shown in Fig. 20, it is difficult to continuously deal with complex situations, including Objective factors such as temperature and humidity when operating on water, which create potential errors and reduce the lifespan of SMG's equipment.

Data is collected and transmitted to the data server, as shown in Fig. 19, and then a cloud computer installed with PaaS processes the data to estimate the RUL of the device, as shown in Fig. 20. The calculation process RUL can be summarized in the following five steps [76]:

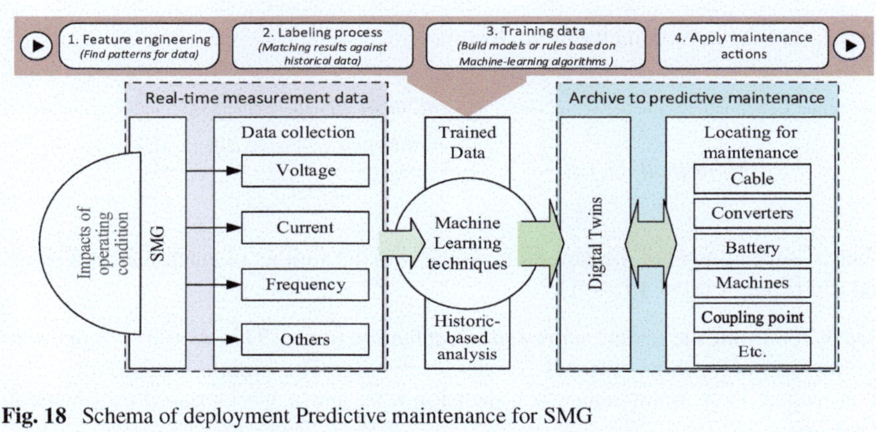

Fig. 18 Schema of deployment Predictive maintenance for SMG

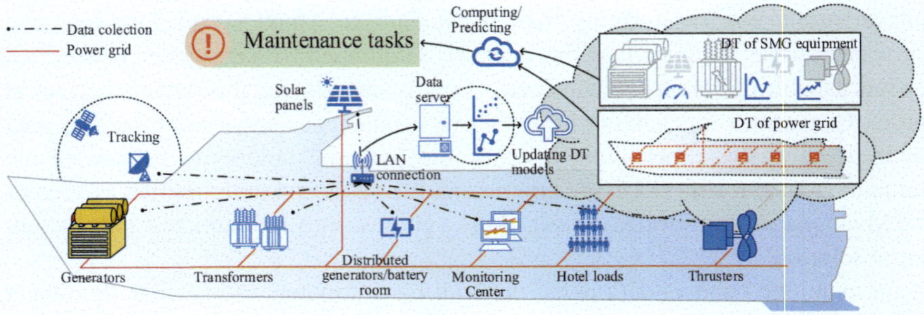

Fig. 19 Deploy digital twins for SMG

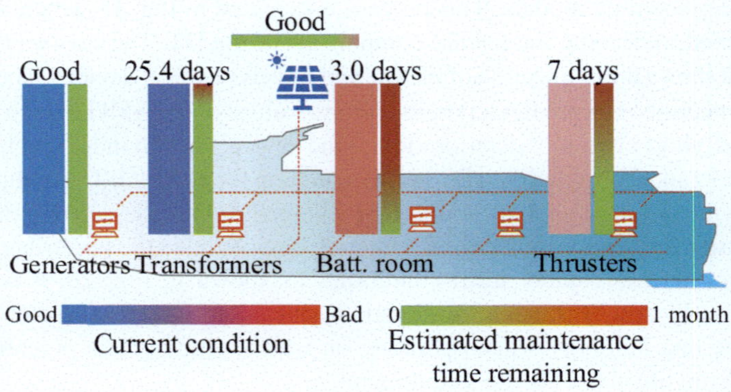

Fig. 20 A Predictive-maintenance-as-a-service interface: RUL data of SMG's equipment

Step 1: Find a feature from collected data, then post-process the extracted feature. Assume that our feature is $x = \left[x_1 \ \ldots \ x_n \right]^T$

Step 2: find the monotonicity of the feature

$$Monotocity(x_i) = \frac{1}{m} \sum_{1}^{m} \frac{\left| \begin{array}{l} number\ of\ positive\ diff\,(x_i) \\ -number\ of\ negaive\ diff\,(x_i) \end{array} \right|}{n-1}$$

where n is the number of measurement points, m is the number of machines monitored and $diff\,(x_i) = x_i(t) - x_i(t-1)$.

Step 3: determine the feature score and select feature fusion. This should be supported by learning algorithms like ML or deep learning. The simple method that can be considered for maintenance is a decision tree, and it works based on practical experience.

Step 4: using statistics to fit Exponential Degradation Models for RUL estimation.

$$h(t) = \phi + \theta \exp\left(\beta t + \epsilon - \frac{\sigma^2}{2}\right)$$

where $h(t)$ is the health indicator as a function of time. ϕ is the intercept term considered as a constant. θ and β are random parameters determining the slope of the model, where θ is lognormal-distributed and β is Gaussian-distributed. At each time step t, the distribution of θ and β is updated to the posterior based on the latest observation of $h(t)$. ϵ is a Gaussian white noise yielding to $N\left(0, \sigma^2\right)$. The $-\frac{\sigma^2}{2}$ term in the exponential is to satisfy the expectation of $h(t)$.

$$E[h(t)\theta, \beta] = \phi + \exp(\beta t)$$

If historical data is available, use the fit method provided by the Exponential Degradation Model to estimate the priors and intercept. However, historical data is not available for this wind turbine-bearing dataset. The prior of the slope parameters are chosen arbitrarily with large variances ($E(\theta) = 1, Var(\theta) = 10^6, E(\beta) = 1, Var(\beta) = 10^6$) so that the model is mostly relying on the observed data. Based on $E[h(0)] = \phi + E(\theta)$, intercept θ is set to -1 so that the model will start from 0 as well.

The relationship between the variation of health indicators and the variation of noise can be derived as

$$\Delta h(t) \approx (h(t) - \phi)\Delta\epsilon(t)$$

Here the standard deviation of the noise is assumed to cause 10% of variation of the health indicator when it is near the threshold. Therefore, the standard deviation of the noise can be represented as $\frac{10\% \cdot threshold}{threshold - \phi}$.

The exponential degradation model also provides a functionality to evaluate the significance of the slope. Once a significant slope of the health indicator is detected, the model will forget the previous observations and restart the estimation based on the original priors.

Step 5: evaluate probability by α bound, assumed to be set to 20%. The probability that the estimated RUL is between the α bound of the true RUL is calculated as a performance metric of the model:

$$\Pr(r^*(t) - \alpha r^*(t) < r(t) < r^*(t) + \alpha r^*(t) | \Theta(\mathrm{t}))$$

where $r(t)$ is the estimated RUL at time t, $r^*(t)$ is the true RUL at time t, $\Theta(\mathrm{t})$ is the estimated model parameters at time t.

Predictions always have errors, and the advantage of ML is that errors can be continuously corrected through the learning process. DT provides additional simulation functionality so that training can occur outside the real model with the best results. Based on ML and DT, new SMG architectures will have adequate preparation as well as many options for the most appropriate and effective maintenance strategies.

5.3 Seaport Microgrids

Seaports are essential for global trade, facilitating the transportation of nearly 90% of goods globally. The role of all-electric ships (AESs) in this landscape is paramount. Guided by the International Maritime Organization's ambition to cut annual greenhouse gas emissions by half by 2050 compared to 2008, several innovations tailored for AES are emerging [77]. Among these, the cold-ironing technology stands out. This technology allows seaport microgrids to supply electricity to moored ships from the shore, presenting a more sustainable solution than traditional diesel engines [77, 78].

However, integrating AES with renewable energy resources (RES) in microgrid brings complex power-system optimization challenges. In the past, microgrids managed voltage separately from berth allocation when dealing with diesel ships [79]. However, with the growing prevalence of AESs, this approach is becoming less viable. Docking several AESs simultaneously can amplify charging demands, potentially destabilizing local voltage levels. This could trigger a need to revisit berthing strategies in light of microgrid's operational boundaries. Consequently, a holistic approach that combines voltage management with berth allocation becomes imperative to obtain their combined benefits [80].

At the following, a cohesive strategy for voltage stabilization and efficient AES berth allocation within seaport microgrids is provided. Figure 21 illustrates this integrated control framework, which melds electrical and communicative elements. This distributed control model's ability to synchronize advantages for system operators and AESs makes it possible [81]—the seaport microgrid layout, emphasizing power sources, such as various RESs and the primary grid. Enhancing port operation efficiency remains the core purpose of port microgrids [82].

The rapid integration of RESs into microgrids and clusters underscores the demand for robust control strategies. Multi-agent-based controls have made it possible to balance power, maintain voltage and frequency stability, and coordinate economic activities in microgrids. The critical differences between microgrids and regular power systems include how they work and behave, mainly because of how power electronic converters work. To counter this, hierarchical control mechanisms have been discussed in [83].

5.3.1 Seaport Smart Grid

Communication networks enable effective control and oversight of power generation and distribution processes. According to the European Regulatory Group for

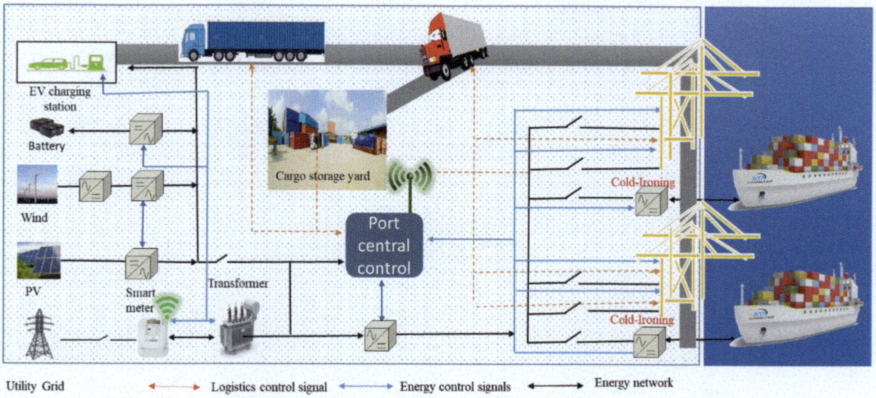

Fig. 21 An illustration of Seaport Microgrid topology [81]

Electrical and Gas, the smart grid may be defined as an electrical network that can efficiently integrate the behaviours and activities of all interconnected users [84]. Consequently, there is an observed improvement in operating efficiency, a decrease in power dissipation, and an increase in levels of quality and security. The core characteristics of an intelligent grid include a focus on customer needs, the capacity to autonomously recover from disruptions, efficient utilisation of assets, resistance to both physical and cyber risks, reliable communication infrastructure, commitment to environmental sustainability, assurance of energy safety, and enhanced dependability. Figure 22 presents a visual representation of the implementation of intelligent energy management within a port setting. It effectively showcases five distinct energy consumption areas: vessels, quay cranes, warehouse and reefers regions, charging stations, and rail-mounted gantry cranes. Using clever communication technology, the aforementioned loads connect with the smart energy management system. The energy sources include a variety of technologies, including solar panels, wind turbines, the utility grid supply, and energy storage devices [85]. The intelligent energy management system efficiently oversees electricity allocation via sophisticated control techniques enabled by communication networks [13]. The [86] thoroughly examines the strategic plan for implementing a smart grid system at the port. The approach encompasses several elements, such as achieving energy homeostasis, analysing equipment load, evaluating the advantages of smart grid technology, and conducting scenario analysis. The steps indicated above are significant in evaluating the diurnal variations of RESs, improving peak load reduction, advancing energy storage systems, and efficiently managing expenditures and pricing structures. Microgrids play a crucial role in facilitating the development of the smart grid. Microgrid refers to a specific subset inside a larger electrical grid, including components derived from the central utility grid. In contrast, an intelligent grid functions on a larger scale within the utility sector, including a wide range of transmission and distribution assets. In contrast, microgrids exhibit a higher degree of localization and can operate independently from the main utility power grid. Port sustainability, as outlined in [87], refers to the

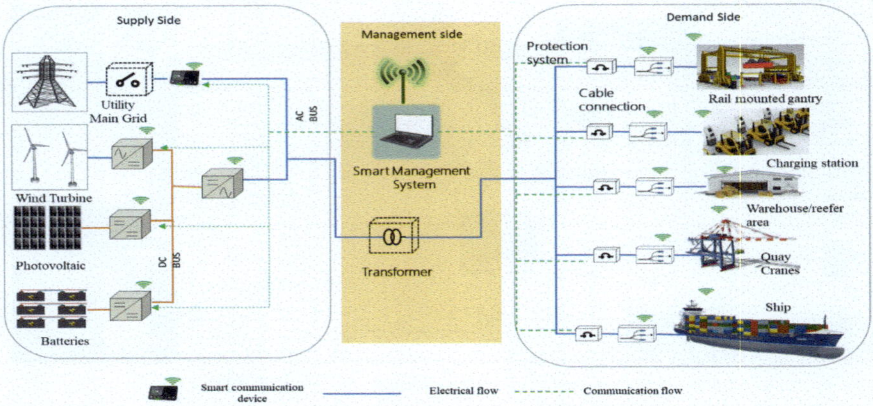

Fig. 22 An illustration of smart energy management at the harbour

state where the port authority takes responsibility and proactive measures to enhance port operations, guided by an ecologically focused development plan and in cooperation with its stakeholders. Implementing the green/smart port idea offers a viable strategy for achieving energy sustainability in port environments [80].

5.3.2 Challenges and Opportunities in Smart Seaports

Leading European ports are delving into blockchain technology to shift from traditional paper-based cargo documentation and customs processing. Such innovations streamline operations and bolster port safety, reduce energy consumption—as evidenced by motion-sensitive terminal lighting—and improve waste management. A case in point is the Port of Valencia, where the introduction of motion-sensitive lighting resulted in an 80% reduction in energy consumption, recouping its investment within two years. To fully harness digital transformation's benefits, ports need to recognize and overcome the inherent technological, operational, and governance challenges. Table 2 enumerates these obstacles related to port digitalization. The ever-evolving transportation sector necessitates that ports refurbish their existing infrastructure. They must harmonize community aspirations with infrastructural growth to sustain maritime transit. The essence of a smart port is its ability to integrate the port ecosystem into the wider infrastructure framework. This ethos is evident in the Copenhagen-Malmo Port, which utilized digital twin technology to re-engage and assimilate the local community into port functions [88]. Numerous ports are consolidating to establish unified, smart ports, meeting the burgeoning demand for port enlargement.

Figure 23 illustrates the daily challenges confronting ports, from infrastructure hurdles to the capricious nature of global trade, financial limitations, regulatory obligations, security imperatives, digital transition, sustainability obligations, and intermittent community endorsement. Given the amplified global focus on maritime

Table 2 Technical challenges to smart seaports

Type	Challenges	Description	References
Technological	Insufficient standardization	Digital technologies might be limited by IT system incompatibility and/or need for interoperability across freight ecosystem operators	[96]
	Data security	Highly networked DT systems may increase data security issues and cyberattack susceptibility	[97]
	Type of technology	The biggest issue for businesses is choosing and implementing the correct digital solutions to support their digital strategy	[98]
Governance and organizational	Being open to experimentation and iteration	An important obstacle to overcome is getting employees to try new things, experiment, iterate, and operate more agilely	[99]
	Organizational culture	Digital technology' disruptive impact is stifled by the organization's structural components, both physical (e.g., means of production) and intangible (e.g., organizational culture)	[100]
	Inertia and resistance	Companies that have been using the same methods of operating for many years tend to be resistant to change. Companies often struggle to alter the attitudes and worldviews of their staff	
	Digital blindness	Companies suffer from both a lack of a feeling of urgency and a lack of a digital vision	[98]
Operational	Trust and corporate secrets	Businesses are reluctant to share data with ecosystem partners because of concern that they may lag behind or be taken advantage of by rival enterprises. So, it's difficult to build trust between companies	[101]
	Resource scarcity	A large financial outlay is essential to guarantee the accessibility of cutting-edge digital tools and resources	[100]
	Missing skills and talent	There is a serious lack of qualified digital professionals. Training and development opportunities are necessary for employees in order to improve both their IT abilities and their process expertise	[100]
	Lack of clarity regarding the economic benefit	Due to fragmented value chain deployment, technical advances and economic advantages are uncertain	[102]

transportation, these challenges are set to deepen. Yet, these issues can transform into significant opportunities with a nimble approach to digital transition. With the escalating significance of the shipping industry, it's evident that conventional infrastructure struggles to match the requirements of maritime and port operations [89]. Real-time forecasting tools, powered by big data, IoT sensors, and AI, can enable ports to precisely predict ship arrivals, thereby mitigating congestion and maximizing capacity utilization. However, managing expansive production capacity against finite facilities amidst a booming global trade remains daunting. Modern technologies are instrumental for ports to enhance their throughput. As an example, the Port of Singapore aspires to evolve into the world's leading automated port, integrating automated vehicles and cranes, indicating a pivotal industry transformation. Digital dashboards offer instantaneous insights, streamlining port operations and enhancing strategic planning [90].

Safety remains paramount in port operations. AI-infused smart port solutions can automate high-risk tasks, protecting personnel. Proactive monitoring mechanisms also empower authorities to mitigate potential hazards. Amidst global sustainability pacts like the Paris Agreement and the 2030 Agenda for Sustainable Development [91], ports grapple with aligning operational gains with eco-responsible practices. A digital-green transition is quintessential, resonating with the shipping industry's ecological trajectory and offering sustainable energy solutions. Nonetheless, intensive port operations can occasionally strain adjacent communities, challenging consistent local support. Thus, smart ports are investing in integrating community feedback into their strategic framework, aspiring to elevate local living standards

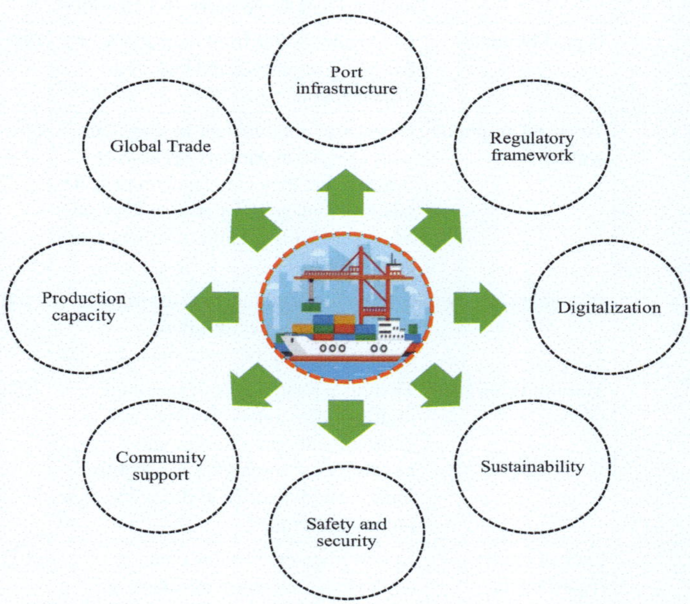

Fig. 23 Challenges to the smart seaports

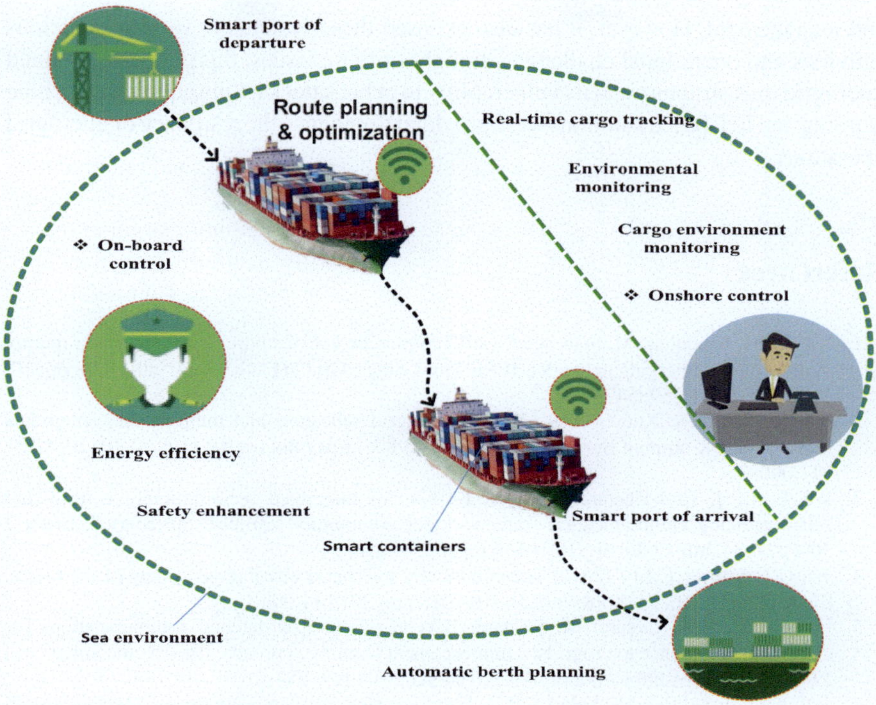

Fig. 24 Emerging applications of smart architecture

and cultivate eco-sensitive ports. Regulatory standards accentuate the imperative for sustained environmental surveillance and compliance [92]. Figure 24 highlights the manifold advantages of port digitization, encompassing safety enhancements, route optimization, and real-time cargo monitoring.

The adoption of digitized, IoT-driven applications is essential for ensuring maritime safety, curbing incidents, and fortifying security. The integration of sensors and M2M system technologies under the intelligent water transportation paradigm [93] is slated to be instrumental for future maritime endeavors. Pioneering route-planning algorithms, instantaneous cargo tracking systems, and environmental surveillance tools are ushering in a transformative maritime era [94, 95].

6 Conclusion

Renewable microgrids, owing to their lower spring reserve, high dynamic loads, and the inherent volatility of renewable generation sources, require precise real-time control and management. The development of Industry 4.0, along with the integration of digital towing and IoT technology, has indeed facilitated this control and management. However, it has also exposed these systems to various malicious activities and operational challenges. In light of these issues, this chapter examined microgrid infrastructure and its vulnerability to cyber-attacks. Subsequently, machine learning and IoT-based solutions were provided to enhance the resilience of microgrid operations.

References

1. Albu MM, Sănduleac M, Stănescu C (2017) Syncretic use of smart meters for power quality monitoring in emerging networks. IEEE Trans Smart Grid 8(1):485–492. https://doi.org/10.1109/TSG.2016.2598547
2. Fang X, Misra S, Xue G et al (2012) Smart grid—the new and improved power grid: a survey. IEEE Commun Surv Tutor 14(4):944–980. https://doi.org/10.1109/SURV.2011.101911.00087
3. Wu L, Li J, Erol-Kantarci M et al (2017) An integrated reconfigurable control and self-organizing communication framework for community resilience microgrids. Electr J 30(4):27–34. https://doi.org/10.1016/j.tej.2017.03.011
4. Wu L, Ortmeyer T, Li J (2016) The community microgrid distribution system of the future. Electr J 29(10):16–21. https://doi.org/10.1016/j.tej.2016.11.008
5. Elsayed M, Erol-Kantarci M, Kantarci B et al (2020) Low-latency communications for community resilience microgrids: a reinforcement learning approach. IEEE Trans Smart Grid 11(2):1091–1099. https://doi.org/10.1109/TSG.2019.2931753
6. Jasiūnas J, Lund PD, Mikkola J (2021) Energy system resilience—a review. Renew Sustain Energy Rev 150:111476. https://doi.org/10.1016/j.rser.2021.111476
7. Rouhani SH, Mojallali H, Baghramian A (2021) An optimized fuzzy sliding based active disturbance rejection control for simultaneous cyber-attack tolerant and demand response participation program 31(12):e13206. https://doi.org/10.1002/2050-7038.13206
8. Rouhani SH, Mojallali H, Baghramian A (2022) Load frequency control in the presence of simultaneous cyber-attack and participation of demand response program 44(10):1993–2011. https://doi.org/10.1177/01423312211068645
9. Nejabatkhah F, Li YW, Liang H et al (2021) Cyber-security of smart microgrids: a survey 14(1):27. PubMed PMID: https://doi.org/10.3390/en14010027
10. Rouhani SH, Abbaszadeh E, Sepestanaki MA et al (2023) Adaptive finite-time tracking control of fractional microgrids against time-delay attacks. IEEE Trans Ind Appl, 1–13. https://doi.org/10.1109/TIA.2023.3312223
11. Larsen MH, Lund MS (2021) Cyber risk perception in the maritime domain: a systematic literature review. IEEE Access 9:144895–144905. https://doi.org/10.1109/ACCESS.2021.3122433
12. Li Z, Shahidehpour M, Liu X (2018) Cyber-secure decentralized energy management for IoT-enabled active distribution networks. J Mod Power Syst Clean Energy 6(5):900–917. https://doi.org/10.1007/s40565-018-0425-1
13. Yigit Y, Nguyen LD, Ozdem M et al (2023) TwinPort: 5G drone-assisted data collection with digital twin for smart seaports. Sci Rep 13(1):12310. https://doi.org/10.1038/s41598-023-39366-1
14. Chowdhury S, Chowdhury SP, Crossley P (2022) Microgrids and active distribution networks
15. Chen X, Zhou J, Shi M et al (2022) Distributed resilient control against denial of service attacks in DC microgrids with constant power load. Renew Sustain Energy Rev 153:111792. https://doi.org/10.1016/j.rser.2021.111792
16. Ghosh S, Chanda CK, Das J (eds) (2021) A comprehensive survey on communication technologies for a grid connected microgrid system. In: 2021 international conference on artificial intelligence and smart systems (ICAIS). IEEE

17. Miller D, Mirzaeva G, Townsend CD et al (2021) The use of power line communication in standalone microgrids. IEEE Trans Ind Appl 57(3):3029–3037. https://doi.org/10.1109/TIA.2021.3057343
18. Yang H, Li Q, Chen W (2019) Microgrid communication system and its application in hierarchical control. In: Smart power distribution systems. Elsevier, pp 179–204
19. Volkova A, Niedermeier M, Basmadjian R et al (2018) Security challenges in control network protocols: a survey. IEEE Commun Surv 21(1):619–639
20. Burbano RAG, Gutierrez MLO, Restrepo JA et al (2019) IED design for a small-scale microgrid using IEC 61850. IEEE Trans Ind Appl 55(6):7113–7121. https://doi.org/10.1109/TIA.2019.2938734
21. Mackiewicz RE (ed) (2006) Overview of IEC 61850 and benefits. In: 2006 IEEE power engineering society general meeting, 18–22 Jun 2006
22. Hussain SMS, Ustun TS, Kalam A (2020) A review of IEC 62351 security mechanisms for IEC 61850 message exchanges. IEEE Trans Industr Inf 16(9):5643–5654. https://doi.org/10.1109/TII.2019.2956734
23. Humble T (2013) Quantum security for the physical layer 51:56–62
24. Farwell JP, Rohozinski RJS (2011) Stuxnet and the future of cyber war 53(1):23–40
25. Case DU (2016) Analysis of the cyber attack on the Ukrainian power grid 388(1–29):3
26. Birleanu FG, Bizon N (2020) Control and protection of the smart microgrids using internet of things: technologies, architecture and applications. In: Microgrid architectures, control and protection methods. Springer, pp 749–770
27. Venkataramanan V, Hahn A, Srivastava A (2019) CP-SAM: cyber-physical security assessment metric for monitoring microgrid resiliency. IEEE Trans Smart Grid 11(2):1055–1065
28. Vosughi A, Tamimi A, King AB et al (2022) Cyber–physical vulnerability and resiliency analysis for DER integration: a review, challenges and research needs. Renew Sustain Energy Rev 168:112794. https://doi.org/10.1016/j.rser.2022.112794
29. Ustun TS, Hussain SMS, Yavuz L et al (2021) Artificial intelligence based intrusion detection system for IEC 61850 sampled values under symmetric and asymmetric faults. Ieee Access 9:56486-56495. https://doi.org/10.1109/Access.2021.3071141. PubMed PMID: WOS:000641940600001; English
30. Ustun TS, Hussain SMS, Ulutas A et al (2021) Machine learning-based intrusion detection for achieving cybersecurity in smart grids using IEC 61850 GOOSE messages. Symmetry-Basel 13(5):826. ARTN 826. https://doi.org/10.3390/sym13050826. PubMed PMID: WOS:000654610200001; English
31. Zhu R, Liu C-C, Hong J et al (2020) Intrusion detection against MMS-based measurement attacks at digital substations. IEEE Access 9:1240–1249
32. Rouhani SH, Su CL, Elsisi M et al (2023) Enhancing cyber-physical resiliency based on meta-heuristic algorithms for microgrids against malicious cyber attacks. In: Razmjooy N, Ghadimi N, Rajinikanth V (eds) Metaheuristics and optimization in computer and electrical engineering: volume 2: hybrid and improved algorithms. Springer International Publishing, Cham, pp 367–379
33. Weiser M (1999) The computer for the 21st century 3(3):3–11
34. Wang F, Hu L, Hu J et al (2017) Recent advances in the internet of things: multiple perspectives. IETE Tech Rev 34(2):122–132. https://doi.org/10.1080/02564602.2016.1155419
35. Amin M (2014) A smart self-healing grid: in pursuit of a more reliable and resilient system [In My View]. IEEE Power Energy Mag 12(1):112–110. https://doi.org/10.1109/MPE.2013.2284646
36. Elma O, Kuzlu M, Zohrabi N (2023) Internet of energy for renewable energy-based decarbonized electrical energy systems, 11. https://doi.org/10.3389/fenrg.2023.1160184. English
37. Goudarzi A, Li Y, Xiang J (2021) Chapter 13—efficient energy management of renewable resources in microgrids. In: Yang Q, Yang T, Li W (eds) Renewable energy microgeneration systems. Academic Press, pp 285–321

38. Ullah A, Azeem M, Ashraf H et al (2021) Secure healthcare data aggregation and transmission in IoT—a survey. IEEE Access 9:16849–16865. https://doi.org/10.1109/ACCESS.2021.3052850
39. Makkar A, Garg S, Kumar N et al (2021) An efficient spam detection technique for IoT devices using machine learning. IEEE Trans Industr Inf 17(2):903–912. https://doi.org/10.1109/TII.2020.2968927
40. Doan QT, Kayes ASM, Rahayu W et al (2020) Integration of IoT streaming data with efficient indexing and storage optimization. IEEE Access 8:47456–47467. https://doi.org/10.1109/ACCESS.2020.2980006
41. Xiao L, Wan X, Lu X et al (2018) IoT security techniques based on machine learning: how do IoT devices use AI to enhance security? IEEE Signal Process Mag 35(5):41–49. https://doi.org/10.1109/MSP.2018.2825478
42. Geneiatakis D, Kounelis I, Neisse R et al (eds) (2017) Security and privacy issues for an IoT based smart home. In: 2017 40th international convention on information and communication technology, electronics and microelectronics (MIPRO); 22–26 May 2017
43. Chang J, Nimer Kadry S, Krishnamoorthy S (2020) Review and synthesis of Big Data analytics and computing for smart sustainable cities 14(11):1363–1370. https://doi.org/10.1049/iet-its.2020.0006
44. Li W, Au MH, Wang Y (2021) A fog-based collaborative intrusion detection framework for smart grid 31(2):e2107. https://doi.org/10.1002/nem.2107
45. Noura M, Atiquzzaman M, Gaedke M (2019) Interoperability in internet of things: taxonomies and open challenges. Mob Netw Appl 24(3):796–809. https://doi.org/10.1007/s11036-018-1089-9
46. Aly M, Khomh F, Guéhéneuc YG et al (2019) Is fragmentation a threat to the success of the internet of things? IEEE Internet Things J 6(1):472–487. https://doi.org/10.1109/JIOT.2018.2863180
47. Antonakakis M, April T, Bailey M et al (eds) (2017) Understanding the mirai botnet. In: 26th USENIX security symposium (USENIX security 17)
48. Lin H, Bergmann NW (2016) IoT privacy and security challenges for smart home environments 7(3):44. PubMed PMID: https://doi.org/10.3390/info7030044
49. Mohri M, Rostamizadeh A, Talwalkar A (2018) Foundations of machine learning. MIT Press
50. Rahman Fahim S, K. Sarker S, Muyeen SM et al (2020) Microgrid fault detection and classification: machine learning based approach, comparison, and reviews 13(13):3460. PubMed PMID: https://doi.org/10.3390/en13133460
51. Schnürch S, Wagner A (2020) Electricity price forecasting with neural networks on EPEX order books. Appl Math Financ 27(3):189–206. https://doi.org/10.1080/1350486X.2020.1805337
52. Razmi P, Ghaemi Asl M, Canarella G et al (2021) Topology identification in distribution system via machine learning algorithms. PLoS ONE 16(6):e0252436. https://doi.org/10.1371/journal.pone.0252436
53. Creswell A, White T, Dumoulin V et al (2018) Generative adversarial networks: an overview 35(1):53–65
54. Sadek SM, Omran WA, Hassan MAM et al (2021) Data driven stochastic energy management for isolated microgrids based on generative adversarial networks considering reactive power capabilities of distributed energy resources and reactive power costs. IEEE Access 9:5397–5411. https://doi.org/10.1109/ACCESS.2020.3048586
55. Gao Y, Hu Y, Li J et al (2020) Modeling and simulation analysis of power dispatch of solar photovoltaic microgrid in paper mill, p 81
56. Du Y, Li F (2020) Intelligent multi-microgrid energy management based on deep neural network and model-free reinforcement learning. IEEE Trans Smart Grid 11(2):1066–1076. https://doi.org/10.1109/TSG.2019.2930299
57. Elsisi M, Mahmoud K, Lehtonen M et al (2021) Reliable Industry 4.0 based on machine learning and IoT for analyzing, monitoring, and securing smart meters 21(2):487. PubMed PMID: https://doi.org/10.3390/s21020487

58. Kumar P, Gupta GP, Tripathi R et al (2023) DLTIF: deep learning-driven cyber threat intelligence modeling and identification framework in IoT-enabled maritime transportation systems. IEEE Trans Intell Transp Syst 24(2):2472–2481. https://doi.org/10.1109/TITS.2021.3122368
59. Ali Z, Terriche Y, Hoang LQN et al (2021) Fault management in DC microgrids: a review of challenges, countermeasures, and future research trends. IEEE Access 9:128032–128054. https://doi.org/10.1109/ACCESS.2021.3112383
60. Ali F, Sarwar S, Shafi QM et al (2023) Securing IoT based maritime transportation system through entropy-based dual-stack machine learning framework. IEEE Trans Intell Transp Syst 24(2):2482–2491. https://doi.org/10.1109/TITS.2022.3177772
61. Du M, Pierrou G, Wang X et al (2021) Targeted false data injection attacks against AC state estimation without network parameters. IEEE Trans Smart Grid 12(6):5349–5361. https://doi.org/10.1109/TSG.2021.3106246
62. Salem AA, Ismail MH, Ibrahim AS (2023) Active reconfigurable intelligent surface-assisted MISO integrated sensing and communication systems for secure operation. IEEE Trans Veh Technol 72(4):4919–4931. https://doi.org/10.1109/TVT.2022.3227319
63. He X, Wang J, Liu J et al (2023) DNS rebinding threat modeling and security analysis for local area network of maritime transportation systems. IEEE Trans Intell Transp Syst 24(2):2643–2655. https://doi.org/10.1109/TITS.2021.3135197
64. Ben Farah MA, Ukwandu E, Hindy H et al (2022) Cyber security in the maritime industry: a systematic survey of recent advances and future trends 13(1):22. PubMed PMID: https://doi.org/10.3390/info13010022
65. Ashraf I, Park Y, Hur S et al (2023) A survey on cyber security threats in IoT-enabled maritime industry. IEEE Trans Intell Transp Syst 24(2):2677–2690. https://doi.org/10.1109/TITS.2022.3164678
66. Oruc A (2020) Claims of state-sponsored cyberattack in the maritime industry
67. Zhang Y, Xiong R, He H et al (2018) Long Short-term memory recurrent neural network for remaining useful life prediction of lithium-ion batteries. IEEE Trans Veh Technol 67(7):5695–5705. https://doi.org/10.1109/TVT.2018.2805189
68. Bolbot V, Kulkarni K, Brunou P et al (2022) Developments and research directions in maritime cybersecurity: a systematic literature review and bibliometric analysis. Int J Crit Infrastruct Prot 39:100571. https://doi.org/10.1016/j.ijcip.2022.100571
69. Chaudhry SA, Irshad A, Khan MA et al (2023) A lightweight authentication scheme for 6G-IoT enabled maritime transport system. IEEE Trans Intell Transp Syst 24(2):2401–2410. https://doi.org/10.1109/TITS.2021.3134643
70. Aslam S, Michaelides MP, Herodotou H (2020) Internet of ships: a survey on architectures, emerging applications, and challenges. IEEE Internet Things J 7(10):9714–9727. https://doi.org/10.1109/JIOT.2020.2993411
71. Button R, Martin BD, Sollinger JM et al (2015) Assessment of surface ship maintenance requirements. Rand Corporation Santa Monica, CA
72. Martin B, McMahon ME, Riposo J et al (2017) A strategic assessment of the future of US Navy ship maintenance, p 53
73. Miraftabzadeh SM, Foiadelli F, Longo M et al (eds) (2019) A survey of machine learning applications for power system analytics. In: 2019 IEEE international conference on environment and electrical engineering and 2019 IEEE industrial and commercial power systems Europe (EEEIC/I&CPS Europe), 11–14 June 2019
74. Amgai R, Shi J, Santos R et al (eds) (2013) Machine learning based diagnosis support for ShipBoard Power Systems controls. In: 2013 IEEE electric ship technologies symposium (ESTS), 22–24 April 2013
75. Accorsi R, Manzini R, Pascarella P et al (2017) Data mining and machine learning for condition-based maintenance. Procedia Manuf 11:1153–1161. https://doi.org/10.1016/j.promfg.2017.07.239
76. Coble JB (2010) Merging data sources to predict remaining useful life–an automated method to identify prognostic parameters

77. Iris Ç, Lam JSL (2019) A review of energy efficiency in ports: Operational strategies, technologies and energy management systems. Renew Sustain Energy Rev 112:170–182. https://doi.org/10.1016/j.rser.2019.04.069
78. Rolán A, Manteca P, Oktar R et al (2019) Integration of cold ironing and renewable sources in the barcelona smart port. IEEE Trans Ind Appl 55(6):7198–7206. https://doi.org/10.1109/TIA.2019.2910781
79. Bierwirth C, Meisel F (2015) A follow-up survey of berth allocation and quay crane scheduling problems in container terminals. Eur J Oper Res 244(3):675–689. https://doi.org/10.1016/j.ejor.2014.12.030
80. Alzahrani A, Petri I, Rezgui Y et al (2021) Decarbonisation of seaports: a review and directions for future research. Energy Strat Rev 38:100727. https://doi.org/10.1016/j.esr.2021.100727
81. Sadiq M, Ali SW, Terriche Y et al (2021) Future greener seaports: a review of new infrastructure, challenges, and energy efficiency measures. IEEE Access 9:75568–75587. https://doi.org/10.1109/ACCESS.2021.3081430
82. Nutkani IU, Loh PC, Blaabjerg F (2013) Distributed operation of interlinked AC microgrids with dynamic active and reactive power tuning. IEEE Trans Ind Appl 49(5):2188–2196. https://doi.org/10.1109/TIA.2013.2262092
83. Meng L, Dragicevic T, Vasquez JC et al (2015) Tertiary and secondary control levels for efficiency optimization and system damping in droop controlled DC–DC converters. IEEE Trans Smart Grid 6(6):2615–2626. https://doi.org/10.1109/TSG.2015.2435055
84. Pham TY (2023) A smart port development: systematic literature and bibliometric analysis. Asian J Shipp Logist 39(3):57–62. https://doi.org/10.1016/j.ajsl.2023.06.005
85. Esmaeili Shayan M, Ghasemzadeh F, Rouhani SH (2023) Energy storage concentrates on solar air heaters with artificial S-shaped irregularity on the absorber plate. J Energy Storage 74:109289. https://doi.org/10.1016/j.est.2023.109289
86. Attanasio G, Battistella C, Chizzolini E (2023) The future of energy management: results of a Delphi panel applied in the case of ports. J Clean Prod 417:137947. https://doi.org/10.1016/j.jclepro.2023.137947
87. Lyridis DV, Prousalidis JM, Lekka AM et al (2023) Holistic energy transformation of ports: the proteus plan. IEEE Electrification Mag 11(1):8–17. https://doi.org/10.1109/MELE.2022.3232923
88. Kaklis D, Varlamis I, Giannakopoulos G et al (2023) Enabling digital twins in the maritime sector through the lens of AI and industry 4.0. Int J Inf Manag Data Insights 3(2):100178. https://doi.org/10.1016/j.jjimei.2023.100178
89. Liu J, Wang X, Chen J (2023) Port congestion under the COVID-19 pandemic: the simulation-based countermeasures. Comput Ind Eng. 183:109474. https://doi.org/10.1016/j.cie.2023.109474
90. Lv Z, Shang W (2023) Impacts of intelligent transportation systems on energy conservation and emission reduction of transport systems: a comprehensive review. Green Technol Sustain 1(1):100002. https://doi.org/10.1016/j.grets.2022.100002
91. Singh T, Solanki A, Sharma SK et al (2022) A decade review on smart cities: paradigms, challenges and opportunities. IEEE Access 10:68319–68364. https://doi.org/10.1109/ACCESS.2022.3184710
92. Sornn-Friese H, Sofev P, Kondratenko K (2023) The port authority as system builder in cross-border regionalization: an exploratory study of port Esbjerg in the development of north sea wind. Marit Transp Res 4:100084. https://doi.org/10.1016/j.martra.2023.100084
93. Gubbi J, Buyya R, Marusic S et al (2013) Internet of Things (IoT): a vision, architectural elements, and future directions. Futur Gener Comput Syst 29(7):1645–1660. https://doi.org/10.1016/j.future.2013.01.010
94. Cheung K-F, Bell MGH (2021) Attacker–defender model against quantal response adversaries for cyber security in logistics management: an introductory study. Eur J Oper Res 291(2):471–481. https://doi.org/10.1016/j.ejor.2019.10.019
95. Yang J-M, Tseng C-M, Tseng PS (2015) Path planning on satellite images for unmanned surface vehicles. Int J Nav Arch Ocean Eng 7(1):87–99. https://doi.org/10.1515/ijnaoe-2015-0007

96. Gong C, Ribiere V (2021) Developing a unified definition of digital transformation. Technovation 102:102217. https://doi.org/10.1016/j.technovation.2020.102217
97. Kechagias EP, Chatzistelios G, Papadopoulos GA et al (2022) Digital transformation of the maritime industry: a cybersecurity systemic approach. Int J Crit Infrastruct Prot 37:100526. https://doi.org/10.1016/j.ijcip.2022.100526
98. Agrawal P, Narain R, Ullah I (2019) Analysis of barriers in implementation of digital transformation of supply chain using interpretive structural modelling approach 15(1):297–317
99. Kane GC, Palmer D, Phillips AN (2017) Achieving digital maturity. MIT Sloan Manag Rev
100. Tijan E, Jović M, Aksentijević S et al (2021) Digital transformation in the maritime transport sector. Technol Forecast Soc Chang 170:120879. https://doi.org/10.1016/j.techfore.2021.120879
101. Vial G (2019) Understanding digital transformation: a review and a research agenda. J Strat Inf Syst 28(2):118–144. https://doi.org/10.1016/j.jsis.2019.01.003
102. Heilig L, Lalla-Ruiz E, Voß S (2017) Digital transformation in maritime ports: analysis and a game theoretic framework. NETNOMICS: Econ Res Electron Netw 18(2):227–254. https://doi.org/10.1007/s11066-017-9122-x

Deep Learning-Based Microgrid Protection

Muhammad Uzair, Li Li, and Syed Basit Ali Bukhari

Abstract Microgrids are rapidly gaining popularity as a reliable solution for localized power generation and distribution. However, ensuring microgrids' reliable operation and protection remains a critical challenge. Traditional protection methods often fail to address the complexities of microgrids, such as integrating renewable energy sources and multiple interconnected systems. This chapter introduces a unique microgrid protection system based on tunable-Q wavelet transform and multi-layered long short-term memory-based deep learning. Besides accurately detecting and classifying the faults, the system can also determine the specific phase affected by the fault, enabling prompt and targeted responses. A tunable-Q wavelet transform is employed to obtain unique patterns from three-phase current signals, and a multi-layered long short-term memory-based deep learning network is applied for fault identification and classification. Accurate fault detection ensures a quick response to faults, minimizing downtime and maximizing microgrid reliability. Secondly, fault classification enables the system to distinguish between different types of faults, which aids in implementing appropriate mitigation strategies. Lastly, the faulted phase identification capability allows for targeted interventions, streamlining maintenance efforts and reducing costs. The performance of the protection system is evaluated on a simulated microgrid model. The results demonstrate the system's ability to accurately detect faults, classify them with high precision, and identify the faulted phase correctly. These findings highlight the potential of deep learning-based microgrid protection as an effective and efficient solution for enhancing microgrids' reliability and operational efficiency.

Keywords Fault detection · Fault classification · Deep learning · Tunable-Q wavelet transform · Long short-term memory

M. Uzair (✉) · L. Li
School of Electrical and Data Engineering, University of Technology Sydney, Ultimo, Australia
e-mail: muhammad.uzair@uts.edu.au

S. B. A. Bukhari
Department of Electrical Engineering, University of Azad Jammu and Kashmir, Muzaffarabad, Pakistan

F. Shahnia and J. M. Guerrero (eds.), *Microgrids and Virtual Power Plants*,
Power Systems, https://doi.org/10.1007/978-981-97-6623-9_9

1 Introduction

The emergence of distributed generation necessitates the microgrid concept, a framework for handling distributed generation sources and their corresponding loads as a cohesive, localized power system. This approach allows the grid to detach from the distribution system during disturbances, isolating its load to maintain service without affecting the transmission grid's integrity [1]. It has gained widespread recognition for its potential to enhance energy resilience and reliability. However, robust operation and protection of microgrids present significant challenges. Traditional protection methods for conventional power systems often fail to address microgrids' unique complexities. In response to these challenges, this chapter introduces a novel microgrid protection system that combines the tunable-Q wavelet transform and multi-layered long short-term memory (multi-LSTM)-based deep learning.

The primary objective of the protection system is to overcome the limitations of traditional methods by offering a more nuanced and accurate approach to fault detection, classification, and phase identification. The system extracts unique patterns from three-phase current signals using the Tunable-Q Wavelet Transform. Integrating a multi-LSTM-based deep learning network further enhances the system's capabilities, allowing for real-time fault identification and classification. The system goes beyond conventional approaches by determining the specific faulted phase.

Various methodologies integrating machine learning (ML) and signal processing techniques have been proposed by numerous researchers to address the challenges associated with fault detection, classification, location and protection coordination in microgrids.

1.1 Machine Learning-Based Protection Methods

Fault detection and classification in microgrids have garnered significant attention, leading to various innovative approaches.

One notable strategy involves combining the Hilbert-Huang Transform (HHT) with ML. In [2], the HHT is used to pre-process post-fault voltage and current signals to extract features. These are then input to the logistic regression classifier for fault detection and the AdaBoost classifier for fault classification. In the same way, [3] has employed a method to address the challenge of fault detection and classification in microgrids by combining HHT for extracting robust features from transient signals while minimizing noise and Brownboost, a sequential ensemble learning method for accurate fault detection and classification.

Likewise, a microgrid fault protection scheme, combining signal processing and data mining, is sugegsted in [4]. It utilizes the multiresolution decomposition of wavelet transform for pre-processing voltage and current signals, computing total harmonic distortion and statistical indices. A random forest-based classifier is trained and tested for fault detection, identification, and classification.

On the contrary, [5] has employed a machine learning-based adaptive microgrid protection method. The approach leverages a hybrid model integrating artificial neural networks and support vector machines for robust microgrid state recognition. Correspondingly, an artificial neural network-based method for fault classification and location in a medium voltage distribution network is discussed in [6]. Fast Fourier Transform is used for feature extraction from frequency spectra of voltage and current samples. At the same time, a multi-layer perceptron neural network with back-propagation is utilized for fault classification and location.

Similarly, a method for microgrid fault detection and classification using multiresolution analysis and various Neural Networks is presented in [7]. Wavelet and Wavelet Packet Transforms are used for fault detection and feature extraction from fault signals. Meanwhile, artificial neural networks, Neuro-Fuzzy, and Wavelet Neural Networks are used for fault classification.

Furthermore, an intelligent microgrid protection approach is presented in [8]. Mathematical morphology is used for feature extraction, and extreme learning machines for fault classification. It extracts features from current signals to train and test the extreme learning machines classifier for primary and backup protection. Moreover, a microgrid protection scheme combining HHT for feature extraction and ML for fault classification is proposed in [9]. The scheme utilizes current signals, decomposes them using empirical mode decomposition, extracts differential features with HHT, and feeds them to three machine learning classifiers, Naive Bayes, support vector machines and extreme learning machines, for fault identification. The results demonstrate that the extreme learning machines-based method surpasses support vector machines and Naive Bayes classifiers to detect faults across various operational modes and network topologies.

Meanwhile, a novel machine-learning approach for protecting low-voltage AC microgrids against faults is introduced in [10]. It extracts unique features from simulated data using novel feature extraction techniques. Random Forest, trained on top-ranked features, is used for fault detection and classification. This method surpasses existing techniques in accurately detecting and classifying faults, besides identifying the affected phase(s). It also demonstrates high sensitivity in protecting the microgrid under different operating conditions and fault scenarios.

1.2 Protection Methods Involving Multi-agent Systems

Combining multi-agent systems and ML makes an intelligent and coordinated approach to AC microgrid protection possible. The collaborative intelligence eliminates the limitations of traditional protection methods, making way for faster response, improved fault isolation, and enhanced system stability. Interestingly, a microgrid protection approach using multi-agent combined with ML is presented in [11] which offers fault detection, protection coordination, adaptive relay settings updates and provides primary and backup protection in grid-connected and islanding modes of the microgrid. The methodology involves designing a multi-agent system framework with distinct agent roles and layers. A meshed microgrid model in Simulink provides fault data for ML training and testing, while AnyLogic software validates individual agent behaviour and interactions. Simulation results confirm the effectiveness of the multi-agent systems algorithm in both microgrid operation modes. Likewise, a protection scheme for microgrids using Q-learning and a multi-agent system is presented in [12]. Each agent, representing a protection relay, is trained using Q-learning to identify and isolate various fault types. Decentralized communication using blockchain technology facilitates information exchange between the agents.

1.3 Deep Learning Based Protection Methods

The reliable operation of AC microgrids is dependent on effective protection against faults. Deep learning emerges as a powerful tool, offering novel fault detection and classification methods, surpassing traditional accuracy and adaptability techniques. An intelligent microgrid protection scheme using S-transform for feature extraction from post-fault currents and a deep belief network for fault detection is presented in [13] which involves training a model on extracted features from simulated fault scenarios. The trained model is then validated on unseen data. Matlab-RTDS is also integrated to assess real-time performance.

Similarly, a wavelet and deep neural network-based protection scheme is used in [14]. The method aims to rapidly identify fault type, phase, and location, aiding microgrid protection and service restoration. The approach utilizes branch current measurements, processed by wavelet transforms, to extract key features fed into deep neural networks for fault analysis. Likewise, a distribution network fault detection method utilizing a deep learning neural network is discussed in [15]. The approach involves decomposing the fault signal using a dB3 wavelet and feeding the calculated sub-band energies into the sparse self-encoding neural network to detect faults.

Additionally, an intelligent protection strategy for microgrids using Convolutional Neural Network (CNN) is developed in [16]. The strategy tackles the challenge of selecting appropriate fault features from three-phase current signals by combining feature extraction and classification. Simulations demonstrate that the method can accurately detect faults in microgrids. On the other hand, the performance of LSTM and CNN models for classifying faults in microgrids is compared in [17]. The study highlights that LSTM excels at fault identification, achieving high accuracy with just a single cycle post-fault voltage signal. Both models were trained on simulated fault data generated in Matlab/Simulink, incorporating real-world noise. While both LSTM and CNN utilized discrete wavelet transform (DWT) and multiresolution analysis (MRA) for pre-processing, the LSTM network proved more effective in accurately classifying various fault types within microgrids.

Finally, an innovative protection scheme based on advanced signal processing and machine learning techniques is described in [18]. The method utilizes Empirical Wavelet Transform (EWT) to decompose the current signals, followed by feature extraction. These features and raw current data are used as input for the LSTM model to identify fault type, phase, and location. EWT has a fixed Q-factor, which may not be best suited for rapidly changing signals, like transients in electrical signals. The Tunable Q-factor Wavelet Transform (TQWT) is more appropriate for analyzing such signals.

The above literature review shows that a feature extraction technique is commonly required to obtain features for training ML and deep learning algorithms. Commonly used feature extraction techniques are HHT, ST, FT and DWT.

A core shortcoming of HHT is the generation of unwanted components in the low-frequency band and its inefficiency in isolating low-energy signal components [19]. Conversely, the S-transform faces spectral leakage due to its Gaussian window, causing inaccurate frequency localization. Wavelet transform, unlike Fourier and short-time Fourier transform, can analyze signals in time–frequency domains, offering an advantage [20]. The primary issue with Wavelet transform-based protection methods is selecting the optimum mother wavelet function. Using various mother wavelets yields varied outcomes [21] that may result in the failure of the protection system. Moreover, the DWT's response depends significantly on the sampling rate and fault inception angle, while its resolution is notably lower in the high-frequency region [22].

Hence, other feature extraction techniques that can overcome the above shortcomings are required.

2 Methodology

This chapter introduces a method for AC microgrids to detect and classify faults by combining two powerful techniques. TQWT is used to extract features from the three-phase current signals, and the multi-LSTM network is used for fault identification and classification.

2.1 Tunable Q-Factor Wavelet Transform

DWT typically adopt a fixed Q-factor of $\sqrt{2}$ for an orthogonal wavelet transformation. Nevertheless, depending on the data, alternative Q-factors may be more advantageous. The TQWT is designed to generate a wavelet MRA characterized by a variable Q-factor. The Q-factor in wavelet transformation is the ratio of the centre frequency to the bandwidth of the filters employed in the transformation process. Notably, the TQWT ensures an accurate reconstruction of the original signal.

2.1.1 Advantages of TQWT Over DWT

- Tailored Signal Analysis: DWT's fixed Q-factor can introduce spectral leakage for signals with varying frequency content. TQWT allows Q-factor selection to match the dominant frequencies, minimizing leakage and improving analysis accuracy, besides avoiding wavelet base function selection issues [23].
- Enhanced Transient Capture: Lower Q-factors in TQWT lead to more comprehensive filters that better capture rapid signal changes like transients. This proves valuable in analyzing signals such as post-fault current signals for fault detection in power systems [24].
- Improved Oscillatory Signal Analysis: Higher Q-factors yield narrower filters, which demonstrate useful for analyzing oscillatory signals [25].
- Flexibility for Evolving Research: As signal processing domains expand, the ability to adapt analysis tools becomes crucial. TQWT's tunability provides researchers with a versatile tool for exploring diverse signals and emerging applications [26].

2.1.2 Frequency-Domain Scaling

An essential component of the TQWT is frequency-domain scaling.

Lowpass Scaling

Lowpass scaling retains low-frequency information.

For a signal $x(n)$, sampled at a rate f_s, its corresponding frequency-domain representation is $X(\omega)$. Performing lowpass scaling by a factor of α (where $0 < \alpha < 1$) essentially compresses the frequency spectrum of $X(\omega)$ by the same factor α. This compression translates to a lower sample rate for the resulting signal $y(n)$, denoted by $Y(\omega)$.

Mathematically, this operation is expressed as:

$$Y(\omega) = X(\alpha\omega) \tag{1}$$

This equation highlights that any frequency component ω in the original signal gets mapped to a new frequency $\alpha\omega$ in the scaled version. Frequencies beyond $\alpha\pi f_s$ are effectively removed, resulting in a lower-frequency signal representation.

Significance of Lowpass Scaling

The ability to control the amount of low-frequency content through this scaling mechanism is crucial for the flexibility of the TQWT. By adjusting the scaling factor α, different types of signals can be analyzed:

- Higher α values: Retain more low-frequency components, suitable for analyzing slowly varying trends or signals with dominant low-frequency content.

- Lower α values: Emphasise higher-frequency content, beneficial for capturing rapid changes or analyzing high-frequency components in the signal.

This tunability allows the TQWT to adapt to diverse signal characteristics, enhancing its effectiveness in various applications.

Highpass Scaling

Another crucial operation within the TQWT is highpass scaling in the frequency domain. Unlike its lowpass counterpart, this technique selectively amplifies high-frequency content and reveals hidden details within the signal.

By performing highpass scaling by a factor of β (where $0 < \beta \leq 1$), the high-frequency components of $X(\omega)$ are essentially magnified by the same factor β. This amplification translates to a new signal $y(n)$, denoted by $Y(\omega)$, with an enhanced emphasis on high frequencies.

Mathematically, this operation is expressed as:

$$Y(\omega) = \begin{cases} X(\beta\pi + (1-\beta)\pi), 0 < \omega < \pi, \\ X(\beta\pi - (1-\beta)\pi), -\pi < \omega < 0. \end{cases} \tag{2}$$

Frequencies below $\beta\pi f_s$ are effectively suppressed, resulting in a high-frequency-focused signal representation.

Significance of Highpass Scaling

The ability to isolate and analyze high-frequency content using this scaling mechanism significantly expands the analytical power of the TQWT. By adjusting the scaling factor β, different scenarios can be analyzed:

- Higher β values: Magnify high-frequency details, useful for uncovering rapid changes, transient events, or high-frequency oscillations within the signal.
- Lower β values: Retain a broader range of higher frequencies, beneficial for studying non-stationary signals or extracting frequency-specific information.

This tunability allows the TQWT to delve deeper into various signal components. It is a valuable tool for diverse applications such as fault detection or signal feature extraction.

2.2 TQWT Algorithm

The TQWT leverages a two-channel filter bank to analyze the signal. During analysis, it creates two subbands:

- Lowpass subband ($v_0(n)$): This carries the low-frequency information and retains a sample rate of α times the original rate ($\alpha \cdot f_s$).

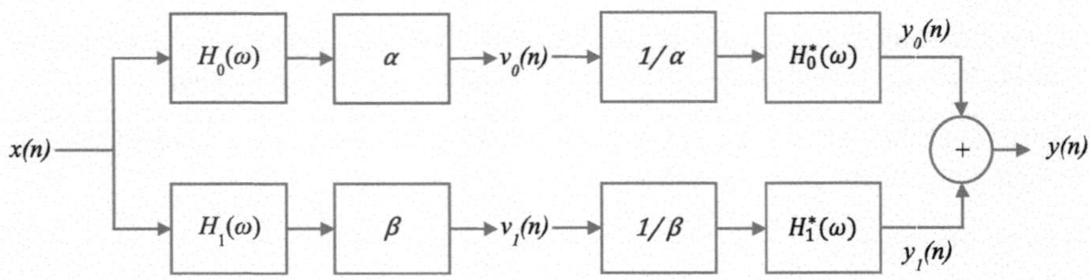

Fig. 1 TQWT two-channel filter bank

- Highpass subband ($v_1(n)$): This captures the high-frequency details with a sample rate of β times the original rate ($\beta \cdot f_s$).

Interestingly, the combined sample rate of these subbands exceeds the original rate, making the filter bank "oversampled" by a factor of $\alpha + \beta$. This oversampling ensures perfect reconstruction of the original signal later in the process (see Fig. 1).

The TQWT is characterized by two principal parameters: the Q-factor and the redundancy.

2.2.1 Redundancy

In the TQWT, $0 < \alpha < 1$ and $0 < \beta \le 1$, $\beta \cdot \alpha^{k-1} \cdot f_s$ is the sample rate at the kth iteration, where f_s is the original sample rate. As more iterations occur, the sample rate converges to $\frac{\beta}{1-\alpha} \cdot f_s$.

Where *redundancy* of the TQWT, r is:

$$r = \frac{\beta}{1-\alpha} \tag{3}$$

The redundancy (r) defines the oversampling factor in the TQWT. It relates the total sample rate across all iterations to the initial rate. The redundancy ensures perfect signal reconstruction later.

A redundancy $r \ge 3$ is recommended to obtain wavelets well-localized in time [27].

2.2.2 Q-Factor

Each iteration (k) employs a specific highpass filter $H_1^k(\omega)$.

This filter's frequency response focuses on a particular interval $[(1 - \beta)\alpha^{k-1}\pi, \alpha^{k-1}\pi]$. The centre frequency fc of the highpass filter acts as the central point of the interval and is given by:

$$f_c = \alpha^k \frac{2-\beta}{4\alpha} \cdot f_s \tag{4}$$

The bandwidth (BW) represents the length of the interval, reflecting the range of frequencies captured by the filter:

$$BW = \frac{1}{4}\beta \cdot \alpha^{k-1} \cdot f_s \tag{5}$$

Finally, the Q-factor combines these measures, providing a single value that describes the filter's frequency selectivity:

$$Q = \frac{f_c}{BW} = \frac{2-\beta}{\beta} \tag{6}$$

2.3 *Multi-LSTM Network*

Recent studies have investigated the pros and cons of using many hidden layers in an LSTM layer compared to multiple LSTM layers. In [28], it was discovered that many shallow LSTM layers with few hidden units can outperform deep LSTM layers with many hidden units. It is also suggested that a shallow layer has fewer than 100 hidden units.

2.3.1 Long Short-Term Memory

The LSTM Network is a recurrent neural network structure that addresses the vanishing and exploding gradient limitations of conventional recurrent neural networks [29]. Unlike standard recurrent neural networks, whose output relies solely on the prior time step, LSTMs incorporate a dedicated memory cell and a multi-gated system. This system, comprising input, forget, update, cell candidate, and output gates, meticulously controls the flow of information into and out of the memory cell, allowing the network to capture and utilize both short-term and long-term dependencies within sequential data.

The memory cell is responsible for retaining long-term memory, while the input gate governs the integration of received data into the memory cell. Conversely, the forget gate controls the degree to which previous memory is discarded from the cell. Additionally, the cell candidate supplements the cell state with new information. At the same time, the output gate acts as a curator, determining the fraction of the memory cell's contents at each time step.

The LSTM model comprises four primary components, each defined by its own set of mathematical equations:

$$f_t = \sigma\left(W_f \cdot \left[H_{t-1}, X_t\right] + b_f\right) \tag{7}$$

$$i_t = \sigma\left(W_i \cdot \left[H_{t-1}, X_t\right] + b_i\right) \tag{8}$$

$$C_t = f_t \odot C_{t-1} + i_t \odot \text{Tanh}\left(W_C \cdot \left[H_{t-1}, X_t\right] + b_C\right) \tag{9}$$

$$o_t = \sigma\left(W_o \cdot \left[H_{t-1}, X_t\right] + b_o\right) \tag{10}$$

$$H_t = o_t \odot \text{Tanh}(C_t) \tag{11}$$

where: f_t, i_t, C_t, and o_t are the forget gate, input gate, cell state, and output gate vectors at time step t, respectively. σ is the sigmoid activation function. $\odot$ represents Hadamard product. W and b are weight matrices and bias vectors for different LSTM components. H_{t-1} is the previous hidden state, and X_t is the input feature vector at time step t (see Fig. 2).

The LSTM layers feature a hidden state size of 64, while dropout layers are employed to regularize the network and mitigate overfitting. With a dropout rate of 0.2, roughly 20% of the neurons in the LSTM layer are randomly deactivated during training, leaving 80% for computation. Introducing dropout layers makes the model less reliant on specific neurons for predictions, enhancing its generalization capability and reducing overfitting risk. This dropout mechanism also fosters the development of a more robust representation of input data by discouraging neuron co-adaptation, thus enhancing performance on unseen data.

In the fully connected layer, 11 output neurons are allocated for ten faults and NF conditions. Finally, the softmax activation function is applied to normalize the output of the fully connected layer, ensuring a probability distribution over the 11 classes. The LSTM network is shown in the following figure.

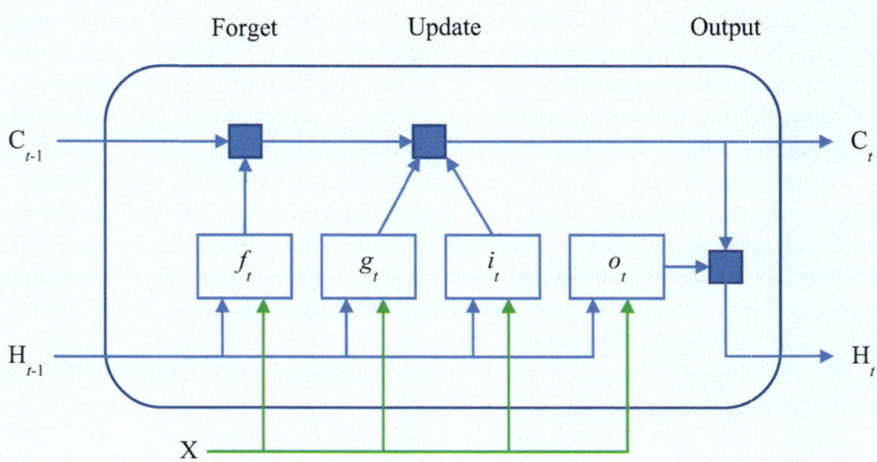

Fig. 2 LSTM cell structure and data flow at time step t

Table 1 Fault cases

Parameters	Count
Fault on different lines (DL1-DL5)	5
Fault location (10, 25, 50, 75 and 95%)	5
Fault resistance (0.01, 10, 50 and 80 Ω)	4
Fault inception angle (0, 45, 90 and 180°)	4
Operational mode (GC/AUTO)	2
Topology (Radial/Meshed)	2
Fault types (LG, LL, LLG, LLL)	10
Simulated fault scenarios	**16000**

2.4 Simulation Setup and Parameters

A standard IEC meshed microgrid model is developed to evaluate a novel protection method. This system operates at 25 kV and incorporates four distributed energy resources, two synchronous machines and two inverter-interfaced units. Each distributed energy resource is connected through a dedicated step-up transformer. The microgrid can switch between meshed and radial configurations using circuit breakers CB-Loop-1 and CB-Loop-2. This adaptability enables testing under various network structures. The system can operate in either grid-connected or islanded mode. The transition between these modes is achieved through the CB located before PCC Bus-1.

A substantial dataset comprising 16,000 cases for fault conditions and 432 cases for non-fault conditions is acquired for training and testing purposes. Current signals in three phases are recorded for each fault and no-fault case (see Tables 1 and 2).

The IEC Microgrid model for data collection and testing is shown in Fig. 3, and Figs. 4 and 5 shows the three-phase current signal for one of the fault and no-fault cases.

Table 2 No-fault cases

Parameters	Count
Load variations	6
Operational mode	2
Topology	2
Capacitor switching	6
Levels of distributed energy resources penetration	3
Simulated no fault scenarios	**432**

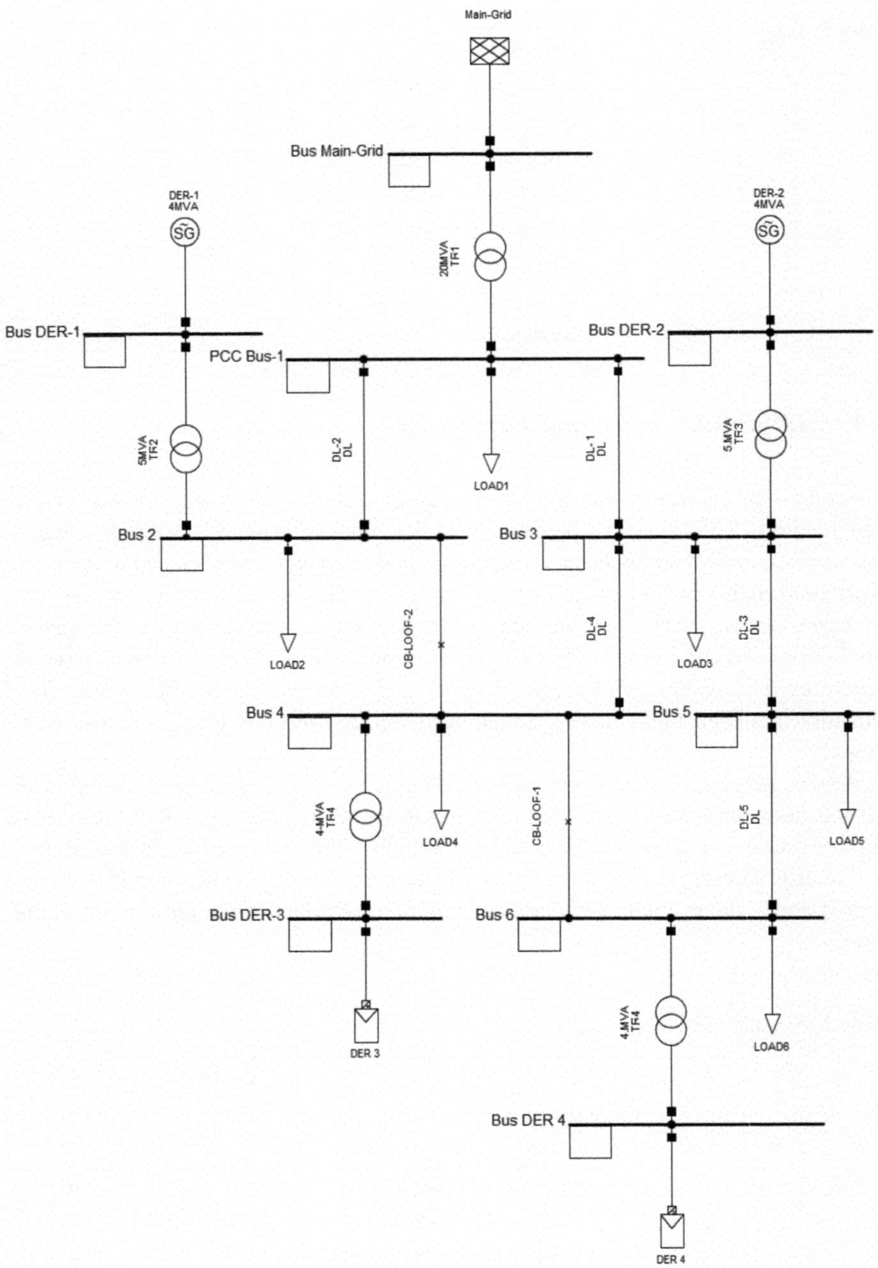

Fig. 3 IEC test microgrid

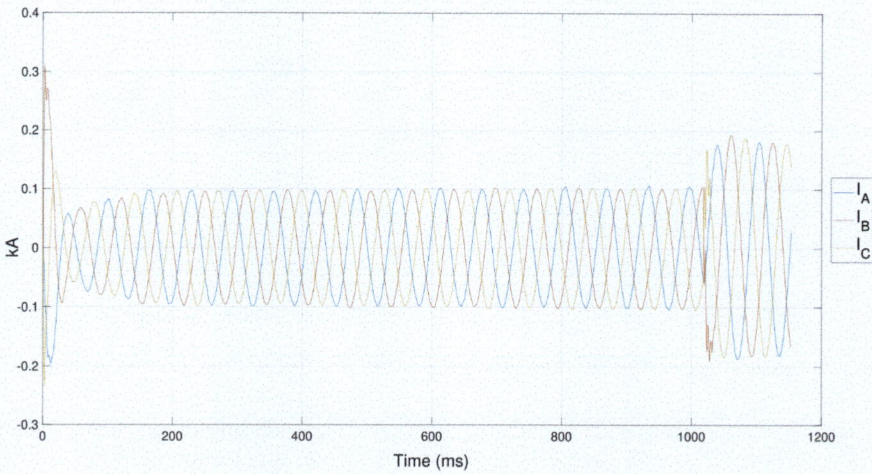

Fig. 4 The three-phase current signal for ABC fault in GC mode on DL1

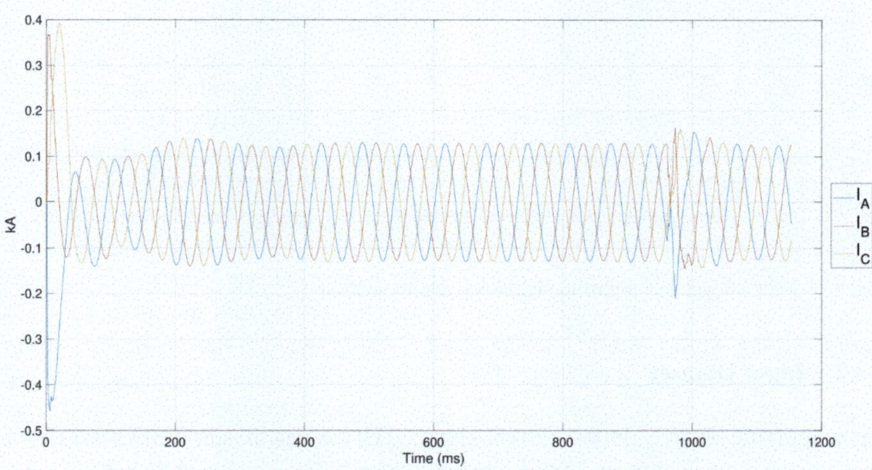

Fig. 5 The three-phase current signal for random no fault conditions

2.4.1 TQWT Multiresolution Decomposition

TQWT 4-level decomposition using a Quality factor of 1. The relative energy in the resulting approximation is close to 98% as shown in Fig. 6, while Figs. 7, 8 and 9 shows signal reconstruction for different decomposition levels and approximation.

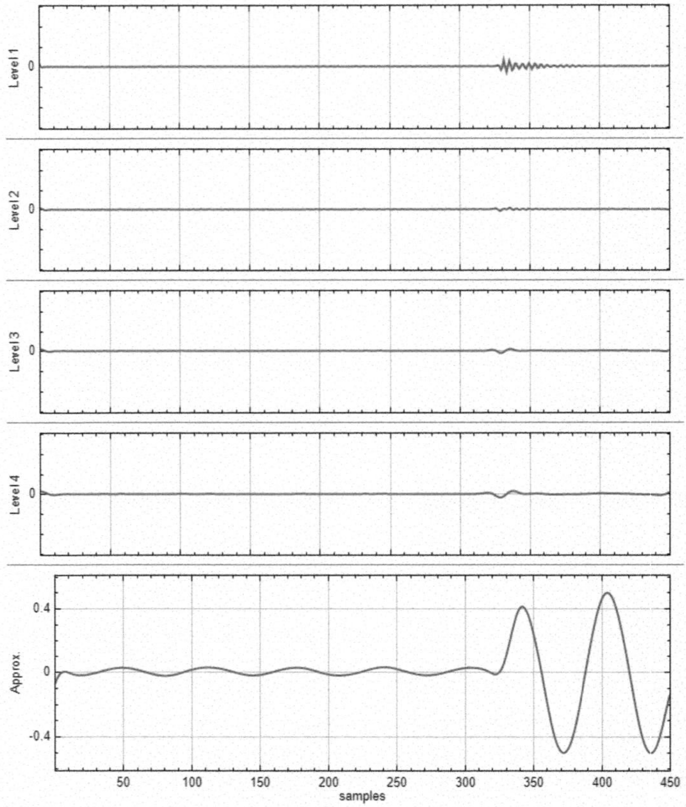

Fig. 6 TQWT 4-level decomposition including approximation

2.4.2 Input Dataset

After applying TQWT on the current signal, THD is calculated. THD serves as a measure of signal distortion, amplifying harmonics and is a crucial indicator of faults. It represents the proportion of total harmonics to the fundamental frequency, applicable to both voltage and current harmonics [30]. The following equation calculates THD_i for current, denoted as THD_i.

$$THD_i = \frac{\sqrt{I_2^2 + I_3^2 + I_4^2 + \ldots\ldots I_n^2}}{I_1} \tag{12}$$

The input dataset is derived from extracting the statistical parameters (such as standard deviation, maximum, and minimum) of total harmonic distortion (THD) (see Fig. 10).

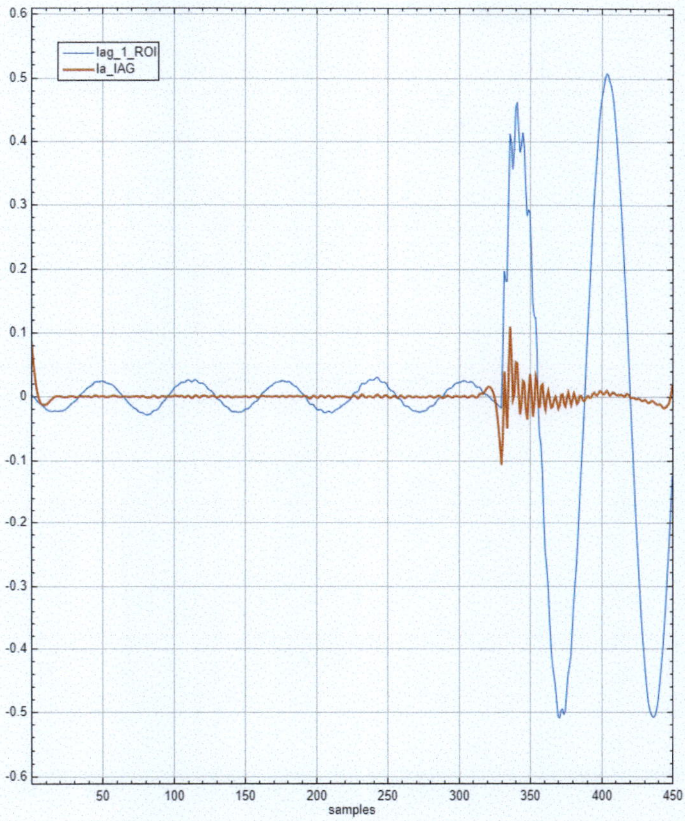

Fig. 7 Signal reconstruction using only the 4-level decomposition

3 Results and Discussion

Employing the dataset generated through a comprehensive simulation of diverse fault and non-fault scenarios, a LSTM network is trained and evaluated. The proposed multi-layered LSTM training progress using an Epoch size of 64, earning rate of 0.01 and a total 15,744 iterations is shown in Fig. 11.

The following training and test confusion matrix exhibit the multiclass classification performance of this network (see Figs. 12 and 13).

The proposed Deep Learning-based method achieves very high accuracy for detecting fault with fault type and faulted phase for both modes of microgrid operation. The results are shown in Table 3.

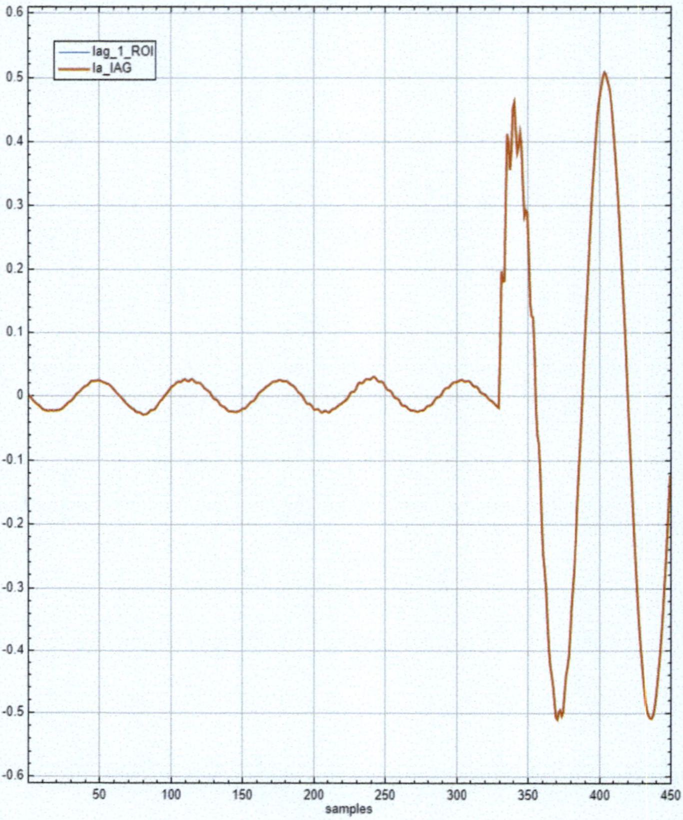

Fig. 8 Signal reconstruction using 4-level decomposition and approximation

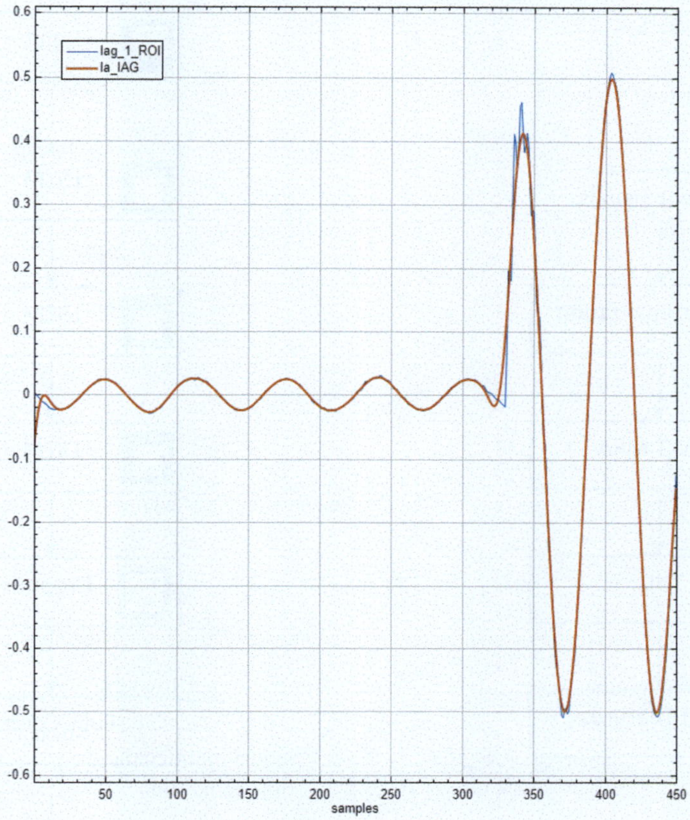

Fig. 9 Signal reconstruction using only the approximation

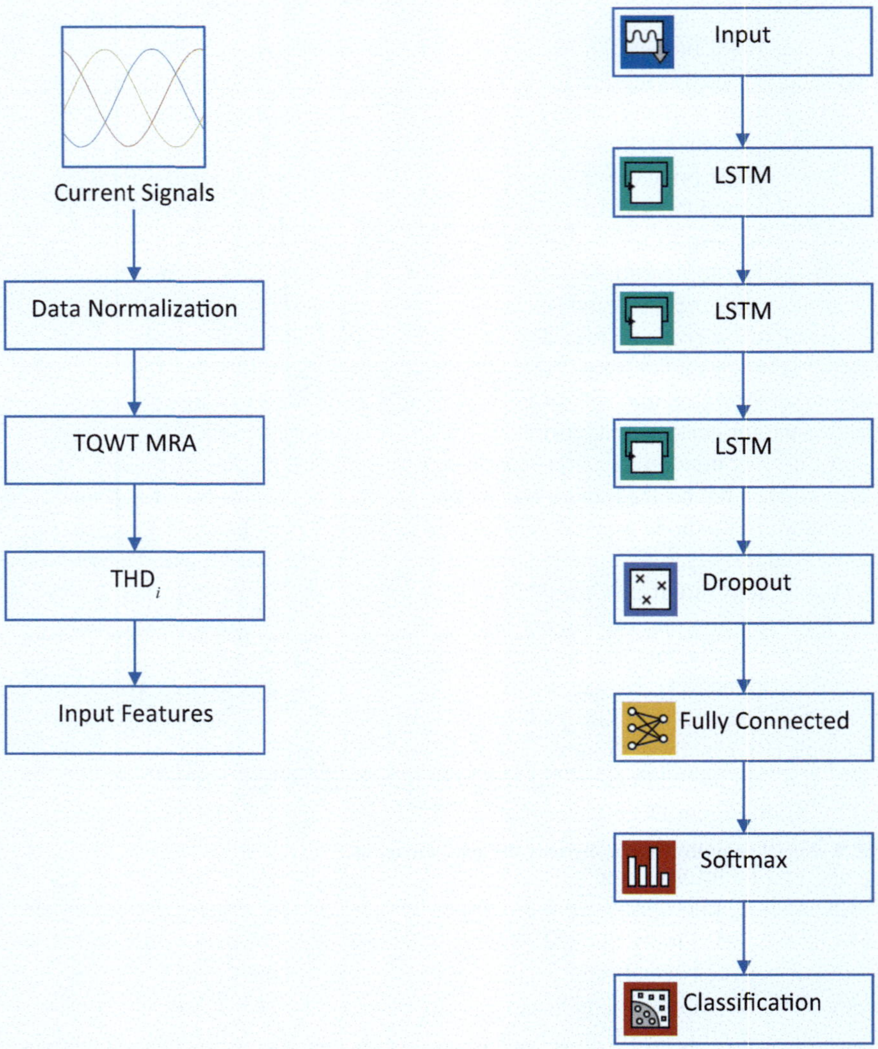

Fig. 10 Flow diagram for fault detection and classification using multi-LSTM layered deep learning network

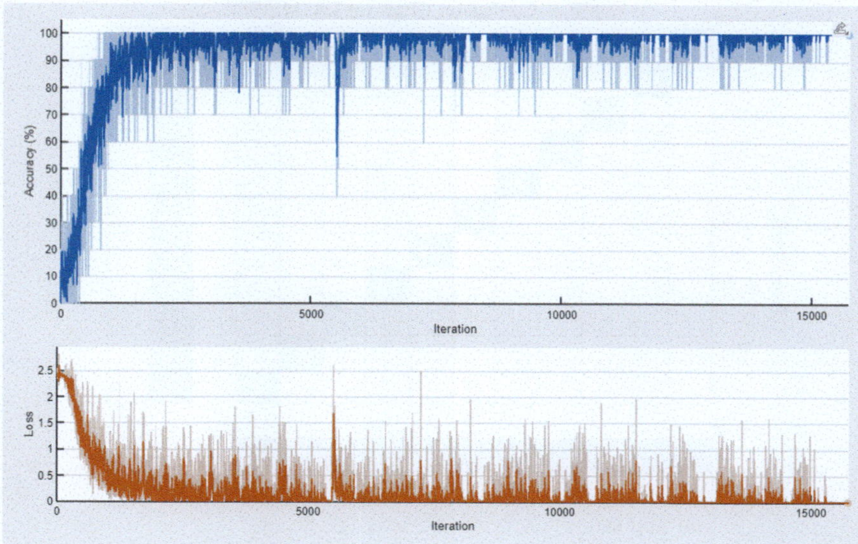

Fig. 11 Proposed multi-LSTM layered deep learning network training progress

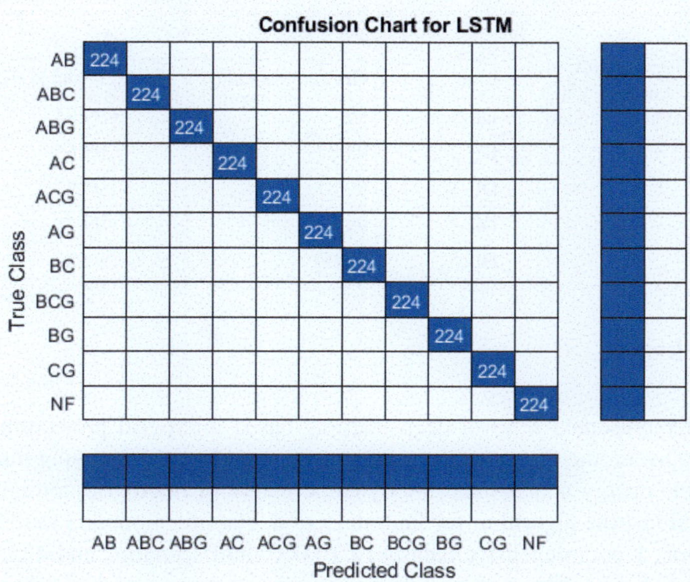

Fig. 12 Training confusion matrix for DL1 in GC mode

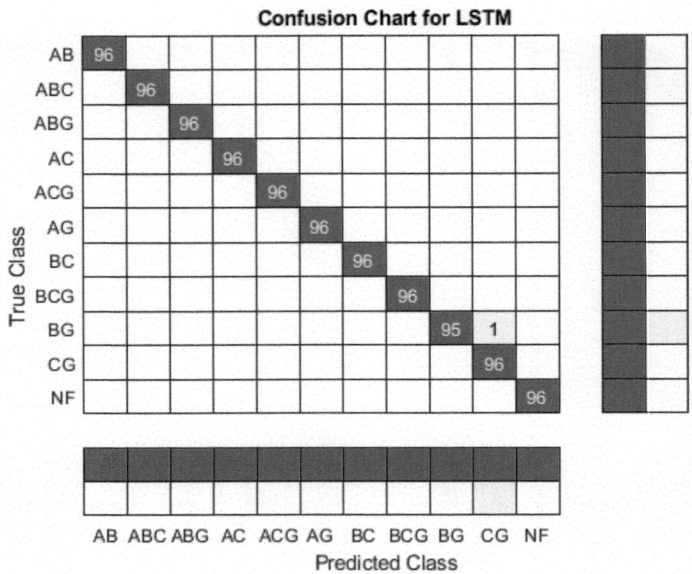

Fig. 13 Test confusion matrix for DL1 in GC mode

Table 3 Fault—no-fault detection—test accuracy in grid-connected and islanded mode

Line	Grid-connected (%)	Islanded (%)
DL-1	99.90	99.84
DL-2	99.87	99.65
DL-3	99.87	99.87
DL-4	99.84	99.78
DL-5	99.90	99.87

4 Conclusion

This chapter presented a novel deep learning-based microgrid protection system designed to overcome the protection challenges posed by the increasing integration of renewable energy sources and the dynamic nature of microgrid configurations. To train and test the algorithm for fault detection, classification, and faulted phase identification, a comprehensive dataset of 16,000 fault scenarios and 432 no-fault conditions was used. Three-phase current signals were acquired for each case, and the combined dataset was partitioned into training (70%) and testing (30%) sets. The proposed system delivers a comprehensive solution by achieving accurate fault detection, classification, and faulted phase identification in a single step. By combining fault detection and classification into a single step significantly reduces computational time and resource requirements compared to traditional, multi-stage methods. This efficiency makes the proposed method more feasible for real-time implementation.

Additionally, the system achieves an over 99% accuracy rate for both modes of microgrid operation, demonstrating its effectiveness in diverse operating conditions.

The findings presented in this chapter enrich the current understanding of deep learning applications in power systems, showcasing a potential pathway for significant advancements in microgrid protection technologies. Future research directions include exploring the impact of different deep learning architectures, incorporating additional signals and real-time validation to confirm the system's real-world efficacy.

References

1. Lasseter RH, Paigi P (2004) Microgrid: a conceptual solution. In: 2004 IEEE 35th annual power electronics specialists conference (IEEE Cat. No. 04CH37551). IEEE
2. Baloch S, Muhammad MS (2021) An intelligent data mining-based fault detection and classification strategy for microgrid. IEEE Access 9:22470–22479
3. Azizi R, Seker S (2021) Microgrid fault detection and classification based on the boosting ensemble method with the Hilbert-Huang transform. IEEE Trans Power Deliv 37(3):2289–2300
4. Baloch S, Samsani SS, Muhammad MS (2021) Fault protection in microgrid using wavelet multiresolution analysis and data mining. IEEE Access 9:86382–86391
5. Lin H et al (2019) Adaptive protection combined with machine learning for microgrids. IET Gener Transm Distrib 13(6):770–779
6. Aslan Y, Yağan YE (2017) Artificial neural-network-based fault location for power distribution lines using the frequency spectra of fault data. Electr Eng 99:301–311
7. Panigrahi BK et al (2018) Detection and classification of faults in a microgrid using wavelet neural network. J Inf Optim Sci 39(1):327–335
8. Mishra M, Panigrahi RR, Rout PK (2019) A combined mathematical morphology and extreme learning machine techniques based approach to micro-grid protection. Ain Shams Eng J 10(2):307–318
9. Mishra M, Rout PK (2018) Detection and classification of micro-grid faults based on HHT and machine learning techniques. IET Gener Transm Distrib 12(2):388–397
10. Uzair M et al (2022) Machine learning based protection scheme for low voltage AC microgrids. Energies 15(24):9397
11. Uzair M et al (2019) A protection scheme for AC microgrids based on multi-agent system combined with machine learning. In: 2019 29th Australasian universities power engineering conference (AUPEC). IEEE
12. Satuyeva B et al (2019) Q-learning based protection scheme for microgrid using multi-agent system. In: 2019 international conference on smart energy systems and technologies (SEST). IEEE
13. Gashteroodkhani OA et al (2019) An intelligent protection scheme for microgrids based on S-transform and deep belief network. In: 2019 IEEE power & energy society general meeting (PESGM). IEEE
14. James J et al (2017) Intelligent fault detection scheme for microgrids with wavelet-based deep neural networks. IEEE Trans Smart Grid 10(2):1694–1703
15. Xi P et al (2017) Fault detection algorithm for power distribution network based on sparse self-encoding neural network. In: 2017 international conference on smart grid and electrical automation (ICSGEA). IEEE
16. Bukhari SBA et al (2020) Convolutional neural network-based intelligent protection strategy for microgrids. IET Gener Transm Distrib 14(7):1177–1185
17. Karan S, Yeh H-G (2020) Fault classification in microgrids using deep learning. In: 2020 IEEE green energy and smart systems conference (IGESSC). IEEE

18. Bukhari SBA et al (2022) Empirical wavelet transform-based intelligent protection scheme for microgrids. Energies 15(21):7995
19. Abdelgayed TS, Morsi WG, Sidhu TS (2017) Fault detection and classification based on co-training of semisupervised machine learning. IEEE Trans Industr Electron 65(2):1595–1605
20. Addison PS (2017) The illustrated wavelet transform handbook: introductory theory and applications in science, engineering, medicine and finance. CRC Press
21. Ngui WK et al (2013) Wavelet analysis: mother wavelet selection methods. Appl Mech Mater 393:953–958
22. Gao RX, Yan R (2010) Wavelets: theory and applications for manufacturing. Springer Science & Business Media
23. Gupta K, Bajaj V, Ansari IA (2023) A support system for automatic classification of hypertension using BCG signals. Expert Syst Appl 214:119058
24. Hu Y et al (2021) Compound fault diagnosis of rolling bearings based on improved tunable Q-factor wavelet transform. Meas Sci Technol 32(10):105018
25. Bayram I, Selesnick IW (2009) Frequency-domain design of over complete rational-dilation wavelet transforms. IEEE Trans Signal Process 57(8):2957–2972
26. Mohamed A et al (2022) Tool condition monitoring for high-performance machining systems—a review. Sensors 22(6):2206
27. Selesnick IW (2011) Wavelet transform with tunable Q-factor. IEEE Trans Signal Process 59(8):3560–3575
28. Lipton ZC et al (2015) Learning to diagnose with LSTM recurrent neural networks. arXiv: 1511.03677
29. Hochreiter S, Schmidhuber J (1997) Long short-term memory. Neural Comput 9(8):1735–1780
30. Danandeh A, Seyedi H, Babaei E (2012) Islanding detection using combined algorithm based on rate of change of reactive power and current THD techniques. In: 2012 Asia-Pacific power and energy engineering conference. IEEE

Cyber-Attacks Detection and Mitigation in Microgrids

Himani Modi, Mohit Kachhwaha, and Deepak Fulwani

Abstract Introducing distributed communication and control in DC microgrids significantly improves the overall performance of the system. However, these developments also raise the possibility of malicious attacks. Denial-of-service (DoS) attacks and false data injection (FDI) attacks are the most common cyber-attacks. DoS attacks disrupt data availability, and FDI attacks tamper with data integrity, which is perilous for the DC microgrid system. These attacks have the potential to create a detrimental effect on the DC microgrid's sensors, actuators, and communication channels. The combination of DoS and FDI attacks on the actuator signal can have a severe impact, endangering the secure functioning of DC microgrid and potentially destabilizing the entire system. To address these concerns, this chapter introduces an attack detection and mitigation method for mixed FDI and DoS attacks. This detection mechanism uses a state-machine-based technique with a dynamic signature function. The dynamic function constantly compares the actual signal of the actuator with its estimated signal. A Linear Unknown Input Functional Observer (LUIFO) can mitigate mixed attacks and provide an estimated actuator signal to the signature function. This detection and mitigation method ascertain the security and stable functioning of the DC microgrid system. The MATLAB simulation results on four distributed generators-based DC microgrid demonstrate the performance of the methods.

Keywords FDI attack · DoS attack · Mixed attack · Cyber-physical system · Linear unknown input functional observer

H. Modi · M. Kachhwaha · D. Fulwani (✉)
Departmentof Electrical Engineering, Indian Institute of Technology Jodhpur (IITJ), Rajasthan 342030, India
e-mail: df@iitj.ac.in

H. Modi
e-mail: modi.1@iitj.ac.in

M. Kachhwaha
e-mail: kachhwaha.1@iitj.ac.in

F. Shahnia and J. M. Guerrero (eds.), *Microgrids and Virtual Power Plants*, Power Systems, https://doi.org/10.1007/978-981-97-6623-9_10

1 Introduction

Modern energy distribution systems need microgrids to integrate different renewable energy sources also known as distributed generators (DGs), loads, controllers, power electronic interfaces, protective devices, and communication systems [1]. Due to the rising adoption of DC sources and loads, including solar photovoltaic systems, fuel cells, batteries, and other alternatives, the use of DC microgrids has expanded recently. Since most of renewable energy sources are DC in nature, fewer power conversions are needed, which leads to increased efficiency and decreased energy losses [2]. There are two fundamental control objectives of DC microgrid voltage regulation and proportional current sharing among DGs. The droop control was incorporated in DC microgrid to provide proportional current sharing among multiple DGs; however, it causes voltage deviation in the DC bus. A centralized secondary controller is implemented, along with droop control, to restore the DC bus voltage. On the other hand, a centralized controller has drawbacks like a single point of failure, less flexibility, and fewer possibilities for expansion [3]. Due to the distributed secondary controller's more robust and reliable features, it has been extensively utilized in recent times to address these problems [4].

DC microgrid is regarded as an advanced cyber-physical system that accomplishes voltage regulation and proportional current sharing by using consensus-based communication. The cyber and physical layers of DC microgrid are shown in Fig. 1. The reliance on the communication network in distributed control systems makes microgrid more susceptible to cyber-attacks [4]. The risk of unauthorized access to actuator signal or sensor data is a serious concern. In the literature survey of DC microgrid, various forms of cyber-attacks were identified, including FDI [5], DOS [6], hijacking [7], replay [8], and man-in-the-middle attacks [9]. The most difficult and destructive cyberattacks are the combination of DoS and FDI attacks [10]. FDI attack on a DC microgrid involves manipulating or injecting incorrect information into the control system of the microgrid to disrupt its normal operation [11]. DoS attacks block information so that it can affect the overall performance of the system [6]. These attacks can disrupt the controller's ability to take necessary actions, which impacts the stability of the DC microgrid. FDI and DoS attacks are destructive in nature; extensive research has been conducted to investigate various protection schemes and their effectiveness. FDI attack detection and mitigation methods are divided into three categories:

- Singal-based methods: This approach focuses on detecting FDI attacks by analyzing signal characteristics and changes in the system [12].
- Model-based methods: System modeling and parameter information are used to defend against FDI attacks. There are several model-based techniques, such as state estimation [13], observer-based [11, 14], and parity-based [15].
- Data-driven based methods: This approach is based on historical data and measurement data to develop protection strategies. Data-driven methods like machine learning and reinforcement learning are implemented on DC microgrid [16].

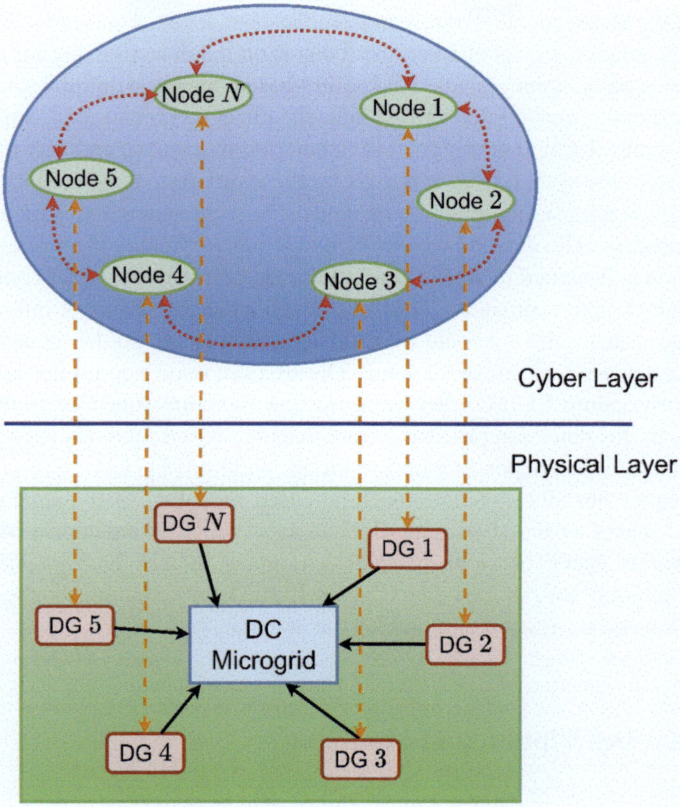

Fig. 1 Cyber layer and physical layer of DC microgrid

Research is currently underway to discover more effective methods of preventing DoS attacks. These include a new sampling period and communication mechanism [6, 17], controllers that are based on attack parameters and resilient control laws [18], an improved state estimation algorithm [19], event-triggered control [20], and software-defined networks [21]. However, in existing works, DoS attacks and FDI attacks in microgrids are separately considered and its prevention is mentioned. But it is possible that attackers can create adverse effects on the performance of the DC microgrid through combination attacks [22]. Previous literature has used limited detection and prevention strategies, including mixed DoS and FDI attacks. In [22], the design of microgrid controllers under mixed attacks was studied, and the theoretical analysis provided an understanding of how various attacks can affect stability; however, the AC microgrid system was taken into consideration in this chapter. An adaptive secondary control-based approach for combination attacks on DC microgrid was suggested in [23, 24]. However, the attack model for FDI was taken as a pulse form.

Most of the above-mentioned research studies have focused on sensor or communication link attacks. In this chapter, the focus is on the detection and mitigation of mixed attacks on actuator signals. An LUIFO-based detection and mitigation technique is used in this chapter that can estimate the linear combination of system states or specific states. LUIFO is designed to handle the parametric and non-parametric uncertainties of the system [25]. It is used for the estimation of the feedback control signal, which is a linear combination of states. The estimated value is helpful in the detection process. The detection is performed using a dynamic signature function. This function is designed to address the difference between the actual control input and the control input estimated by LUIFO. When an attack occurs, it influences the actual value, causing it to deviate from its estimated value, so the signature value increases and crosses its threshold value. The attacked node's controller is switched to the corresponding LUIFO after verifying the signature function to avoid false detections. In this detection process, a state machine-based model is used to switch among different states depending upon various conditions [11].

This chapter introduces a state-machine-based modeling of dynamic signature function to detect mixed, FDI, and DoS attacks, which is an improvement over existing works. Since previous research on mixed attacks has not specifically addressed actuator attacks, this chapter uses LUIFO for actuator signal estimation and detection and mitigation of attack on the actuator.

2 System Description and Modeling

Microgrid controls have various functions, such as voltage regulation, proportional load sharing among DGs, improved reliability, and reduced risk of blackouts. DC microgrid's overall control structure is shown in Fig. 2. DC microgrid incorporates multiple DG units, which consist of a DC source that is coupled via a DC-DC boost converter. DC microgrid incorporates multiple DG units, which consist of a DC source that is coupled via a DC-DC boost converter. An additional term for the DC source and boost converter combination is a node of a multi-agent system. Through line resistance, this node is connected to the DC bus. Additionally, the DC loads are connected to the DC bus, as shown in Fig. 2. This structure allows a precise analysis of the effectiveness of the DC microgrid control systems in real-world scenarios.

2.1 Control Layers

As seen in Fig. 2, the DC microgrid system makes use of a hierarchical control system with a distributed control approach at the secondary layer. Multiple levels of control and coordination are incorporated into this hierarchical control structure, which enables effective management and optimization of the DGs [26]. The three layers of DC microgrid control are the primary, secondary, and tertiary control layer.

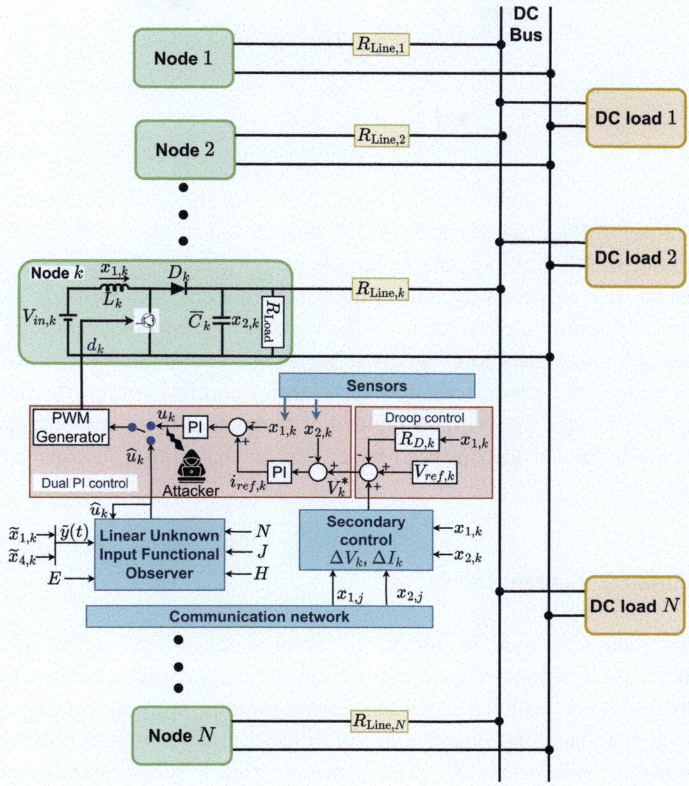

Fig. 2 Overall structure of DC microgrid

2.1.1 Primary Control Layer

Each node has its own primary controller, which is also known as lower-level control. To achieve the preliminary control objectives, voltage regulation, proportional power sharing, and quick reaction to converter dynamics, it comprises dual PI control and droop control at this control layer. Figure 2 illustrates the architecture of the dual PI control and droop control. The dual-loop PI control configuration comprises a faster inner current control loop with a higher bandwidth and a relatively slower outer voltage control loop with a lower bandwidth [11]. The reference provided to the PI controller is given by [27]

$$V_k^* = V_{ref,k} - R_{D,k}x_{1,k} + \Delta V_k + \Delta I_k \tag{1}$$

The dual PI voltage and current control loops generate an actuator signal, denoted as u_k. The equations for this signal are given below. The actuator signal is compared with a high-frequency (f_{sw}) sawtooth carrier signal to produce pulse width modulation (PWM) signals, which is provided to the switching of the boost converter.

$$u_k = \left(\frac{k_{ic}}{s} + k_{pc}\right)\left(i_{ref,k} - x_{1,k}\right) \tag{2}$$

$$i_{ref,k} = \left(\frac{k_{iv}}{s} + k_{pv}\right)\left(V_k^* - x_{2,k}\right) \tag{3}$$

where V_k^* is the desired output voltage, $V_{ref,k}$ represents the global reference voltage, $x_{1,k}$ is the output voltage, $x_{2,k}$ represents the inductor current, and $R_{D,k}$ denotes the droop resistance of kth node. k_{pc} and k_{ic} are the proportional and integral terms of the current PI controller, respectively. The output of current control loop is d_k. k_{pv} and k_{iv} are the proportional and integral terms of the voltage PI controller, respectively. The output of the voltage control loop is denoted as $i_{ref,k}$. Yet, the droop control method encounters challenges with current sharing accuracy and DC voltage deviation [28]. To address these issues, the primary controller needs to receive the appropriate set points from the higher-level controller based on the DC microgrid's overall system performance.

2.1.2 Secondary Control Layer

The secondary control layer is indeed a crucial aspect of managing power flow between nodes in DC microgrids. It helps to maintain appropriate voltage levels and enhance power quality, which is essential for ensuring the smooth functioning of the system. In the distributed secondary control layer, each DG must establish communication with its neighboring DGs. The communication between DGs is modeled using graph theory where each DG is a node, and the communication link between them is an edge of the graph. The kth node communicates voltage and current information with its neighbors, sending $\left[\overline{V}_k, i_k^{pu}\right]$ and receiving $\left[\overline{V}_j, i_j^{pu}\right]$ from the jth neighbouring node as shown in Fig. 3. This is accomplished by using an adjacency matrix of graph theory $\mathcal{A} = a_{kj} \in \mathbb{R}^{(n\times n)}$. The adjacency matrix is defined as [29]:

$$a_{kj} = \begin{cases} > 0 \ if \left(x_k, x_j\right) \in E \\ 0 \quad Otherwise \end{cases} \tag{4}$$

where ϵ is an edge connecting two nodes, the variables x_k and x_j represent the voltage and current information of the local node and neighboring nodes, respectively. Furthermore, the graph's Laplacian matrix is determined to be $\mathcal{L} = Z^{in} - \mathcal{A}$. where $Z^{in} = \sum_{i\in n} a_{kj}$ is a matrix with incoming information. For the Laplacian matrix, the row sum of all elements equals zero [29]. The correction terms for the secondary controller of a DC microgrid are:

$$\overline{V}_k(t) = V_k + \int_0^t \sum_{j\in N} a_{kj}\left(\overline{V_j}(\tau) - \overline{V_k}(\tau)\right)d\tau \tag{5}$$

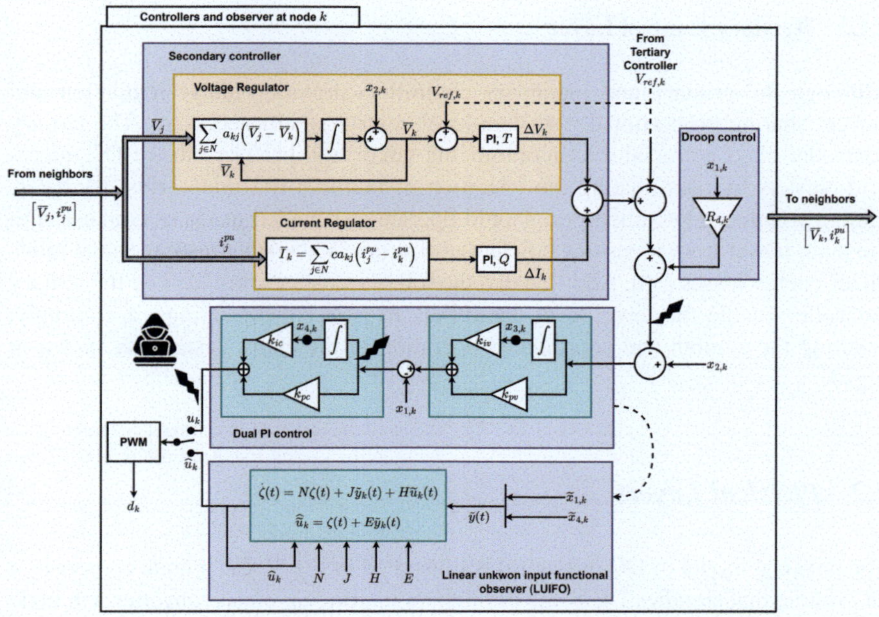

Fig. 3 Controllers and observer structure

$$\Delta V_k = T_P\big(V_{ref} - \overline{V_k}(t)\big) + T_I \int \big(V_{ref} - \overline{V_k}(\tau)\big)d\tau \tag{6}$$

$$\overline{I}_k = \sum_{j\in N} ca_{kj}\left(i_j^{pu} - i_k^{pu}\right) \tag{7}$$

$$\Delta I_k = Q_P\overline{I}_k + Q_I \int \overline{I}_k \tag{8}$$

where T_P and T_I are the proportional and integral gains of the PI control of the voltage regulator, T. Q_P and Q_I are the proportional and integral gains of the PI control of the current regulator, Q. The voltage regulator updates its own estimated value $\overline{V}_k$ by processing the neighbors' estimated value $\overline{V}_j$ and local voltage measurement $\overline{V}_k$. The current regulator compares the local per-unit current $i_k^{pu} = \frac{i_k}{i_k^{base}}$ with its neighbor's per-unit value $i_j^{pu} = \frac{i_j}{i_j^{base}}$, and this mismatch $\overline{I}_k$ is fed to the PI controller. The voltage regulator and current regulator provide ΔV_k and ΔI_k to the primary controller, respectively.

2.1.3 Tertiary Control Layer

Although the primary and secondary controllers provide voltage regulation and current sharing proportionally, system-level control is still necessary. The tertiary controller plays a crucial role in optimizing the overall performance of DC microgrid clusters by coordinating the operation of multiple DC microgrids. By determining whether a DC microgrid should operate in islanded mode or connect with the main power grid, it ensures efficient utilization of available resources and minimizes energy losses [30]. Additionally, the reference values provided by the tertiary controller enable the lower-level controllers to adjust their settings, accordingly, ensuring the smooth and coordinated operation of the entire system, as shown in Fig. 3.

2.2 *Physical Layers*

The physical model of DC microgrid is illustrated in Fig. 2. Each node comprises a DC source that supplies DC loads through an interfacing boost converter. *k*th node of a microgrid is defined by the following set of equations [31]:

$$L_k \dot{x}_{1,k} = V_{in,k} - (1 - u_k)x_{2,k} \tag{9a}$$

$$\overline{C}_k \dot{x}_{2,k} = (1 - u_k)x_{1,k} - x_{2,k}/R_k \tag{9b}$$

$$\dot{x}_{3,k} = V_{ref,k} - x_{2,k} \tag{9c}$$

$$\dot{x}_{4,k} = -x_{1,k} + k_{pv,k}\left(V_{ref,k} - x_{2,k}\right) + k_{iv,k}x_{3,k} \tag{9d}$$

where $V_{in,k}$ is converter input voltage, u_k is the control input. R_k, L_k, and $\overline{C}_k$ are the resistive load, inductor, and capacitor. $x_{1,k}$ and $x_{2,k}$ are inductor current and capacitor voltage, respectively. $x_{1,k}$ is also termed as i_k^{pu} in this chapter, it is noted that both have the same meaning. $x_{3,k}$ and $x_{4,k}$ are integrals of voltage and current, representing flux state and charge state, respectively as shown in Fig. 3. All small changes or disturbances occurring in the system states, as well as in the control input, are represented as perturbations [11]. The following equations mathematically express it:

$$x_{n,k} = X_{n,k} + \tilde{x}_{n,k};\ n = 1,2,3,4 \tag{10a}$$

$$u_k = U_k + \tilde{u}_k \tag{10b}$$

Linearizing and separating the DC and AC parts of (9a–9d), the AC part of each state and the control input are represented as follows:

$$\dot{\tilde{x}}_k = A_k\tilde{x}_k + B_k\tilde{u}_k, \tilde{y}_k = C_k\tilde{x}_k \tag{11}$$

where the small-signal state vector, system, input and output matrices are $\tilde{x}_k = \left[\tilde{x}_{1,k}\tilde{x}_{2,k}\tilde{x}_{3,k}\tilde{x}_{4,k}\right]$, A_k, B_k, and C_k as [31]:

$$A_k = \begin{bmatrix} 0 & \frac{(U_k-1)}{L_k} & 0 & 0 \\ \frac{(U_k-1)}{C_k} & \frac{-1}{R_k C_k} & 0 & 0 \\ 0 & -1 & 0 & 0 \\ -1 & -k_{pv,k} & k_{iv,k} & 0 \end{bmatrix}; B_k = \begin{bmatrix} \frac{x_{2,k}}{L_k} \\ \frac{-x_{1,k}}{C_k} \\ 0 \\ 0 \end{bmatrix}; \tag{12}$$

$$C_k = \begin{bmatrix} 1\,0\,0\,0 \\ 0\,1\,0\,0 \end{bmatrix}; \tag{13}$$

$$\tilde{u}_k = -\left[\, k_{pc}\ k_{pv}k_{pc}\ -k_{iv}k_{pc}\ -k_{ic}\,\right]\tilde{x}_k \tag{14}$$

where U_k, $X_{1,k}$, and $X_{2,k}$ are duty cycle, inductor current, and DC bus voltage at the desired operating points. These matrices will be useful for the design of LUIFO.

3 Modeling of Cyber-Attacks

This chapter focuses on mitigation methods of FDI and DOS-based mixed attacks. Each DG's converter requires an actuator signal for the switching of IGBT. Consider that the attacker can manipulate or disrupt the actuator signal by mixed attack. In this section, the modeling of DoS attacks and FDI attacks is first illustrated, and then mixed cyber-attack modeling is addressed.

3.1 DoS Attack

The DoS attack can block the channel for certain periods, resulting in a lack of actuator signal for the PWM generator, which is illustrated in Fig. 3. The mathematical representation of a DoS attack takes place in two-time sequences. One time sequence is for when the DoS attack happens: $\rho_D = \{s_1, s_2, \ldots, s_k, ..\}$. The second time sequence is for the duration of the attack: $\rho_I = \{q_1, q_2, \ldots, q_k, \ldots\}$ [24]. Since DoS attacks often don't last continuously, that exists $s_k, q_k \to \infty$ and $s_{k+1} - s_k > q_k > 0$. Based on this, the set $\Gamma_D(s, t)$ is where the attacker disables the communication network, causing the actuator signal to be absent at that time, and $\Gamma_N(s, t)$ is the set where there is no attack. Then, a piecewise function describes the characteristics of the DoS attack on the communication channel [32].

$$D_k = \begin{cases} 1\ for\ t \in \Gamma_D(s,t) \\ 0\ for\ t \in \Gamma_N(s,t) \end{cases} \tag{15}$$

The mathematical expression for the compromised actuator signal caused by DoS for the kth converter is $\overline{u}_{k,D} = (1 - D_k) * u_k$. Where D_k represents the presence of a DoS attack on the kth node.

3.2 FDI Attack

In an FDI attack, false data can be inserted to change actuator or sensor information and cause the DC microgrid system to be unstable and unreliable. In this article, the FDI attack manipulates the actuator signal of the converter with the addition of a random signal. The FDI attack on the kth node's actuator is expressed as $\overline{u}_{k,F} = u_k + \gamma_k * u_{a,k}$. where $\gamma_k = 1$ means there is an FDI attack and $\gamma_k = 0$ means there is no attack [11].

3.3 Mixed Attack

The combination of FDI and DoS attacks is a mixed attack, which can severely affect the DC microgrid system [23]. Mathematically, the actuator signal during the mixed attack is

$$\overline{u}_{k,M} = (1 - D_k) * u_k + \gamma_k * u_{a,k} \tag{16}$$

Mixed attack occurs when $D_k = 1$ and $\gamma_k = 1$. The modified kth node power converter model, which includes the attack signal, is as follows:

$$L_k \dot{x}_{1,k} = V_{\text{in},k} - \left(1 - \overline{u}_{k,M}\right) x_{2,k} \tag{17a}$$

$$\overline{C}_k \dot{x}_{2,k} = \left(1 - \overline{u}_{k,M}\right) x_{1,k} - x_{2,k}/R_k \tag{17b}$$

$$\dot{x}_{3,k} = V_{ref,k} - x_{2,k} \tag{17c}$$

$$\dot{x}_{4,k} = -x_{1,k} + k_{pv,k}\left(V_{ref,k} - x_{2,k}\right) + k_{iv,k} x_{3,k} \tag{17d}$$

Upper control layers are vulnerable to attacks due to their reliance on communication. Consider that an attacker attacks the primary controller with a single attack or multiple attacks [11]. This could lead to a complete takeover of the entire control system. Here, the attacks on the control input, which is also called the actuator signal, that is given to the PWM generator of the boost converter are considered. However, sensor attacks and communication link attacks indirectly affect the actuator signal, u_k, as shown in Fig. 3.

4 Linear Unknown Input Functional Observer

Observers utilize the plant's input and output signals to create an estimation of the plant's state. Luenberger was the first to introduce and develop this approach, which is known as the Luenberger observer [33]. Later in the early stages of development, observers for linear systems can estimate the state variable when unknown inputs are affecting the system dynamics, which is known as the unknown input observer [34]. The term "unknown inputs" refers to external disturbances or inputs that affect the system but are not measured directly [35]. The unknown input observer estimates only the state variable, but for the estimation of a linear combination of the system states LUIFO is applicable [36].

The basic idea behind the use of observers for cyber-attack detection is to form residuals from the difference between the actual system outputs and the estimated outputs using an observer. Once an attack occurs, the residuals are expected to react by becoming greater than a prespecified threshold. In this chapter, cyber-attacks are considered as unknown inputs. LUIFO can provide an estimate of the feedback control signal; it is represented as a linear combination of states in the presence of unknown inputs [25]. We may express (11) as:

$$\dot{\tilde{x}}_k(t) = A_k\tilde{x}_k(t) + B_k\tilde{u}_k(t) + Fd(t)$$

$$\tilde{y}_k(t) = C_k\tilde{x}_k(t) + Gd(t)$$

$$z(t) = L\tilde{x}_k(t) \tag{18}$$

where $\tilde{x}_k(t) \in \mathbb{R}^n$ is a state vector, $\tilde{u}_k(t) \in \mathbb{R}^m$ is the input vector, $\tilde{y}_k(t) \in \mathbb{R}^p$ is the output vector, and $d(t) \in \mathbb{R}^q$ is the unknown input vector. The vector that has to be estimated is $z(t) \in \mathbb{R}^r$. The DC microgrid model above calculates the known constant matrices A, B, C, F, and G with the proper dimensions. The rank of matrix L should be r, and it should be appropriately designed to satisfy the observer's requirements. The main objective of the design of the LUIFO is to estimate the control input signal $z(t) = \tilde{u}_k(t)$. The term $Fd(t)$ is employed to model the uncertainties and inherent mismatch in the DC microgrid model. The term $Gd(t)$ refers to the occurrence of cyber-attacks and any faults that may arise in the actuator, measurement, and communication channels. The LUIFO is structured as follows [25].

$$\dot{\zeta}(t) = N\zeta(t) + J\tilde{y}_k(t) + H\tilde{u}_k(t) \tag{19a}$$

$$\hat{\tilde{u}}_k(t) = \zeta(t) + E\tilde{y}_k(t) \tag{19b}$$

where $\hat{\tilde{u}}_k(t) \in \mathbb{R}^r$ is the estimation of $\tilde{u}_k(t)$, and $\zeta(t) \in \mathbb{R}^r$ is the observer's state vector. The matrices $N, J, H,$ and E are constants, and their values need to be determined accurately. Error $e(t)$ and matrix ϕ are defined as

$$e(t) := \tilde{u}_k(t) - \hat{\tilde{u}}_k(t) = L\tilde{x}_k(t) - \hat{\tilde{u}}_k(t) = \phi\tilde{x}_k(t) - \zeta(t) - EGd(t) \tag{20}$$

$$\phi := L - EC \tag{21}$$

Taking the derivative of the error $e(t)$,

$$\begin{aligned} \dot{e}(t) &= L\dot{\tilde{x}}_k(t) - \dot{\tilde{u}}_k(t) = Ne(t) + (\phi A - N\phi - JC)\tilde{x}_k(t)(\phi B - H)\tilde{u}_k(t) \\ &+ (\phi F + NEG - JG)d(t) + (EG)\dot{d}(t) \end{aligned} \tag{22}$$

The dynamical system given in (19a, 19b) is a functional observer of the DC microgrid system given in (18) if the abovementioned matrices need to satisfy the conditions outlined in Theorem 1. Moreover, the error $e(t)$ asymptotically approaches zero [25].

Theorem 1 *Let the pair* (A_k, B_k) *be controllable. The jth order observer* (19a, 19b) *will estimate* $u(t)$ *if and only if the following conditions hold.*

$$N \text{ is Hurwitz} \tag{23a}$$

$$\phi F + NEG - JG = 0 \tag{23b}$$

$$\phi A - N\phi - JC = 0 \tag{23c}$$

$$H = \phi B \tag{23d}$$

$$EG = 0 \tag{23e}$$

***Proof*.**

The stability proof of the LUIFO observer is given in [37].

A robust functional observer should be developed to mitigate the cyber-attacks; hence, it's necessary to determine the matrices N, J, H, and E that satisfy (23a–23e).

Given that the matrix L is assumed to be of full row rank, let's define matrix P_1 as a $(n-r) \times n$ matrix,

$$\begin{bmatrix} L \\ P_1 \end{bmatrix} = \begin{bmatrix} H_1 & E_1 \end{bmatrix}^{-1} \tag{24}$$

From (21) and conditions (23a–23e), matrix N is derived,

$$N = LAH_1 - \begin{bmatrix} E & K \end{bmatrix} \begin{bmatrix} CA \\ C \end{bmatrix} H_1 \tag{25}$$

where $\begin{bmatrix} E & K \end{bmatrix}$ is an unknown matrix that conforms to the following criteria.

$$\begin{bmatrix} E & K \end{bmatrix} w_1 = w_2 \tag{26}$$

where

$$[w_1] = \begin{bmatrix} CAE_1 & CF & G \\ CE_1 & G & 0 \end{bmatrix},$$

$$[w_2] = \begin{bmatrix} LAE_1 & LF & 0 \end{bmatrix}$$

and $K = J - NE$, The matrix J can be derived from this equation. For the existence of (26), The necessary condition is

$$rank \begin{bmatrix} w_1 \\ w_2 \end{bmatrix} = rank[w_1] \tag{27}$$

It turns out that the general solution to (26) is found as

$$\begin{bmatrix} E & K \end{bmatrix} = w_2 w_1^+ + Z(I - w_1 w_1^+) \tag{28}$$

where w_1^+ is the Moore–Penrose inverse of the w_1 matrix, and Z is an arbitrary matrix of adequate dimension. The E matrix can be derived from the above equation. From (25) and (27), matrix N is calculated as

$$N = \mu - Z\lambda \tag{29}$$

where

$$\mu = LAH_1 - w_2 w_1^+ \begin{bmatrix} CA \\ C \end{bmatrix} H_1$$

$$\lambda = (I - w_1 w_1^+) \begin{bmatrix} CA \\ C \end{bmatrix} H_1$$

The LUIFO matrices $N, J, H,$ and$E$ can be obtained through Eqs. (23)–(29). Once derived, these matrices enable the estimation of the actuator signal. This estimated value plays an essential role in the subsequent steps of cyber-attack detection and mitigation within the system.

5 Attack Detection and Mitigation Technique Using LUIFO

The first step towards preventing a cyber-attack is to detect the attack. It is assumed that the attack on the actuator input signal requires an actual actuator value u_k and LUIFO-based estimated actuator value $\hat{u}_k$. The dynamic signature function used in the detection process is presented as

$$S_k = \alpha_k \int_0^t \beta_k^{|u_k - \hat{u}_k|} d\tau \tag{30}$$

where α_k and β_k are positive constants. Illustrated in Fig. 4 is a state machine diagram providing a comprehensive overview of the detection technique. After initializing all the variables $(C_k, S_k, O, r, \textit{and } v_k = 0)$, the procedure begins with entering state $E1$. Basically, $O(t)$ and $r(t)$ are clock functions. The switch function, denoted by v_k, controllers send the actual actuator signal to the converter when $v_k = 0$ and the estimated signal when $v_k = 1$. In state $E1$, the value of the signature function S_k is calculated dynamically according to (30). In the absence of attacks, u_k and $\hat{u}_k$ have almost the same values, so the S_k value does not increase much; it will reset with a period of time T_O using the timer $O(t)$. When the kth converter is under attack, u_k deviates from its estimated value, $\hat{u}_k$. According to (30), the signature function value S_k starts increasing at the time of attack. When S_k is greater than its predetermined threshold value, $S_{k,\max}$, the algorithm enters state $E2$. the counter C_k is configured to count the occurrences of S_k crossing $S_{k,\max}$. If the value of counter C_k is lower than the prespecified $C_{k,\max}$ value, the algorithm transitions into state $E3$, which indicates the absence of an attack. The system remains in its current state until the timer function $O(t)$ resets the value of S_k, at $S_k = 0$ it returns to state $E1$. If counter C_k crosses $C_{k,\max}$, the system has a detrimental impact due to attacks, resulting in the algorithm transitioning to state $E4$. As the switch v_k is set to 1, the counter C_k resets, which indicates the attack is detected and LUIFO is activated for the particular node. Additionally, a timer function $r(t)$ resets the timer to prevent false attack detection by resetting it with a T_c period. The actuator signal shifts to the estimated value $\hat{u}_k$. The aforementioned detection and mitigation technique has the capability to mitigate DoS, FDI, and mixed attacks effectively.

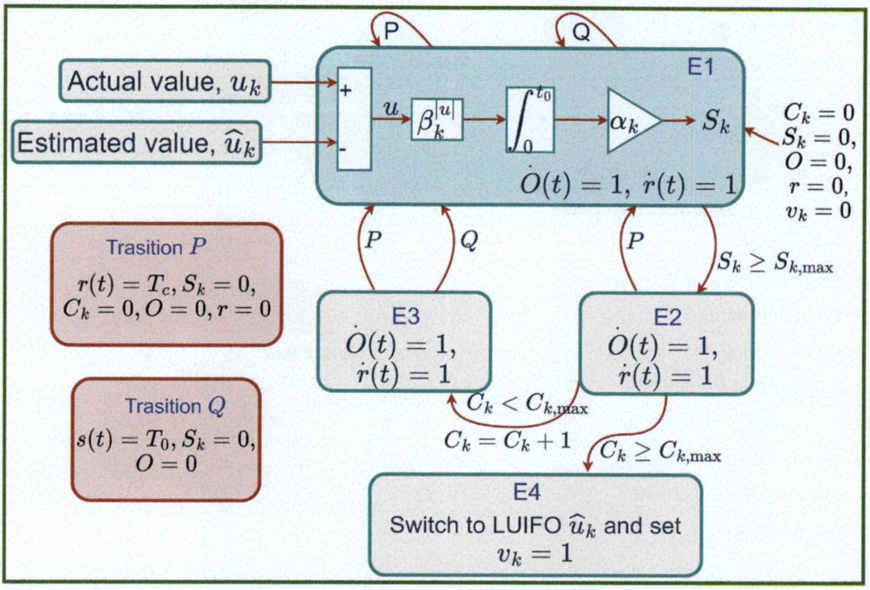

Fig. 4 Dynamic signature function-based detection method

6 Simulation Results

The detection and mitigation method is validated through a four-node DC microgrid setup on MATLAB Simulink. The connection of the nodes and its primary and secondary controllers is shown in Figs. 2 and 3. All the parameters of DC microgrid are shown in Table 1. LUIFO is designed for individual nodes, as mentioned in Sect. 4. The parameters required for LUIFO are:

$$A_k = \begin{bmatrix} 0 & -250 & 0 & 0 \\ 10638 & -443.26 & 0 & 0 \\ 0 & -1 & 0 & 0 \\ -1 & -0.0657 & 91.6 & 0 \end{bmatrix}, B_k = \begin{bmatrix} 60000 \\ -10638 \\ 0 \\ 0 \end{bmatrix}, C_k = \begin{bmatrix} 1\,0\,0\,0 \\ 0\,0\,0\,1 \end{bmatrix}$$

$$L = \begin{bmatrix} -0.0022 & -0.0054 & 6.005 & -32.85 \end{bmatrix}, F = \begin{bmatrix} 1\,0 \\ 0\,1 \\ 1\,1 \\ 1\,1 \end{bmatrix}, G = \begin{bmatrix} 1\,0 \\ 0\,0 \end{bmatrix},$$

$$N = [-374.78], E = \begin{bmatrix} -0.058 & -26.786 \end{bmatrix}, K = \begin{bmatrix} -26.786 & -364.508 \end{bmatrix}, \text{ and}$$

$$J = \begin{bmatrix} -0.0047 & 9.674 \end{bmatrix} * 10^3.$$

Table 1 Parameters of DC microgrid setup

Parameters	Symbol	Value
Input voltage	$V_{in,k}$	60 V
Output reference voltage	$V_{ref,k}$	120 V
Inductance in all boost converters	L_k	2 mH
Capacitance in all boost converters	$\overline{C}_k$	100 µF
Line resistance	R_{Line}	0.2 Ω
Resistive load	R_k	10–14 Ω
Nominal duty ratio	U_k	0.5
Droop resistance	$R_{D,1}, R_{D,2}, R_{D,3}, R_{D,4}$	0.5 Ω, 1 Ω, 0.5 Ω, 1 Ω
Primary current PI controller	k_{pc}, k_{ic}	0.0826, 244
Primary voltage PI controller	k_{pv}, k_{iv}	0.0657, 91.4
Secondary current regulator	Q_P, Q_I	1, 50
Secondary voltage regulator	T_P, T_I	5, 10
Adjacency matrix	$\mathcal{A}$	$\begin{bmatrix} 0&1&0&1 \\ 1&0&1&0 \\ 0&1&0&1 \\ 1&0&1&0 \end{bmatrix}$
Detection process constants	α_k β_k	9000 30
Signature function threshold	$S_{k,\max}$	128
Counter threshold	$C_{k,\max}$	3
Switching frequency of boost converters	f_{sw}	20 kHz
Power rating	P	1 kW

The overall loading on DC microgrid is 1 kW, and four DGs share it equally. As shown in Fig. 5, all the DGs operate steadily at first. Initially, $u_k = \hat{u}_k$, the actuator signal and its estimated value are same for all converters. At time $t = 0.05$ s, a mixed attack is initiated on node 2. Mixed attack is a combination of a random FDI attack that adds false data to the actuator signal in the range of $[-0.35, 0.35]$ and a DoS attack. The sequence of the DoS attack is given in Fig. 5; DoS value 1 means the attack blocks the information and 0 means there is no DoS attack. The mixed attack significantly impacts the operational effectiveness of DC microgrid. As illustrated in Fig. 5, there is evident destabilization observed in parameters such as the actuator signal, as well as the voltage and current waveforms. Actuator value u_2 varies from its estimated value $\hat{u}_2$; with that signature function of node 2, S_2 starts increasing drastically, and it crosses its threshold value $S_{2,\max}$ every subsequent period. The counter raises by 1 ($C_2 = C_2 + 1$) when $S_2 > S_{2,\max}$ periodically. The mitigation process begins when C_2 exceeds the predetermined threshold value $C_{2,\max}$; at that point, the actuator value u_2 switches to the observer value $\hat{u}_2$. It is observed that signal u_2 is still compromised by mixed attacks, but LUIFO mitigates the effect of the attack on the system. Furthermore, the system experiences a load variation of 40% at $t = 0.14$ s and $t = 0.20$ s in a sequence of 1000 W → 1400 W → 1000 W. Clearly, load changes are handled promisingly, and current sharing is also maintained.

LUIFO-based attack detection and mitigation techniques can detect and mitigate single attacks as well. A DoS attack on the actuator signal targets node 2 at $t = 0.05$ s, as shown in Fig. 6. When an attack occurs, the detection process starts with it, and the signature value S_2 increases. Periodically, when $S_2 > S_{2,\max}$, the counter increases; when $C_2 > C_{2,\max}$, u_2 switches to $\hat{u}_2$. After attack detection, the system becomes steady again. Figure 7 shows the detection and mitigation process of a random FDI attack. In both cases, the overall system is also stable with load changes.

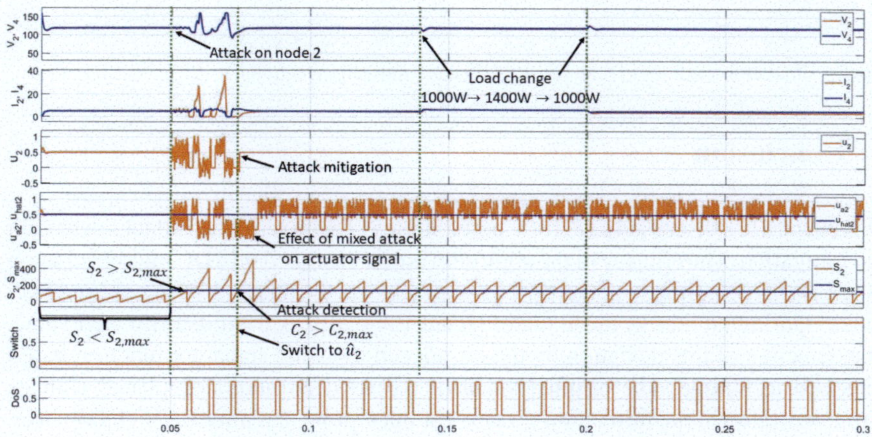

Fig. 5 Simulation study of mixed attack detection and mitigation technique

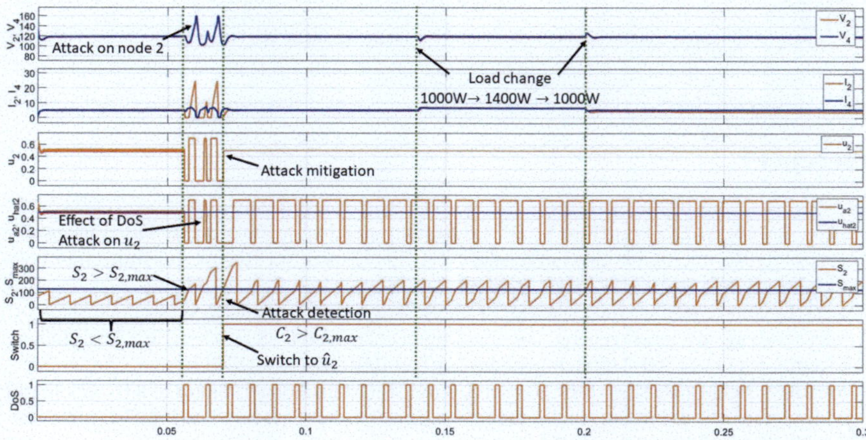

Fig. 6 Simulation study of DoS attack detection and mitigation technique

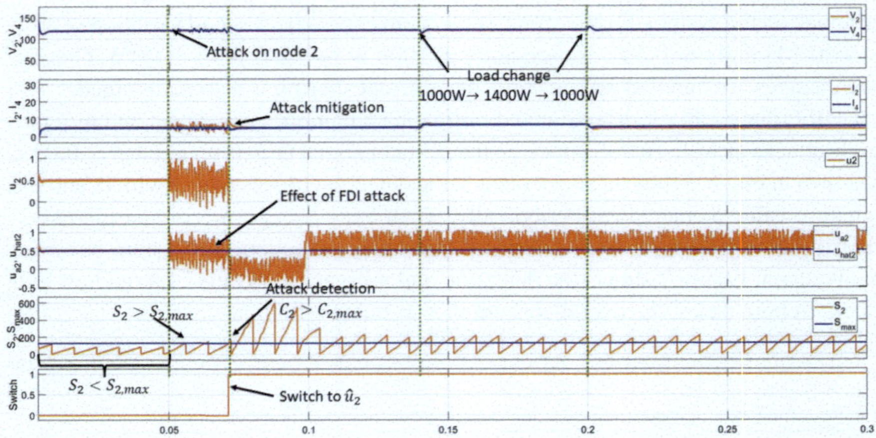

Fig. 7 Simulation study of FDI attack detection and mitigation technique

7 Conclusion

This chapter has discussed a LUIFO- based strategy for mixed cyber-attacks detection and mitigation. The efficacy of the algorithm is demonstrated on a DC microgrid setup in MATLAB Simulink with three types of attacks: mixed attack, FDI attack, and DoS attack. During an attack, the actuator signal value differs from its estimated value. The detection of this deviation by the introduced dynamic signature function-based detection technique. The original value of the attacked node is replaced with its estimated value in the mitigation process. Studies show that after LUIFO took over, it becomes possible to handle the load changes.

Acknowledgement This work was supported by the Ministry of Electronics and Information Technology, India under the project "Cyber-Physical Systems" under Project S/MeitY/SKS/20210139.

References

1. Jiang Y, Yang Y, Tan S-C, Hui SY (2021) Distributed sliding mode observer-based secondary control for DC microgrids under cyber-attacks. IEEE J Emerg Sel Top Circuits Syst 11(1):144–154. https://doi.org/10.1109/JETCAS.2020.3046781
2. Deng C, Guo F, Wen C, Yue D, Wang Y (2022) Distributed resilient secondary control for DC microgrids against heterogeneous communication delays and DoS attacks. IEEE Trans Industr Electron 69(11):11560–11568. https://doi.org/10.1109/TIE.2021.3120492
3. Mehdi M, Jamali SZ, Khan MO, Baloch S, Kim CH (2020) Robust control of a DC microgrid under parametric uncertainty and disturbances. Electr Power Syst Res 179:106074. https://doi.org/10.1016/J.EPSR.2019.106074

4. Lu J, Zhang X, Hou X, Wang P (2022) Generalized extended state observer-based distributed attack-resilient control for DC microgrids. IEEE Trans Sustain Energy 13(3):1469–1480. https://doi.org/10.1109/TSTE.2022.3162757
5. Liu M, Zhao C, Deng R, Cheng P, Chen J (2022) False data injection attacks and the distributed countermeasure in DC microgrids. IEEE Trans Control Netw Syst 9(4):1962–1974. https://doi.org/10.1109/TCNS.2022.3181483
6. Li Y, Meng W, Fan B, Zhao S, Yang Q (2022) Distributed aperiodic control of multibus DC microgrids with DoS-attack resilience. IEEE Trans Smart Grid 13(6):4815–4827. https://doi.org/10.1109/TSG.2022.3180502
7. Sahoo S, Peng JCH, Mishra S, Dragicevic T (2020) Distributed screening of hijacking attacks in DC microgrids. IEEE Trans Power Electron 35(7):7574–7582. https://doi.org/10.1109/TPEL.2019.2957071
8. Taher MA, Tariq M, Behnamfar M, Sarwat AI (2023) Analyzing replay attack impact in DC microgrid consensus control: detection and mitigation by Kalman-filter-based observer. IEEE Access 11:121368–121378. https://doi.org/10.1109/ACCESS.2023.3317799
9. Jena S, Padhy NP, Guerrero JM (2023) Multi-layered coordinated countermeasures for DC microgrid clusters under man in the middle attack. IEEE Trans Ind Appl https://doi.org/10.1109/TIA.2023.3308557
10. Kayem AVDM, Wolthusen SD, Meinel C (2018) Smart micro-grid systems security and privacy. Springer, Cham. Accessed 10 Jan 2024. http://www.springer.com/series/5576
11. Kachhwaha M, Modi H, Nehra MK, Fulwani D (2024) Resilient control of DC microgrids against cyber attacks: a functional observer based approach. IEEE Trans Power Electron 39(1):459–468. https://doi.org/10.1109/TPEL.2023.3326308
12. Zhou Q, Shahidehpour M, Alabdulwahab A, Abusorrah A (2020) A cyber-attack resilient distributed control strategy in islanded microgrids. IEEE Trans Smart Grid 11(5):3690–3701. https://doi.org/10.1109/TSG.2020.2979160
13. Kurt MN, Yilmaz Y, Wang X (2018) Real-time detection of hybrid and stealthy cyber-attacks in smart grid. IEEE Trans Inf Forensics Secur 14(2):498–513. https://doi.org/10.1109/TIFS.2018.2854745
14. Tan S, Xie P, Guerrero JM, Vasquez JC, Han R (2022) Cyberattack detection for converter-based distributed dc microgrids: observer-based approaches. IEEE Ind Electron Mag 16(3):67–77. https://doi.org/10.1109/MIE.2021.3059996
15. Tan S, Xie P, Guerrero JM, Vasquez JC (Mar 2022) False data injection cyber-attacks detection for multiple DC microgrid clusters. Appl Energy 310. https://doi.org/10.1016/j.apenergy.2021.118425
16. Zhang H, Yue D, Dou C, Hancke GP (May 2022) Resilient optimal defensive strategy of micro-grids system via distributed deep reinforcement learning approach against FDI attack. IEEE Trans Neural Netw Learn Syst, 1–11. https://doi.org/10.1109/TNNLS.2022.3175917
17. Lian Z, Guo F, Wen C, Deng C, Lin P (2021) Distributed resilient optimal current sharing control for an islanded DC microgrid under DoS attacks. IEEE Trans Smart Grid 12(5):4494–4505. https://doi.org/10.1109/TSG.2021.3084348
18. Hu S, Yang F, Gorbachev S, Yue D, Kuzin V, Deng C (2022) Resilient control design for networked DC microgrids under time-constrained DoS attacks. ISA Trans 127:197–205. https://doi.org/10.1016/J.ISATRA.2022.02.022
19. Li X, Jiang C, Du D, Li W, Fei M, Wu L (2023) A novel state estimation method for smart grid under consecutive denial of service attacks. IEEE Syst J 17(1):513–524. https://doi.org/10.1109/JSYST.2022.3171751
20. Lu J, Jiang H, Hou X, Wang P (2023) Distributed edge-based event-triggered control for voltage restoration and current sharing in DC microgrids under DoS attacks. IEEE Trans Industr Electron 70(8):8053–8063. https://doi.org/10.1109/TIE.2022.3224153
21. Danzi P, Angjelichinoski M, Stefanović Č, Dragičević T, Popovski P (2018) Software-defined microgrid control for resilience against denial-of-service attacks. IEEE Trans Smart Grid 10(5):5258–5268. https://doi.org/10.1109/TSG.2018.2879727

22. Zhou Q, Shahidehpour M, Alabdulwahab A, Abusorrah A, Che L, Liu X (2021) Cross-layer distributed control strategy for cyber resilient microgrids. IEEE Trans Smart Grid 12(5):3705–3717. https://doi.org/10.1109/TSG.2021.3069331
23. Liu XK, Wen C, Xu Q, Wang YW (2021) Resilient control and analysis for DC microgrid system under DoS and impulsive FDI attacks. IEEE Trans Smart Grid 12(5):3742–3754. https://doi.org/10.1109/TSG.2021.3072218
24. Liu XK, Wang SQ, Chi M, Liu ZW, Wang YW (2024) Resilient secondary control and stability analysis for DC microgrids under mixed cyber attacks. IEEE Trans Industr Electron 71(2):1938–1947. https://doi.org/10.1109/TIE.2023.3262893
25. Haes Alhelou H, Golshan MEH, Hatziargyriou ND (2019) A decentralized functional observer based optimal LFC considering unknown inputs, uncertainties, and cyber-attacks. IEEE Trans Power Syst 34(6):4408–4417. https://doi.org/10.1109/TPWRS.2019.2916558
26. Shafiee Q, Dragičević T, Vasquez JC, Guerrero JM (2014) Hierarchical control for multiple DC-microgrids clusters. IEEE Trans Energy Convers 29(4):922–933. https://doi.org/10.1109/TEC.2014.2362191
27. Abhinav S, Modares H, Lewis FL, Davoudi A (2019) Resilient cooperative control of DC microgrids. IEEE Trans Smart Grid 10(1):1083–1085. https://doi.org/10.1109/TSG.2018.2872252
28. Zhang J, Sahoo S, Peng JCH, Blaabjerg F (2021) Mitigating concurrent false data injection attacks in cooperative DC microgrids. IEEE Trans Power Electron 36(8):9637–9647. https://doi.org/10.1109/TPEL.2021.3055215
29. Sahoo S, Mishra S, Peng JCH, Dragicevic T (2019) A stealth cyber-attack detection strategy for DC microgrids. IEEE Trans Power Electron 34(8):8162–8174. https://doi.org/10.1109/TPEL.2018.2879886
30. Meng L et al (2017) Review on control of DC microgrids and multiple microgrid clusters. IEEE J Emerg Sel Top Power Electron 5(3):928–948. https://doi.org/10.1109/JESTPE.2017.2690219
31. Kachhwaha M, Modi H, Nehra M, Fulwani D (2024) Robust observer-based defense strategy against actuator and sensor cyber-attacks in DCMGs. IEEE Transact Ind Inform 1–10. https://doi.org/10.1109/TII.2024.3412211
32. Deng C, Wen C (2020) Distributed resilient observer-based fault-tolerant control for heterogeneous multiagent systems under actuator faults and DoS attacks. IEEE Trans Control Netw Syst 7(3):1308–1318. https://doi.org/10.1109/TCNS.2020.2972601
33. Luenberger DG (1971) An introduction to observers. IEEE Trans Automat Contr 16(6):596–602. https://doi.org/10.1109/TAC.1971.1099826
34. Bhattacharyya SP (1978) Observer design for linear systems with unknown inputs. IEEE Trans Automat Contr 23(3):483–484. https://doi.org/10.1109/TAC.1978.1101758
35. Yang G, Herrera L, Yao X (2023) Detection of false data injection and series arc faults in DC microgrids based on unknown input observers, pp 1155–1159. https://doi.org/10.1109/ECCE53617.2023.10362745
36. Trinh H, Fernando T (2012) Functional observers for dynamical systems. Lecture notes in control and information sciences, vol 420, pp 1–221https://doi.org/10.1007/978-3-642-24064-5_1
37. Darouach M, Darouach M (2004) Functional observers for systems with unknown inputs. https://www.researchgate.net/publication/368528784

Peer-To-Peer Trading Among Microgrid Prosumers in Local Energy Markets

Ziqing Zhu, Siqi Bu, and Qian Hu

Abstract This chapter investigates the transformative potential of Peer-to-Peer (P2P) trading in local energy markets, emphasizing the role of distributed energy resources in facilitating efficient market operations and fostering sustainable energy practices. The text first provides a foundational understanding of local energy markets, highlighting their definition and significance, and introduces the emergent concept of P2P energy trading among prosumers, with a focus on the microgrid and nanogrid levels. It underscores the advantages of such trading, including improved energy efficiency, enhanced grid reliability, and the promotion of renewable energy sources. Further, the chapter delves into the specifics of P2P ancillary service trading, with particular attention to frequency regulation support among microgrid and nanogrid prosumers, exploring its operational benefits and contribution to grid stability. Advancing the discussion, the text introduces a novel aspect of P2P markets—the carbon emission auction trading within local energy spheres. This section probes the theoretical and practical implications of integrating carbon emission considerations into energy trading and examines the market mechanisms through which microgrid prosumers might interact within this innovative paradigm. The chapter is structured into two principal sections. The first addresses P2P energy and ancillary service trading among nanogrid prosumers within a microgrid setting, focusing on real-time market operations for energy balancing and frequency regulation. The second section examines the interconnections between P2P energy, ancillary service, and carbon emission quota trading among multiple microgrid prosumers, presenting advanced modeling techniques and algorithms such as the multi-agent deep deterministic policy gradient for strategy optimization and risk mitigation. The chapter concludes with a synthesis of the explored concepts, reinforcing the significance of P2P trading in advancing the decarbonization of local energy markets and its potential for incentivizing the adoption of green technologies. It offers insights into

Z. Zhu · S. Bu (✉)
Department of Electrical and Electronics Engineering, Hong Kong Polytechnic University, Kowloon, Hong Kong, China
e-mail: siqi.bu@polyu.edu.hk

Q. Hu
Department of Physics, Hong Kong Baptist University, Kowloon Tong, Hong Kong, China

F. Shahnia and J. M. Guerrero (eds.), *Microgrids and Virtual Power Plants*,
Power Systems, https://doi.org/10.1007/978-981-97-6623-9_11

the market structures, strategic behaviors of prosumers, and the envisioned impact on the overarching energy landscape.

Keywords Peer-to-Peer (P2P) Trading · Local energy markets · Energy trading · Ancillary services · Carbon emission trading · Distributed Energy Resources (DERs)

1 Introduction

In the evolving landscape of power systems, local energy markets (LEMs) have emerged as pivotal platforms for the transaction of electrical energy within confined geographical boundaries. Fundamentally, an LEM can be defined as a marketplace where various stakeholders, including residential, commercial, and industrial entities, commonly known as prosumers, engage in the exchange of electricity [1]. These stakeholders possess the dual capability to produce (generate) and consume (utilize) energy, primarily derived from distributed energy resources (DERs). The architecture of LEMs is predicated upon the decentralization of energy generation, and it operates parallel to, or entirely independent from, the traditional centralized grid.

The significance of LEMs lies in their potential to revolutionize energy distribution networks by enhancing efficiency, resilience, and sustainability [2]. By fostering local energy trading, LEMs incentivize the integration of renewable energy sources, thereby expediting the transition towards a low-carbon economy. They provide a dynamic and flexible framework that allows participants to capitalize on their excess generation capacity by facilitating direct transactions, which can lead to a reduction in transmission losses and overall energy costs. Furthermore, LEMs can contribute to the stabilization of the grid by offering demand-side management capabilities, which are essential in the face of fluctuating supply from renewable sources. In essence, LEMs are instrumental in aligning the objectives of modern energy policies with the technological advancements in smart grid infrastructure. They serve as a testament to the shifting paradigm where end-users become integral components of the energy value chain, influencing the patterns of energy production, consumption, and distribution. The proliferation of LEMs is anticipated to be a cornerstone in the creation of sustainable and self-sufficient communities, underscoring their paramount significance in the grand tapestry of energy management and policy-making.

1.1 Peer-To-Peer (P2P) Energy Trading in LEMs

Peer-to-peer (P2P) energy trading represents a transformative model within LEMs, characterized by the direct exchange of electrical power between prosumer entities [3]. This model diverges from the traditional, centralized paradigm of energy distribution, advocating for a bottom-up approach in which individual energy units can trade excess energy production autonomously. A prosumer [4] in this context is an agent who both produces and consumes energy, often harnessing renewable energy sources such as solar or wind power, and thus actively participates in the energy market not only as a consumer but also as a supplier. The benefits of P2P energy trading are multifaceted. It empowers prosumers by providing them with control over their energy production and consumption, leading to optimized energy costs and enhanced energy independence. Economically, it enables the monetization of surplus energy, facilitating a new revenue stream for prosumers while promoting renewable energy adoption through potential financial incentives. Environmentally, it encourages a reduction in greenhouse gas emissions by optimizing the use of locally generated renewable energy. Additionally, P2P trading fosters resilience in energy systems by distributing energy generation, thereby mitigating the risks associated with centralized energy production, such as large-scale outages [5].

Within the spectrum of P2P energy trading, microgrid and nanogrid prosumers represent two critical hierarchical elements [6]. Microgrid prosumers are constituents of a localized grid system capable of operating both connected to and independent from the larger utility grid. These microgrids incorporate multiple generation sources and consumption loads, providing energy autonomy and reliability to a small community or a specific area like a university campus or a hospital complex. Nanogrid prosumers, on the other hand, operate at a more granular level. A nanogrid typically refers to a smaller, self-contained network that services a single building or a specific residential or commercial entity. The nanogrid concept embodies the epitome of localized energy management, offering an even more individualized approach to energy generation and consumption. These prosumers can optimize their energy usage to an even finer degree, leading to increased operational efficiency and cost savings. Incorporating the concepts of microgrid and nanogrid prosumers into the P2P energy trading framework highlights the scalability and versatility of decentralized energy systems. By acknowledging the distinct roles and capabilities of different prosumer levels, P2P energy trading within local energy markets can be tailored to accommodate the diverse needs and capacities of various energy stakeholders, paving the way for a more resilient, efficient, and sustainable energy landscape.

P2P ancillary service trading [7] among microgrid and nanogrid prosumers constitutes an innovative extension of P2P energy transactions, where the exchange transcends mere energy to encompass vital support services essential for maintaining grid stability and efficiency. This trading paradigm enables microgrids and nanogrids to not only trade electricity but also offer ancillary services—such as frequency regulation and voltage support—to the power system. Frequency regulation, a primary type of ancillary service, involves the rapid balancing of supply and demand to maintain the grid's frequency within the prescribed boundaries. It is crucial for the prevention of power system failures that could lead to blackouts. Voltage support, or reactive power support, on the other hand, refers to the management of voltage levels across the grid to ensure that the power delivery is both efficient and within safe operational limits. Both services are integral to the modern power systems, ensuring reliability and security of the electricity supply.

The benefits of P2P ancillary service trading are substantial [8]. It allows for a more flexible and resilient grid by leveraging DERs for grid support, which is especially useful given the intermittent nature of renewable energy sources. Microgrid and nanogrid prosumers can contribute to ancillary services, offering a decentralized approach to grid management. By providing frequency regulation, prosumers can help maintain a stable and constant power frequency, even as the grid integrates a growing share of variable renewable energy sources. This service is particularly valuable in a landscape increasingly dominated by renewables, where traditional frequency regulation resources are becoming less adequate. Voltage support provided by prosumers can mitigate the challenges associated with the variable nature of distributed generation. Through reactive power compensation, prosumers can help maintain voltage levels, reducing the strain on the grid and enhancing power quality for all users. This service is beneficial for managing the local voltage profiles and for reducing losses in distribution networks, thus optimizing grid operation. In the context of this chapter, the focus is predominantly on the trading of frequency regulation support among microgrid and nanogrid prosumers. Frequency regulation is a dynamic and complex service that necessitates rapid response to grid conditions, making it an interesting case study for P2P ancillary service trading. The chapter will delve into the mechanisms by which these prosumers can engage in the market, the technologies enabling such participation, and the implications of their involvement for the broader energy ecosystem.

Carbon emission auction trading [9] in local energy markets is an emergent transaction type that represents the intersection of energy distribution and environmental stewardship. This innovative trading mechanism involves the auctioning of carbon emission quotas (CEQs) to energy producers and consumers within a local market, effectively integrating a carbon pricing element into the energy trading landscape. CEQs essentially cap the amount of greenhouse gases that a prosumer can emit, requiring the purchase of additional quotas if their emissions exceed the predefined limit. The foundational idea behind this approach is to incorporate the externality costs of carbon emissions into the market dynamics, thereby providing economic incentives for reducing greenhouse gas outputs. The benefit of incorporating a carbon emission auction system into local energy markets lies in its potential to drive investment in cleaner energy sources and technologies. By assigning a cost to carbon emissions, prosumers are motivated to optimize their energy production and consumption, favoring low-carbon or carbon–neutral methods. Another benefit is the facilitation of a market-driven approach to achieving regulatory emissions targets. Rather than imposing flat rates or mandates for emission reductions, an auction system allows for flexibility and market-based price discovery for carbon prices, potentially leading to more cost-effective decarbonization strategies.

It is important to emphasize that carbon emission auction trading within local energy markets is a nascent concept and has not yet been widely implemented in actual industry practices. Currently, this trading type is in a preliminary exploration stage, which this chapter will address. The exploration is particularly relevant given the increasing urgency of climate change mitigation and the ongoing search for effective market-based solutions to reduce carbon emissions. While carbon markets exist on national and international levels, the innovation here lies in the application and adaptation of these mechanisms to the local energy market context. This presents a unique set of challenges and opportunities, such as integrating local renewable energy sources, aligning with broader energy market regulations, and ensuring equitable participation among diverse prosumers. In this chapter, we embark on an exploratory journey to investigate how carbon emission auction trading can be structured in local energy markets, the potential it holds for catalyzing a shift towards a more sustainable energy future, and the implications for policy, technology, and market operations.

Given the aforementioned context of P2P trading in LEM with different market paradigms and participants, in this chapter, we will focus on two important research topics: the mechanism design, and the market equilibrium estimation. The mechanism design is to determine the rules for trading, while a good mechanism design facilitates the trading with both economic efficiency and fairness. The market equilibrium estimation is to investigate the behaviors of market participants in a specified market, and can be used to quantitively assess the performance of market mechanisms, thereby providing valuable information for market operators to optimize the market mechanism design. The related research regarding to these two topics are summarized in the following two subsections.

1.2 P2P Trading Mechanism Design

To promote efficient usage of local resources, P2P trading as an emerging approach, enabling participants of the network to share their own resources with one another, has become a research hotspot [10, 11]. Notably, an increasing number of studies that integrate P2P trading mechanism into the microgrid energy market has appeared recently. Ref. [12] introduced a four-layer system architecture of P2P energy trading for grid-connected microgrid, consisting of business layer, control layer, ICT layer and power grid layer, while the business layer was the main focus of this section and a P2P energy trading platform 'Elecbay' was introduced. In [13], prosumers were grouped into virtual microgrids and the social benefits of P2P energy trading in virtual microgrids were optimized by using a Stackelberg game. A aggregated battery control framework was described in [14] for P2P energy sharing in a community microgrid. In [15], P2P energy trading for a community microgrid was built based on the game-theoretic model, where the interaction between sellers and buyers is modelled as Stackelberg game and the pricing competition among sellers is modelled as a noncooperative game. [16] was focused on the P2P trading price in the microgrid market and the trading prices were acquired based on the market participations.

Nevertheless, above studies only consider 'energy' as the main transactive commodity in the P2P microgrid energy market. As a matter of fact, the coalitions of small-scale prosumers can provide flexible ancillary service, which means the transactive commodities should not be only limited to energy [17]. According to the report of National Grid UK, frequency response can be categorized into two types of service: dynamic and non-dynamic. Dynamic frequency response is a continuously provided service to control the normal second by second changes, whilst non-dynamic frequency response is usually referred to the discrete service triggered at a defined frequency deviation [18]. In the microgrid, entities such as dispatchable generation units, energy storage systems and demand-responsible loads can adjust their power generation or demand to flexibly offer the frequency response service [19, 20]. A comprehensive P2P trading mechanism for the microgrid should include ancillary service as a small but critical part of the entire trading process.

Some works have incorporated ancillary service into energy markets in the distribution or transmission level. A day-ahead scheduling strategy was presented in [21] for the integrated community energy system that participates in the joint energy and ancillary service markets. But this joint market was still implemented in the level of wholesale markets. In [17], microgrids were considered to be active prosumers to provide energy and ancillary service to the distribution network. Although the incorporation of joint energy and ancillary service exchanges was discussed, the focus of this research lies in the optimal management of the distribution system. An incentive contract between independent system operator and frequency regulation service providers was designed in [22] for purchasing energy and ancillary service at the same time. The energy and frequency reserve were optimally allocated among regulation service producers. However, the contract design between two parties was the focus of the section, without further comprehensive consideration in the view of the market design.

In fact, an appropriate P2P marketing mechanism for the microgrid should be put in place to facilitate not only the energy trading but the frequency response provision in an economic manner. The frequency response service required by the system operator should be able to gain the profit for the providers. However, it is founded that in most of the existing literatures in terms of the design of microgrid energy market, the frequency regulation as an essential tradable ancillary service to maintain the local balance of generation-demand and frequency stability, has not been addressed carefully. The regulation effect which could be closely related to the market mechanism has not been clearly investigated. To fill the research gap, it is imperative to design a hybrid microgrid energy management system that enables prosumers in the microgrid to trade energy and frequency regulation service simultaneously, economically coping with the balancing and frequency stability issue in the islanded microgrid.

1.3 P2P Market Equilibrium Analysis

In the joint P2P market, producer and consumer microgrids will form pairs via match-making conducted by the market operator based on their bidding prices in a merit-order manner, and microgrids will compete with each other for the priority of being matched, while considering the potential cost in the carbon emission auction market (CEAM). Intuitively, microgrids cannot obtain the "optimal" bidding strategy at the very beginning of the bidding game, because the "optimality" of one's strategy is highly dependent on other rivals' strategies which are unknown and would vary in a competitive bidding game. Hence, microgrids will learn to "be adaptive to" each other by modifying their bidding strategies. The modification of microgrids' bidding strategy is a long-term process, and will converge to a steady state in which each microgrid cannot obtain more profit by modifying its bidding price, subject to the fixed bidding price of other rival microgrids. Such a steady state is the so-called Nash Equilibrium Point (NEP) [23], i.e., the set of optimal strategies with maximized profits in response to their rivals' fixed strategies. The estimation and analysis of NEP is the key to maximizing microgrids' financial profits and optimizing market design [24]. On the one hand, the optimality of each microgrid's bidding strategy is dependent on rival microgrids, and therefore one microgrid needs to make its bidding strategies adaptive to other rival microgrids' decisions. On the other hand, the market operator may investigate the impact of market designs on the social welfare and fairness of competition via the analysis of all possible NEPs, to facilitate the incentive compatibility [25] of market operation. References [26, 27, 28, 29] have contributed to the NEP estimation in P2P markets using conventional mathematical programming methods method. The most important deficiencies of these methods are twofold. Firstly, only the non-strategic bidding is considered, while the dynamic bidding strategy modification procedure is neglected. Secondly, it is based on the very strong assumption that each agent knows the global information of the game, i.e., other agents' decisions, which is the private information that would not be available in real-world markets. Hence, the obtained NE must be deviated significantly from that of the real markets. Ref. [30] introduces the Q-learning-based Reinforcement Learning (RL) method to tackle the first limitation, while the privacy protection concern is still not addressed. This limitation has been effectively addressed in [31, 32] by Multi-Agent Deep Deterministic Policy Gradient (MADDPG) method as a fully distributed agent-based RL simulation without any requirement of other rival agents' private information.

Meanwhile, microgrids are faced with risks when participating in the joint P2P EM and CEAM. Firstly, the uncertainty of microgrids' net load will result in the deviation between scheduling and actual energy procurement, leading to the real-time imbalance with financial losses and security concerns [33]. Secondly, the costs of purchasing CEQs with uncertain prices in the CEAM will further complicate the bidding decisions, and the NEP in such a joint market with uncertainties will therefore be different from that of a conventional EM. To the authors' best knowledge, few attempts have been made to incorporate the risk mitigation while deploying RL to estimate the NEP. In [34], the "robust programming" method is adopted to estimate the "robust NE" in the worst-case scenario, while this method is extremely conservative with limited applicability in real-world electricity market. Compared with robust programming, stochastic programming with conditional value-at-risk (CVaR) has been widely deployed in [35, 36, 37] to achieve a balance between profit maximization and risk mitigation.

2 P2P Market Mechanism Design

In this section, the market mechanism design of P2P trading is investigated, with a focus on the P2P energy and ancillary service trading among nanogrid prosumers within microgrid, and the market design methodology can be further extended to other P2P markets in LEM with different stakeholders (e.g., multiple microgrids, etc.).

As aforementioned, the presence of geographically dispersed DERs, the microgrid can be organized into a number of nanogrids which provides an efficient approach to aggregating DERs for the participation of ancillary service with high level of efficiency and controllability. In this regard, this section presents a multi-market nanogrid trading framework for simultaneous real-time imbalance elimination and frequency regulation procurement, based on the architecture of P2P trading and communication. The framework comprises three consecutive markets: P2P bilateral energy market, balancing market and ancillary service market, where the capacity of participated nanogrids can be coordinately traded and reserved for the real-time frequency regulation service. Optimal offering strategies are designed to maximize the profits of producer nanogrids and meanwhile to tackle the capacity allocation tradeoffs among three markets. When it is approaching to the real-time, the market will be closed with completion of balancing, and selected nanogrids with reserved capacity for the ancillary service will provide the real-time frequency regulation in a fully distributed manner. A parameter so called engagement factor is designed based on the market clearing results to determine the deployment of frequency regulation service of each participated nanogrids. Subsequently, the pinning-based distributed secondary control upon P2P information sharing among nanogrids is implemented to regulate the frequency and ensuring the power sharing. The results of case studies indicate the feasibility of the introduced framework and its good performance, while different remuneration mechanisms are compared for discussion of their pros and cons.

2.1 Preliminaries

2.1.1 Microgrid Partition and Set-Up

To enhance the reliability and flexibility, a microgrid can be grouped into a set of sub-networks, i.e., networked microgrid [38] 39. As it is not feasible for an individual small-scale DER to play a significant role in the market trading and to make impact on the real-time operation due to its limited capacity, in this section, an islanded microgrid is partitioned into several sub-networks, so called nanogrids. DERs that are geographically close to each other can be grouped into one nanogrid and each nanogrid functions as a single entity for the energy trading and frequency regulation hereinafter.

Each nanogrid locally contains uncontrollable loads and two types of DERs: non-dispatchable DERs (e.g., PV, wind turbine) and dispatchable DERs (e.g., gas turbines, energy storage devices). The nanogrid can be categorized into two types: consumer nanogrids (N_c) and producer nanogrids (N_p). In a specific trading timeslot, if there is a power deficiency in a nanogrid, it becomes a consumer nanogrid; if there is a power surplus in a nanogrid, then it becomes a producer nanogrid. A single nanogrid is the main trading and regulation entity hereafter in the section. Figure 1 shows a schematic diagram of an islanded microgrid consisting multiple nanogrids, where the important network components are displayed in the diagram. Apart from nanogrids, there exist other virtual agents and smart ICT-based infrastructures that facilitate the market implementation, which are introduced below.

Communication Router

Every market participant in the microgrid has a communication router for the information sending and receiving. It is the fundamental infrastructure to enable P2P energy trading and information sharing among nanogrids.

Database

A cloud server is embedded in the top level of the nanogrid energy management system, providing a date repository for the extraction of historical data of nanogrids and relevant components in the microgrid. The trading information for each trading would also be stored in the database.

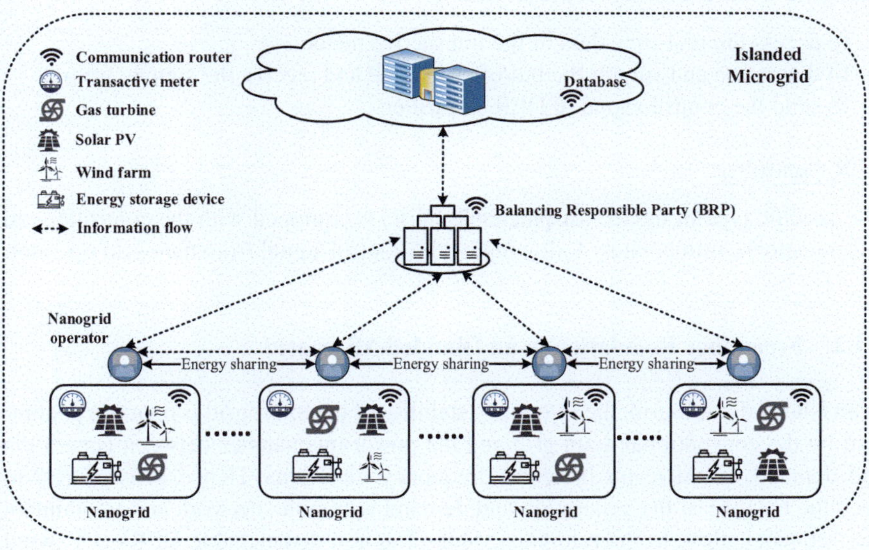

Fig. 1 An islanded microgrid consisting of multiple nanogrids with different DERs for P2P energy and ancillary service trading

Transactive Meter

Each nanogrid is equipped with a transactive meter to record the energy data, e.g. real-time demand and available power capacity, which is monitored and managed by the nanogrid operator.

Balancing Responsible Party

Balancing responsible party (BRP) is a profit-oriented stakeholder to coordinate P2P transactions among nanogrids and is responsible for the power balance of the whole microgrid. It has following duties:

- To receive the bids/offers from nanogrid operators who are willing to trade in different markets
- To match the offers and bids of bilateral market according to their submitted prices
- To clear the balancing and ancillary service market based on their submitted bids/offers, in order to fulfill the requirements of imbalance elimination and frequency regulation capability.

Nanogrid Operator

The main goal of the nanogrid operator is to locally optimize the profit and ensure the secure operation of the nanogrid. nanogrid operator could be a virtual agent or platform that has the specific knowledge of the conditions of the nanogrid, including the rated capacity of the generation units, storage devices and the loads inside the nanogrid. It has following responsibilities:

- To access the real-time data of the transactive meter
- To determine and submit the bid/offer to BRP and receive the signals from it
- To send the control signal to DER controllers.

DER Controller

The specific type of device (dispatchable DER) is equipped with the controller and it can response immediately following by the control signal from nanogrid operator.

2.1.2 Frequency Regulation in an Islanded Microgrid

In an islanded microgrid, the frequency stability of the microgrid is no more guaranteed by the conventional main grid and the power mismatches between generation and demand would result in severe frequency violations. Hence, it is critical to keep the balance of the generation and demand within the network and to maintain the frequency close to the reference value. The non-dispatchable DERs (PV, wind turbines) operating in a grid-following mode can provide the maximum available power, which are seen as the negative loads in the section. The dispatchable DERs (gas turbines, energy storage devices), operated as voltage source inverters in a grid-forming mode, can participate in the system frequency regulation and power sharing to satisfy the network demand. The power sharing among nanogrids is conducted by implementing primary droop control, which is shown as:

$$f_i = f^* - k_{eq,i} P_i \tag{1}$$

where f^* is the reference value of frequency; f_i and P_i correspond to the measured output frequency and power of nanogrid i, respectively; $k_{eq,i}$ denotes the equivalent droop coefficient of nanogrid i which is the sum of the droop gains of all dispatchable DERs in nanogrid i. The main idea of the droop mechanism is to mimic the response capability of synchronous generators, where the system frequency can be regulated down or up when the power output is increased or decreased [40]. Once load demand changes, the power output of grid-forming DERs would be adjusted instantaneously based on their droop characteristics to maintain the power balance. However, conventional primary droop control would cause frequency deviated from the reference value [5]. A secondary control is required to mitigate the frequency deviation. As such, the droop function is now formulated as:

$$f_i = f^* - k_{eq,i} P_i + \Delta f_i \tag{2}$$

where Δf_i is the frequency correction term, which is provided by the secondary control to ensure the frequency can be regulated to the reference value. A distributed secondary control based on the local measurements and neighboring information sharing is compatible with the P2P nanogrid energy and ancillary service trading under the P2P communication structure. A market-enabled distributed secondary control considering the economic incentives for the frequency regulation will be introduced in Sect. 2.5.

2.2 Mechanisms of Multi-market Framework for Nanogrid P2P Energy and Ancillary Service Trading

2.2.1 Overview of Multi-market Framework

A framework including three consecutive markets can be presented to facilitate the cost-effective real-time operation in the microgrid, addressing the issue of demand and generation balancing, while offering potential transaction opportunities for nanogrids to provide both energy deployment and real-time frequency regulation service. Figure 2 shows a sequential implementation procedure of this multi-market framework. The entire market trading procedure is conducted in a 15-min timeline. After the market gate closure, the responsibility of balancing and frequency regulation which is determined during the market trading will be deployed in the subsequent hourly real-time operation. Such a 15-min trading timeline is feasible for nanogrids

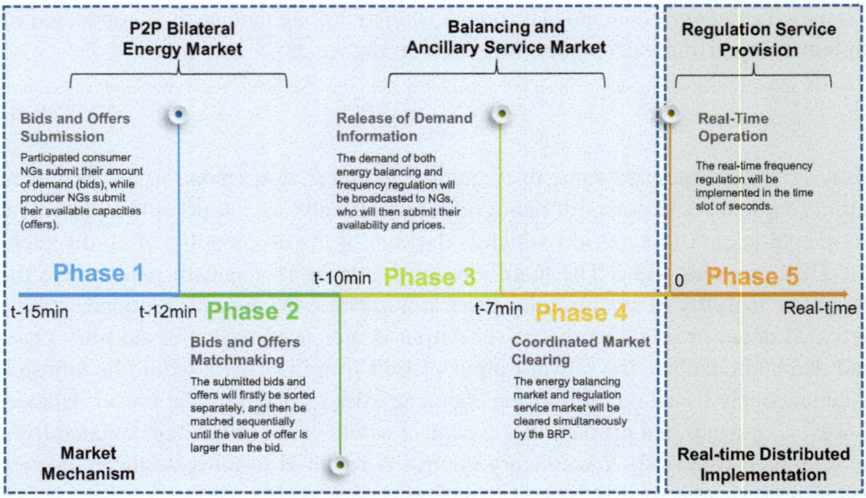

Fig. 2 Sequential implementation procedure of a multi-market nanogrid P2P energy management framework

consisting of flexible small-scale generators with fast response capability and without pre-determined annually/monthly/weekly start-up and shut-down schedule.

The markets will be sequentially coordinated by the BRP. In the bilateral P2P trading market (Phase 1 and 2), each nanogrid will mitigate the power imbalance resulting from inaccuracy of prediction by participating in the bilateral P2P transaction, in which the matchmaking procedure is conducted by the BRP. In the balancing and ancillary service market (Phase 3&4), BRP will organize a centralized bidding procedure and coordinated market clearing for deployment of both energy and regulation service, to enable available nanogrids participating in: (1) further elimination of the remaining imbalance due to failure of matchmaking at the Phase 2; and (2) real-time frequency regulation service following the control signal. This transaction procedure could also be considered as a P2P trading between nanogrids and BRP. Details in terms of market mechanisms are discussed in the following subsections.

2.2.2 Optimal Offering Strategy in P2P Bilateral Energy Market

From the perspective of producer nanogrids, the 'tradeoff' that needs to be tackled is: how much capacity should be offered to the bilateral transactions and subsequent balancing and ancillary service market respectively, to achieve the maximum total profit? The expected unit price in the balancing and ancillary service market would apparently be higher than the bilateral market, as it is much more approaching to the market closure time, and the 'willingness' of buyers to purchase such service would therefore be strengthened. However, devoting all of the available capacity to the subsequent markets is also not the best decision, as producers cannot precisely foresee the actual demand of network balancing and ancillary service, and it would be possible that the value of submitted offers is much larger than the actual demand. According to above descriptions, the optimal offering strategy for producer nanogrids to allocate their available capacity to different markets can be formulated as follows:

$$\max\left\{\begin{array}{c}\mu_{S1,i}\left[\lambda_{S1,i}(t)-C_i(t)\right]q_{S1,i}(t)+\\+\mu_{S2,i}^{B}\left[\lambda_{S2,i}^{B}(t)-C_i(t)\right]q_{S2,i}^{B}(t)+\\\mu_{S2,i}^{R}\left[(\lambda_{S2,i}^{R}(t)-C_i(t))\tau_i+\lambda_{S2,i}^{B}(t)(1-\tau_i)\right]q_{S2,i}^{R}(t)\end{array}\right\}, i\in N_P \tag{3}$$

s.t.

$$q_{S1,i}(t)\geq 0 \tag{3a}$$

$$q_{S2,i}^{B}(t)\geq 0 \tag{3b}$$

$$q_{S2,i}^{R}(t)\geq 0 \tag{3c}$$

$$q_{NG,i}^{min}\leq q_{S1,i}(t)+q_{S2,i}^{B}(t)+q_{S2,i}^{R}(t)\leq q_{NG,i}^{max} \tag{3d}$$

where $\lambda_{S1,i}(t)$, $\lambda_{S2,i}^{B}(t)$ and $\lambda_{S2,i}^{R}(t)$ are the capability price offered by nanogrid i in P2P bilateral market, balancing market and ancillary service market in period t, respectively; $C_i(t)$ is the cost function of the generation; $q_{S1,i}(t)$ is the trading capacity in the bilateral energy trading market, $q_{S2,i}^{B}(t)$ and $q_{S2,i}^{R}(t)$ are respectively, expected balancing capacity in the balancing market and regulation capacity in the ancillary service market in period t; $q_{NG,i}^{min}$ and $q_{NG,i}^{max}$ are the minimum and maximum capacity limit of nanogrid i; $\mu_{S1,i}$, $\mu_{S2,i}^{B}$ and $\mu_{S2,i}^{R}$ denote the *successful rate of bidding* of producer nanogrid i in each market, respectively. The *successful rate of bidding* indicates the expected ratio of actually allocated capacity and the submitted amount of capacity, which is a non-negative value in the range of [0,1]. It can be statistically determined based on the historical trading data that are stored in the database of the microgrid.

The objective function (3) shows the expected maximal profit of a producer nanogrid comprises three profitable transactions in different markets. The first and second item of (3) are the expected profit acquired in the bilateral market and balancing market, respectively. The third item, including two types of settlements, quantifies the expected profit acquired from the regulation service provision. τ_i in the third item denotes the *probability of the regulation capacity to be deployed* by nanogrid i, which can be extracted from the historical data that also are stored in the database. Its first part denotes the expected revenue of capacity to be actually deployed for frequency regulation service, in which the trading price of ancillary market will be paid. The rest of capacity that is not deployed will be remunerated for the trading price of the balancing market as a compensation for the opportunity cost, which is shown in the second part of this item.

It should be noted that, each operator of producer nanogrid will determine the amount of traded capacity in three markets based on above offering strategy once the market is open. Although three optimal solutions $q_{S1,i}$, $q^B_{S2,i}$ and $q^R_{S2,i}$ will be obtained by this strategy, only $q_{S1,i}$ as the devoted capacity for bilateral trading which will be actually offered in Phase 1 in the P2P bilateral energy market. Once the optimal capacity for P2P trading is confirmed, nanogrid operator will submit this offer to BRP. Then BRP will conduct the matchmarking procedure to clear the market in Phase 2, as described in the following Subsection. The tradeoff between devotion to energy balancing market and ancillary service market will be further tackled again after the information of demand being released by the BRP, i.e. at the beginning of the balancing market opening in Phase 3.

2.2.3 Matchmaking Procedure in P2P Bilateral Energy Market

In Phase 1, nanogrid operators will submit their available energy (producer nanogrid) or demands (consumer nanogrid) to the transaction platform managed by the BRP, who will thereafter examine the submission to avoid any violations. The producers and consumers will be sorted separately in the order of submitted offers (from the lowest to the highest) and bids (from the highest to the lowest). For bids and offers with same values, those with earlier submission time will be treated with priority. Once upon the 'queues' of producers and consumers is constructed, the BRP will serve as the dealmaker to implement the matchmaking procedure in Phase 2, which can be described as follows:

- The producer with the lowest offers will be matched with the consumer with the highest bids.
- The producer with the second lowest offers will be matched with the consumer with the second highest bids.
- The rest of matches can be made in the same manner, until the highest value of the rest of bids is less than the lowest value of the rest offers.
- The final transaction price of each matched producer–consumer pair will be the average value of their offers and bids.
- Transactions with amount of power exchange that exceed the maximum capacity of tie-lines between the producer and the consumer will be suspended.

2.2.4 Mechanisms of Balancing and Ancillary Service Market

At the beginning of Phase 3, i.e., 10 min prior to the real-time operation, BRP will organize the centralized bidding and coordinated clearing of energy balancing market and frequency regulation service market. In the balancing market, the amount of remaining power imbalance q_B of the entire microgrid will be broadcasted by BRP to nanogrids, while available nanogrids will submit their applicable amount of capacity and price to BRP. In this section, only power deficiency is considered for the power imbalance that requires the participation of producer nanogrids. As for the unbalanced situation of power surplus, power generation of renewables such as PV and wind farms will be curtailed accordingly, which is not the focus in the section. To be noted, the cost of purchasing energy balancing should be paid by the imbalanced consumer nanogrids instead of BRP, as they are responsible for the failure of matchmaking because of overly high bidding price. Meanwhile, for the ancillary service market, the amount of expected requirement capacity q_F for frequency regulation will be broadcasted to nanogrids while available nanogrids will also submit their applicable amount of capacity and price. The requirement of the regulation service is based on the net-load forecast, determined as the $110^{\%}$ of the largest predicted net-load variation within a timeslot [41]. As the producer nanogrid is the main focus in this section, only up-regulation is considered. For the scenario of generation exceeding load demand, the additional unexpected renewable generation will be curtailed directly, as the cost of curtailment is much less than purchasing the down-regulation service [?].

The BRP, as the consumer of balancing and ancillary service, will simultaneously clear the balancing and ancillary service market by minimizing the total cost, while the requirement of capacity to maintain network balancing and frequency regulation in the real-time must be fulfilled. Three types of capacity remuneration mechanism (CRM) are considered for the payments of producers in the balancing and ancillary service market: marginal price (MP), pay-as-bid (PAB) and pay as the opportunity cost (PAO), which are introduced below.

- MP: The nanogrids with winning bids after the market clearing will be paid by the market marginal price. This is a currently one of the most common CRMs.
- PAB: The nanogrids with winning bids after the market clearing will be paid by the price they procure for the unit of capacity.
- PAO: For the remuneration of ancillary service deployment, the policy of PAO, in which the subsidy price is directly determined by the BRP according to the calculated opportunity cost for ancillary service provision, was implemented in several existing markets to avoid possible strategic arbitrage of service providers. In this section, the reserved capacity for the ancillary service will be remunerated by the marginal price determined in the balancing energy market.

The case study will illustrate how three types of CRM will impact the total producer profit in this framework.

2.2.5 Optimal Offering Strategy in Balancing and Ancillary Service Market

For these two markets, producer nanogrids will tackle the tradeoff between devoting their available capacity to balancing market and ancillary service market to maximize their benefits. The optimal offering strategy for producers to allocate their available capacity to these two markets can be formulated as follows:

$$max\left\{\begin{array}{c}[\lambda^{B}_{S2,i}(t)-C_i(t)]q^{B^*}_{S2,i}(t)+\\ [(\lambda^{R}_{S2,i}(t)-C_i(t))\tau_i+\lambda^{B}_{S2,i}(t)(1-\tau_i)]q^{R^*}_{S2,i}(t)-\\ C_{opp,i}\end{array}\right\}, i\in N_P \tag{4}$$

$$C_{opp,i}=\underbrace{[\lambda^{B}_{S2,i}(t)-C_i(t)]q^{B^*}_{S2,i}(t)}_{\text{term 1}}-\underbrace{[(\lambda^{R}_{S2,i}(t)-C_i(t))\tau_i+\lambda^{B}_{S2,i}(t)(1-\tau_i)]q^{R^*}_{S2,i}(t)}_{\text{term 2}} \tag{4a}$$

s.t.

$$q^{min}_{NG,i}-q_{S1,i}\leq q^{B^*}_{S2,i}+q^{R^*}_{S2,i}\leq q^{max}_{NG,i}-q_{S1,i} \tag{4b}$$

$$q^{B^*}_{S2,i}\geq q_B \tag{4c}$$

$$q^{R^*}_{S2,i}\geq q_F \tag{4d}$$

The first item in (4) represents the expected profit acquired in the balancing market. The second item describes how much profit can be gained by providing frequency regulation service. The third term $C_{opp,i}$ reflects the *opportunity cost* of reserved capacity for the provision of frequency regulation service instead of participating in the balancing market, which can be acquired by (4a). If the capacity reserved for the ancillary service is all traded in the balancing market, the profit would be represented by term 1 in (4a), while the actual profit of the ancillary service is described by term 2 in (4a). The *opportunity cost* can be represented by the difference of these two. Constraint (4b) declares that the total devoted capacity of each nanogrid should be within the minimum and maximum value of availability. Constraint (4c) and (4d) indicate that the demand of energy balancing and frequency regulation requirement must be fulfilled.

In practice, in terms of various existing formations of ancillary service market, most of them are specifically for transmission system which are organized and operated by the system operators or transmission organizations [42]. For example, a web-based service so called frequency response price submission that allows frequency response providers to submit holding prices on a monthly basis has been conducted by National Grid UK [18]. As the frequency regulation is the service that is more approaching to the real-time, this framework provides opportunities for potential players to participate in the market with near to real-time procurements, which can encourage more prosumers to get involved into the markets. As time goes by, the regulation service market would have higher priority comparing to the other markets, which therefore leads to a higher price to reserve the capacity for the regulation service. The price $\lambda^{R}_{S2,i}$ submitted by the producer nanogrid for the ancillary service provision essentially must cover the holding cost and capacity cost. Holding cost is the expected payments made to producer nanogrid by BRP to cover the costs when the nanogrid chooses to provide the frequency regulation capacity cost, while capacity costs are the payments by BRP to remunerate the producers the amount of capacity delivered when providing frequency regulation.

2.3 *Market-Based Real-Time Distributed Frequency Regulation*

Under the P2P communication infrastructure, small fluctuations in terms of the frequency due to the mismatch between supply and demand can be automatically and constantly balanced by controlling the power dispatch of nanogrids. In the nanogrid energy management framework, nanogrids that are providing frequency regulation service should follow the control signal from the distributed hierarchical control structure and change the output accordingly based on the previous market trading results. Given a certain limited capacity, how active a single nanogrid participate in the regulation service provision is closely related to how much capacity is determined to devote in the ancillary service market based on the optimal offering strategies in the previous markets.

To ensure the economic efficiency of real-time implementation, the devoted capacity for the ancillary service is incorporated into the control action for potentially participated nanogrids. Specifically, the nanogrid with largest reserved capacity has the highest incentive to provide the service. Here a parameter *engagement factor* φ_i is introduced to reveal the participation level of available nanogrids in the regulation. It is determined by the ratio between devoted capacity of frequency regulation to the rated capacity of the nanogrid, which is defined as

$$\varphi_i = 1 + q_{S2,i}^{R^*} / q_{NG,i}^{max} \tag{5}$$

The *engagement factor* is embedded into (2) and a new droop characteristic indicating the priority of the participated nanogrids can be reformulated as:

$$f_i = f^* - \varphi_i k_{eq,i} P_i + \Delta f_i \tag{6}$$

The frequency correction term Δf_i can be compensated a secondary control structure. The pinning-based synchronization control protocols are developed for distributed secondary control. To achieve the frequency reference f^*, each participated nanogrid needs to exchange the information with its neighboring nanogrids. Assuming the total number of participated nanogrids is *n*, the frequency disagreement dynamic between neighboring nanogrids via P2P information exchanging is shown as follows:

$$\dot{\Delta f_i} = -c\left[\sum_{j=1, j\neq i}^{n} a_{ij}(f_i - f_j) + d(f_i - f^*)\right] - c\sum_{j=1, j\neq i}^{n} a_{ij}\left(\varphi_i k_{eq,i} P_i - \varphi_j k_{eq,j} P_j\right) \tag{7}$$

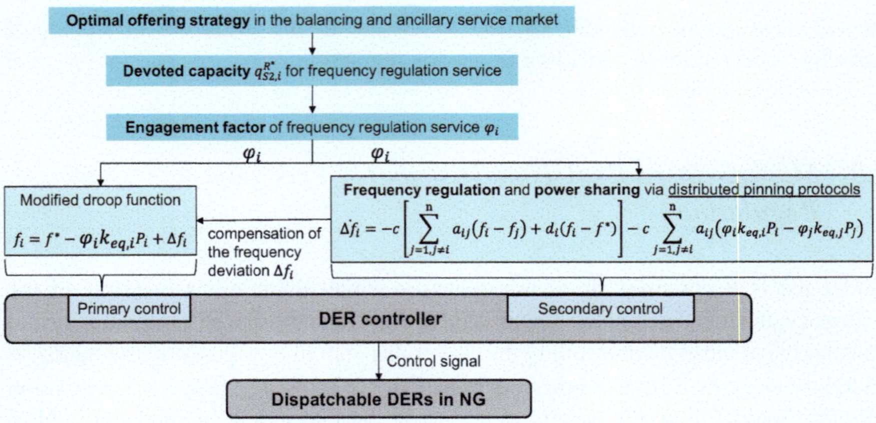

Fig. 3 Implementation of pinning-based distributed frequency regulation considering the engagement level of each nanogrid

where c is the control gain; a_{ij} indicates the communication connection between nanogrid i and nanogrid j: if nanogrid i can receive the information from nanogrid j, then$a_{ij} = 1$, otherwise it is 0; $d > 0$ is the pinning gain of the pinning nanogrid which can access the reference frequency signalf^*,, otherwise it is 0. By utilizing the control protocol (6), within a certain iteration steps, pinning nanogrid will regulate its output frequency to the reference value while the non-pinning nanogrids will minimize the disagreement with their neighbors until the frequency of all nanogrids achieve to the reference state. Figure 3 shows the real-time distributed frequency regulation procedure based on the previous market trading results.

Next, the controllability of the protocols is proved. Let $E_i = f_i - f^*$ as the frequency regulation error, the error dynamics can be written as follows:

$$\dot{E}_i = \sum_{j=1,j\neq i}^{N} ca_{ij}\left(E_j - E_i\right) - d_i E_i \tag{8}$$

The following Lyapunov candidate is used to derive the stability properties of the pinning-based control:

$$\mathrm{V} = \frac{1}{2}\sum_{i}^{n} {E_i}^T E_i \tag{9}$$

Then it follows that

$$\dot{V} = \sum_{i}^{n} {E_i}^T \dot{E}_i \tag{10}$$

$$= \sum_{i}^{n} E_i^T \left\{ \sum_{j=1, j \neq i}^{N} ca_{ij}(E_j - E_i) - dE_i \right\} \tag{10a}$$

$$\leq \sum_{i}^{n} E_i^T \sum_{j=1, j \neq i}^{N} ca_{ij}(E_j - E_i) - \sum_{i}^{n} d_i \|E_i\|^2 \tag{10b}$$

$$\leq \sum_{i}^{n} \|E_i^T\| \sum_{j=1, j \neq i}^{N} ca_{ij} \|E_j - E_i\| - \sum_{i}^{n} d_i \|E_i\|^2 \tag{10c}$$

$$\leq \sum_{i}^{n} \|E_i^T\| \sum_{j=1, j \neq i}^{N} ca_{ij}(\|E_j\| + \|E_i\|) - \sum_{i}^{n} d_i \|E_i\|^2 \tag{10d}$$

$$= |\boldsymbol{E}|^T (\boldsymbol{B} - \boldsymbol{D})|\boldsymbol{E}| \tag{11}$$

where $\boldsymbol{E} = [E_1, E_2, \ldots, E_N]^T$ is the error vector; $\boldsymbol{B} = \left[ca_{ij}\right]$ is the coupling configuration matrix of the communication network; $\boldsymbol{D} = diag(d_1, d_2, \ldots, d_N)$ is the pinning gain matrix. Now it is observed that if $(\boldsymbol{B} - \boldsymbol{D})$ is the Hurwitz matrix where the real part of eigenvalues are strictly negative, then the asymptotic stability of the pinning-based secondary frequency control can be achieved, i.e., the frequency can be regulated to the reference state by given the proper pinning gain and communication connection.

2.4 Case Study

In this subsection, three cases are designed to test and validate the nanogrid multi-market framework in terms of the market mechanisms and the real-time frequency regulation. The physical model of nanogrids and frequency regulation protocols are created in the MATLAB/Simulink. The programs for the optimal offering strategy are built in MATLAB and solved using YALMIP. For all cases, an islanded microgrid consisting of seven nanogrids are considered and it is assumed that four of nanogrids are producers while the rest of nanogrids are consumers. The main objective of each case is shown as follows:

(1) Case I: To demonstrate the sequential implementation procedure of market trading within a single timeslot.
(2) Case II: To verify the effectiveness of market-based real-time distributed frequency regulation.
(3) Case III: Based on the predicted net-load curve in an eight-timeslot, to demonstrate the amount of allocated capacity in three market and to examine the total profit of the producer nanogrid under three CRMs.

2.4.1 Case I

The parameters of each producer nanogrid in different markets are illustrated in Table 1 for Case I. For the multi-market trading framework, 15 min prior to the real-time operation, P2P bilateral market is organized to enable prosumer nanogrids trade their energy based on their requirements. According to the optimal solutions from the optimal sharing strategy elaborated in Eq. (3), each nanogrid operator of producer nanogrids submits the amount of devoted capacity along with the selling price to the transactive platform managed by BRP. Meanwhile, each consumer nanogrid submits its required amount of capacity along with the potential purchasing price. The matchmaking process is conducted by BRP as introduced in Subsection 2.3.3, where lowest offer is matched with the highest bid first at the first place. The submitted bids and offers and the determined P2P trading among nanogrids are demonstrated in Fig. 4. The total submitted energy of the four producer nanogrids is 0.97 MWh, which is less than the consumer requirement of 1.4 MWh. Therefore, all the submitted offers are successful, while NG7 with the highest bidding price fails to be matched. The remaining imbalance of 0.43 MWh will be addressed in the next phase with a potentially higher cost. In the P2P energy market, the final transaction price is the average value of offer and bid for the matched producer–consumer pair. Then the potential profit (in $) obtained by each producer nanogrid in our hour is calculated as 17.64, 25.81, 15.6 and 36.6 respectively.

In the subsequent balancing and ancillary service market, the producer nanogrids will devote their remained available quantities to the market, following the optimal allocation strategy elaborated in Eq. (4). Table 2 shows the overall trading results for each producer nanogrid. The balancing requirement is the remaining 0.43 MWh from previous P2P energy market, and the frequency regulation requirement is 0.71 MW. The BRP will then clear these two markets simultaneously based on the merit-order, while the bidding with lower price will be accepted with priority, until the requirement is fulfilled. In this case, NG1 and NG4 are selected to provide energy balancing with

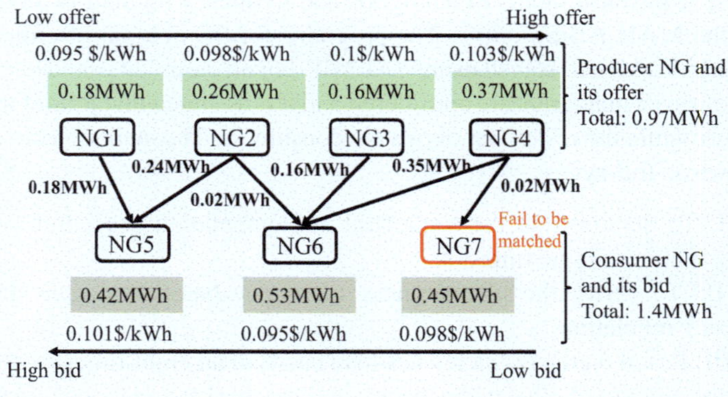

Fig. 4 P2P energy trading among nanogrids based on their offers and bids in the P2P bilateral energy market

Table 1 Trading parameters of producer nanogrid in different markets

Case I	Rating (MW)	P2P bilateral market						Balancing and ancillary service market		
		Price ($/kWh)			Successful rate of bidding			Price ($/kWh)	Price ($/kW)	Probability
		$\lambda_{S1,i}$	$\lambda_{S2,i}^{B}$	$\lambda_{S2,i}^{R}$	$\mu_{S1,i}$	$\mu_{S2,i}^{B}$	$\mu_{S2,i}^{R}$	$\lambda_{S2,i}^{B}$	$\lambda_{S2,i}^{R}$	τ_i
NG1	0.8	0.095	0.099	0.105	0.80	0.72	0.64	0.099	0.111	0.82
NG2	1.0	0.098	0.102	0.111	0.82	0.66	0.52	0.104	0.117	0.79
NG3	1.0	0.100	0.108	0.110	0.75	0.62	0.58	0.112	0.116	0.81
NG4	1.2	0.103	0.098	0.114	0.78	0.65	0.53	0.102	0.119	0.82

Table 2 Trading results of producer nanogrid in different markets

Case I	P2P bilateral market				Balancing and ancillary service market			
	Consumer requirement: 1.4 MWh				Balancing requirement: 0.43 MWh			
					Frequency regulation service requirement: 0.71 MW			
	Devoted (MWh)	Allocated (MWh)	Remained (MW)	Profit ($)	Energy devoted (MWh)	Ancillary service devoted (MW)	Energy allocated (MWh)	Ancillary service allocated (MW)
NG1	0.18	0.18	0.62	17.64	0.21	0.41	0.21	0.41
NG2	0.26	0.26	0.74	25.81	0.3	0.44	0	0
NG3	0.16	0.16	0.84	15.6	0.41	0.43	0	0.3
NG4	0.37	0.37	0.83	36.66	0.39	0.44	0.22	0

successful bidding amount of 0.21 MWh and 0.22 MWh, while NG1 and NG3 are selected to provide frequency regulation service with successful bidding amount of 0.41 MW and 0.3 MW. In this regard, the engagement factor can be acquired for the following real-time operation, which are $\varphi_{NG1} = 1 + 0.41/0.8 = 1.51$ and $\varphi_{NG3} = 1+0.3/1 = 1.3$. In the real-time operation, such the engagement factor will be used to formulate the primary and secondary control protocols for the frequency regulation.

2.4.2 Case II

In Case II, the effectiveness of market-based distributed frequency regulation is verified. Figure 5 shows the P2P communication link among nanogrids in the test islanded microgrid, where the local information can be transmitted and received bidirectionally between neighbouring nanogrids. For the pinning-based secondary

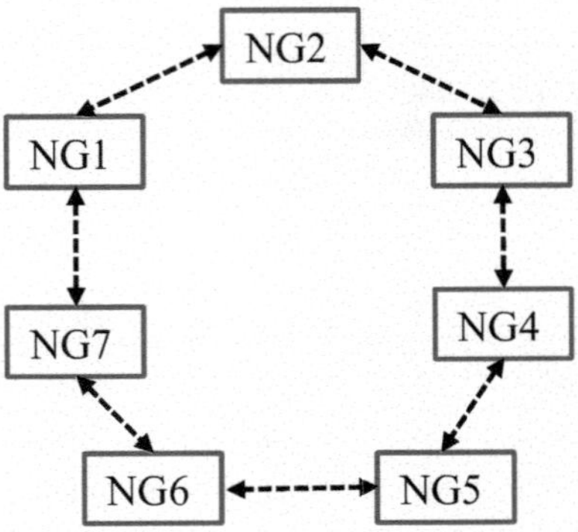

Fig. 5 Communication graph of nanogrids in the islanded microgrid

protocol, the control gain c is set to 100 and the pinning gain d is set to 10 which ensures the stability of the control protocol. The frequency reference is set to 50 Hz.

As soon as the power imbalance occurs in the islanded microgrid, the frequency is deviated from the reference value and consequently the frequency regulation will be automatically initiated. For the real-time frequency regulation in this section, how much capacity will be provided, and which nanogrid will participate in the regulation service provision is directly associated with the trading results in the market. Figure 6 shows the real-time power dispatch of each prosumer nanogrid according to the trading results in Case I. During 0-2s, the entire network operates in the steady-state, where the power output of each prosumer nanogrid is approximately along with the sum of allocated capacity in the P2P energy and balancing energy market shown in Table 2: NG1 (0.39 MW), NG2 (0.26 MW), NG3 (0.16 MW) and NG4 (0.59 MW). At $t = 2s$, in response to the frequency drop due to a sudden load increase of 0.63 MW, NG1 and NG3 instantly adjust their power output to provide frequency regulation service as they are the committed nanogrids determined in the trading stage of the ancillary service market. The power sharing is conducted based on their engagement factors which are integrated in Eq. (6) and (7). It is observed that power output of NG1 and NG3 is increased by 0.35 MW and 0.28 MW respectively, which is consistent with the allocated capacity for the ancillary service market. Figure 7 shows the frequency response of each producer nanogrid, where the frequency can be regulated fast to the reference value. By contrast, Fig. 8 shows the power sharing without the consideration of the market trading, purely based on the pre-defined droop gain, where each producer nanogrid contributes to the regulation proportionally.

2.4.3 Case III

Figure 9 illustrates the predicted net load profile (total load consumption minus total generation) throughout eight timeslots. In Case III, based on the predicted net load profile, the allocated capacity in P2P bilateral energy market, balancing energy market and ancillary service market is investigated respectively. The frequency regulation service requirement for each timeslot is 110% of the net load fluctuation of

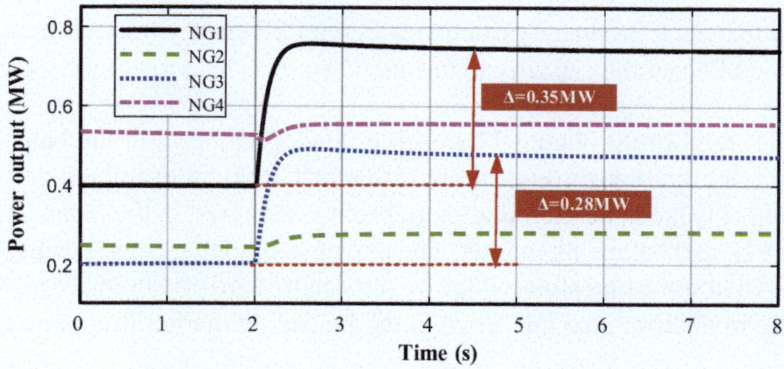

Fig. 6 Frequency regulation service provision by NG1 and NG3 considering the market trading

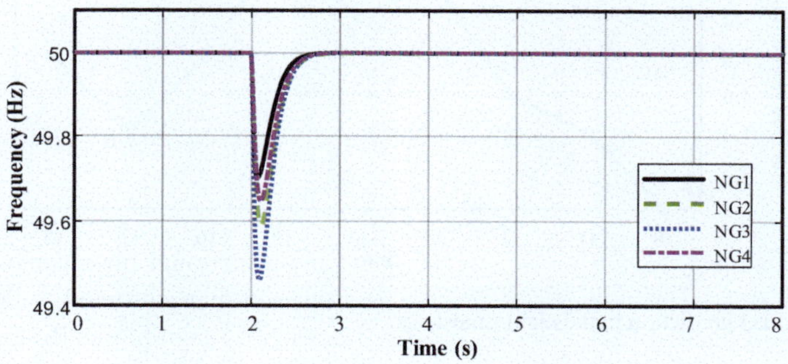

Fig. 7 Frequency response by implementing market-based distributed frequency regulation

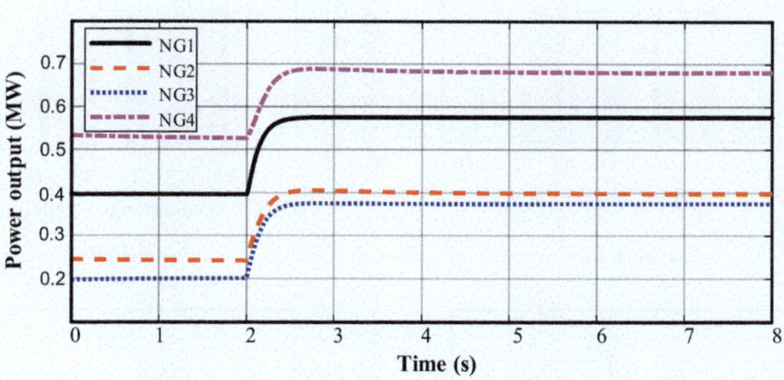

Fig. 8 Power sharing without the consideration of market trading

the current timeslot and next timeslot. For example, the required capacity in the first timeslot is $(t_1 - t_0) \times 1.1 = 1.4 - 1 = 0.44$MW. As only up-regulation is considered, the frequency regulation requirement is zero if the net load is predicted to be reduced, i.e., in the 4th, 6th and 7th timeslot the regulation requirement is zero. Figure 10 shows the results for each producer nanogrid in eight timeslots. It is observed that each producer participates in the P2P energy trading market, while none of them allocate the capacity to the ancillary service market in the 4th, 6th and 7th timeslot.

Moreover, total profit obtained by each producer nanogrid in the balancing and ancillary service market through three different CRMs is examined respectively, shown in Fig. 11. It can be seen that more profit is acquired in the balancing market than in the P2P and balancing market. The reason behind that is the ancillary service is not required in some timeslots, which directly cause zero procurement for the ancillary service. In addition, the unit price in the balancing market is apparently higher

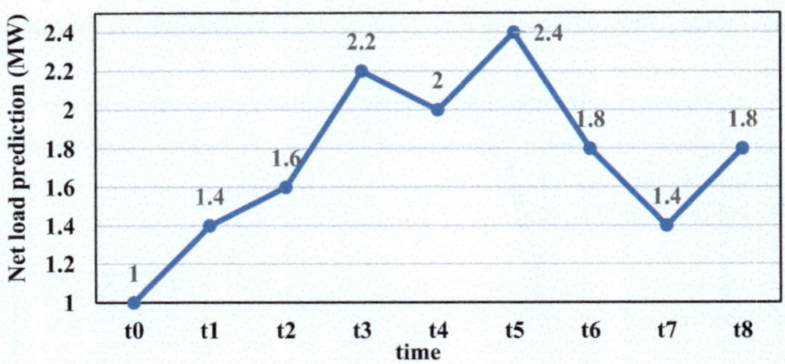

Fig. 9 Net load prediction in the eight timeslots

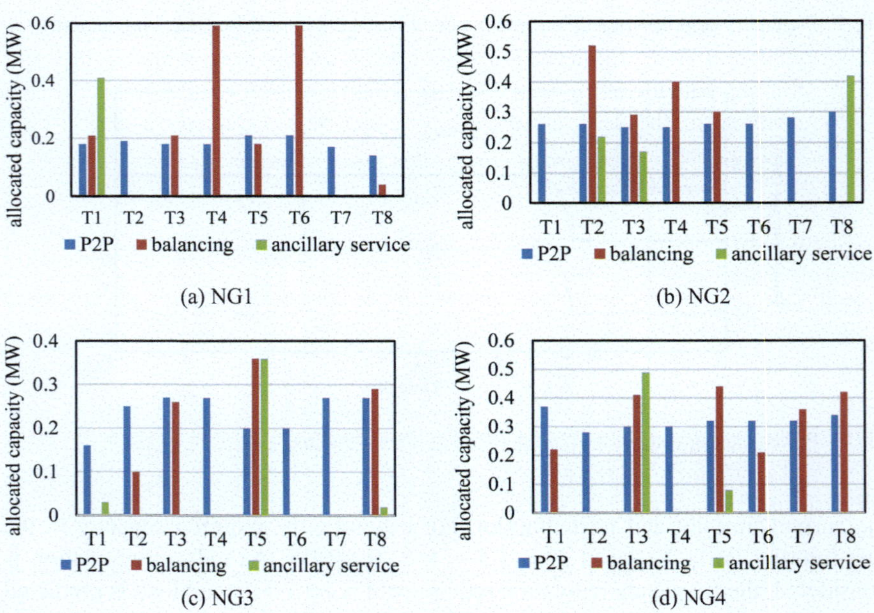

Fig. 10 Allocated capacity of each nanogrid in three market throughout eight timeslots

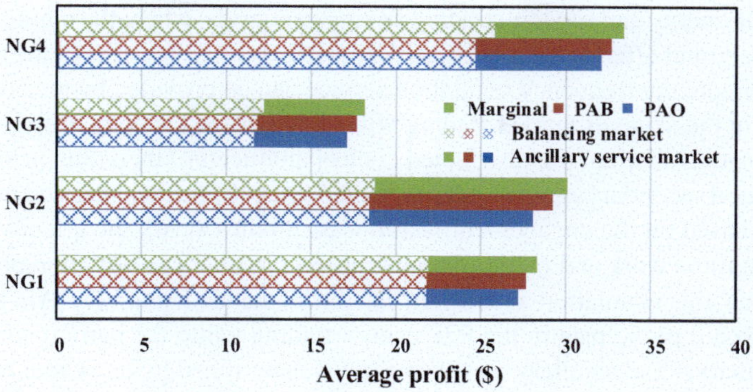

Fig. 11 Average profit within eight timeslots under different CRMs

than that in the P2P market, as it is more approaching to the real time. It is noticeable that different nanogrids received different profits with the same approach of remuneration, which is due to their different bidding strategies. Those who submitted a lower price will be more likely to be selected as successful bidding, and therefore their obtained benefits would be higher.

It should also be noted that, different remuneration mechanisms will have different impact on the profit of producer nanogrids. If they are remunerated by the marginal price of the market, than they will receive more benefits, which would potentially contribute to their willingness of devoting their capacity to the market to facilitate real-time network operation. However, this also means the BRP will pay more cost for the purchase of service procurement. In comparison, the PAB approach will lead to less financial cost in the perspective of BRP, and the producer nanogrids will be fairly remunerated according to their willingness of bidding, so this is the most widely-adopted method in existing markets. The difference between PAB and PAO is the payment method for ancillary service deployment. By adopting this method, the BRP could directly pay the producer nanogrids with the actual opportunity cost instead of overly-high bidding price, which could effectively reduce the investment. However, the willingness of nanogrids' devotion would be affected as they can only receive the payback same as devoting the capacity to the energy market, which could result in potential lack of capacity for frequency regulation in the real-time operation.

In this section, a multi-market trading framework is used to achieve the goal of maximal economic benefit of the producer nanogrid, while the frequency regulation service can be provided economically based on the market trading results. Specifically, under this framework and in light of high-performing P2P communication infrastructures, nanogrids can be coordinated not only to trade energy with one another, but to maintain the power balance and frequency within a microgrid by providing reserved capacity for the ancillary service. The entire framework is organized from the perspective of the potential producers in the market. P2P bilateral energy trading market, balancing market and ancillary service market is considered to facilitate the energy and capacity trading among nanogrids, where the market mechanisms and optimal offering strategies are designed carefully. In terms of the real-time implementation, the frequency regulation service can be provided by the committed nanogrids in the stage of market trading with the reserved capacity. Relying on the P2P information sharing, the distributed control architecture including primary and pinning-based secondary control is implemented to enable the incentivized frequency regulation based on the engagement factor. Case studies verify the effectiveness of this trading framework and market-based real-time frequency regulation protocols.

Based on the simulation results, it can be concluded that: (1) All producer nanogrid would participate in the P2P energy trading while the participation in the balancing market and ancillary service market is determined by the actual balancing requirement and predicted frequency regulation requirement; (2) Comparing to the frequency regulation without the consideration the market incentives, the market-based frequency regulation enables a dedicated power sharing for the regulation service provision with the same regulation effect. (3) The remuneration method of MP will positively contribute to the willingness of nanogrids' participation in the markets, while the PAO method could effectively reduce the financial cost of BRP.

3 P2P Market Equilibrium Estimation Using Multi-Agent Reinforcement Learning

In this section, the market equilibrium estimation methodology using multi-agent reinforcement learning is introduced, with a focus on the joint P2P electricity market (EM) and CEAM among prosumer microgrids in the LEM of distribution network, which a promising paradigm to facilitate the participation of DERs and incentivize the decarbonization. This multi-agent reinforcement learning-based market equilibrium estimation method can also be extended to other market paradigms, providing reference for market operators to facilitate the market design optimization.

In this market, microgrids will modify their bidding strategies to be adaptive to other rival microgrids' for profit maximization. Such modification will converge to the Nash Equilibrium Point (NEP), where each microgrid cannot obtain more profits by modifying its strategy subject to the fixed strategy of other rival microgrids. In this section, the NEP under such a joint market paradigm is investigated, in which microgrids will trade electricity in the EM and purchase CEQs in the CEAM. In addition, microgrids must adjust their bidding strategies considering penalties due to deviations between day-ahead scheduling and real-time procurement caused by uncertainties of net load, as well as the price fluctuation in the CEAM. The NEP is estimated by a MADDPG algorithm, and the risk mitigation is achieved by incorporating CVaR constraint. The computational performance and effectiveness of risk mitigation of the suggested algorithm, and the obtained NEP in the joint EM and CEAM, are analyzed in the case studies.

3.1 Problem Formulation

3.1.1 Market Design of Joint P2P EM and CEAM

In this market paradigm, microgrids are market participants trading electricity in a bilateral manner, while they are required to purchase CEQ in the CEAM on a daily basis. The distribution system operator (DSO) is served as the market operator, being responsible for the match-making of transactions, market operation and management, as well as the network balancing. The generalized architecture of this market is depicted in Fig. 12, and the time frame of market operation is presented in Fig. 13, with detailed descriptions as follows.

(1) ***Microgrids as Prosumers***: Based on the sufficiency of electricity, microgrids are herein classified into two categories: producer microgrids (to sell electricity in the P2P market) and consumer microgrids (to purchase electricity in the P2P market). They are jointly defined as "Prosumer MGs" because their sufficiency of electricity is time-varying, i.e., a producer microgrid at time t could become a consumer microgrid at time t + 1. The objective of MGs is to maximize the total profit by participating in the electricity and CEQ trading with strategic bidding decisions, including the available/insufficient capacity and the bidding

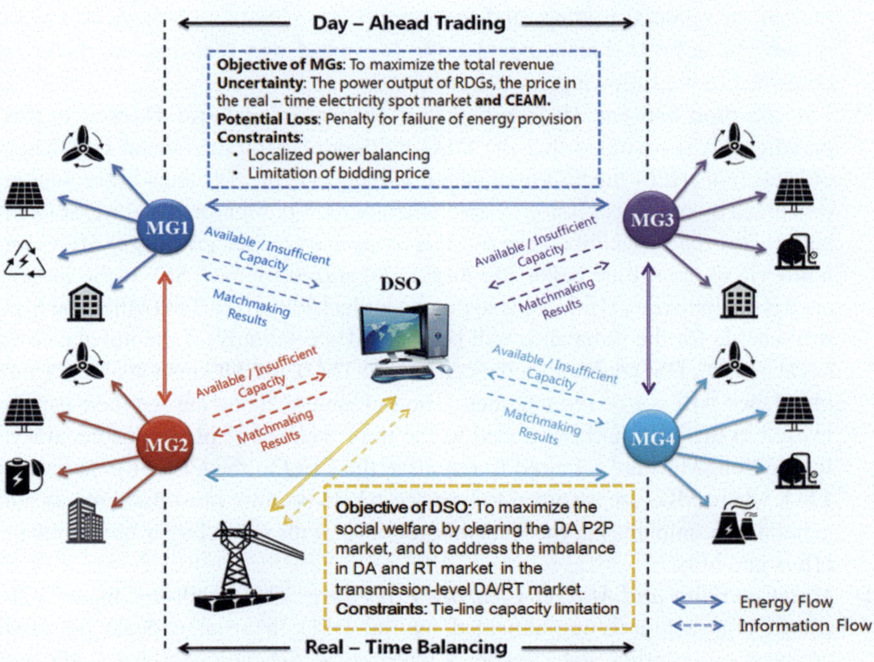

Fig. 12 Framework of joint EM and CEAM paradigm

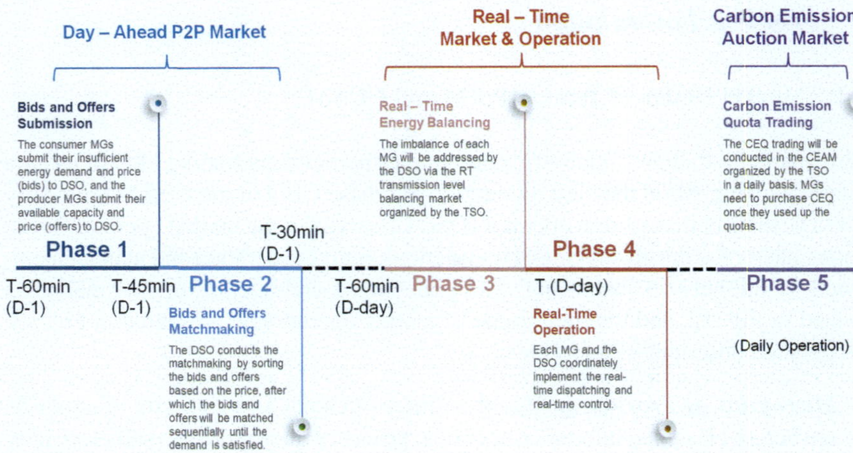

Fig. 13 Time-frame of the joint market

price, and they need to consider both the localized power balancing and bidding price limitations as hard constraints when making decisions.

(2) ***DSO as Market Operator***: The responsibilities of DSO are twofold: (1) to operate the P2P market, including the collection of bids and offers, the matchmaking of bilateral trading, and the monitoring of bidding behaviors; (2) to secure the network balancing in both day-ahead and real-time markets, as described in the following.

(3) ***Coordination between Distribution and Transmission Level Market***: In this paradigm, it is assumed that the DSO will address the day-ahead imbalance of MGs *(caused by the failure of match-making)* via the day-ahead spot market organized by the transmission system operator (TSO). Meanwhile, the DSO will address the real-time imbalance of MGs *(caused by the uncertainty of MGs' net load)* via the real-time balancing market organized by the TSO. If the imbalance is positive (i.e., DSO needs to purchase electricity from TSO), then the MG responsible for the imbalance will be charged accordingly; if the imbalance is negative (i.e., DSO needs to sell electricity to TSO), the MG responsible for the imbalance will not be remunerated. The rationale is to encourage their participation in the P2P market, as well as the improvement of prediction accuracy. In addition, MGs are required to purchase the CEQ in the CEAM held by the TSO, where MGs are assumed to be price-takers because most participants are generation companies at the transmission level with much larger quantities of offers and bids.

(4) ***Match-making and Market Clearing Procedure***: The match-making of P2P energy transaction is implemented by the DSO to arrange deals between producer microgrids and consumer microgrids. Suppose that the DSO has received a set of bids $\left\{P^{t,d}_{MG,bid,p}, \lambda^{t,d}_{MG,bid,p}\right\}, p = 1,2, \ldots, m$ from p th consumer

microgrid and a set of offers $\left\{P^{t,d}_{MG,off,q}, \lambda^{t,d}_{MG,off,q}\right\}$, $q = 1,2,\ldots,n$ from the q th producer microgrid. Then, the match-making will be conducted following the steps as shown below:

Step 1:To rank the bids in descending order of the price: $\lambda^{t,d}_{MG,bid,1} \geq \lambda^{t,d}_{MG,bid,2} \geq \cdots \geq \lambda^{t,d}_{MG,bid,m}$.

Step 2:To rank the offers in ascending order of the price: $\lambda^{t,d}_{MG,off,1} \leq \lambda^{t,d}_{MG,off,2} \leq \cdots \leq \lambda^{t,d}_{MG,off,n}$.

Step 3:To form the pairs:$(\lambda^{t,d}_{MG,bid,1}, \lambda^{t,d}_{MG,off,1})$,$(\lambda^{t,d}_{MG,bid,2}, \lambda^{t,d}_{MG,off,2})$, ...,$(\lambda^{t,d}_{MG,bid,s}, \lambda^{t,d}_{MG,off,s})$ where $\sum_{p=1}^{s} P^{t,d}_{MG,bid,m} = \mathrm{Min}\left\{\sum_{q=1}^{n} P^{t,d}_{MG,off,q}, \sum_{p=1}^{m} P^{t,d}_{MG,bid,m}\right\}$.

In other words, the match-making is terminated when all the bids for load demand are satisfied or all the offers for energy provision are used up.

Step 4:Determine the transaction price of each pair:

$$\lambda^{t,d}_{MG,j} = \frac{\lambda^{t,d}_{MG,bid,j} + \lambda^{t,d}_{MG,off,j}}{2}, j = 1, 2, \ldots, s. \tag{12}$$

Step 5:If $\sum_{q=1}^{n} P^{t,d}_{MG,off,q} \leq \sum_{p=1}^{m} P^{t,d}_{MG,bid,m}$, i.e., the total offers for energy provision is less than the total load demand, the remaining load demand will be addressed by the DSO via the real-time balancing market, in which the cost for purchasing balancing resources will be charged to the corresponding consumer microgrids.

The "market clearing scheme" which constitutes both the "match-making rule" and the "pricing rule" can guarantee the maximization of total social welfare with consideration of network balancing and mitigation of load curtailment, as shown in Fig. 14. After ranking offers of consumer MGs in descending order and bids of producer MGs in ascending order, the intersection of the stacked bids curve and offers curve denotes the "ideal transacted amount of electricity" and the "ideal marginal price". The wording "ideal" means that the social welfare is maximized if only the infra-marginal deals are implemented, because all producer MGs get more payment than they bid, and all consumer MGs pay less than they offer. The Eq. (1) denotes a pricing method for each pair as a compromise after bargaining, and the resulting social welfare is equal to the "pay as bid" scheme. However, the remaining demands of consumers must also be considered, which will lead to satisfying these demands via the P2P market will be much more convenient and cheaper than purchasing electricity in the transmission-level market [43], though these extra-marginal deals will inevitably lead to negative social welfare. Therefore, the total social welfare of this scheme is still maximized considering the mitigation of load curtailment.

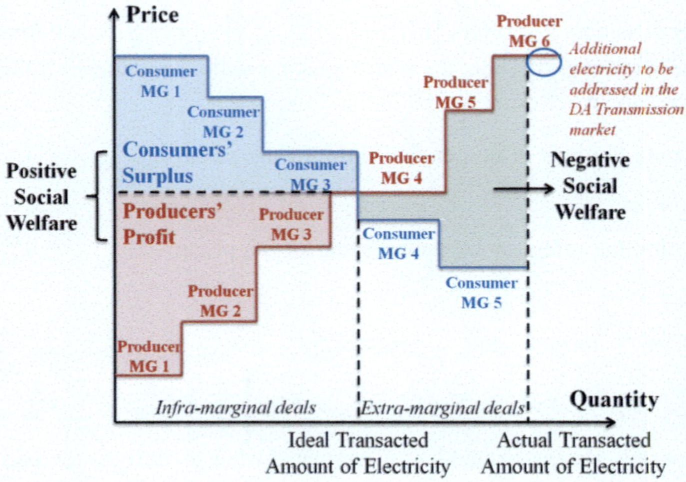

Fig. 14 Illustration of market clearing rules

3.1.2 Strategic Bidding Decisions of MGs with Uncertainties

Based on the market design, the MGs need to consider the following uncertainties while making their bidding decisions:

(1) ***Uncertainty of MGs' inherent net load***: Due to the uncertainty of MGs' inherent RDGs and load demand, the net load prediction error will lead to deviations between day-ahead scheduling in the P2P market and real-time procurement. Such deviation will be addressed by the DSO in the real-time balancing market, and the corresponding balancing cost will be charged to the corresponding MG, and lead to the risk of potential financial losses.
(2) ***Uncertainty of price in the Transmission-level market***: The aggressive bidding decision (i.e., high prices for offers and low prices for bids) may lead to higher profits for MGs; however, it may also lead to failure of match-making in the day-ahead P2P market, and the remaining imbalance must be eliminated by the DSO in the day-ahead transmission-level market. Considering the price fluctuation in the transmission-level market, the MGs may encounter less profit/higher cost compared with a conservative bidding decision in the day-ahead P2P market.
(3) ***Uncertainty of price in the CEAM***: MGs also need to purchase CEQ in the CEAM as a price-taker based on their actual emissions and allocated quotas. However, the price of CEQ is also varying with time and difficult to be predicted, and therefore adds more uncertainties to MGs' profits.

3.1.3 CVaR-Constrained Optimal Bidding Strategy of MG Prosumers

Based on the aforementioned market paradigm and uncertainties, the optimal bidding strategy of MGs can be formulated as a stochastic dynamic programming model, with the objective of maximizing the total profit in the joint P2P EM and CEAM, as shown in (13). Note that notation $R(A|B)$ denotes that the profit R is determined by the decision variable A, which is affected by the factor B.

$$\text{Max} \sum_{t \in T} \left\{ \begin{array}{l} \delta_{MG}^{t,d} \left[\begin{array}{c} \underbrace{R_{MG,k}^{t,d}\left(P_{MG,k}^{t,d}, \lambda_{MG,k}^{t,d} \middle| P_{MG,k}^{t,d-1}, P_{k,act}^{t,d-1}, \lambda_{MG,k}^{t-1,d}, P_{Load,total}^{t,d}\right)}_{\text{① Revenue in P2P Market}} \\ -\underbrace{\sigma_k^{t,d} C_{MG,k}^{t,d}\left(P_{k,allo}^{t,d}\right)}_{\text{② Generation Cost}} - \underbrace{\sigma_k^{t,d} C_{CEAM,k}^{t,d}\left(P_{k,allo}^{t,d}\right)}_{\text{③ Emission Cost}} - \underbrace{C_{pen,k}^{t,d}\left(P_{MG,k,err}^{t,d}, \lambda_{spot}^{t,d}\right)}_{\text{④ Penalty}} \end{array} \right] \\ -\left(1-\delta_{MG}^{t,d}\right) \left[\underbrace{C_{MG,load}^{t,d}\left(P_{Load,k}^{t,d}, \lambda_{MG,k}^{t,d}\right)}_{\text{⑤ Cost for purchase}} + \underbrace{C_{pen,k}^{t,d}\left(P_{MG,k,err}^{t,d}, \lambda_{spot}^{t,d}\right)}_{\text{⑥ Penalty}} \right] \end{array} \right\} \tag{13}$$

$$\underline{P_{local,k,min}} \leq P_{MG,k,total}^{t,d} - P_{MG,k}^{t,d} \leq \overline{P_{local,k,max}} \tag{13a}$$

$$\underline{\lambda_{thre,min}} \leq \lambda_{MG,k}^{t-1,d} / \lambda_{MG,k}^{t,d} \leq \overline{\lambda_{thre,max}} \tag{13b}$$

$$C_{CVaR,MG,k}^{t,d}\left(P_{RDG,k,err}^{t,d}, P_{Load,k,err}^{t,d}\right) \leq C_{risk,k,max} \tag{13c}$$

where $\delta_{MG}^{t,d}$ is the binary variable indicating MG purchase/sell energy from/to DSO in d-day, $R_{MG,k}^{t,d}$ and $C_{MG,k}^{t,d}$ denotes the revenue and cost of the k th MG, $P_{k,allo}^{t,d}$ denotes the allocated capacity for each MG, and $\sigma_k^{t,d} = P_{k,allo}^{t,d-1} / P_{k,act}^{t,d-1}$ indicates the estimated rate of actual procurement. For each round of bidding, each MG submits its available capacity $P_{MG,k}^{t,d}$ and bidding price $\lambda_{MG,k}^{t,d}$ for consecutive 24 h, based on the broadcasted total demand $P_{Load,total}^{t,d}$ and estimated $\sigma_k^{t,d}$, following the localized power balancing limitation (13-a), bidding price limitation (13b), and CVaR constraint (13-c). Specifically, given the threshold value of loss $\xi = C_{risk,max}$ and random variable X, the CVaR of X for exceeding ξ at the lowest β-quantile can be defined as:

$$C_{CVaR,k}^t = C_{VaR,k}^t + \frac{1}{(1-\beta)} \mathbb{E}[X - C_{VaR,k}^t]^+ \tag{14}$$

where $[X - C_{VaR,k}^t]^+ = \max\{0, (X - C_{VaR,k}^t)\}$, and the VaR is defined as $C_{VaR,k}^t = \inf\{\xi : \mathbb{P}\{X \leq \xi\} \geq \beta\}$.

The sum of term ③, ④ and ⑥ in (13) denote the random variable X, indicating the risk faced by MGs. Term ③ denotes the estimated cost of purchasing CEQ in the CEAM, and this cost is affected by both the actual generation and the real-time price of CEQ. Term ④ and ⑥ denote potential penalties $C_{pen,k}^{t,d}$ for balancing cost of the imbalance capacity $P_{MG,k,err}^{t,d}$ due to:

(1) The positive deviation between day-ahead allocated capacity $P_{k,allo}^{t,d}$ and real-time actual procurement, because of the prediction error of the net load.
(2) The lack of energy supply due to the failure of match-making, because of the excessively aggressive bidding price. Such penalty is imposed by the DSO to purchase energy from the main grid for real-time balancing with spot market price $\lambda_{spot}^{t,d}$.

3.1.4 Markov Decision Process with Uncertain Reward (MDPUR) Formulation

As shown in (13), the decision making of MGs apparently inherits the Markov Property, i.e., the decision in time slot t is affected by states in $d-1$. Hence, (13) could be intuitively re-formulated as a Markov decision process with elements elaborated as follows:

- ***State*** ($s_{t,d}^{k,h} \in \mathcal{S}_{t,d}^{k}$): The state refers to the set of MGs' observations, including the current load demand $P_{Load,total}^{t,d}$ broadcasted by the DSO, and the MGs' forecasted net load $P_{Load,k}^{t,d}$.
- ***Action*** ($a_{t,d}^{k,h} \in \mathcal{A}_{t,d}^{k}$): The action refers to the set of MGs' bidding decisions at the current time slot t, including the purchase/sell status $\delta_{MG}^{t,d}$, capacity $P_{MG,k}^{t,d}$, and price $\lambda_{MG,k}^{t,d}$.
- ***Reward*** ($r_{t,d}^{k,h} \in \mathcal{R}_{t,d}^{k}$): The reward refers to the received profit of MGs at the current time slot t after the actual procurement of electricity and transaction of CEQ.

However, as indicated in (13), the reward after the actual procurement of energy and transaction of CEQ is non-observable to MGs and implicit because of the uncertain match-making results, penalties, and the real-time price of the CEQ. Hence, the aforementioned Markov decision process should be considered as the MDPUR, with the uncertain reward $\overset{\sim}{\mathcal{R}}_t^k$ specified as follows:

$$\overset{\sim}{\mathcal{R}}_t^k = \delta_{MG}^{t,d}[R_{MG,k}^{t,d} - \sigma_k^{t,d}(C_{CEAM,k}^{t,d} + C_{MG,k}^{t,d}) - C_{pen,k}^{t,d}] - \left(1 - \delta_{MG}^{t,d}\right)(C_{MG,load}^{t,d} + C_{pen,k}^{t,d}) \quad (14)$$

The objective of this MDPUR model is to maximize the cumulative uncertain reward $\sum_{t \in T} \overset{\sim}{\mathcal{R}}_t^k$ by adjusting the policy function $\mu^k(a_{t,d}^k | s_{t,d}^k)$ while satisfying constraints (13a) to (13c). However, this MDPUR model cannot be solved by existing multi-agent reinforcement learning algorithms because of the uncertainty of reward function and the implicit CVaR constraint. The CCMA-DDPG Algorithm will be deployed to tackle this problem, as elaborated in the next subsection.

3.2 *CVaR-Constrained Multi-agent Deep Deterministic Policy Gradient (CCMA-DDPG) Algorithm Implementation*

To solve the MDPUR model, each MG should be enabled to continuously update its bidding strategy based on the observations without information on other MG's bidding strategies, to maximize the cumulative reward while ensuring the CVaR not exceeding the imposed constraint. Hence, in the algorithm, the CVaR constraint is

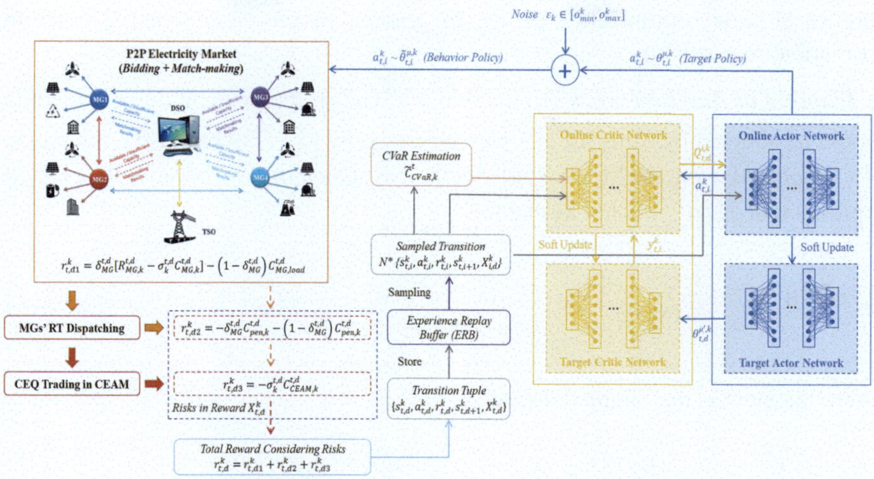

Fig. 15 Framework of the CCMA—DDPG algorithm

incorporated into a fully-distributed MADDPG framework as depicted in Fig. 15, in which the general methodology of MADDPG and its architecture of Actor-Critic (elaborated in Sect. 3.3.1) are adopted with important modifications (in Sect. 3.3.2) to enable the distributed training and CVaR-based risk mitigation. Finally, some important tricks deployed in this algorithm to improve the computational performance are briefly introduced in Sect. 3.3.3.

3.2.1 Framework of MADDPG Algorithm

MADDPG is a well-known off-policy RL algorithm to obtain the optimal policy function $\mu_{t,d}^k(a_{t,d}^k|s_{t,d}^k)$ that maximizes the reward $\mathcal{R}_{t,d}^k$ obtained by the agent. The rationale for deploying MADDPG as the baseline in this section includes: (1) the optimization of policy function in a continuous state and action space, instead of the discrete space which would result in the curse of dimensionality; (2) the fully distributed training framework that enables MG agents to implement the autonomous bidding strategy optimization without any information of other rival MGs' confidential bidding strategies. The objective of this algorithm is to optimize the "performance objective" of each MG agent:

$$J_\beta^k(\mu_{t,d}^k) = \mathbb{E}_{\mathcal{R}_{t,d}^k \sim \mu_{t,d}^k, a_{t,d}^k \sim \mu_{t,d}^k}[Q_{t,d}^k(s_{t,d}^k, a_{t,d}^1, a_{t,d}^2, \ldots, a_{t,d}^p)] \tag{15}$$

where $k = 1,2, \ldots, p$, $Q_{t,d}^k$ is the Q-function without explicit formulation. Hence, the MADDPG algorithm is developed based on the actor-critic architecture, in which "actor" and "critic" refer to two convolutional neural networks (CNNs) with the purpose of policy optimization of $\mu_{t,d}^{k}$ by gradient ascent method and Q-function estimation, respectively.

1) Training of Actor Network: To maximize $J_{\beta}^{k}(\mu_{t,d}^{k})$, the gradient descent method should be adopted by using a CNN parameterized by $\theta_{t,d}^{\mu,k}$. For each MG, the policy $\mu_{t,d}^{k}$ is updated using the Monte-Carlo method as shown in (17), because the explicit expression of $J_{\beta}^{k}(\mu_{t,d}^{k})$ is not available:

$$\nabla_{\theta_{t,d}^{\mu,k}} J_{\beta}(\mu_{t,d}^{k}) \approx \frac{1}{N}\sum\nolimits_{n\in N}\left\{\nabla_{\theta_{t,d}^{Q,k}} Q_{t,d}^{k}(s_{t,d}^{k}, a_{t,d}^{1}, a_{t,d}^{2}, \ldots, a_{t,d}^{p}) \bullet \nabla_{\theta_{t,d}^{\mu,k}} \mu_{t,d}^{k}(a_{t,d}^{k}|s_{t,d}^{k})\right\} \tag{17}$$

where the input is the sampled dataset

$$\tau = \{s_{t,d}^{k}, a_{t,d}^{1}, a_{t,d}^{2}, \ldots, a_{t,d}^{p}, r_{t,d}^{k}, s_{t,d+1}^{k}\} \tag{18}$$

The training process of this CNN is actually the procedure of optimizing the $J_{\beta}^{k}(\mu_{t,d}^{k})$, indicated by $\nabla_{\theta_{t,d}^{\mu,k}} \mu_{t,d}^{k}(a_{t,d}^{k}|s_{t,d}^{k})$. The parameter of actor network $\theta_{t,d}^{\mu,k}$ will be updated by the Adam Optimizer based on the results of (15) and learning rate η_k:

$$\theta_{t,d+1}^{\mu,k} = \theta_{t,d}^{\mu,k} + \eta_k \nabla_{\theta_{t,d}^{\mu,k}} J_{\beta}(\mu_{t,d}^{k}) \tag{19}$$

However, the value of $Q_{t,d}^{k}$ is conventionally obtained by the iterative Bellman Equation [44] which is also implicit. This problem is tackled by adopting another CNN (known as the critic network) parameterized by $\theta_{t,d}^{Q,k}$ to estimate $Q_{t,d}^{k}$, as indicated in the term $\nabla_{\theta_{t,d}^{Q,k}} Q_{t,d}^{k}$.

2) Training of Critic Network: The training of critic network is similar to the supervised learning, using the back-propagation approach by minimizing the mean square error (MSE) to estimate the Q-function:

$$L_k(\theta_{t,d}^{Q,k}) = \frac{1}{N}\sum\nolimits_{n\in N}(y_{t,d}^{k} - Q_{t,d}^{k}(s_{t,d}^{k}, a_{t,d}^{1}, a_{t,d}^{2}, \ldots, a_{t,d}^{p}))^2 \tag{20}$$

where $Q_{t,d}^{k}$ is the estimated Q-function estimated by the critic network parameterized by $\theta_{t,d}^{Q,k}$, and $y_{t,d}^{k}$ is the "label function":

$$y_{t,d}^{k} = \overset{\sim k}{\mathcal{R}}_{t,d} + \gamma \max Q_{t,d+1}^{k}(s_{t,d+1}^{k}|\theta_{t,d}^{Q,k}) \tag{21}$$

where $\overset{\sim k}{\mathcal{R}}_{t,d}$ is the immediate reward, and $Q_{t,d+1}^{k}$ denotes the Q-value of implementing the current policy $\mu_{t,d}^{k}$ in the same hour of $d+1$ day, indicating the long-term reward with discount factorγ. This Q-value is estimated based on the Q-function parameterized by $\theta_{t,d}^{Q,k}$. In summary, the critic network is trained by updating $\theta_{t,d}^{Q,k}$

to minimize the MSE between the estimated Q-value $Q^k_{t,d}$ and label function $y^k_{t,d}$ as the "real" Q-value. However, as indicated in (20) and (21), the minimization of MSE is based on the label function$y^k_{t,d}$, which is iteratively dependent on$\theta^{Q,k}_{t,d}$, i.e., the parameter intended to be optimized. This problem will lead to the instability of training, which can be tackled by adopting another "Target Network" copied from the original "Main Critic Network" per fixed steps of update:

$$\theta^{Q\prime,k}_{t,d} \leftarrow \rho\theta^{Q\prime,k}_{t,d} + (1 - \rho_k)\theta^{Q,k}_{t,d} \tag{22}$$

where $\theta^{Q\prime,k}_{t,d}$ denotes the parameter of target network, and $\rho_k \in (0,1)$ is a hyper-parameter indicating the rate of update.

3.2.2 Modified CCMA-DDPG Algorithm

To enable the aforementioned baseline MADDPG to be fully-distributed and CVaR-constrained, two modifications are implemented herein and elaborated as follows.

Firstly, as indicated in (20), the update of each MG agent's critic network requires the availability of other rival MGs' bidding strategies, which is inconsistent with the privacy protection in real-world market operations. To remove this assumption, the k th MG is enabled to estimate other j th ($j = 1,2,\ldots, K - 1$) rival MGs' bidding strategies [45] $\{\widehat{\mu}^1_{t,d}, \widehat{\mu}^2_{t,d}, \ldots, \widehat{\mu}^j_{t,d}\}$ parameterized by $\widehat{\theta}^{\mu,j}_{t,d}$, which is updated by each MG via maximizing the likelihood function:

$$L^j_k\left(\widehat{\theta}^{\mu,j}_{t,d}\right) = -\mathbb{E}\left[\log\widehat{\mu}^j_{t,d}(a^j_{t,d}|\widehat{\theta}^{\mu,j}_{t,d}) + \varpi_k H(\widehat{\mu}^j_{t,d})\right] \tag{23}$$

where $H\left(\widehat{\mu}^j_{t,d}\right) = -\widehat{\mu}^j_{t,d}(a^j_{t,d}|\widehat{\theta}^{\mu,j}_{t,d})\sum_{a^j_{t,d}}\log\widehat{\mu}^j_{t,d}(a^j_{t,d}|\widehat{\theta}^{\mu,j}_{t,d})$ is the entropy regularizer [45] with coefficient ϖ_k to facilitate the exploration of environment, and $\log\widehat{\mu}^j_{t,d}(a^j_{t,d}|\widehat{\theta}^{\mu,j}_{t,d})$ denotes the likelihood function (i.e., estimated policy function of the j th MG) based on the sampled dataset of actions $\{a^1_{t,d}, a^2_{t,d}, \ldots, a^j_{t,d}\}$.

Secondly, the label function $y^k_{t,d}$ is modified with the incorporation of the CVaR constraint as a penalty, which is inspired by the Lagrangian Relaxation procedure, where $\lambda^k_{t,d}$ is equivalent as the "Lagrangian multiplier", to be updated in the following approach [46]:

$$y^k_{t,d} = \overset{\sim k}{\mathcal{R}}_{t,d} + \gamma\max_{s^k_{t,d+1}} Q^k_{t,d+1}\left(s^k_{t,d+1}|\theta^{Q,k}_{t,d}\right) - \lambda^k_{t,d}C^{t,d}_{CVaR,k} \tag{24}$$

where

$$\lambda^k_{t,d} \leftarrow \max\left\{0,\ \lambda^k_{t,d-1} + \sigma^k\left(C^{t,d}_{CVaR,k} - C_{risk,max}\right)\right\} \tag{25}$$

By imposing such a penalty term, the policy that results in CVaR values exceeding the limitation will be disregarded by deducting the corresponding Q-values. The final problem to be tackled is the estimation of $C^{t}_{CVaR,k}$, which is also an unknown function without explicit formula. This value is approximated by adopting the following theorem:

Theorem 1 Given that $\{\omega_i\}_{i=1}^{N}$ representing N samples based on $\omega_i \sim X$, then the CVaR can be estimated as follows:

$$\tilde{C}^{t}_{CVaR,k} = \tilde{C}^{t}_{VaR,k} + \frac{1}{(1-\beta)}\sum\nolimits_{i=1}^{N}(\omega_i - \tilde{C}^{t}_{VaR,k})^{+} \tag{26}$$

where $\tilde{C}^{t}_{VaR,k} = \inf\{\xi : \frac{1}{N}\sum_{i=1}^{N}\mathbb{I}\{\omega_i \le \xi\} \ge \beta\}$, and $\mathbb{I}\{\omega_i \le \xi\}$ is an indicator function [47]. In other words, if samples ω_i are ranked in the ascending order $\omega_{[1]} \le \omega_{[2]} \le \cdots \le \omega_{[n]}$, then the VaR can be empirically estimated as $\omega_{[n\beta]}$.The sample of loss $X_k^{t,d}$ is obtained by

$$X_k^{t,d} = \delta^{t,d}_{MG} C^{t,d}_{pen,k} + \left(1 - \delta^{t,d}_{MG}\right) C^{t,d}_{pen,k} + \delta^{t,d}_{MG} \cdot C^{t,d}_{CEAM,k} \tag{27}$$

after each round of bidding.

3.2.3 Tricks for Computational Performance Improvement

Based on the CCMA-DDPG architecture elaborated in the previous subsections, two effective tricks are deployed herein as supplementary countermeasures to address bottlenecks for the computational performance improvement of the algorithm.

1) Double Q-Learning: The over-estimation of Q-value is a common problem of RL algorithms benchmarked by the DDPG algorithm [48]. To address this problem, two critic networks separated from each other will be trained with different initialized network parameters:

$$L_k^i(\theta^{Q_i,k}_{t,d}) = \frac{1}{N}\sum\nolimits_{n\in N}(y^k_{t,d} - Q^{i,k}_{t,d}(s^k_{t,d}, a^j_{t,d}))^2, i = 1,2 \tag{28}$$

and the smaller estimated Q-value will be taken to update the label function $y^k_{t,d}$:

$$y^k_{t,d} = \overset{\sim k}{\mathcal{R}}_{t,d} + \gamma \min\nolimits_{i=1,2} \max\nolimits_{s^k_{t,d+1}} Q^{i,k}_{t,d+1} - \lambda^k_{t,d} C^{t,d}_{CVaR,k} \tag{29}$$

Note that these two critic networks share a common $y^k_{t,d}$.

2) Target Policy Smoothing: Apart from aforementioned over-estimations of Q-value, there still exist several important reasons of estimation bias, including:

- The non-stationarity of multiple MGs' bidding process: For instance, from the aspect of the k th MG, the sub-optimal bidding strategy of other j rival MGs will lead to a biased evaluation of the Q-value.
- The biased estimation of CVaR: Although the potential financial loss due to these uncertainties is estimated by the CVaR term in the label function $y_{t,d}^{k}$, the CVaR estimation at the early stage of training will be with significant bias because of the lack of sampled dataset. The label function $y_{t,d}^{k}$ with biased CVaR will further result in bias of Q-value estimation.

To mitigate the effect of biased Q-value on the policy update (i.e., trapping into the local optimality), the trick of target policy smoothing [49] is implemented herein, by adding noise to the action generated by the target policy $\theta_{t,d}^{\mu,k}$:

$$a_{t,d}^{k} \leftarrow a_{t,d}^{k}\left(s_{t,d}^{k}|\theta_{t,d}^{\mu,k}\right) + \varepsilon_k, \varepsilon_k \in [o_{min}^{k}, o_{max}^{k}] \tag{30}$$

where o_{min}^{k} and o_{max}^{k} denote the threshold value of the noise. By adding such a noise, the exploration to the environment will be further enriched, and therefore the problem of local optimum will be effectively addressed. The action can be considered to be generated by the stochastic policy (also called as behavior policy) $\{\overset{\sim\mu,1}{\theta_{t,d}}, \overset{\sim\mu,2}{\theta_{t,d}}, \ldots, \overset{\sim\mu,p}{\theta_{t,d}}\}$, which is a combination of deterministic policy (also called as target policy) $\{\theta_{t,d}^{\mu,1}, \theta_{t,d}^{\mu,2}, \ldots, \theta_{t,d}^{\mu,p}\}$ and the imposed noise. The entire workflow of this algorithm is summarized as follows.

3.2.4 Discussions of Existence and Uniqueness of NEP by Implementation of CCMA—DDPG

The existence of NEP in any *non-cooperative normal-form* game has been strictly proven in [50]. The bidding game used in this section is competitive (i.e., *non-cooperative*), with finite MGs and finite set of actions (i.e., *normal-form*). Note that we assume the action set of each MG agent is finite because the space of bidding price and capacity decisions always have intervals (e.g., 0.1 kWh, 0.1 RMB/kWh, etc., which varies in different markets.).

The uniqueness of NEP, however, to the best of the authors' knowledge, cannot be proven in this bidding game model. Existing methodology for proof of NEP uniqueness include the following: (1) Rosen et al. [51], which requires the utility function to be concave; (2) Luenberger et al. [52], in which the explicit form of utility function must be known; (3) Moulin [53], in which the utility function must be twice continuously differentiable and pseudo-concave; and (4) to use the Lemke–Howson algorithm [54] to compute all the NEPs. However, the utility function of each MG agent is stochastic and therefore implicit, and the game is in the form of n-player, n-strategy. Hence, the algorithm is the most applicable method to estimate the NEP in such a game model.

The principle of the algorithm to reach the NEP is elaborated as follows. Each MG agent improves its policy function (i.e., the bidding strategy) by deterministic policy gradient, as shown in (6)–(8). If the gradient of all the agents are zero, it indicates that all the agents have found their optimal strategies and there is no motivation to change their strategies, which is equivalent to the definition of NEP.

Algorithm 1 Workflow of CCMA-DDPG Algorithm

Input:
1)The network configuration parameters and line capacity limitations;
2)The rated capacity (upper bound of the capacity limitation) of controllable generators in each MG;
3)The bidding price limitation;
4)The net load profile of each MG, with uncertainty set indicating the probabilistic of possible scenarios;
5)The carbon emission factors and quota allocation of each MG;
6)The set of hyper-parameters for training
Output:
The converged policy of bidding decisions of each MG
1: Parameter Initialization: $\theta_{t,d}^{Q_i,k}, \theta_{t,d}^{\mu,k}, \lambda_{t,d}^{k}, .$
For each episode:
For $t = 1 : T$:
For MG agent $1 : k$:
 # Interaction with the environment
2: The actor network of each MG agent generates action $a_{t,d}^{k}$ based on the behavior policy $\theta_{t,d}^{\mu,k}$ and noise ε_k
3: Each MG submits their actions to the DSO, who returns the P2P trading match-making results to MGs
4:Generate the random prediction error $\tilde{E}_{t,d}^{k}$ which denotes the prediction error of MGs' net load
5:Each MG performs the real-time dispatching using the method introduced in [24]
6:Each MG purchases CEQs in the CEAM based on the actual generation and the remained CEQ allowance
6:Each MG agent computes the reward $r_k^{t,d}$ considering the loss $X_k^{t,d}$ with penalties imposed by the DSO
7:The actor network of each MG restores the transition $\{s_{t,d}^{k}, a_{t,d}^{k}, r_{t,d}^{k}, s_{t,d+1}^{k}, X_{t,d}^{k}\}$ to the ERB
 # Policy improvement via *network training*
8:A random transition is selected from ERB as a mini-batch of training dataset
9:Update the CVaR estimation $C_{CVaR,k}^{t,d}$ using (26)
10:Update the Lagrangian multiplier $\lambda_{t,d}^{k}$ using (25)
11:Compute the gradient of Critic using (28) and (29)
12:Update $\theta_{t,d}^{Q_i,k}$ using the Adam Optimizer
13:Estimate the policy gradient of Actor using (17)
14:Update $\theta_{t,d}^{\mu,k}$ using the Adam Optimizer
End for MG agent
End for time step
15:Update $\theta_{t,d}^{\mu\prime,k}$ and $\theta_{t,d}^{Q\prime_i,k}$ using (22)
End for episode

3.3 Case Study

3.3.1 Test System and Input Data

In this case study, a test market consisting of 5 distributed MG agents in a modified IEEE 33-Bus Network is adopted as shown in Fig. 16, with the following input data to be generated as indicated in Algorithm 1. These MG agents are classified as producers, prosumers and consumers, with their representative characteristics elaborated in Table 3. The hyper-parameters for network training are summarized in Table 4. The rated capacity and generation cost of different generators are referred to [55]. The initialized CEQ allowance is directly set as zero to clearly demonstrate the impact of CEAM. The bidding price limitation in the P2P EM is (100, 150) RMB/MWh. The emission factor of each type of generator is extracted from [56]. The prediction error of power output of wind turbines, PVs and the load demand is generated based on the probability density function provided in [57] with consideration of probability density function correlations among DERs and the load demand. The real-time dispatching of each MG follows the general methodology in [55]. The price of CEQ in the CEAM is extracted from [58].

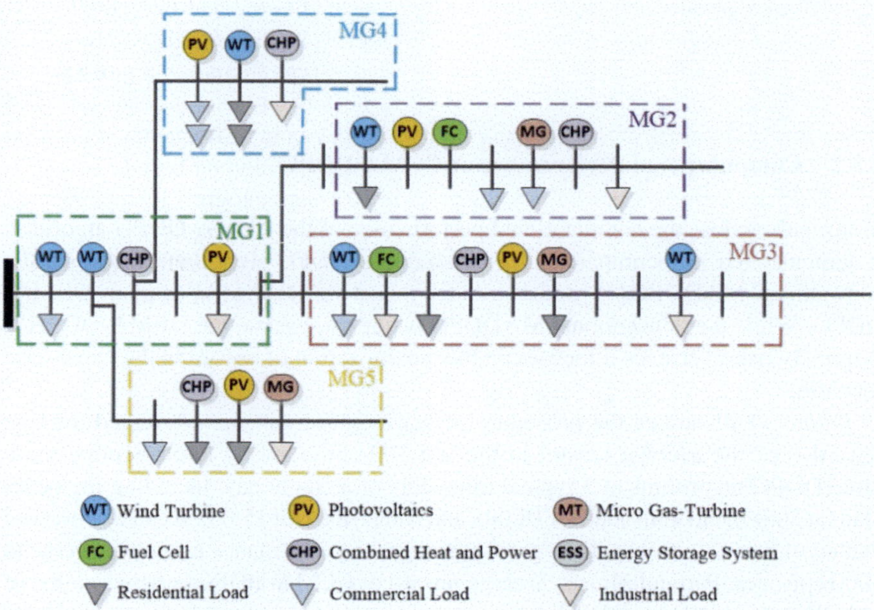

Fig. 16 Modified IEEE 33-bus network with MGs

Table 3 Types and characteristics of MG agents

MG agent	Characteristics	Type
1	Large amount of RDGs	Producer
2	Large amount of controllable DGs	Producer
3	Large amount of RDGs and controllable DGs	Prosumer
4	Large amount of RDGs	Consumer
5	Large amount of controllable DGs	Consumer

Table 4 Neural network architectures and hyper-parameters

Neural network	Actor	Critic
No. of hidden layers	2	2
No. of neurons	[58, 59]	[58, 59]
Activation function	Sigmoid	ReLU
Learning rate	1e-6	1e-6
Delayed update frequency	1	4
Optimizer	Adam	Adam
Training parameters	Replay buffer size	1e + 4
	Replay start size	100
	Batch size	16
	Discount factor	0.99

3.3.2 Computational Performance of CCMA-DDPG

In this subsection, the computational performance of the CCMA-DDPG algorithm is demonstrated with comparisons of two baseline DDPG algorithms, including the conventional MADDPG [45] and distributed robust MADDPG [34], namely DRMA-DDPG. While the conventional MADDPG is not risk-averse, the DRMA-DDPG is extremely risk-averse, as it focuses on the maximization of reward in the worst-case scenario.

Figure 17 illustrates the procedure of convergence and the converged average reward (i.e., the average reward in the last 100 episodes) by implementing these three DDPG algorithms at 3 typical time-slots of a single day, including the peak-load (at 20:00), medium-load (at 18:00), and valley-load (at 15:00). It can be observed that all of these algorithms have good convergence performance as they converge in 8000 episodes. Particularly, the average reward of CCMA-DDPG is between that of MADDPG and DRMA-DDPG.

For a more in-depth comparison, Fig. 18 shows the converged MP in a consecutive 24 h by implementing these algorithms. As both CCMA-DDPG and DRMA-DDPG are risk-averse, the corresponding bidding price is apparently lower than that of MADDPG, because their bidding strategies are much more conservative. Furthermore, it can be observed that the CCMA-DDPG is a compromise between pursuing

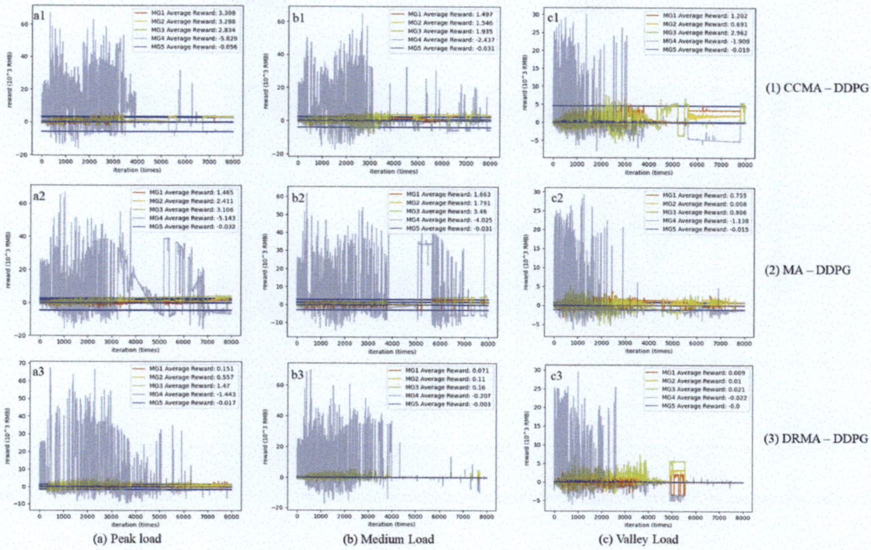

Fig. 17 Convergence performance of different algorithms

higher reward and risk mitigation, as shown in Table 5, in which the average reward of MGs in 24 h is listed. The reason is the CVaR-based risk mitigation does not lead to a solution that is robust to all the possible scenarios, i.e., the risk in the worst case scenario must be mitigated, which is the core idea of DRMA-DDPG. Instead, the CCMA-DDPG algorithm enables MGs adjust the bidding strategy based on their own willingness of undertaking the risk, i.e., they can determine the threshold CVaR indicating the risk they cannot tolerate. Hence, the obtained NEP will enable all participated MGs with more rewards than the DRMA-DDPG, and better risk mitigation performance than the MADDPG.

3.3.3 Bidding Strategy Analysis of MG Agents

In this subsection, the impact of uncertainties (as mentioned in Subsection 3.2) and incorporation of CEAM on MG agents' bidding strategy preference is analyzed. To exhaust the possible scenarios, the CCMA-DDPG-based simulation was repeated 100 times, and the converged bidding price of each MG in consecutive 24 h is illustrated as a box-plot in Fig. 19. Each box indicates the information of the minimum, the lower and upper quartile, and the maximum bidding prices among these 100 times of simulation. Figure 20 shows the average bidding price of producers, prosumers and consumers MGs, from which the following are observed:

(1) Both the producers and consumers prefer the price-taker strategy (i.e., to submit a lower bidding price around 120 RMB/MWh) in the valley-load time (1:00 to 9:00, 22:00 to 24:00), and prefer the price-maker strategy (i.e., to submit a higher

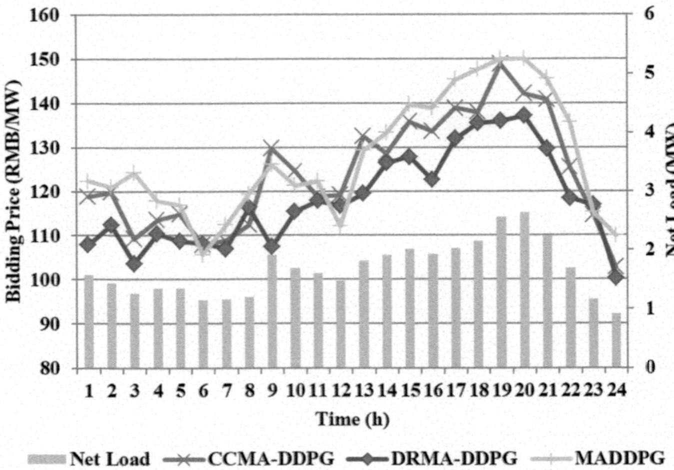

Fig. 18 Converged MP by implementation of different algorithms

Table 5 Average reward of MGs in 24 Hours (10^3 RMB)

	MG1	MG2	MG3	MG4	MG5
CCMA-DDPG	0.684	0.596	0.219	-0.474	-0.395
DRMA-DDPG	0.232	0.227	0.098	-0.526	-0.447
MADDPG	0.395	0.373	0.182	-0.315	-0.471

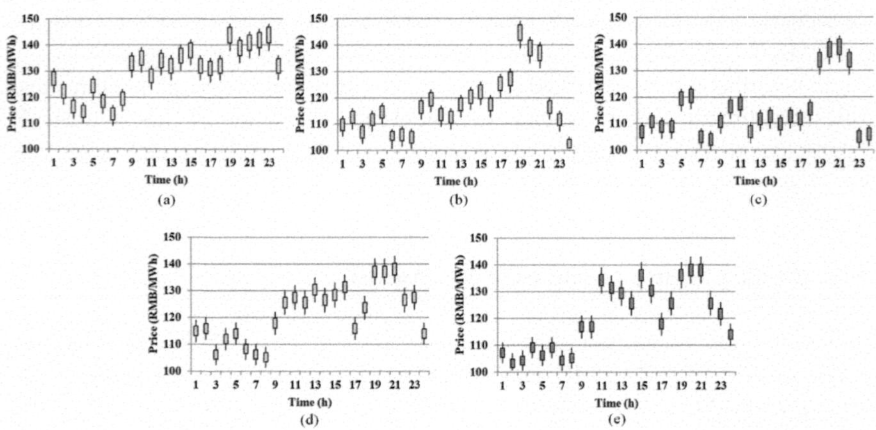

Fig. 19 Box-plot for bidding prices of different MGs in 24 h. **a** Bidding price of MG1; **b** Bidding price of MG2; **c** Bidding price of MG3; **d** Bidding price of MG4; **e** Bidding price of MG5

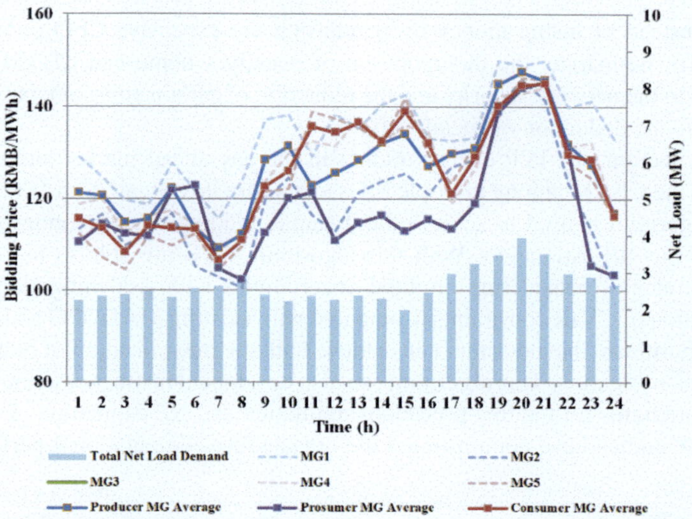

Fig. 20 Bidding price of different MGs in 24 h

bidding price around 140 RMB/MWh) in the peak-load (19:00 to 21:00). This is consistent with the supply–demand-price correlations, which indicates that the price would increase with the rise of demand.

(2) While the RDG output is at the peak-time, i.e., 10:00 to 15:00, the producers and consumers will also increase their bidding prices to the level of around 130 RMB/MWh. The reason is that the penalty for deviation will significantly deduct the MGs' reward, and therefore MGs prefer to submit a higher price, for a less allocated capacity with higher remuneration. Notably, for the prosumer MG3, it prefers price-taker strategy with much lower bidding price (around 110 RMB/MWh) in this period, because it possesses large amount of controllable DGs which are not affected by the penalty, and the CEQ price is also at the lower level with less impact on MG3's reward.

To further investigate the impact of penalty for deviation and CEAM incorporation on MGs' bidding strategy, the following supplementary simulations are conducted. Firstly, the penalty is reset from its 100% rated value (i.e., 500 RMB/MWh) to 10% rated value (i.e., 50 RMB/MWh), and the resulted bidding price modification is shown in Figs. 21 and 22. It is obvious that, with the deduction of imposed penalty, both the prosumers' and customers' bidding price are decreased, especially in the period of 10.00 AM to 15.00 PM. Although the impose of penalty may potentially stimulate MGs for an accurate net load prediction, it would also uplift the energy trading price in the P2P market and lead to a deducted total social welfare. Secondly, the bidding procedure is simulated with and without CEAM incorporation, and the resulted bidding price modification is shown in Figs. 23 and 24. It can be observed that the incorporation of CEAM also increases the producers' and customers' bidding price, because they must ensure that trading in the P2P market could bring positive rewards instead of losing money by purchasing the expensive CEQ quota in the CEAM. To summarize, the incorporation of penalty scheme and CEAM may be beneficial to the network security and the reduction of carbon emission, but it would also lead to the deduction of social welfare.

In this section, the P2P energy trading in the day-ahead market and the CEQ trading in the CEAM among multiple MGs agents are investigated, and the CCMA-DDPG algorithm is used to achieve the dynamic bidding strategy optimization of MGs and the NE estimation. With this algorithm, MGs are able to modify their bidding strategies considering multiple uncertainties with risk mitigation via the incorporation of CVaR constraints. Compared with existing MADDPG and DRMA-DDPG algorithms, the algorithm can achieve both the maximization of total reward and the risk mitigation. This algorithm can also be adopted in other emerging energy trading paradigms for bidding procedure simulation and NE estimation. The designated case studies have demonstrated the outstanding computational performance

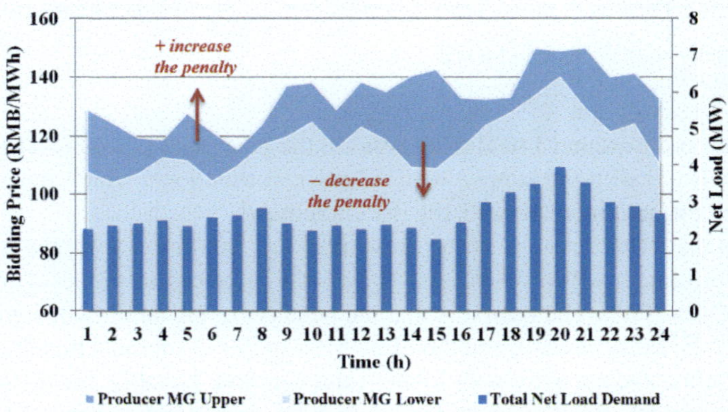

Fig. 21 Impact of penalty on bidding price of producer MGs

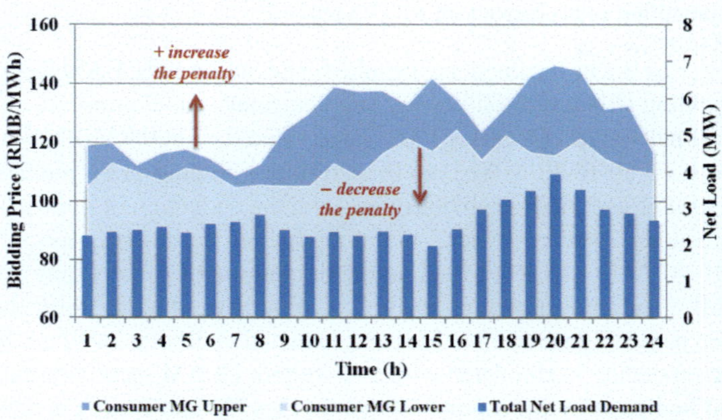

Fig. 22 Impact of penalty on bidding price of consumer MGs

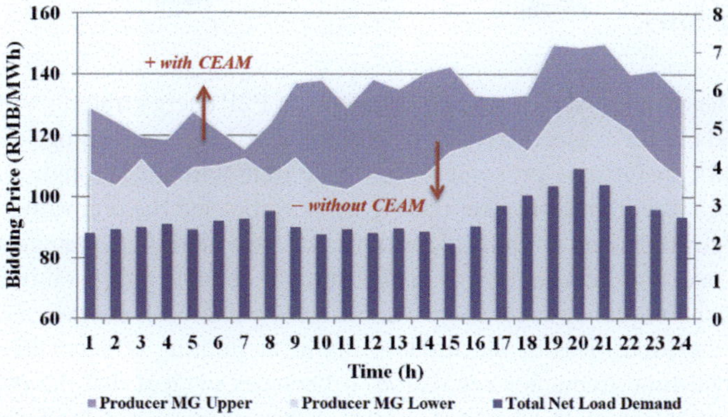

Fig. 23 Impact of CEAM on bidding price of producer MGs

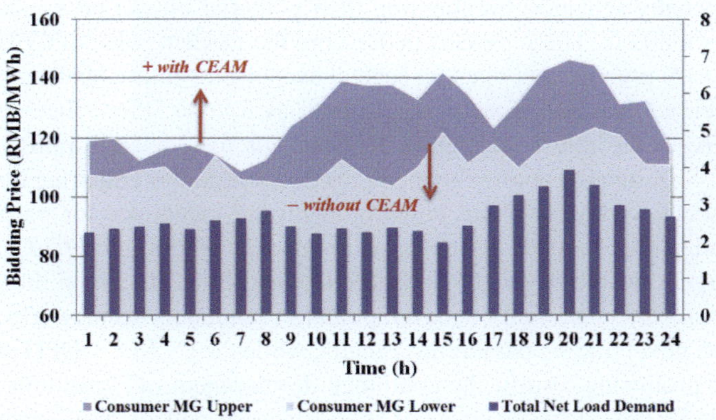

Fig. 24 Impact of CEAM on bidding price of consumer MGs

of this algorithm. In addition, it indicates that the incorporation of penalty scheme and CEAM may lead to the deduction of social welfare in the market. Hence, the mechanism design of deviation mitigation and CEAM incorporation, especially the parameter setting for the imposed penalty and CEQ price limitation, is in pressing need to be further investigated.

4 Conclusions

This chapter has examined the intricate dynamics of P2P energy trading within local energy markets, with a specific focus on the roles of MGs and nanogrids as both producers and consumers—prosumers—in these transactions. Through a comprehensive exploration of multi-market trading frameworks and the innovative application of deep reinforcement learning algorithms, we have identified strategies that not only maximize economic benefits but also ensure the reliable provision of ancillary services such as frequency regulation. The insights garnered from the case studies demonstrate the robustness of the frameworks in facilitating efficient P2P energy trading and real-time frequency regulation. It becomes evident that the integration of market mechanisms with advanced control architectures allows for a synergistic operation of nanogrids within a microgrid, thereby enhancing stability and efficiency. Nonetheless, the exploration has also surfaced complexities associated with CEQ trading and penalty schemes, highlighting their potential impact on social welfare within energy markets. These aspects underscore the nascent stage of CEQ trading in actual industry practice and open up several avenues for future investigation.

Some potential future prospects are outlined as follows. First, there is a need for in-depth research into the design of mechanisms for deviation mitigation and CEQ trading. Specifically, optimizing parameters for penalties and price limitations to enhance social welfare without compromising the incentives for participation and decarbonization. Second, as the concepts of P2P trading and CEQ auctions are still emergent in industry settings, future studies should focus on the practical implementation challenges and the development of regulatory frameworks that can facilitate these novel trading paradigms. Finally, while the CCMA-DDPG algorithm demonstrates promising results, the continued development of algorithms that can effectively handle the stochastic nature of energy markets and integrate various types of market players is crucial.

References

1. Lv T, Ai Q (2016) Interactive energy management of networked microgrids-based active distribution system considering large-scale integration of renewable energy resources. Appl Energy 163:408–422
2. Olivares DE et al (2014) Trends in microgrid control. IEEE Trans Smart Grid 5(4):1905–1919
3. Guerrero JM, Vasquez JC, Matas J, De Vicuña LG, Castilla M (2011) Hierarchical control of droop-controlled AC and DC microgrids—a general approach toward standardization. IEEE Trans Ind Electron 58(1):158–172
4. Hu J, Bhowmick P (2020) A consensus-based robust secondary voltage and frequency control scheme for islanded microgrids. Int J Electr Power Energy Syst
5. Khayat Y et al (2020) On the secondary control architectures of AC microgrids: an overview. IEEE Trans Power Electron 35(6):6482–6500
6. Morstyn T, Farrell N, Darby SJ, McCulloch MD (2018) Using peer-to-peer energy-trading platforms to incentivize prosumers to form federated power plants. Nat Energy 3(2):94–101

7. Monacchi A, Elmenreich W (2016) Assisted energy management in smart microgrids. J Ambient Intell Humaniz Comput 7(6):901–913
8. Mengelkamp E, Gärttner J, Rock K, Kessler S, Orsini L, Weinhardt C (2018) Designing microgrid energy markets: a case study: the Brooklyn Microgrid. Appl Energy 210:870–880
9. Lu Z, Bai L, Wang J, Wei J, Xiao Y, Chen Y (2022) Peer-to-peer joint electricity and carbon trading based on carbon-aware distribution locational marginal pricing. IEEE Trans Power Syst, 1, https://doi.org/10.1109/TPWRS.2022.3167780
10. Tushar W, Saha TK, Yuen C, Smith D, Poor HV (2020) Peer-to-peer trading in electricity networks: an overview. IEEE Trans. Smart Grid 11(4):3185–3200
11. Lin J, Pipattanasomporn M, Rahman S (2019) Comparative analysis of auction mechanisms and bidding strategies for P2P solar transactive energy markets. Appl Energy 255:113687
12. Zhang C, Wu J, Zhou Y, Cheng M, Long C (2018) Peer-to-peer energy trading in a microgrid. Appl Energy 220(March):1–12
13. Anoh K, Maharjan S, Ikpehai A, Zhang Y, Adebisi B (2020) Energy peer-to-peer trading in virtual microgrids in smart grids: a game-theoretic approach. IEEE Trans Smart Grid 11(2):1264–1275
14. Long C, Wu J, Zhou Y, Jenkins N (2018) Peer-to-peer energy sharing through a two-stage aggregated battery control in a community Microgrid. Appl Energy 226(April):261–276
15. Paudel A, Chaudhari K, Long C, Gooi HB (2019) Peer-to-peer energy trading in a prosumer-based community microgrid: a game-theoretic model. IEEE Trans Ind Electron 66(8):6087–6097
16. An J, Lee M, Yeom S, Hong T (2020) Determining the Peer-to-Peer electricity trading price and strategy for energy prosumers and consumers within a microgrid. Appl Energy 261:114335
17. Wu Y, Shi J, Lim G, Fan L, Molavi A (2020) Optimal management of transactive distribution electricity markets with co-optimized bidirectional energy and ancillary service exchanges. IEEE Trans Smart Grid 3053(c):1
18. Grid N (2013) GC022—frequency response workgroup report
19. Yuen C, Oudalov A, Timbus A (2011) The provision of frequency control reserves from multiple microgrids. IEEE Trans Ind Electron 58(1):173–183
20. Chau TK, Yu SS, Fernando T, Iu HHC (2018) Demand-side regulation provision from industrial loads integrated with solar PV Panels and energy storage system for ancillary services. IEEE Trans Ind Inform 14(11):5038–5049
21. Zhou Y, Wei Z, Sun G, Cheung KW, Zang H, Chen S (2018) A robust optimization approach for integrated community energy system in energy and ancillary service markets. Energy 148:1–15
22. Rayati M, Sheikhi A, Ranjbar AM (2019) Optimal contract design for purchasing from frequency regulation service providers with private information. IEEE Trans Power Syst 34(3):2445–2448
23. Anoh K, Maharjan S, Ikpehai A, Zhang Y, Adebisi B (2020) Energy peer-to-peer trading in virtual microgrids in smart grids: a game-theoretic approach. IEEE Trans Smart Grid 11(2):1264–1275. https://doi.org/10.1109/TSG.2019.2934830
24. Ali L, Muyeen SM, Bizhani H, Simoes MG (2022) Economic planning and comparative analysis of market-driven multi-microgrid system for peer-to-peer energy trading. IEEE Trans Indus Appl 58(3):4025–4036, https://doi.org/10.1109/TIA.2022.3152140
25. Chen L, Liu N, Li C, Wang J (2021) Peer-to-peer energy sharing with social attributes: a stochastic leader-follower game approach. IEEE Trans Industr Inf 17(4):2545–2556. https://doi.org/10.1109/TII.2020.2999328
26. Amin W et al (2020) A motivational game-theoretic approach for peer-to-peer energy trading in islanded and grid-connected microgrid. Int J Electr Power Energy Syst 123:106307, https://doi.org/10.1016/j.ijepes.2020.106307
27. Li J, Ye Y, Papadaskalopoulos D, Strbac G (2021) Computationally efficient pricing and benefit distribution mechanisms for incentivizing stable peer-to-peer energy trading. IEEE Internet Things J 8(2):734–749, https://doi.org/10.1109/JIOT.2020.3007196
28. Wang L et al (2022) Security constrained decentralized peer-to-peer transactive energy trading in distribution systems. CSEE J Power Energy Syst 8(1):188–197. https://doi.org/10.17775/CSEEJPES.2020.06560

29. Khorasany M, Paudel A, Razzaghi R, Siano P (2021) A new method for peer matching and negotiation of prosumers in peer-to-peer electricity markets. IEEE Trans Smart Grid 12(3):2472–2483. https://doi.org/10.1109/TSG.2020.3048397
30. He L, Liu Y, Zhang J (2021) Peer-to-peer energy sharing with battery storage: energy pawn in the smart grid. Appl Energy 297:117129. https://doi.org/10.1016/j.apenergy.2021.117129
31. Qiu D, Ye Y, Papadaskalopoulos D, Strbac G (2021) Scalable coordinated management of peer-to-peer energy trading: a multi-cluster deep reinforcement learning approach. Appl Energy 292:116940. https://doi.org/10.1016/j.apenergy.2021.116940
32. Chen T, Bu S, Liu X, Kang J, Yu FR, Han Z (2022) Peer-to-peer energy trading and energy conversion in interconnected multi-energy microgrids using multi-agent deep reinforcement learning. IEEE Trans Smart Grid 13(1):715–727. https://doi.org/10.1109/TSG.2021.3124465
33. Hosseini SA, Toubeau J-F, De Grève Z, Vallée F (2020) An advanced day-ahead bidding strategy for wind power producers considering confidence level on the real-time reserve provision. Appl Energy 280, https://doi.org/10.1016/j.apenergy.2020.115973
34. Zhu Z, Chan KW, Xia S, Bu S (2022) Optimal bi-level bidding and dispatching strategy between active distribution network and virtual alliances using distributed robust multi-agent deep reinforcement learning. IEEE Trans Smart Grid, 1, https://doi.org/10.1109/TSG.2022.3164080
35. Yang H, Zhang S, Qiu J, Qiu D, Lai M, Dong Z (2017) CVaR-constrained optimal bidding of electric vehicle aggregators in day-ahead and real-time markets. IEEE Trans Industr Inf 13(5):2555–2565
36. AlAshery MK, Xiao D, Qiao W (2020) Second-order stochastic dominance constraints for risk management of a wind power producer's optimal bidding strategy. IEEE Trans Sustain Energy 11(3):1404–1413
37. Khaloie H, Anvari-Moghaddam A, Contreras J, Siano P (2021) Risk-involved optimal operating strategy of a hybrid power generation company: a mixed interval-CVaR model. Energy (Oxford) 232:120975, https://doi.org/10.1016/j.energy.2021.120975
38. Galvan E, Mandal, P Sang Y (2020) Networked microgrids with roof-top solar PV and battery energy storage to improve distribution grids resilience to natural disasters. Int J Electr Power Energy Syst 123
39. Wang Y, Rousis AO, Strbac G (2020) On microgrids and resilience: a comprehensive review on modeling and operational strategies. Renew Sustain Energy Rev 134. Elsevier Ltd
40. Bevrani H, Ise T, Miura Y (2014) Virtual synchronous generators: a survey and new perspectives. Int J Electr Power Energy Syst 54:244–254
41. Zhou Z, Levin, T Conzelmann G (2016) Survey of U.S. ancillary services markets, p 59
42. Oureilidis K et al. (2020) Ancillary services market design in distribution networks: review and identification of barriers. Energies 13(4)
43. Gupta M, Vaishya SR, Abhyankar AR (2022) Facilitating DER participation in wholesale electricity market through TSO-DSO coordination. Energy Conv Econ 3(4):201–213. https://doi.org/10.1049/enc2.12063
44. Sutton RS, Barto AG (1998) Reinforcement learning: an introduction. MIT Press, Cambridge, Mass
45. Lowe R, Wu Y, Tamar A, Harb J, Abbeel P, Mordatch I (2020) Multi-agent actor-critic for mixed cooperative—competitive environments. Proc AAAI Conf Artif Intell 34(5)
46. Reddy, Sai Koti D et al (2019) Risk averse reinforcement learning for mixed multi-agent environments. Proceedings of the 18th international conference on autonomous agents and multiagent systems
47. Kolla RK, L.A P, Bhat SP, Jagannathan K (2019) Concentration bounds for empirical conditional value-at-risk: the unbounded case. Oper Res Lett 47(1):16–20
48. Silver D et al (2014) Deterministic policy gradient algorithms. Proc Mach Learn Res, 387–395
49. Fujimoto S, van Hoof H, Meger D (2018) Addressing function approximation error in actor-critic methods. Proc Mach Learn Res 80:1587–1596
50. Nash J (1950) Equilibrium points in n-person games. Proc Nat Acad of Sci PNAS 36(1):48–49, https://doi.org/10.1073/pnas.36.1.48

51. Rosen JB (1965) Existence and uniqueness of equilibrium points for concave N-person games. Econometrica 33(3):520–534. https://doi.org/10.2307/1911749
52. Luenberger D (1978) Complete stability of noncooperative games. J Optim Theory Appl 25(4):485–505. https://doi.org/10.1007/BF00933516
53. Moulin H (1980) On the uniqueness and stability of Nash equilibrium in non-cooperative games. Applied stochastic control in econometrics and management science
54. Rosenmuller J (1971) On a generalization of the lemke-howson algorithm to noncooperative N-person games. SIAM J Appl Math 21(1):73–79. https://doi.org/10.1137/0121010
55. Zhu Z, Wing Chan K, Bu S, Zhou B, Xia S (2021) Real-Time interaction of active distribution network and virtual microgrids: market paradigm and data-driven stakeholder behavior analysis. Appl Energy 297:117107
56. Lin B, Jia Z (2019) Impacts of carbon price level in carbon emission trading market. Appl Energy 239:157–170. https://doi.org/10.1016/j.apenergy.2019.01.194
57. Ajoulabadi A, Ravadanegh SN, Behnam M (2020) Flexible scheduling of reconfigurable microgrid-based distribution networks considering demand response program. Energy 196
58. Freitas CJP, da Silva PP (2015) European Union emissions trading scheme impact on the Spanish electricity price during phase II and phase III implementation. Utilities policy 33:54–62
59. Melgar-Dominguez OD, Pourakbari-Kasmaei M, Lehtonen M, Sanches Mantovani JR (2020) An economic-environmental asset planning in electric distribution networks considering carbon emission trading and demand response. Electric Power Syst Res 181:106202, https://doi.org/10.1016/j.epsr.2020.106202

Embedding Regulatory Frameworks in Microgrids Management

Helena Martín and Jordi de la Hoz

Abstract This chapter demonstrates the key importance of embedding the regulatory framework in the microgrid models, for either their optimal sizing and/or management or for its technoeconomic assessment. Using a case study focused on a basic microgrid for domestic self-consumption, a conceptual regulatory framework is embedded in the model for its energy management optimization. The obtained results are compared to those stemming from disregarding regulatory constraints, as is frequently observed in the literature, which leads to significantly inaccurate results.

Keywords Regulatory framework · Electricity tariff · Energy management · Optimization

1 Introduction

A huge body of research has been conducted in the last decades on microgrids (MGs), mainly focused in technical aspects of these systems. Some of these studies also evaluate the economic performance of MGs, or address their optimal sizing and/or management including economic considerations in the objective function. Nevertheless, the constraints imposed by the regulatory frameworks under which those MGs will operate are rarely taken into account in the analysis, which may lead to misleading investment decisions [1, 2].

MGs are usually tailor-made solutions, meeting site-specific requirements with particular configurations. This diversity has been identified as a barrier for the wide development of the necessary regulatory frameworks needed for fostering their deployment [3]. Economic and safety concerns for the utility electric system to which MGs may connect have also played a role in delaying the development of the regulatory frameworks [4].

H. Martín (✉) · J. de la Hoz
Department of Electrical Engineering, Barcelona East School of Engineering (EEBE), Polytechnic University of Catalonia (UPC), Eduard Maristany 16, 08019 Barcelona, Spain
e-mail: m.helena.martin@upc.edu

F. Shahnia and J. M. Guerrero (eds.), *Microgrids and Virtual Power Plants*, Power Systems, https://doi.org/10.1007/978-981-97-6623-9_12

Emerging regulatory efforts can be found in certain USA states such as California, Connecticut, Hawaii, Maine and Puerto Rico, which have defined MGs in statute, either establishing funding mechanisms, service tariffs, grid interconnection requirements or access to rights-of-way [5]. In most world areas, however, a clear legal definition of the MG concept cannot be found in the energy regulations, although some MG typologies could be ruled by the provisions designed for certain decentralized energy systems [6, 7]. In the case of the European Union, that could be the case for the Closed Distribution Systems, the Citizens Energy Communities or the Renewable Energy Communities, which are figures integrated in the European Union's energy regulatory framework [6]. Any case, the lack of knowledge or the uncertainty about the legal provisions to apply are deterrent factors for a wider deployment of MGs and for the inclusion of the related constraints in their assessment.

In the context of increasing electricity prices, MGs can be a reliable alternative for grid electricity consumers to reduce the electricity supply costs (ESCs) while contributing to the decarbonization process. To understand how MGs might be key in contributing to the reduction of the ESCs, understanding the structure of these costs of regulatory nature is mandatory. The task of explaining these costs will be undertaken in the following sections of this chapter.

2 Electricity Supply Contracts and Electricity Supply Costs

Grid-connected electricity consumers differ regarding their consumption level and contractual arrangements for purchasing electricity. Business and residential consumers enter into a supply contract with an electric energy retailer. In contrast, industrial consumers may buy electricity directly at the wholesale market or conduct a power purchase agreement with an energy retailer with more competitive prices. Consequently, electricity consumers can be subject to various contractual arrangements for electricity supply, with location-dependent taxation rules and other particularities.

Focusing on the case of medium and small consumers, with contracted powers not exceeding certain regulatorily determined threshold, the retail contracts that determine their ESC_i within a year i typically charge separately the cost of the demanded power DPC_i and the cost of the consumed energy CEC_i. Disregarding taxes and other particularities, the ESC_i is then determined as:

$$ESC_i = DPC_i + CEC_i \tag{1}$$

These contracts can adopt two main structures, namely, fixed electricity price or pass-through electricity price [8]. In the case of a fixed price contract and monthly billing, the ESC_i would be determined as follows:

$$ESC_i = \sum_{m=1}^{12} PT_{i,m} * CP_{i,m} + \sum_{m=1}^{12} ET_{i,m} * CE_{i,m} \quad (2)$$

where for each month m of a specific year i, the cost of the demanded power $DPC_{i,m}$ is calculated as the product of the power term $PT_{i,m}$ (¤/kW) by the contracted power $CP_{i,m}$ (kW). Likewise, the cost of the consumed energy $CEC_{i,m}$ is determined by the product of the energy term $ET_{i,m}$ (¤/kWh) by the consumed electricity $CE_{i,m}$ (kWh), where the symbol ¤ stands for the unspecified currency of application.

In this regard, Fig. 1 graphically depicts the structure of the fixed electricity price contracts, showing at its left side how the ESC_i is computed and at its right side how the CEC_i is determined on a monthly basis.

Concerning the fixed price contracts another possible option is to have several fixed price values for different times of the day or periods. In this case, the monthly contracted power CP_{i,m,P_p} and consumed electricity CE_{i,m,P_E} are priced according to the time of the day. Thus, in the case of having two periods for the power term $P_P = \{P_{P1}, P_{P2}\}$ and three periods for the energy term $P_E = \{P_{E1}, P_{E2}, P_{E3}\}$, the ESC_i would result as follows:

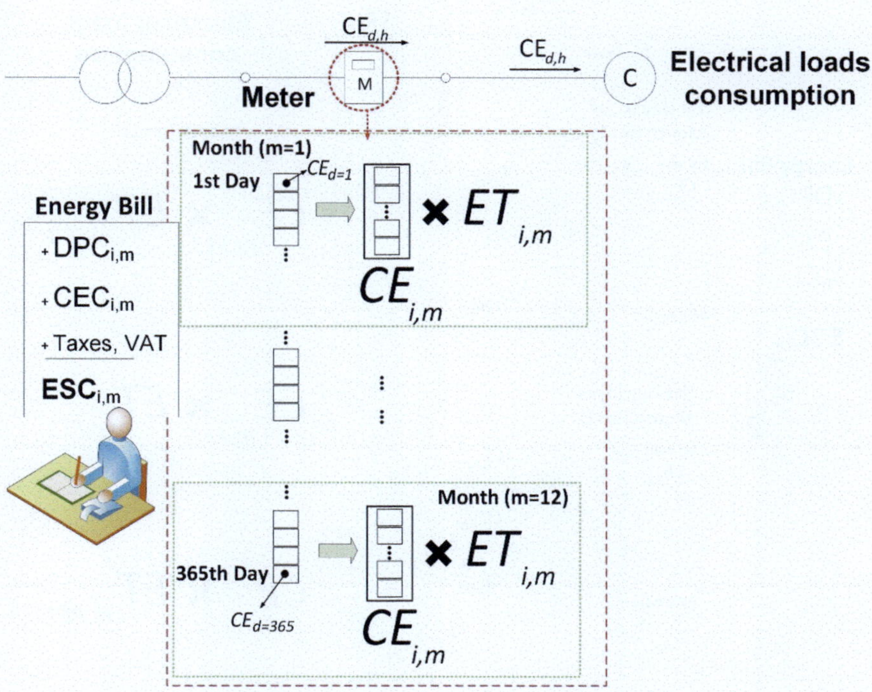

Fig. 1 Energy supply cost $ESC_{i,m}$ structure and consumed energy cost $CEC_{i,m}$ determination on a monthly basis

$$ESC_i = \sum_{m=1}^{12} \sum_{P_p=P_{P1}}^{P_{P2}} PT_{i,m,P_p} * CP_{i,m,P_p} + \sum_{m=1}^{12} \sum_{P_E=P_{E1}}^{P_{E3}} ET_{i,m,P_E} * CE_{i,m,P_E} \quad (3)$$

Figure 2 illustrates the billing process for CE_{i,m,P_E}. The hourly consumed electricity $CE_{i,m,d,h}$ is added up based on its particular consumption period P_E. Thus, at the end of each day, there is a daily amount of consumed energy during each of the established periods CE_{i,m,d,P_E}. Finally, at the end of the month *m*, the total consumed energy CE_{i,m,P_E} for each period is calculated adding the former daily values and is billed according to the different energy terms for each period ET_{i,m,P_E}. The billing process for CP_{i,m,P_p} follows an analogous approach.

On the other hand, under the pass-through price structure, electricity consumers may be charged a price indexed to the spot market $Pm_{i,d,h}$ of an hour *h* of a day *d*. In addition, the cost of the consumed energy $CEC_{i,m}$ may also include a margin for the electricity retailer $EMC_{i,m}$ and a proportional cost to the energy losses in the network $ELC_{i,m}$. Assuming as above two periods for the power term and monthly billing, with *Nd(m)* being the number of days in a month *m*, the ESC_i would be depicted as:

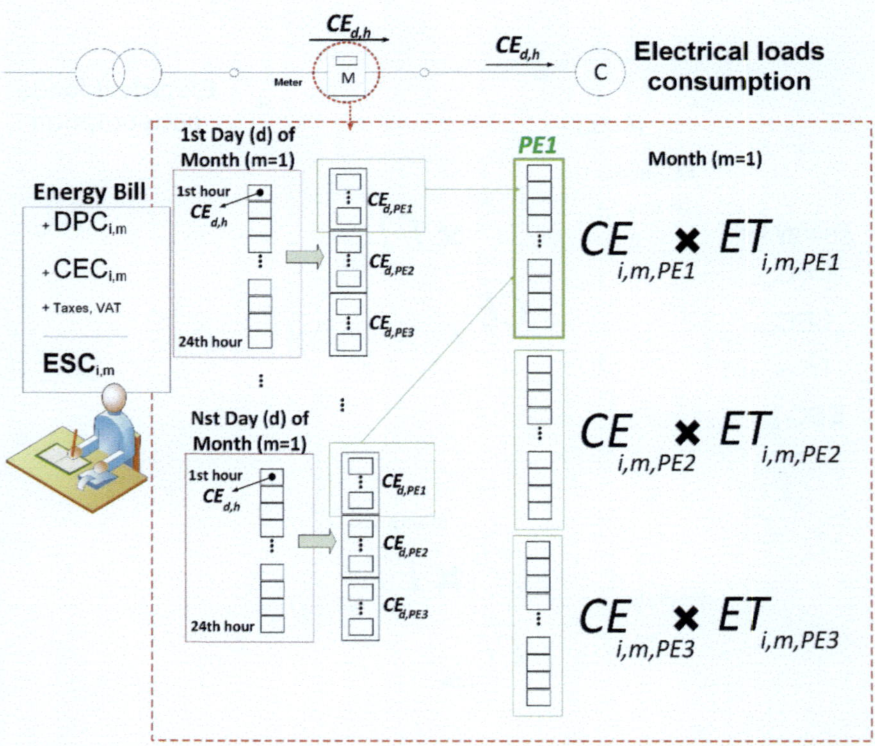

Fig. 2 Billing process for the consumed energy CE_{i,m,P_E} in energy supply contracts with different fixed price periods

$$ESC_i = \sum_{m=1}^{12} \sum_{P_p=P_{P1}}^{P_{P2}} PT_{i,m,P_p} * CP_{i,m,P_p} + \sum_{m=1}^{12} \left\{ \left(\sum_{d=1}^{Nd(m)} \sum_{h=1}^{24} Pm_{i,d,h} * CE_{i,d,h} \right) + EMC_{i,m} + ELC_{i,m} \right\} \quad (4)$$

Other retail electricity supply contract types may be considered as lying between the basic structures described before [9].

Electricity consumers often lack knowledge about what is included in their ESCs, why they are depicted as they are and how they are calculated. Nevertheless, understanding the energy consumers electricity tariff structure is necessary for discriminating between the different existing commercial offers and selecting the one in their best interest.

For clarifying such questions, it should be noted that the electricity retailer faces various significant costs when providing energy to a customer. The first one corresponds to the electricity acquisition cost EAC_i, i.e., the cost of purchasing energy in advance for its subsequent supply to the consumer.

In addition, the electricity retailer is usually in charge of managing the consumer's access to the network, as well as satisfying its access tariffs and charges. These payments are addressed to fulfil the regulated electricity costs (RECs), namely, those related to transportation and distribution networks, and other system costs as renewable energy incentives, nuclear waste treatment, etc. The amount of the RECs depends on the consumer's voltage level and rated power and are fixed by the regulator.

The REC related to the amount of demanded power REC_P_i, depends on the product of a power term PT_REC by CP and the REC related to the amount of consumed energy REC_E_i, depends on the product of an energy term ET_REC by CE. Continuing with the example of two periods for the power term and three periods for the energy term, the REC_i faced by the retailer could be depicted as follows:

$$REC_i = REC\_P_i + REC\_E_i = \sum_{m=1}^{12} \sum_{P_p=P_{P1}}^{P_{P2}} PT\_REC_{i,m,P_p} * CP_{i,m,P_p} + \sum_{m=1}^{12} \sum_{P_E=P_{E1}}^{P_{E3}} ET\_REC_{i,m,P_E} * CE_{i,m,P_E} \quad (5)$$

Furthermore, the energy supplier incurs operating expenses $O\&M\_C_i$ linked to its business activity and also faces the cost proportional to the energy losses in the network ELC_i. Thus, the cost of the energy retailer $Cost_i$ for supplying electricity to a consumer would be:

$$Cost_i = EAC_i + REC_i + O\&M\_C_i + ELC_i \quad (6)$$

The operating revenue $Revenue_i$ of the energy retailer for supplying electricity to a determined customer is equal to the ESC_i of that customer. Consequently, the earnings before interest, taxes, depreciation, and amortization $EBITDA_i$ of the energy retailer are expressed as:

$$EBITDA_i = Revenue_i - Cost_i = ESC_i - Cost_i \tag{7}$$

Evincing in Eq. (6) the cost of the demanded power DPC_i and the cost of the consumed energy CEC_i that compose ESC_i as well as the different elements of $Cost_i$ shown in (6), the $EBITDA_i$ is formulated as:

$$EBITDA_i = DPC_i + CEC_i - EAC_i - REC_i - O\&M\_C_i - ELC_i \tag{8}$$

The $EBITDA_i$ can be itemized based on the type of product delivered, namely, ($EBITDA\_P_i$) for the guaranteed power and ($EBITDA\_E_i$) for the supplied energy. Continuing with the example of two periods for the power term and three periods for the energy term, for whatever energy price contract type it holds:

$$\begin{aligned} EBITDA\_P_i = & \sum_{m=1}^{12} \sum_{p_P=P_{P1}}^{P_{P2}} PT_{i,p_P,m} * CP_{i,p_P,m} \\ & - \sum_{m=1}^{12} \sum_{p_P=P_{P1}}^{P_{P2}} PT\_REC_{i,p_P,m} * CP_{i,p_P,m} \end{aligned} \tag{9}$$

$$\begin{aligned} EBITDA\_E_i = & \sum_{m=1}^{12} CEC_{i,m} \\ & - \sum_{m=1}^{12} \left(EAC_{i,m} + \sum_{p_E=P_{E1}}^{P_{E3}} ET\_REC_{i,p_E,m} * CE_{i,p_E,m} + O\&M\_C_{i,m} + ELC_{i,m} \right) \end{aligned} \tag{10}$$

It is clear that the energy retailer must set the power term $PT_{i,p_P,m}$ and the either fixed or pass-through energy term within $CEC_{i,m}$ to meet its economic expectations while ensuring competitiveness in the electricity retail market:

$$\sum_{m=1}^{12} \sum_{p_P=P_{P1}}^{P_{P2}} PT_{i,p_P,m} * CP_{i,p_P,m} \geq \sum_{m=1}^{12} \sum_{p_P=P_{P1}}^{P_{P2}} PT\_REC_{i,p_P,m} * CP_{i,p_P,m} \tag{11}$$

$$\sum_{m=1}^{12} CEC_{i,m}$$

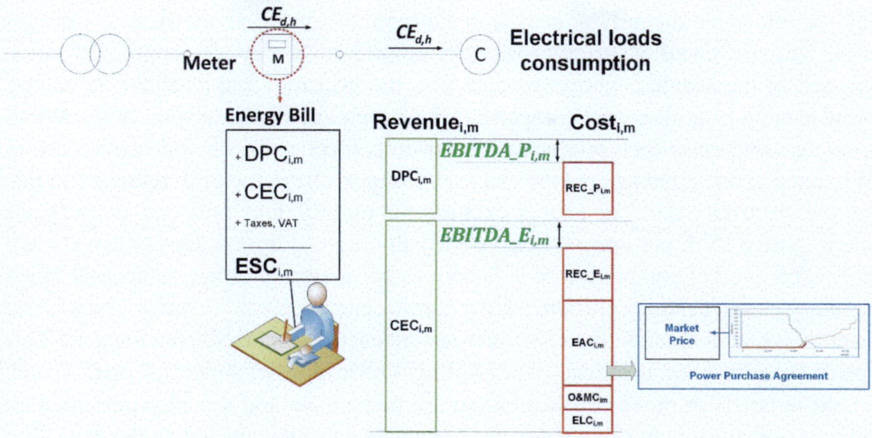

Fig. 3 Components of $Revenue_{i,m}$, $Cost_{i,m}$ and $EBIDTA_{i,m}$ of the electricity retailer

$$\geq \sum_{m=1}^{12} \left(EAC_{i,m} + \sum_{p_E=P_{E1}}^{P_{E3}} ET\_REC_{i,p_E,m} * CE_{i,p_E,m} + O\&M\_C_{i,m} + ELC_{i,m} \right) \tag{12}$$

Figure 3 graphically illustrates the revenues, costs and EBIDTA components of the electricity retailer.

3 Electricity Self-Consumption Came to the Rescue

In a context of high electricity prices, self-consumption may reduce the ESCs by replacing purchases of electricity from the grid with power generated on-site by the consumers. The role that self-consumption may play in securing the energy supply and contributing to the decarbonization process is also well acknowledged, particularly regarding the case of distributed solar photovoltaic (PV) facilities. To reflect this, recent regulatory efforts and ambitious programs have been undertaken in wide regions of the world for further fostering self-consumption. For instance, in the revised European Union Directive for the promotion of energy from renewable sources it has been legislated a great speed-up in the permit-granting procedures for PV self-consumption facilities in all Member States [10] and the Chinese National Energy Administration has issued a program for expanding PV at rooftops that in 2022 alone accomplished above 51 GW of new capacity additions [11].

Self-consumption policies are location-dependent, and within a location they may vary in time. Beyond the basic legal right to self-consume, there may exist a wide variety of differences between its implementation rules. In this regard, the revenues for the self-consumed electricity can merely be the savings in the electricity bill or may include some incentives, as bonuses, premiums or green certificates. Charges on the self-consumed electricity for financing grid costs may also apply. The value assigned to the surplus energy injected into the grid may imply either an energy compensation or a monetary compensation or even any compensation at all. When an energy compensation is produced the consumer receives an allowance of credits in kWh that can be netted against the electricity imports from the grid, reducing in this way the electricity bill. The energy exchange is usually done on a one-to-one basis (one imported kWh per one exported kWh), although additional fees or taxes could apply. This type of compensation is known as net-metering or net-balance. When a monetary compensation is produced, the surplus energy is sold to the utility at a rate lower than that charged to the consumer for the energy imported from the grid. This type of compensation is known as net-billing. Other types of support schemes, such as feed-in tariffs or green certificates can too be applied, and it is also possible that any compensation is perceived for the surplus electricity injected to the grid. The typical maximum time frames for the compensation of the credits received for the surplus energy can be either annual, monthly or real-time (15 min or shorter). Special charges may apply in case of using storage systems. There may also exist limits for the maximum size and distance of the production and the consumption equipment, the type of stakeholders and the criteria for energy sharing between participants [12, 13].

Given the aforementioned great casuistry, it is not strange that in most techno-economic assessments found in the academic literature the regulatory constraints in force in a determined location are not fully considered in the analysis [2]. This is particularly true regarding the possible charges and taxes that can apply to the self-consumed and/or surplus electricity or to the inclusion of storage systems in the scheme. However, it must be taken into account that these the regulatory constraints frequently disregarded in the analyses are policy instruments for shaping the deployment path of distributed facilities for self-consumption with a non-despicable impact on the profitability outcomes and consequently on the investment decisions.

4 Philosophical and Methodological Management Approach in a Self-Consumption Context

Although a variety of definitions for a MG can be found in the academic literature, certain traits which are common to all of them have been identified, i.e., the existence of locally controlled small-scale generation and loads and possibly energy storage units, a point of common coupling with the main grid allowing for either grid-tied or islanded operation, and typical implementation at the low voltage distribution level [7]. Particularly, the key concept differentiating a MG from other distributed renewable energy systems is its ability to behave as a locally controlled entity towards the centrally controlled main grid, e.g., showing autonomous capabilities in the event of the main grid failure. This control faculty may also empower the MG for the provision of complex services or for coordinating with other entities in the network [1].

In this section, a conceptual framework has been developed in order to illustrate the need of embedding the electricity tariff structure and the regulatory framework constraints in the energy management of MGs. For this purpose, a simple system consisting of PV generation equipment and electric batteries for supplying a domestic low voltage demand having an electricity supply contract has been selected, as shown in Fig. 4.

The selected MG shown in Fig. 4 meets the requirements enunciated before and has well defined legal and economic regulatory frameworks. Modern inverters incorporate backup capabilities for transitioning to islanding operation in case of grid failure, and regarding other sophisticated control characteristics, e.g., auxiliary services provision, it is not a matter of technical infeasibility but of usual legal impediment for carrying out such activities.

For the sake of generality, the regulatory framework for self-consumption applied to the MG in Fig. 4 includes a variety of the characteristics enumerated in Sect. 3, which are summarized in Table 1.

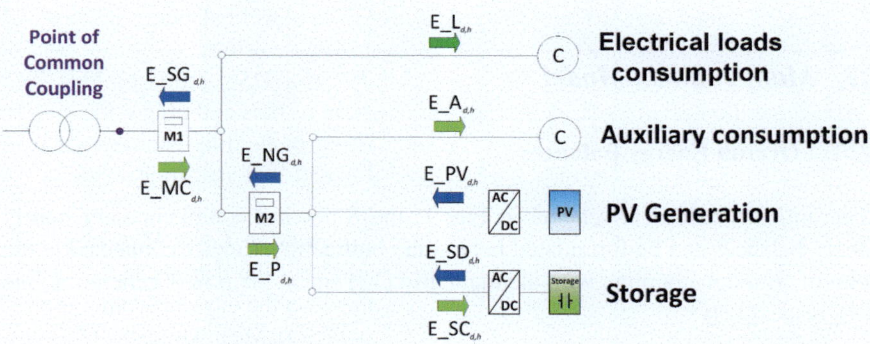

Fig. 4 Conceptual electrical scheme of the considered MG

Table 1 Conceptual regulatory framework characteristics

Self-consumed energy	Revenues from self-consumed energy	Monetary savings	√
		Other incentives	X
	Charges on self-consumed energy		√
Surplus energy	Revenues from surplus energy	Net-metering	X
		Net-billing	√
		No revenues	√
	Access tariff and tax on the revenues	Net-billing	√
		No revenues	X
Other characteristics	Additional charges	On main grid consumption	√
		On storage	√

4.1 Methodology and Source of Data

Following a mathematical programming approach, the optimal management of the system depicted in Fig. 4 has been carried out, minimizing its total daily operating cost. Unlike the common trend of modelling at most the fixed and/or variable operating costs, the energy acquisition cost and the possible revenue for energy selling, here the electricity tariff structure and the constraints introduced by the considered regulatory framework have been embedded in the analysis.

Given the varied casuistry of regulatory frameworks for self-consumption, a conceptual one has been selected incorporating characteristics commonly found in most regulations. Besides contemplating the option for adhering to either a net-billing scheme or receiving no revenue at all for the surplus energy, it includes particular provisions for charging the self-consumed electricity and the use of storage, applying access tariffs and taxes for the surplus electricity injected into the grid, as well as a detailed characterization of an electricity tariff structure. This comprehensive formulation may then be easily adapted to the particular regulatory framework of interest.

4.2 Mathematical Model

4.2.1 Overall Energy Balances

Regarding the electrical scheme in Fig. 1, the following complementary hourly energy balances can be formulated behind the bidirectional meter connected at the point of common coupling with the main grid (13) and at the power generation line (14), respectively:

$$E\_MC_{d,h} + E\_NG_{d,h} = E\_L_{d,h} + E\_P_{d,h} + E\_SG_{d,h} \tag{13}$$

$$E\_P_{d,h} + E\_PV_{d,h} + E\_SD_{d,h} = E\_NG_{d,h} + E\_A_{d,h} + E\_SC_{d,h} \tag{14}$$

where $E\_MC_{d,h}$ is the energy consumption from the main grid, $E\_NG_{d,h}$ is the net energy generation, $E\_L_{d,h}$ is the energy consumption of the electrical loads, $E\_P_{d,h}$ is the energy consumption of the power line after meter 2, $E\_SG_{d,h}$ is the energy surplus generation injected into the grid, $E\_PV_{d,h}$ is the PV generation, $E\_SD_{d,h}$ is the storage energy discharge, $E\_A_{d,h}$ is the auxiliary energy consumption of the PV equipment and $E\_SC_{d,h}$ is the storage energy charge, all of them for an hour h of a day d. All the former and the following variables are defined as non-negative, unless otherwise stated.

Simultaneous flow in Fig. 4 across meter M1 of $E\_MC_{d,h}$ and $E\_SG_{d,h}$ within the same time step is avoided in (15)–(17) by preventing the joint activation of their associated binary variables $b\_MC_{d,h}$ and $b\_SG_{d,h}$, being M a big number that can be physically interpreted as the maximum energy flow allowed by the line and/or its protection devices. The same is done in (18)–(20) regarding meter M2 for $E\_P_{d,h}$ and $E\_NG_{d,h}$ by means of their associated binary variables $b\_P_{d,h}$ and $b\_NG_{d,h}$.

$$E\_MC_{d,h} \leq M \cdot b\_MC_{d,h} \tag{15}$$

$$E\_SG_{d,h} \leq M \cdot b\_SG_{d,h} \tag{16}$$

$$b\_MC_{d,h} + b\_SG_{d,h} = 1 \tag{17}$$

$$E\_P_{d,h} \leq M \cdot b\_P_{d,h} \tag{18}$$

$$E\_NG_{d,h} \leq M \cdot b\_NG_{d,h} \tag{19}$$

$$b\_P_{d,h} + b\_NG_{d,h} = 1 \tag{20}$$

4.2.2 PV Generation

A simplified modelling of the PV generation equipment has been adopted, consisting in defining the maximum PV energy generation $E\_PV_{max\,d,h}$ in (21) as the product of the peak power of the PV equipment P_PV_p, the generation profile and the duration of the time step Δh, and then enforcing in (22) that it is not exceeded by the actual PV energy generation $E\_PV_{d,h}$.

$$E\_PV_{maxd,h} = P\_PV_p \cdot profile_{d,h} \cdot \Delta h \tag{21}$$

$$E\_PV_{d,h} \leq E\_PV_{maxd,h} \tag{22}$$

4.2.3 Storage

The stored energy $E\_S_{d,h}$ is defined in (23) and is constrained to not exceed the maximum energy capacity $E\_S_{max\,d,h}$ of the battery in (24). The state of charge $SoC_{d,h}$ is defined in (25), an initial condition for $E\_S_{d,h}$ is imposed in (26) and a lower $SoC_{d,h}$ at the end of the day than the initial SoC_0 is prevented in (27). The charge $P\_SC_{d,h}$ and the discharge $P\_SD_{d,h}$ powers are defined in (28)–(29), respectively. In (30)–(31) the binary variable $b\_status_{d,h}$ is activated when $SoC_{d,h}$ is comprised between the threshold SoC_{th} and the upper boundary SoC_{max}, while it is deactivated when SoC_h ranges between the lower boundary SoC_{min} and the threshold SoC_{th}. Then, if

$b\_status_{d,h}$ is activated $P\_SC_{d,h}$ and $P\_SD_{d,h}$ are limited in (32)–(33) by the more stringent charge and discharge rates P_SC_{th} and P_SD_{th}, whereas these rates increase up to P_SC_{max} and P_SD_{max} otherwise. Simultaneous flow of $P\_SC_{d,h}$ and $P\_SD_{d,h}$ within the same time step is avoided in (34)–(36) by preventing the joint activation of their associated binary variables $b\_SC_{d,h}$ and $b\_SD_{d,h}$.

$$E\_S_{d,h} = E\_S_{d,h-1} + E\_SC_{d,h} \cdot \eta_C - E\_SD_{d,h}/\eta_D \tag{23}$$

$$E\_S_{d,h} \leq E\_S_{max} \tag{24}$$

$$SoC_{d,h} = E\_S_{d,h}/E\_S_{max} \tag{25}$$

$$E\_S_{d,h-1} = SoC_0 \cdot E\_S_{max},\, h = 1 \tag{26}$$

$$\sum_{h=1}^{23} \left(SoC_{d,h+1} - SoC_{d,h} \right) \geq 0 \tag{27}$$

$$P\_SC_{d,h} = E\_SC_{d,h}/\Delta h \tag{28}$$

$$P\_SD_{d,h} = E\_SD_{d,h}/\Delta h \tag{29}$$

$$SoC_{d,h} \leq SoC_{th} + (SoC_{max} - SoC_{th}) \cdot b\_status_{d,h} \tag{30}$$

$$SoC_{d,h} \geq SoC_{min} + (SoC_{th} - SoC_{min}) \cdot b\_status_{d,h} \tag{31}$$

$$P\_SC_{d,h} \leq P\_SC_{th} + (P\_SC_{max} - P\_SC_{th}) \cdot \left(1 - b\_status_{d,h}\right) \tag{32}$$

$$P\_SD_{d,h} \leq P\_SD_{th} + (P\_SD_{max} - P\_SD_{th}) \cdot \left(1 - b\_status_{d,h}\right) \tag{33}$$

$$P\_SC_{d,h} \leq P\_SC_{max} \cdot b\_SC_{d,h} \tag{34}$$

$$P\_SD_{d,h} \leq P\_SD_{max} \cdot b\_SD_{d,h} \tag{35}$$

$$b\_SC_{d,h} + b\_SD_{d,h} = 1 \tag{36}$$

4.2.4 Detailed Energy Balances

Once the PV generation and the storage models have been presented, the overall energy balances are complemented by those in (37)–(45), where the source and the sink of the different components of the energy flows in (13)–(14) are made explicit. The notation employed for these components consists of the letter *E* (energy) followed by a letter identifying the energy source and a letter representing the energy sink, where *M* stands for the main grid, *L* for the electrical loads, *S* for the storage, *PV* for the PV equipment and *A* for its auxiliary consumption.

$$E\_MC_{d,h} = E\_M\_L_{d,h} + E\_M\_A_{d,h} + E\_M\_S_{d,h} \tag{37}$$

$$E\_SG_{d,h} = E\_PV\_M_{d,h} + E\_S\_M_{d,h} \tag{38}$$

$$E\_P_{d,h} = E\_M\_A_{d,h} + E\_M\_S_{d,h} \tag{39}$$

$$E\_NG_{d,h} = E\_PV\_L_{d,h} + E\_PV\_M_{d,h} + E\_S\_L_{d,h} + E\_S\_M_{d,h} \tag{40}$$

$$E\_L_{d,h} = E\_M\_L_{d,h} + E\_PV\_L_{d,h} + E\_S\_L_{d,h} \tag{41}$$

$$E\_A_{d,h} = E\_M\_A_{d,h} + E\_PV\_A_{d,h} + E\_S\_A_{d,h} \tag{42}$$

$$E\_PV_{d,h} = E\_PV\_L_{d,h} + E\_PV\_A_{d,h} + E\_PV\_M_{d,h} + E\_PV\_S_{d,h} \tag{43}$$

$$E\_SC_{d,h} = E\_M\_S_{d,h} + E\_PV\_S_{d,h} \tag{44}$$

$$E\_SD_{d,h} = E\_S\_L_{d,h} + E\_S\_A_{d,h} + E\_S\_M_{d,h} \tag{45}$$

4.2.5 Electricity Tariff Structure and Regulatory Framework for Self-Consumption

In this section, the daily total operating cost of the modelled MG is determined, taking into account the electricity tariff structure and the constraints introduced by the conceptual regulatory framework for self-consumption that has been selected.

Following (2), the daily cost of the demanded power DPC_d is calculated in (46), assuming in this case that there is a single period for the power term and that its value matches that of the daily regulated electricity cost related to the demanded power REC_P_d:

$$DPC_d = REC\_P_d = PT\_REC_d \cdot CP \tag{46}$$

Likewise, it has been chosen to explicitly calculate the daily regulated electricity cost related to the consumed energy REC_E_d, rather than considering that it is recovered through the electricity purchase final price $P_{epf\,d,h}$. Thus, following (5):

$$REC\_E_d = \sum_h ET\_REC_{d,h} \cdot E\_MC_{d,h} \tag{47}$$

Next, the daily purchased energy cost PEC_d is calculated in (48) as the hourly sum of the product of $P_{epf\,d,h}$ by the energy consumption from the main grid $E\_MC_{d,h}$, considering here that $P_{epf\,d,h}$ already includes concepts such as the margin for the retailer and a cost proportional to the energy losses in the network that were made explicit in (4):

$$PEC_d = \sum_h \left(P_{epf\,d,h} \cdot E\_MC_{d,h}\right) \tag{48}$$

Then, the daily cost of the consumed energy CEC_d is determined in (49):

$$CEC_d = PEC_d + REC\_E_d \tag{49}$$

The electricity tariff structure described up to this point by (46)–(49) would be coincident with that corresponding to conventional consumers, but according to the characteristics of the conceptual self-consumption framework depicted in Table 1, several other charges are applied.

The total daily charges cost TCC_d have been divided into fixed and variable charges costs as well as charges cost for other system services. On the one side, there is a daily fixed charges cost FCC_d calculated as the product of the fixed charges power term PT_FC_d by a specific power for the application of the charges CAP. The CAP has been defined as the sum of the maximum value of $E\_MC_{d,h}$ plus the maximum value of $E\_NG_{d,h}$. This second term would be removed for the case of MGs without storage:

$$FCC_d = PT\_FC_d \cdot CAP_d \tag{50}$$

$$CAP_d = max\left(E\_MC_{d,h}/\Delta h\right) + \left(max\left(E\_NG_{d,h}/\Delta h\right)\right)\big|_{\exists storage} \tag{51}$$

On the other side, there is a daily variable charges cost VCC_d dependent on the variable charges energy term $ET\_VC_{d,h}$ by the sum of $E\_MC_{d,h}$ plus the hourly self-consumed electricity $E\_SelfC_{d,h}$:

$$VCC_d = \sum_h ET\_VC_{d,h} \cdot \left(E\_MC_{d,h} + E\_SelfC_{d,h}\right) \tag{52}$$

where $E\_SelfC_{d,h}$ is computed as:

$$E\_SelfC_{d,h} = E\_NG_{d,h} - E\_SG_{d,h} \tag{53}$$

Daily additional charges for other system services $OSSCC_d$ have also been considered, determined as the product of the other system services charges energy term ET_OSSC by the daily self-consumed electricity $E\_SelfC_{d,h}$:

$$OSSCC_d = ET\_OSSCC \cdot \sum_h E\_SelfC_{h,d} \tag{54}$$

Finally, the daily total charges cost TCC_d is computed:

$$TCC_d = FCC_d + VCC_d + OSSCC_d \tag{55}$$

Then, the daily ESC_d before taxes is:

$$ESC_d = DPC_d + CEC_d + TCC_d \tag{56}$$

Next, the daily cost ETC_d of applying an electricity tax ET on the daily ESC_d before taxes is determined:

$$ETC_d = ET \cdot ESC_d \tag{57}$$

The daily valued added tax cost $VATC_d$ is computed applying the VAT tax on the sum of the daily ESC_d before taxes plus the daily ETC_d plus the daily meter rental cost MRC_d:

$$VATC_d = VAT \cdot (ESC_d + ETC_d + MRC_d) \tag{58}$$

Lastly, the final daily energy supply cost $FESC_d$ results from:

$$FESC_d = ESC_d + ETC_d + MRC_d + VATC_d \tag{59}$$

On the other hand, in the considered conceptual regulatory framework for self-consumption, two different modalities are envisioned. In the first modality, the surplus energy receives no revenues, while in the second one it receives the hourly mean electricity market price $P_{mm\,d,h}$, which can be expressed as:

$$Revenue_d = (Modality - 1) \cdot \sum_h \left(P_{mm_{d,h}} \cdot E\_SG_{d,h}\right),\ Modality = \{1,2\} \tag{60}$$

It has been considered, however, that consumers under the second modality have also the role of electricity producers. In the following they will be referred to as producers, and must support a daily access tariff cost ATC_d calculated as the access tariff energy term ET_AT by the daily surplus energy injected into the main grid:

$$ATC_d = (Modality - 1) \cdot ET\_AT \cdot \sum_h E\_SG_{h,d}, Modality = \{1,2\} \quad (61)$$

Additionally, a tax RT on the revenues from the surplus energy injected into the main grid must be paid, raising the cost RTC_d:

$$RTC_d = RT \cdot Revenue_d \quad (62)$$

The daily operation and management costs $O\&M\_C_d$ of the MG have also been considered, distinguishing between the fixed and the variable costs of PV, i.e., $O\&M\_FC\_PV_d$ and $O\&M\_VC\_PV_d$, and between the fixed and the variable costs of the storage, i.e., $O\&M\_FC\_S_d$ and $O\&M\_VC\_S_d$:

$$O\&M\_C_d = O\&M\_FC\_PV_d + O\&M\_VC\_PV_d + O\&M\_FC\_S_d + O\&M\_VC\_S_d \quad (63)$$

These costs have been calculated as:

$$O\&M\_FC\_PV_d = FC\_PV_d \cdot P\_PV_p \quad (64)$$

$$O\&M\_VC\_PV_d = VC\_PV \cdot \sum_h E\_PV_{h,d} \quad (65)$$

$$O\&M\_FC\_S_d = FC\_S_d \cdot S_{max}/\Delta h \quad (66)$$

$$O\&M\_VC\_S_d = VC\_S \cdot \sum_h \left(E\_SC_{h,d} + E\_SD_{h,d}\right) \quad (67)$$

where FC_PV_d and VC_PV are the unit fixed and variable costs related to O&M of the PV equipment and FC_S_d and VC_S are those corresponding to the storage.

In the end, the daily final O&M cost $F\_O\&M\_C_d$ of the MG results from considering all the former defined costs and revenues:

$$F\_O\&M\_C_d = FESC_d - Revenue_d + ATC_d + RTC_d + O\&M\_C_d \quad (68)$$

Figure 5 jointly shows for comparison purposes the ESC components of both a conventional consumer in its upper side and of a prosumer equipped con the analyzed MG in its lower side, reflecting the greater complexity of the latter case.

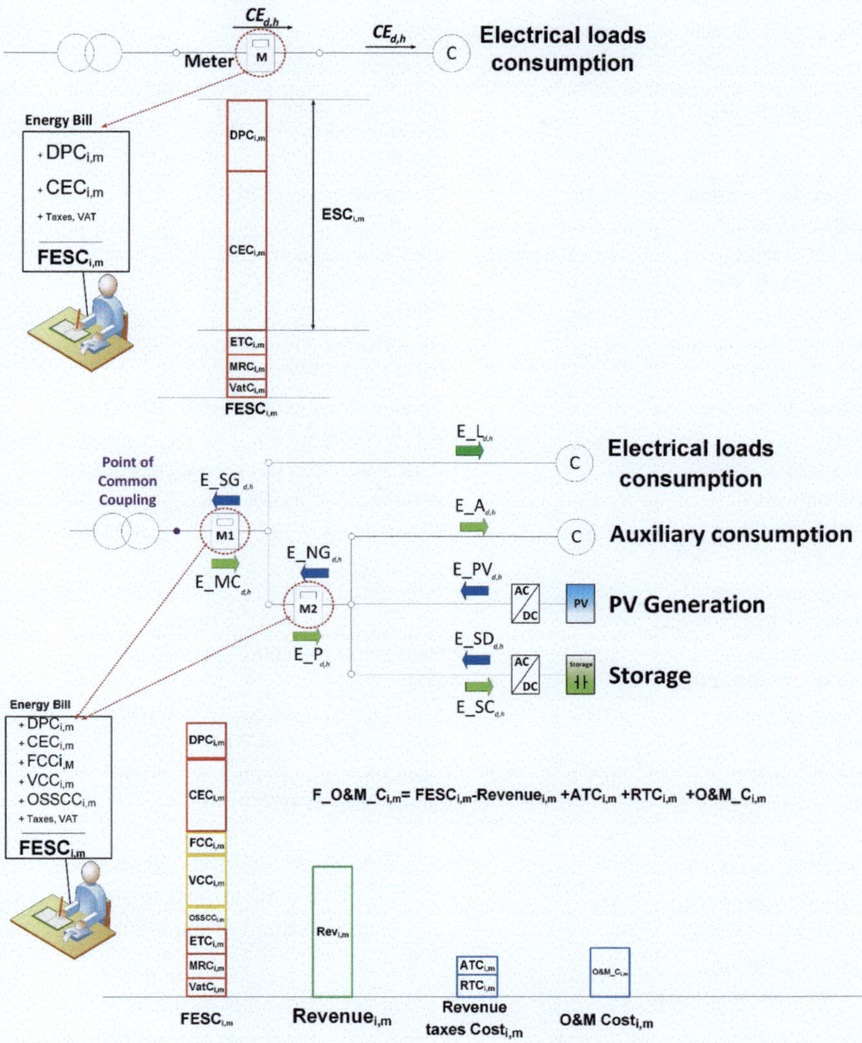

Fig. 5 Representation of the $FESC_{i,m}$ components for the conventional consumer (upper side) and for the prosumer equipped with the considered MG (lower side)

5 Case Study

5.1 Definition

The daily final O&M cost $F\_O\&M\_C_d$ determined in (68) for the MG shown in Fig. 4 has been selected as the objective function to minimize, parameterizing the model in (13)–(67) with the values shown in Table 2 and Fig. 6.

Table 2 MG physical and regulatory parameters

Physical parameters		Regulatory parameters	
PV peak power (P_PV_p) [kW]	2.5	Daily regulated electricity cost power term (PT_REC_d) [€/kW]	0.10
PV auxiliary consumption ($E\_A_{h,d}$) [kWh]	0.01	Contracted power (CP) [kW]	3.5
Daily PV O&M fixed cost (FC_PV_d) [€/kW]	0.09890410959	Regulated electricity cost energy term ($ET\_REC_{h,d}$) [€/kWh]	0.044027
PV O&M variable cost (VC_PV) [€/kWh]	0	Fixed charges power term (PT_FC_d) [€/kW·day]	0.02231268219
Storage charge efficiency (η_C)	1	Variable charges energy term (ET_VC) [€/kWh]	0.043187
Storage discharge efficiency (η_D)	1	Other system services energy term (ET_OSSC) [€/kWh]	0.005344
Storage maximum energy capacity (E_S_{max}) [kWh]	3.5	Electricity tax (ET)	0.0511269632
Storage maximum state-of-charge (SoC_{max})	1	Value added tax (VAT)	0.21
Storage threshold state-of-charge (SoC_{th})	0.96	Meter rental cost (MRC_d) [€/day]	0.074
Storage minimum state-of-charge (SoC_{min})	0.2	Access tariff on surplus energy (ET_AT) [€/kWh]	0.0005
Storage initial state-of-charge (SoC_0)	0.5	Tax on the revenues for the surplus energy (RT)	0.07
Storage maximum charge rate (P_SC_{max}) [kW]	1		
Storage threshold charge rate (P_SC_{th}) [kW]	0.1		
Storage maximum discharge rate (P_SD_{max}) [kW]	1		
Storage threshold discharge rate (P_SD_{th}) [kW]	0.1		
Daily Storage O&M fixed cost (FC_S_d) [€/kW·day]	0.01671232877		
Storage O&M variable cost (VC_S) [€/kWh]	0.00049		

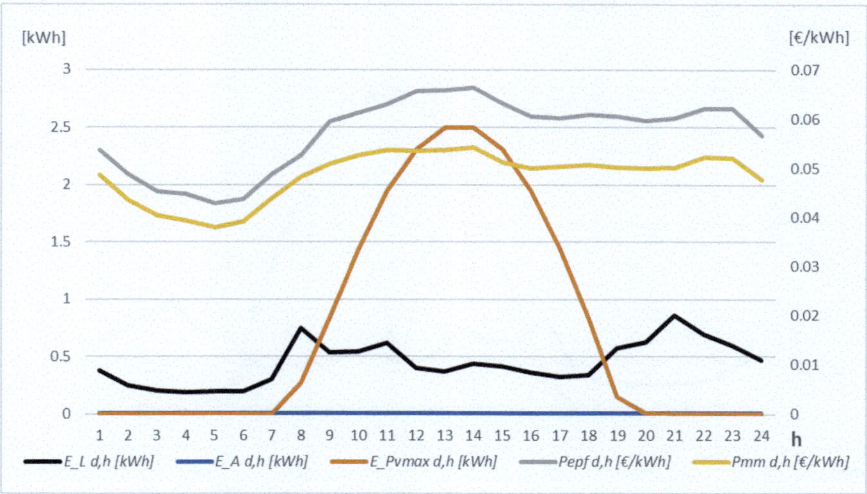

Fig. 6 Hourly values of the electrical loads $E\_L_{d,h}$ (black) and PV auxiliary $E\_A_{d,h}$ (dark blue) energy consumptions, maximum PV energy generation $E\_PVmax_{d,h}$ (orange), electricity purchase final price $P_{epf\,d,h}$ (grey) and mean electricity market price $P_{mmf\,d,h}$ (yellow)

Four different scenarios have been considered, namely, no revenue for the surplus energy injected to the main grid without storage (case 1) and with storage (case 2) and the net-billing arrangement without storage (case 3) and with storage (case 4).

Additionally, in order to highlight the key importance of embedding the electricity tariff structure and all the regulatory constraints for self-consumption within the MG model, a different objective function $Simplified\_F\_O\&M\_C_d$, frequently found in the academic literature has also been considered, namely:

$$Simplified\_F\_O\&M\_C_d = PEC_d - Revenue_d + O\&M\_C_d \tag{69}$$

Comparing the common objective function in (69) to that in (68), it can be noticed that in (69) it is only considered the purchase energy cost PEC_d, disregarding the demanded power cost DPC_d in (46), the regulated energy cost REC_E_d in (47), all the total charges cost TCC_d in (55), the electricity tax cost ETC_d in (57) and the VAT cost $VATC_d$ in (58), as well as the access tariff cost ATC_d in (61) and the revenue tax cost RTC_d in (62) for the surplus generation injected into the main grid.

5.2 Results and Discussion

Figure 7 shows the actual PV energy generation $E\_PV_{d,h}$ for each of the considered cases, along with the electrical loads $E\_L_{d,h}$ and the PV auxiliary $E\_A_{d,h}$ energy consumptions and the maximum PV energy generation $E\_PVmax_{d,h}$. As can be seen,

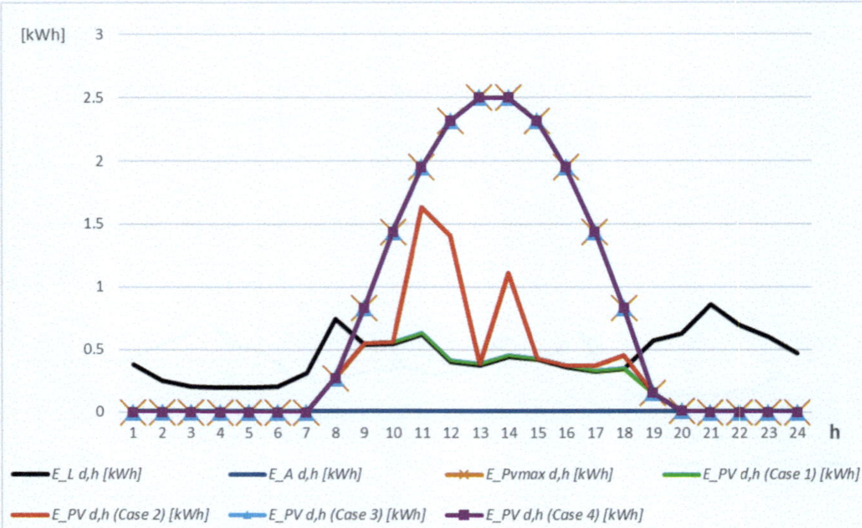

Fig. 7 Hourly values of the actual PV energy generation $E\_PV_{d,h}$ for case 1 (green), case 2 (red), case 3 (light blue with triangular marker) and case 4 (violet with square marker), along with the electrical loads $E\_L_{d,h}$ (black) and PV auxliliary equipment $E\_A_{d,h}$ (dark blue) energy consumption and maximum PV energy generation $E\_PVmax_{d,h}$ (orange with cross marker)

the actual PV energy generation $E\_PV_{d,h}$ in case 1 matches the sum of the electrical loads $E\_L_{d,h}$ and the PV auxiliary $E\_A_{d,h}$ energy consumptions whenever possible, since no revenue is perceived for the surplus injection into the main grid. However, in case 2 $E\_PV_{d,h}$ exceeds the sum of $E\_L_{d,h}$ plus $E\_A_{d,h}$ in certain hours for battery charging. In cases 3 and 4 under the net-billing arrangement, $E\_PV_{d,h}$ matches $E\_PVmax_{d,h}$, allocating the PV generation exceeding the sum of $E\_L_{d,h}$ and $E\_A_{d,h}$ to surplus injection in case 3 and to both surplus injection and battery charging in case 4.

Figure 8 shows the several daily costs defined in (46)–(68) for each of the four considered cases, using green coloured bars. For comparison purposes, those costs also affecting conventional consumers have also been represented by black coloured bars. When the incomplete objective function in (69) is employed for the optimum energy management of the MG, the resulting costs are shown with red bars.

Of particular interest are the several last costs shown towards the right-hand side of each subfigure within Fig. 8. On the one side, there have been represented the actual $F\_O\&M\_C_d$ that the electricity retailer will bill to the prosumer and the *Cost savings* regarding the conventional consumer, either calculated with the optimal energy flows stemming from (68) (green bars) or with the sub-optimal energy flows resulting from (69) (red bars). On the other side, there have also been depicted the *Simplified_F_O&M_*C_d and the *Expected cost savings* obtained using (69) (red bars).

First of all, looking at Fig. 8 it can be observed that, as expected, the final energy supply cost of the conventional consumer ($FESC_d$, black bars) is greater than the

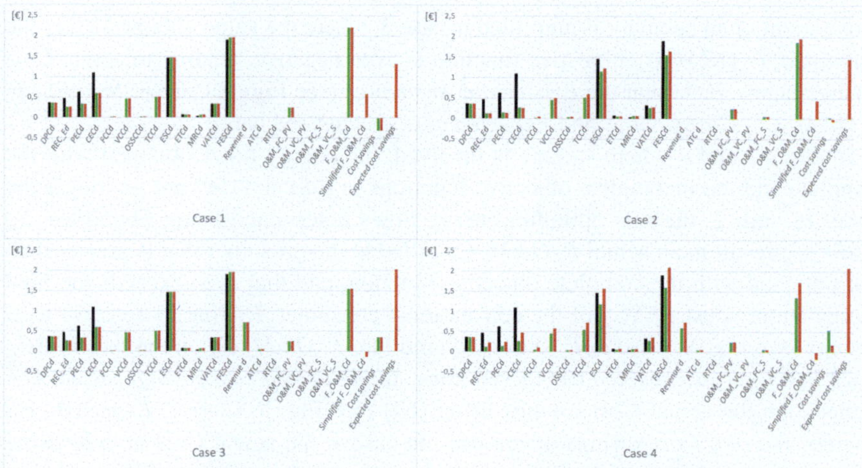

Fig. 8 Breakdown of daily costs for the conventional consumer (black bars) and the prosumer with either the complete (green bars) or the incomplete (red bars) formulation of the daily final O&M cost $F\_O\&M\_C_d$

final cost corresponding to the prosumer ($F\_O\&M\_C_d$, green bars) in all cases except in case 1. For the given size of the PV equipment and the electric load consumption level, the variable charges cost VCC_d and the fixed O&M PV cost $O\&M\_FC\_PV_d$ paid by the prosumer in case 1, as well as the absence of $Revenue_d$ result in negative *Cost savings* (see Fig. 8, Case 1, *Cost savings,* green bar). In case 2 it is obtained a positive value for *Cost savings* with the assistance of the storage, even though that the associated fixed $O\&M\_FC\_S_d$ and variable $O\&M\_VC\_S_d$ costs arise. The values of *Cost savings* progressively increase in cases 3 and 4, respectively, receiving a $Revenue_d$ under the net-billing arrangement.

It can also be observed that in the cases 1 and 3 without storage, the actual $F\_O\&M\_C_d$ and the *Cost savings* for the prosumer are the same regardless the use of the either the complete or the incomplete objective functions (see Fig. 8, Cases 1 and 3, $F\_O\&M\_C_d$, *Cost savings*, green and red bars). In these cases, the management strategy for the energy consumption from the main grid E_MC_d and the self-consumed energy $E\_SelfC_{d,h}$ is identical, i.e., to self-consume whenever possible and to inject the surplus into the main grid when a revenue is received. Consequently, the value that the electricity retailer will calculate and bill for all the disregarded terms in (69) will be at the end the same, since they depend on identical energy flows. Yet, although the actual final cost $F\_O\&M\_C_d$ that the prosumer is billed is the same, quite another thing are the expectations created at the elaboration of the feasibility study if the $Simplified\_F\_O\&M\_C_d$ objective function was employed. In case 1, the value determined using the $Simplified\_F\_O\&M\_C_d$ objective function is around 25% of the actual $F\_O\&M\_C_d$ value, resulting in a prediction of positive *Expected cost savings* while negative *Cost savings* are obtained (see Fig. 8, Case 1, $Simplified\_F\_O\&M\_C_d$, *Cost savings* and *Expected cost savings*, red bars). The above-mentioned behavior is amplified in case 3, where the value obtained using the *Simplified_F_O&M_*C_d objective function is even negative, leading to actual *Cost savings* below 20% than the *Expected cost savings* (see Fig. 8, Case 3, *Simplified_F_O&M_*C_d, *Cost savings* and *Expected cost savings*, red bars).

In cases 2 and 4 where storage is involved, the energy flows obtained with the complete and the incomplete objective functions in (68) and (69) are no longer the same. In case 2, the non-optimum energy flows determined using *Simplified_F_O&M_*C_d result in an actual *F_O&M_*C_d billed to the prosumer that is greater than that obtained with the complete objective function, and that even exceeds the final energy supply cost $FESC_d$ of the conventional consumer, leading to negative *Cost savings* (see Fig. 8, Case 2, *F_O&M_*C_d, red bar, *F_O&M_*C_d, green bar, $FESC_d$ black bar, *Cost savings*, red bar). Even more, the *Expected cost savings* are 5000% higher than the actual *Cost savings* using (69) (see Fig. 8, Case 2, *Expected cost savings, red bar, Cost savings*, green bar). In case 4, the actual *Cost savings* using (69) stand around 30% of those that could have been obtained using (68), and only 8% of the *Expected cost savings* (see Fig. 8, Case 4, *Cost savings*, red bar, *Cost savings*, green bar, *Expected cost savings, red bar*).

6 Conclusion

This chapter discussed the key importance of accurately embedding the electricity tariff structure and the regulatory framework for self-consumption in the MG models. To this end, the typical electricity supply contracts have been explained as well as their main cost components. Likewise, the most frequently found characteristics in the regulatory frameworks for self-consumption have been identified. Then, a simple MG electrical scheme has been selected and a conceptual regulatory framework for self-consumption has been embedded in its physical model. The MG energy management has been optimized selecting as objective function the daily final O&M cost, and the obtained results have been compared to those stemming from the use of an objective function that disregards regulatory constraints, which is a usual approach in the literature. The performed analysis demonstrates that neglecting the regulatory constraints lead to significant inaccurate results, that may lead to misleading investment decisions.

References

1. Gholami M, Muyeen SM, Mousavi SA (2024) Optimal sizing of battery energy storage systems and reliability analysis under diverse regulatory frameworks in microgrids. Energy Strateg Rev 51:101255. https://www.sciencedirect.com/science/article/pii/S2211467X23002055
2. de la Hoz J, Martín H, Alonso A, Luna AC, Matas J, Vasquez JC, Guerrero JM (2019) Regulatory-framework-embedded energy management system for microgrids: the case study

of the Spanish self-consumption scheme. Appl Energy 251:11337. https://www.sciencedirect.com/science/article/abs/pii/S0306261919310487
3. King DE (2008) Theregulatory environment for interconnected electric power micro-grids: insights from state regulatory officials. Carnegie Mellon Electricity Industry Center Working Paper CEIC-05-08. https://www.cmu.edu/ceic/assets/docs/publications/working-papers/ceic-05-08.pdf
4. Zinaman O, Eto J, Marshall-Garcia B, Joo JY, Jeffers R, Schneider K (2021) White paper: enabling regulatory and business models for broad microgrid deployment. National Renewable Energy Laboratory, Golden, CO. NREL/TP5R00–84818. https://www.energy.gov/sites/default/files/2022-09/7-Enabling%20Regulatory%20and%20Business%20Models%20for%20Broad%20Microgrid%20Deployment.pdf
5. Shea D (2022) Microgrids: state policies to bolster energy resilience. National Council of State Legislatures. https://www.ncsl.org/energy/microgrids-state-policies-to-bolster-energy-resilience
6. Behrendt J (2023) Microgrids and EU law: three microgrid models to solve one regulatory puzzle. Energy Policy 177:113483. https://www.sciencedirect.com/science/article/pii/S030142152300068X
7. Wouters C (2015) Towards a regulatory framework for microgrids—the Singapore experience. Sustain Cities and Soc 15:22–32. https://www.sciencedirect.com/science/article/pii/S2210670714001152
8. Zachmann G, Hirth L, Heussaff C, Schlecht I, Mühlenpfordt J, Eicke A (2023) The design of the European electricity market. Current proposals and ways ahead. Policy Department for economic, scientific and quality of life policies, directorate-general for internal policies, European parliament. https://www.europarl.europa.eu/RegData/etudes/STUD/2023/740094/IPOL_STU(2023)740094_EN.pdf
9. de Meulemeester B (2017) White paper: energy contract types. E&C Consultants. Belgium. https://www.eecc.eu/energy-contract-types
10. Directive (EU) 2023/2413 of the European parliament and of the council of 18 October 2023. European Parliament, Council of the European Union. Official Journal of the European Union, 31 October 2023, L series, 2023/2413. http://data.europa.eu/eli/dir/2023/2413/oj
11. Hemetsberger W, Schmela M, Cruz-Capellan T, SolarPower Europe (2023) Global market outlook for solar power 2023–2027. https://www.solarpowereurope.org/insights/outlooks/global-market-outlook-for-solar-power-2023-2027/detail
12. Masson G, Briano JI, Baez MJ (2016) Review and analysis of pv self-consumption policies. Report IEA-PVPS T1-28:2016, photovoltaic power systems programme, IEA International Energy Agency. https://iea-pvps.org/wp-content/uploads/2020/01/IEA-PVPS_-_Self-Consumption_Policies_-_2016_-_2.pdf
13. CEER Report (2019) Regulatory aspects of self-consumption and energy communities. C18-CRM9_DS7-05-03. https://www.ceer.eu/documents/104400/6509669/C18-CRM9_DS7-05-03_Report+on+Regulatory+Aspects+of+Self-Consumption+and+%20Energy+Communities_final/8ee38e61-a802-bd6f-db27-4fb61aa6eb6a?version=1.1

Virtual Power Plant Participation in Australian Wholesale Electricity Markets

Julius Susanto

Abstract Over the last decade, Australia has witnessed unprecedented growth in its distributed energy resource (DER) installed capacity base, growing from around 3 GW in 2013 to nearly 19 GW in 2023. Alongside this massive expansion in DER capacity has come the opportunity for DER to be aggregated to form virtual power plants (VPP) that can be coordinated and controlled by a central operator. By harnessing DER at scale, VPPs can interact with wholesale electricity markets either (i) directly through energy and ancillary services trading or (ii) indirectly by flexibly adjusting imports and exports in response to market prices. This chapter provides a review of how VPPs currently participate in wholesale electricity markets in Australia, and how they could participate in the future.

Keywords Virtual power plants · Electricity markets · Australia

1 Overview

Virtual power plants (VPPs) are defined as a portfolio of distributed energy resources (DER), including responsive loads, that are aggregated into a single controllable entity, akin to a transmission-connected generator [1]. In this article, VPPs refer to centrally controlled aggregations of residential, commercial and industrial behind-the-meter DER such as:

- Local generation (e.g. rooftop solar)
- Local energy storage (e.g. battery systems, etc.)
- Flexible demand/controllable loads (e.g. hot water systems, electric vehicles, air conditioning, smart appliances, etc.)

J. Susanto (✉)
School of Engineering and Energy, Murdoch University, Perth, Australia
e-mail: susanto@ieee.org

F. Shahnia and J. M. Guerrero (eds.), *Microgrids and Virtual Power Plants*, Power Systems, https://doi.org/10.1007/978-981-97-6623-9_13

Individual DER devices are pooled together and aggregated into a VPP using communications technology to monitor and control individual DER from a central VPP platform. In Australia, the communication and control between DER device and VPP platform is generally achieved through either:

(1) The DER device's inherent communications capabilities (typically communicating to the VPP platform via the public internet using the customer's Wi-Fi network), or
(2) A third-party device ("gateway") that interfaces between the DER device and the VPP platform. Gateway devices can communicate to the VPP platform using the customer's Wi-Fi or using an alternative communications link such as mobile radio (4G/5G). The key advantage of using a gateway device is that multiple DER devices from different manufacturers can be monitored and controlled from a single gateway.

The VPP can then be used to facilitate DER participation in wholesale electricity markets (both energy and ancillary services markets), provided there is an avenue for VPPs to participate (Fig. 1).

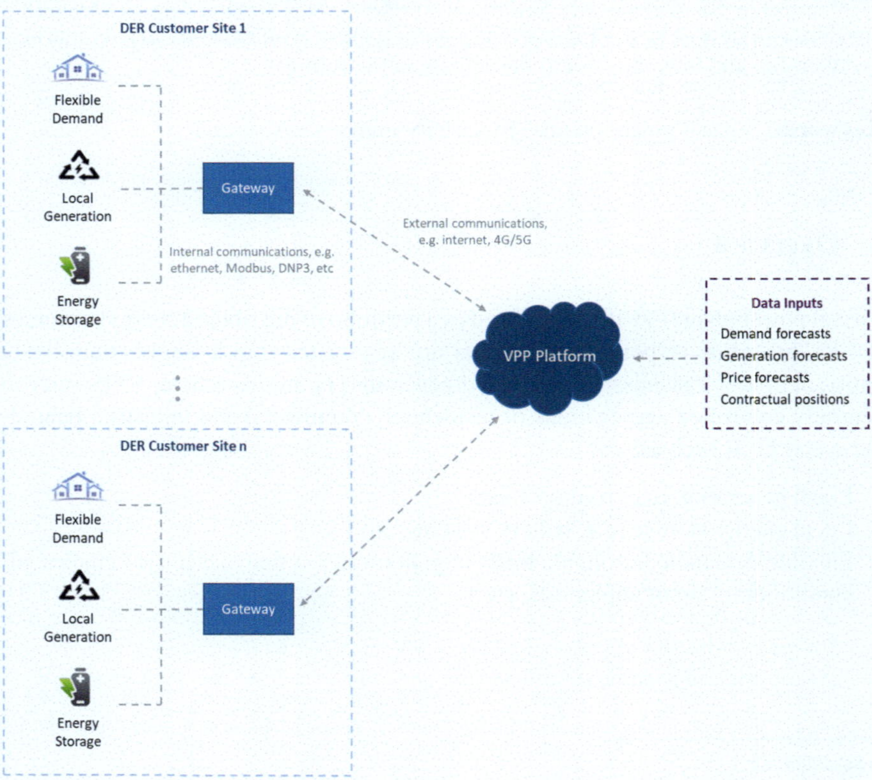

Fig. 1 High-level structure of a VPP

Over the last decade, Australia has witnessed unprecedented growth in its DER installed capacity base, growing from around 3 GW in 2013 to nearly 19 GW in 2023. Following closely on the heels of this massive expansion in DER capacity has been the rise of VPPs, now with more than 1 GW of DER capacity enrolled in a VPP program.

This chapter explores how VPPs in Australia participate in wholesale electricity markets and is organised as follows: Sect. 2 provides an overview of existing VPPs in Australia and the wholesale markets that they inhabit. Section 3 presents the different models currently available for VPPs to participate in wholesale markets. Section 4 discusses a VPP's interactions with other DER control mechanisms that it may be exposed to. Section 5 presents specific VPP case studies, providing details on the mechanisms used for market participation. Section 6 discusses the future of wholesale market participation of VPPs. Finally, Sect. 7 offers conclusions and a summary of the chapter.

2 Overview of VPPs in Australia

VPPs are active in Australia's two major wholesale electricity markets: (i) the National Electricity Market (NEM) that serves approximately 10.7 m customers in the eastern and southern states of New South Wales, Queensland, South Australia, Tasmania and Victoria, as well as the Australian Capital Territory, and (ii) the Wholesale Electricity Market (WEM), that serves over 1m customers in the southwest part of Western Australia.

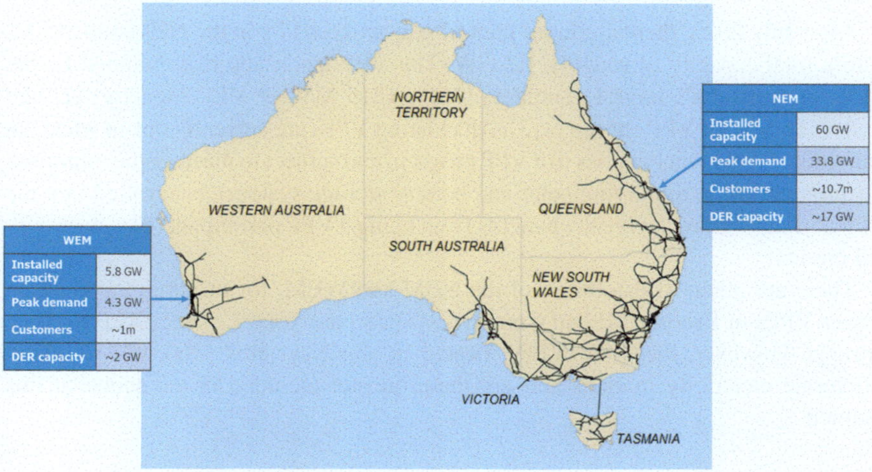

The wholesale market arrangements and regulatory structures of the NEM and WEM are quite different, with differing features that have implications for VPP uptake and market participation. The table below provides some of the key differences that are pertinent to VPPs:

Feature	NEM	WEM	Implications for VPPs
Market design	Energy only market	Energy + capacity markets	The market price cap in the NEM energy only market is significantly higher than in the WEM, which provides sharper real-time pricing signals for VPPs trading on the real-time energy market. However, VPPs could potentially unlock another value stream in the form of Reserve Capacity credits
Retail competition	Full retail competition	Retail competition only allowed for customers that consume >50 MWh per year	The vast majority of residential customers in the WEM can only be served by the state-owned retailer (Synergy), which can limit the scope of what a third-party residential VPP can achieve in the WEM
Forward contracting	ASX energy futures market and bilateral contracting	Bilateral contracting (~85% of energy traded) and day ahead market (~5% of energy traded)	Forward contracting and hedging mechanisms are more transparent in the NEM, which can provide VPPs with greater flexibility in their commercial design and offerings
DER incentives	Multiple state-based schemes offering different feed-in tariffs and incentives	Single DER feed-in tariff (Distributed Energy Buyback Scheme) and no other state-based incentives for solar PV or BESS	There are a greater variety of incentive schemes in the NEM, which can lead to more flexibility and regional diversity in VPP offerings

As at July 2023, there are more than fifteen active VPPs in the NEM and WEM, with a total capacity of roughly 1.2 GW. The VPP landscape is dominated by the VPPs of Australia's largest gentailers Origin (815 MW of VPP capacity) [2] and AGL (316 MW of VPP capacity) [3]. All existing VPPs are currently opt-in where an individual DER owner applies to a VPP program and subject to meeting the minimum requirements (e.g. qualifying solar and battery storage system), is enrolled into the VPP and can start receiving the benefits from being a VPP participant (e.g. electricity bill credits).

There are other VPPs outside of the NEM and WEM, for example the Horizon Power VPPs in Broome [4] and Carnarvon [5] and the Solar Connect VPP in Alice Springs. However, these are VPPs located in small systems (<60 MW) without wholesale electricity markets and are therefore not included in the scope of this chapter.

VPP name	Registered participant	Market	Market registration	Estimated size (MW)
AGL "bring your own solar battery"	AGL Sales Pty Ltd	NEM	Market customer	316
Amber SmartShift	Amber Electric Pty Ltd	NEM	Market customer	No publicly available data
Arcstream/Nectr	Hanwha Energy Retail Australia Pty Ltd	NEM	Market customer	No publicly available data
Ausgrid VPP trial	Ausgrid	NEM	DNSP	3.4 [finished]
Discover energy	Discover Energy Pty Ltd	NEM	Market customer Market small generator aggregation	1 [based on FCAS market registration]
EnergyAustralia	EnergyAustralia Pty Ltd	NEM	Market customer	No publicly available data (but no longer accepting customers to the VPP program)
Members energy	Energy Locals Pty Ltd	NEM	Market customer	No longer exists (from 4 October 2023)
Origin loop	Origin Energy Electricity Ltd	NEM	Market customer	815
Powershop "Charge Force"	Powershop Australia Pty Ltd	NEM	Market customer	No publicly available data
Reposit	Amalgamated energy services	NEM	Demand response—ancillary services Also participates via diamond energy and powershop (market customers)	10
Shine Hub	Shine Hub Pty Ltd	NEM	Demand response—ancillary services	5 [based on FCAS market registration]
Simply energy	Simply energy	NEM	Market customer	4 [based on FCAS market registration]
Sonnen	Sonnen Australia Pty Ltd	NEM	Demand response—ancillary services	3 [based on FCAS market registration]
Tesla	N/A	NEM	Participates via Energy locals (market customer)	~35 (SA VPP, 7000 customers)
SolarHub	ActewAGL Retail	NEM	Market customer	6
Enel X	Enel X	NEM	Demand response	350

(continued)

(continued)

VPP name	Registered participant	Market	Market registration	Estimated size (MW)
Rheem smart network	Rheem Australia	NEM	Demand response	~5 (2,400 hot water systems, 200 air conditioners and 200 pool pumps)
Project symphony	Synergy	WEM	Market customer	No publicly available data
Plico	N/A	WEM	Supplementary reserve capacity	3
Enel X	N/A	WEM	NCESS	120

3 Wholesale Market Participation Models

There are currently a number of models for VPPs to participate in Australian wholesale electricity markets:

(1) Intermediary market customer
(2) Ancillary services provider
(3) Demand side response provider
(4) Registered aggregator
(5) Direct VPP market participation

These participation models are described in more detail in the following subsections.

3.1 Intermediary Market Customer

In this model, a VPP participates in the wholesale energy market indirectly via an intermediary Market Customer (a buyer of energy from the wholesale market). The VPP is not directly involved in the scheduling process (i.e. it cannot be a price setter), but is rather a price taker in the market and can only indirectly affect spot price outcomes by its impact on system demand.

Within a certain range, the VPP is flexible enough to control its aggregate net demand from the wholesale market, which means that the VPP can respond to energy spot prices in real-time. However, the VPP's degree of price sensitivity depends on how much exposure it has to the energy spot price:

- **Spot price pass through**: the VPP is fully exposed to the energy spot price.
- **Hedged exposure**: the VPP has some exposure to the energy spot price, but is also protected by hedging instruments such as cap contracts.

- **Fully integrated with Market Customer**: the VPP is fully integrated with its intermediary Market Customer (e.g. the VPP is owned by the Market Customer) and is exposed to the same spot price risks as the Market Customer. Examples in the NEM are the VPPs owned by the large gentailers AGL and Origin.

One of the drawbacks of this model is that VPP participation in the wholesale market is not transparent, and thus the volume of price-responsive behaviour from VPPs in aggregate cannot be quantified. From the system operator perspective, this makes forecasting system demand more challenging as VPP behaviour is neither visible nor completely predictable in real-time [6].

Another key implication of this participation model is that there is no fundamental geographical or size limit for a single VPP. If desired, a single VPP could cover an entire region or multiple regions. There are potentially material ramifications for both transmission and distribution networks as at scale, a sufficiently large VPP could arbitrarily alter transmission and distribution level power flows and cause (or relieve) network congestion. The behaviour of such a VPP may not be entirely predictable either since it may be operating in real-time relative to its intermediary Market Customer's net trading position (which is not visible to the market). However, the use of dynamic operating envelopes (DOEs) by distribution network service providers (DNSPs), which imposes dynamic limits on the output of DER to manage distribution network congestion and voltages in real-time, would dampen the impact that a single VPP can have on network power flows.

This participation model is most viable in the NEM, where there is full retail competition and allows for Market Customers to allow all classes of customers to participate in a VPP. This is more difficult in the WEM, where retail competition is limited to larger customers that consume more than 50 MWh per year (e.g. typically only commercial and industrial users). Small use customers (<50 MWh) can only be served by the state-owned electricity retailer (Synergy), which means that the majority of residential customers cannot participate in the wholesale market via a VPP unless enrolled in the Synergy VPP pilot (Project Symphony).

Moreover, the high energy market price cap in the NEM (set to AU$16,600/MWh on 1 July 2023) arguably sharpens the incentives for VPPs to trade on the energy spot market by allowing them to take advantage of their flexibility during periods of congestion and/or supply-side scarcity. On the other hand, the energy market price cap in the WEM considerably lower (currently set at less than AU$1,000/MWh) owing to the existence of a reserve capacity market, which may dampen incentives for VPPs to participate in the energy market (relative to the NEM). On the other hand, aggregators can potentially be eligible for reserve capacity credits (via demand response), which unlocks an additional value stream that is not available in the NEM.

3.2 Ancillary Services Provider

In the NEM, VPPs can participate directly in the wholesale ancillary services markets such as the Frequency Control Ancillary Services (FCAS) markets. VPPs can be accredited for ancillary services provision via an intermediary market participant (e.g. a Market Customer) or can directly register under either the "Demand Response Service Provider—Ancillary Service Load" or "Market Small Generator Aggregator" registration classes. There is currently no registration class specifically for VPPs to provide ancillary services.

VPPs can offer into the NEM FCAS markets in the same way as other FCAS providers, i.e. offering up to 10 price-quantity pairs for each 5-min trading interval. The performance specifications for service delivery are described in the AEMO Market Ancillary Service Specification (MASS), which has specific provisions for aggregated DER in terms of measurement, control and verification [7].

As at November 2023, there were eight market participants that have VPPs accredited for FCAS provision with a total capacity of 58 MW [8]. The majority of these participants had registered as part of the AEMO NEM Virtual Power Plant Demonstrations project that took place between 2019 and 2021, where one of the project objectives was to trial direct participation of VPPs in the contingency FCAS markets [9]. This trial proved successful and the AEMO MASS was revised in December 2021 to allow other VPPs (beyond the trial participants) to be accredited for FCAS provision [10].

Participant	DUID	Region	L1	L6s	L60	L5	R1	R6	R60	R5
AGL	VSSAE1V1	SA			6	6			6	6
Amalgamated energy services [reposit]	ASNAES1	NSW	1	9	10	8	1	9	9	10
Discover energy	ASSDE1	SA			1	1			1	1
Energy locals	ASSEL1	SA		12	12	12		12	12	12
Energy locals	ASNSEL2	NSW		1	1	1		1	1	1
Energy locals	ASVEIL2	VIC		1	1	1		1	1	1
Energy locals	VSSEL1V1	SA		12	12	12		12	12	12
Energy locals	VSNEL2S1	NSW		1	1	1		1	1	1
Energy locals	VSVEL2S1	VIC		1	1	1		1	1	1
Hydro Tas	VSQHT1V1	QLD			1	1			1	1
Shine Hub	DRSHVS01	SA			3	4			3	2
Shine Hub	DRSHVN01	NSW			1	2			1	1
Shine Hub	VSSSH1S1	SA		1	1	1		1	1	1
Simply Energy	VSSSE1V1	SA		4	4	4		4	4	4
Sonnen	VSNSN1V1	NSW			1	1			1	1
Sonnen	ASSSN1	SA			2	2			2	2

In the WEM, there is currently no means for a VPP to register and be accredited for ancillary services provision, though contingency frequency response provision was trialled as part of the Project Symphony pilot project.

3.3 Demand Side Response Provider

Demand side participation in the energy market is a straightforward use case for VPPs as within a certain operating range, a VPP can decrease (or increase) its aggregate net consumption from the grid in response to market price signals or a system operator command. In Australia, there are currently two approaches for eliciting demand side response from VPPs: (i) via spot market demand response mechanisms, and (ii) through contracts.

Spot market demand response: in the NEM, a rule change from June 2020 established a new mechanism for direct participation of demand response in the wholesale energy market called the Wholesale Demand Response Mechanism (WDRM) [12]. With this new mechanism, responsive loads can bid a price-quantity pair into the energy spot market to reduce their demand, i.e. offering in "negawatts". Demand response facilities can also participate in the NEM ancillary service markets. In principle, WDRM is generic and allows for VPPs to participate, e.g. either by actually reducing aggregate consumption or offsetting the demand with behind-the-meter energy storage.

Contracted demand response: in the WEM, there is no equivalent mechanism like WDRM to participate directly in the energy spot market. However, there have been Non-Cooptimised Essential System Services (NCESS) contracts awarded to the Enel X VPP to provide demand flexibility services under peak and minimum demand scenarios [13]. In December 2022, AEMO also contracted with the Plico VPP to provide supplementary reserve capacity for the 2022/23 Summer period.

3.4 Registered Aggregator

In the NEM, VPPs can also elect to participate directly in the wholesale market by registering as a Market Small Generator Aggregator (MSGA). However, the current rules require that each customer site in the VPP have a separate connection point and NEM compliant metering [11]. The NEM metering requirements do not consider the typical device-level metering in DER inverters as compliant, so customer sites would be required to install separate metering. Given the other VPP participation options available, this MSGAs are generally not seen as an attractive option.

The WEM has a similar facility registration class as the MSGA called *Small Aggregations*, which is defined as aggregations "comprising distribution connected technologies at a single electrical location" [14]. The reference to a single electrical location (e.g. substation) may limit the geographic extents that a single VPP can encompass, i.e. coverage is limited to distribution feeders electrically connected to the nominated connection point. This registration class has similar issues as the MSGA (e.g. compliant metering requirements) and is not currently used by any VPPs in the WEM.

3.5 Direct VPP Market Participation

Given the lack of transparency and visibility of the intermediate Market Customer model and some of the complexities of VPP participation in the other models, there is a desire (particularly from the system operator) to facilitate direct market participation for VPPs. Direct market participation means that VPPs are included in the central dispatch process and are treated like other scheduled generation and storage facilities in the market. In this model, VPPs are required to submit bids and offers into the market (which can apply to both energy and ancillary services markets) and are involved in the price formation process, i.e. VPPs can be market price setters. In the US, this model is equivalent to the one described in FERC Order 2222 [15].

There are currently no mechanisms for direct VPP market participation in the NEM or WEM, but there are ongoing trials and pilot projects underway to validate and test the concept and systems required to achieve this.

There are several trial projects in the NEM, for example Project EDGE and Project Converge. Each project is trialing different models for VPP participation in the wholesale market. For example, Project EDGE is trialing direct bidding and scheduling of VPPs in the wholesale energy market, while Project Converge is routing bids via a Distribution System Operator (DSO). There is currently only one VPP pilot project in the WEM ("Project Symphony"), which is a state government-led initiative that is testing direct bidding and scheduling in the energy market, as well as frequency control markets.

Note that none of these trial and pilot projects are actually integrated into their respective wholesale markets and are essentially off-market simulations, i.e. they may integrate into a production instance of the market dispatch engine, but do not affect actual market prices or dispatch.

AEMO is also exploring a voluntary "Scheduled Lite" mechanism for the NEM, a component of which aims to lower the barriers for VPPs to directly participate in the market. In the high-level design, a VPP can participate as a single entity provided that it comprises DER that are connected within a single geographic region ("zone") and comply with the technical requirements (e.g. minimum aggregated portfolio threshold, DER metering requirements, etc.) [16].

4 Interactions with Other DER Control Mechanisms

There are several other control mechanisms that apply to aggregated DER, which are currently either in place or being trialled in the NEM and WEM. These mechanisms operate in parallel with the wholesale market, for example to manage distribution network congestion (such as dynamic operating envelopes and dynamic network pricing) or manage system security (such as the DER backstop mechanism).

VPPs participating in the wholesale market must consider the impact of these mechanisms on their ability of a VPP to interact with the wholesale market. For example, dynamic operating envelopes effectively function as local constraints on how much DER connected to a specific feeder can import or export.

A selection of the current control mechanisms is described in more detail in the following subsections.

4.1 Dynamic Operating Envelopes (DOEs)

Dynamic Operating Envelopes (DOE) are dynamically calculated time-varying technical limits for DER exports and imports to and from the grid that reflect real-time local network conditions (e.g. voltages, loading, power quality, etc.) [17]. DOEs were first introduced as part of a 2019 VPP trial in South Australia, with the ultimate goal of unlocking unused distribution network capacity in operational timeframes [18] (Fig. 2).

DOEs are calculated by the Distribution System Operator (DSO) or Distribution Network Service Provider (DNSP) and then sent to individual DER customer devices (e.g. PV inverters or home energy management systems). DOEs are typically calculated at distribution feeder level, which then apply to all DER on the feeder in a uniform manner.

For a VPP, a DOE simply represents another constraint on the aggregate available capacity of the VPP, which must be managed alongside other constraints such as

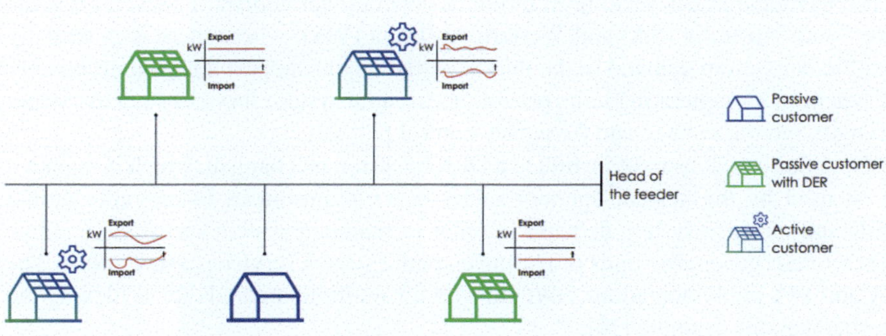

Fig. 2 Illustration of static and dynamic operating envelopes [19]

solar resource availability, behind-the-meter load, battery storage state of charge, etc. In effect, DOEs constrain the maximum volume (in both directions) that a VPP can feasibly trade in the wholesale market.

4.2 Dynamic Network Pricing

Dynamic network pricing is a market-based alternative to DOEs that was first explored through Ausgrid's Project Edith trial [20]. Rather than imposing technical limits on DER import and exports, dynamic network pricing provides time-varying price signals reflecting the cost of using the distribution network at any given time (e.g. based on the level of congestion at the time).

Like DOEs, dynamic network pricing is intended to maximise utilisation of existing distribution network capacity, but through decentralised decision-making intermediated by network price incentives instead of technical limits. For example, when there are high levels of distribution network congestion (such as during a Summer evening peak), the price for imports (load) will be high, while there will be a corresponding negative price or reward for exports (generation).

For a VPP, dynamic network pricing is akin to a market spot price input into the VPP trading engine. The VPP trading engine is designed to optimise for maximum profit (or least cost), and the dynamic network pricing inputs are additional terms in the optimisation algorithm. There are obvious implications to solve times depending on the complexity of the VPP trading engine and the number dynamic network prices being fed into the algorithm. A further complication is the forecasting of dynamic network prices, which may be challenging if prices are volatile.

4.3 DER Backstop Mechanism

The considerable take up of rooftop solar in Australia since 2010 has begun to cause technical issues relating to minimum operational demand in several regions like South Australia (SA) and Western Australia (WA), where there is potentially insufficient system demand in the middle of the day to support a minimum number of synchronous generating units providing essential system services such as system strength, inertia, voltage and frequency control [21, 22].

Both these regions have instituted a DER backstop mechanism, which is intended to be used by the system operator under rare circumstances to remotely curtail DER and increase system demand in order to maintain system security, i.e. allow synchronous generating units providing essential system services to stay online. The SA and WA implementations both allowed for multiple methods for achieving the mechanism, for example [26]:

- Sending a Demand Response Mode (DRM) signal to the DER inverter to disconnect it or stop it from generating any power
- Sending a signal to an electricity meter or central protection unit to disconnect the DER device
- Direct communication and control of the inverter (e.g. via the internet or SCADA) to disconnect the inverter or reduce power exports to the network to zero

For a VPP, the DER backstop mechanism can be viewed similar to a DOE with an export level of 0 MW. The VPP needs to have visibility of the backstop mechanism signals being activated, but then can then factor the adjusted capacity of affected DER in the aggregate VPP capacity available for trading.

5 Case Studies

5.1 NEM Retail VPPs

The VPPs owned and operated by the major Market Customers in the NEM make up the largest share of VPP capacity in Australia. For example, Origin Energy stated that their VPP "Origin Loop" was 815 MW in capacity across over 276,000 connected services (as at 30 June 2023), including 200 MW of flexible large business loads and more than 400 electric vehicles [2]. AGL's VPP has a capacity of 316 MW (as at 30 June 2023) [3], expanding from the original 5 MW VPP that was first started in South Australia in 2016 [23].

Retail VPPs are opt-in and typically allow residential or commercial customers to enrol into the VPP via one of two options: (1) by bringing their own solar and battery system (provided that it meets the VPP's minimum requirements), or (2) by purchasing a compliant solar and battery package from the VPP. Retail VPP programs offer energy bill savings to customers in return for ceding the control of battery charge and discharge cycles to the VPP.

As mentioned earlier, retail VPPs are opaque in that their aggregate behaviour is not directly visible to the market in real-time. The VPP algorithms, including the objective functions, are proprietary and not made available to the public. Given the opt-in nature of VPP programs, there is clearly value for customers from participation, but it is difficult to quantify how much of that value is accrued to the VPP owner vs distributed to the DER owners.

5.2 NEM VPP Demonstration Project

The NEM VPP demonstration project was an Australian Renewable Energy Agency funded project led by AEMO that ran over a 24 month period from 2019 to 2021.

The project was designed to demonstrate the capability of VPPs to respond to wholesale energy and FCAS market prices (value stacking), as well as provide sufficient operational visibility to AEMO, improve the customer experience and understand the cybersecurity threats from VPPs. It was intended the findings from the demonstration project would help to inform any changes required to regulatory and operational frameworks. Eight (8) VPPs and over 7,000 customers participated in the VPP demonstration project.

The project did not test VPP integration into the wholesale energy market, and VPPs in the demonstration were treated as unscheduled resources. The project monitored whether VPPs behaved in a price-responsive manner, but found that price-responsiveness was inconsistent and unpredictable as drivers and algorithms were both varied and opaque across VPPs.

However, one of the project's major achievements was to successfully test VPP integration into the NEM Contingency FCAS markets, allowing the VPPs to fully participate in all Contingency FCAS markets, e.g. submitting market offers and providing the service(s) when cleared. By the end of the project, the market share of VPPs in the Contingency FCAS markets had reached 3% [9]. Most VPPs from the trial continue to participate in the Contingency FCAS markets today.

The technical performance of VPPs providing Contingency FCAS in the demonstration was encouraging, with VPPs generally able to reliably and effectively deliver their cleared Contingency FCAS volumes when required. VPPs in the demonstration were observed to make conservative offers into the market to mitigate risk of under-delivery. This often led to over-delivery during actual contingency events, for example the following responses from two VPPs to a generator contingency on 13 January 2021 [9]:

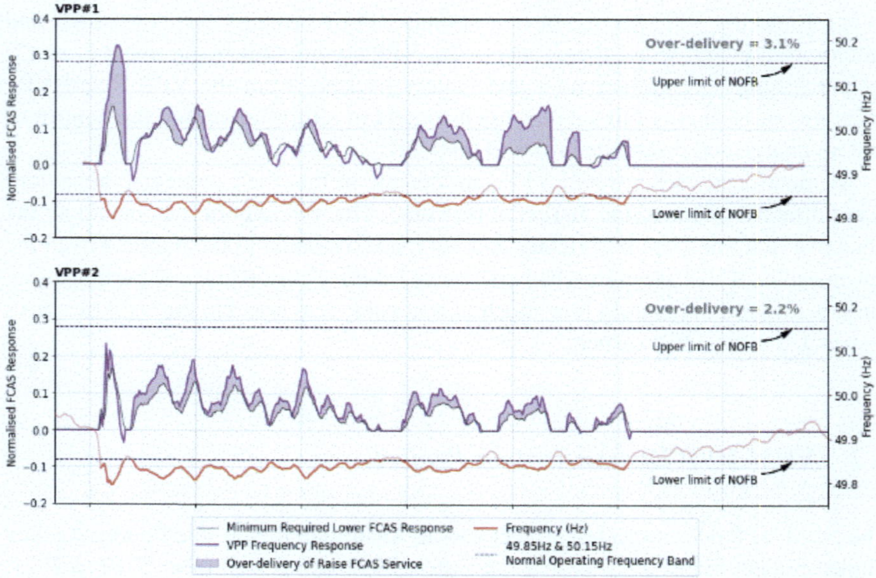

5.3 *Project EDGE and Project Symphony*

Project EDGE and Project Symphony are VPP orchestration trial/pilot projects in the NEM and WEM respectively. Whereas the previous NEM VPP demonstration project was more focused on a narrower set of goals such as value stacking, Contingency FCAS market participation and operational visibility, both Project EDGE and Project Symphony have a broader scope, including (but not limited to) testing the following functionality in an integrated and end-to-end manner:

- Test direct participation of VPPs with wholesale market dispatch and market price formation
- Test dynamic operating envelopes during market dispatch via the DSO
- Test DSO-level visibility and local network services

In testing the end-to-end functionality of direct wholesale market participation (for both energy and ancillary services), it was found that VPPs were able to submit credible bi-directional offers and be viably scheduled in the market. As noted earlier, both VPP projects were off-market simulations, however, both exhibited actual physical responses to simulated market clearing outcomes. For example, the pair of figures below show the behaviour of Project EDGE and Project Symphony in response to price incentives:

Project EDGE [24]

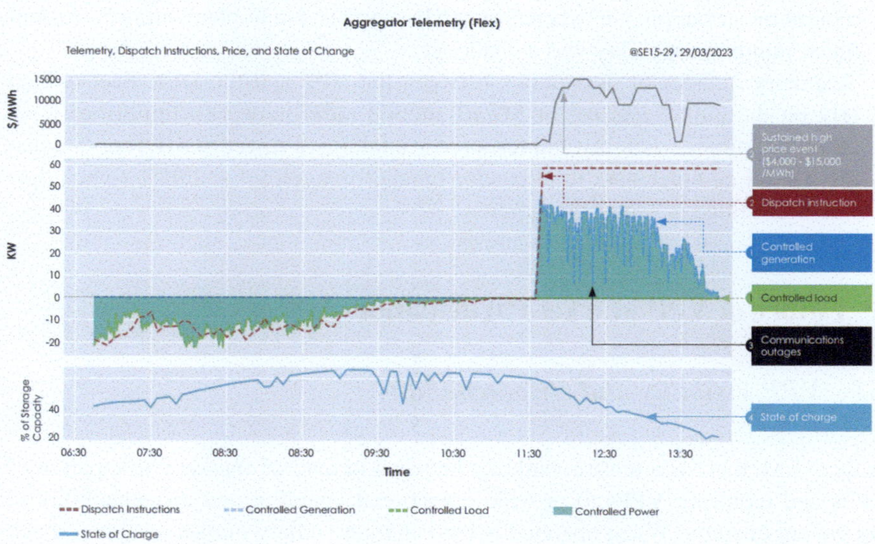

Project Symphony [25]

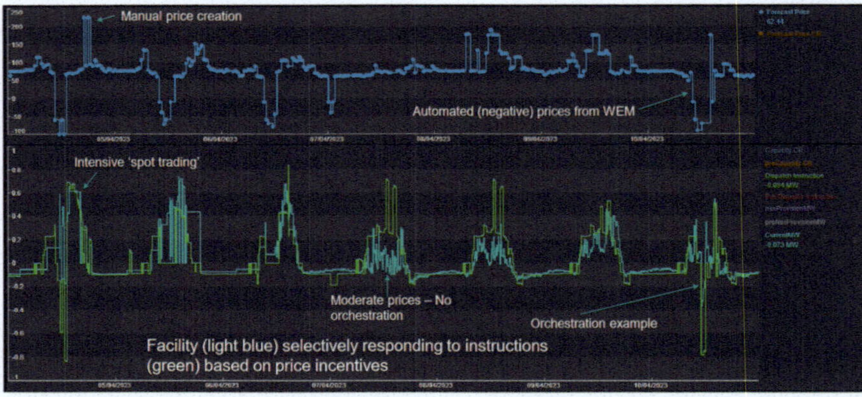

Despite the promising results, there were a number of technical performance issues identified, for example [24]:

- Accurate DER asset and portfolio forecasting remains difficult, and this has implications for dispatch compliance and conformance to DOEs. Project EDGE found that there were improvements to portfolio level forecasting and dispatch compliance with increasing VPP size.
- Market dispatch engines (like the engines used in the NEM and WEM) tend to require linear ramping between dispatch basepoints, but implementing consistent linear ramping capability was a challenge.
- Telemetry outages or disruptions between the VPP and the DER fleet (which rely on the public internet or 3G/4G mobile radio network) diminishes forecasting accuracy, visibility and resource availability. Project EDGE reported that the overall completeness in telemetry ranged between 90 and 95% (when data was transmitted at 1-min resolutions).

6 Future of VPP Market Participation

6.1 VPP Visibility and Dispatchability

In the NEM, there is a tension building between unfettered market participation of VPPs and requiring VPPs to provide operational visibility and dispatchability to the system operator. When operated behind an intermediary market customer, large VPPs (>30 MW) do not have to comply with the same rules and requirements as comparably sized generators and energy storage, e.g. dispatch compliance, metering requirements, SCADA visibility and control, curtailment due to constraints, etc. This gives the VPP more operational flexibility, but at the expense of participation in the price determination process and a lack of operational visibility by the system operator.

The concern for the system operator is that large-scale VPP behaviour that is not visible to the market and system operator can create distortions in system demand forecast accuracy and pose system security risks [9, 24]. However, full scheduling of VPPs via the market dispatch engine may not yet be at a mature enough stage to roll out at scale.

Both the NEM VPP demonstration project and Project EDGE indicate that developing reliable forecasting of available VPP capacity (to the level of sophistication required for dispatchability) remains a challenge. Project EDGE reported that the VPPs in its trial had more than 25% absolute forecast error 5 min before a dispatch interval (measured as the sum of net connection point flows of all participating VPP customers), with errors increasing to 50% for forecasts 48 h before the dispatch interval. Conformance to dispatch targets was relatively poor as a result, though it did show some improvement as the trial went on [24].

Network congestion and voltage stability risks at the transmission-distribution interface points can also emerge from large scale VPP behaviour. However, these can potentially be mitigated through the use of DOEs or other DER control mechanisms, which can prevent large swings in aggregate DER behaviour at a specific transmission node.

The conundrum for the system operator is that there are few, if any, incentives for VPPs to voluntarily provide real-time operational visibility and/or forecast activity, and certainly not to the level of measurement accuracy and communications reliability required of comparably sized generators. The high-level design of the Scheduled-Lite proposal suggested incentive options for providing visibility, such as access to market pre-dispatch schedules, reductions in non-energy cost allocation, capability-type payments during critical times and eligibility to participate in ancillary service markets [16]. However, it remains unclear whether any of these incentives outweigh the costs of firstly implementing the systems required and then meeting the compliance obligations. As VPPs continue to scale, operational visibility may end up being a mandatory requirement.

Finally, it is noted that this tension is more muted in the WEM due to the dominant position of the state-owned retailer Synergy in controlling VPP access and participation (with the support of the state government).

6.2 VPP-Led Market Participants

Existing VPPs tend to either be led by Market Customers (like the AGL and Origin VPPs) or technology providers that are affiliated with Market Customers (like Tesla and Reposit Power). However, there is an emerging trend for technology providers to themselves be registered market participants (for example, solar panel and battery manufacturer Hanwha has a subsidiary company that is registered as a Market Customer in the NEM).

Unlike traditional energy retailers, VPP-led companies can sell energy services to customers with VPPs at the heart of their service offerings, e.g. customers are enrolled into the VPP by default. Alternatively, a VPP-led company could offer an ecosystem of energy products (for example DER, battery storage, EV, home energy management, etc.) where the retail electricity plan is perhaps ancillary to the actual products being sold. This could potentially lead to a proliferation of new consumer offerings, for example retail energy subscriptions structured in a similar way as internet or mobile phone plans.

6.3 Harnessing Electric Vehicles and Vehicle-To-Grid Technology

Despite a huge installed capacity base of rooftop solar in Australia (nearly 19 GW), the expansion of total VPP capacity is fundamentally limited by the penetration of distributed energy storage and controllable (discretionary) loads at the residential and commercial level, which has only seen relatively steady annual growth. However, the large-scale adoption of electric vehicles and vehicle-to-grid (V2G) technology has the potential to massively expand VPP capacity [27].

Given the relatively large battery storage capacity of a typical EV (40–100 kWh), which is significantly larger than the capacity of a typical residential battery storage system (5–20 kWh), a VPP's capacity (and flexibility) can be materially increased by enrolling EVs, particularly those with bi-directional V2G chargers. There will be complexities with integrating and managing EVs in a VPP, not least because of their primary use for transportation (as opposed to energy). For example, EV owners will likely need incentives to participate in a VPP, given the risk that the EV is not charged when it is needed as a transport vehicle [28].

7 Conclusion

This chapter provided an overview of the range of VPP participation models currently available and being trialled in Australia, including an exploration of the delicate balance that needs to be struck between making it attractive to harness DER at scale and keeping the system secure and operable. In a fast moving and emerging area, Australia is arguably at the forefront of VPP adoption, enrolment and participation in wholesale electricity markets. The lessons learned in Australia can serve as a guide for other jurisdictions looking to integrate large-scale DER aggregations into the grid.

References

1. Pudjianto D, Ramsay C, Strbac G (2007) Virtual power plant and system integration of distributed energy resources. IET Renew Power Gener 1(1):10–16. https://doi.org/10.1049/iet-rpg:20060023
2. Origin Energy (2023) Origin energy annual report 2023. Company Annual Report. https://www.originenergy.com.au/wp-content/uploads/54/Origin_2023_Annual_Report.pdf
3. AGL (2023) AGL annual report 2023. Company Annual Report. https://www.agl.com.au/content/dam/digital/agl/documents/about-agl/investors/2023/230810-agl-energy-limited-annual-report-2023-4-4-asx.pdf
4. Horizon Power (2021) Smart sun: summary pilot report findings. https://web.horizonpower.com.au/media/6843/horizonpower_smartsun_pilot_report_2021.pdf
5. Horizon Power (2020) Carnarvon DER trials: lessons learnt report #1. https://arena.gov.au/assets/2020/03/horizon-power-carnarvon-der-lessons-learnt-report-1.pdf
6. AEMO (2017) Visibility of distributed energy resources: future power system security program. Technical Report. https://www.aemo.com.au/-/media/files/electricity/nem/security_and_reliability/reports/2017/aemo-fpss-program-visibility-of-der.pdf
7. AEMO (2023) Market ancillary service specification. Technical Specification, v8.1. https://aemo.com.au/-/media/files/stakeholder_consultation/consultations/nem-consultations/2023/primary-freq-resp-norm-op-conditions/market-ancillary-services-specification-v81.pdf?la=en
8. AEMO (2023) NEM registration and exemption list. https://aemo.com.au/-/media/files/electricity/nem/participant_information/nem-registration-and-exemption-list.xlsx?la=en. Accessed 2 Nov 2023
9. AEMO (2021) AEMO NEM virtual power plant demonstrations—knowledge sharing report #4. Technical Report. https://aemo.com.au/-/media/files/initiatives/der/2021/vpp-demonstrations-knowledge-sharing-report-4.pdf?la=en
10. AEMO (2021) Amendment of the market ancillary service specification—DER and general consultation—final report and determination. Final Determination. https://aemo.com.au/-/media/files/stakeholder_consultation/consultations/nem-consultations/2021/mass/final-determination/final-determination.pdf?la=en
11. AEMO (2021) Small generation aggregators in the NEM. Fact Sheet. https://aemo.com.au/-/media/files/electricity/nem/participant_information/registration/small-generation-aggregator/small-generator-aggregator-fact-sheet.pdf?la=en
12. AEMC (2020) Wholesale demand response mechanism. NER Rule Change. https://www.aemc.gov.au/rule-changes/wholesale-demand-response-mechanism
13. AEMO (2023) Tenders and expressions of interest for NCESS—reliability services 2024–26 (WA). https://aemo.com.au/consultations/tenders/tenders-and-expressions-of-interest-for-ncess-reliability-services-wa. Accessed 10 Nov 2023
14. AEMO (2023) WEM reform: wholesale electricity market design summary. Technical Report. https://aemo.com.au/-/media/files/initiatives/wem-reform-program/wem-reform-market-design-summary.pdf, version 2.0
15. FERC (2020) Participation of distributed energy resource aggregations in markets operated by regional transmission organizations and independent system operators. Docket No. RM18–9–000; Order No. 2222
16. AEMO (2022) Draft high level design for scheduled Lite. Consultation Paper. https://aemo.com.au/-/media/files/initiatives/scheduled-lite/consultation-paper-draft-high-level-design-for-scheduled-lite.pdf?la=en
17. DEIP (2022) Dynamic operating envelopes working group outcomes report. Technical Report, Distributed Energy Integration Program (DEIP). https://arena.gov.au/assets/2022/03/dynamic-operating-envelope-working-group-outcomes-report.pdf
18. SA Power Networks (2021) Advanced VPP grid integration—knowledge sharing report. Technical Report. https://arena.gov.au/assets/2021/05/advanced-vpp-grid-integration-final-report.pdf

19. Project EDGE (2023) Project EDGE: fairness in DOE objective functions—executive summary report. Technical Report. https://aemo.com.au/-/media/files/initiatives/der/2023/the-fairness-in-dynamic-operating-envelope-executive-summary.pdf?la=en
20. Ausgrid (2023) Project Edith knowledge share report—network support: a comparison of current and emerging solutions. Technical Report. https://cdn.ausgrid.com.au/-/media/Documents/About-us/Future-Grid/Edith/Project-Edith-Knowledge-Sharing-Report-2.pdf?rev=5045c84d1ec241a5ac81c89546549828
21. AEMO (2020) Minimum operational demand thresholds in South Australia. Technical Report. https://aemo.com.au/-/media/files/electricity/nem/planning_and_forecasting/sa_advisory/2020/minimum-operational-demand-thresholds-in-south-australia-review.pdf?la=en&hash=BBB27149A93B9259C63B47A8ECDB0B6E
22. Wilkinson S, Maticka M, Lue Y, John M (2021) The duck curve in a drying pond: the impact of rooftop PV on the Western Australian electricity market transition. Utilities Policy 71. https://doi.org/10.1016/j.jup.2021.101232
23. AGL (2020) Virtual power plant in South Australia—final milestone report. Technical Report. https://arena.gov.au/assets/2020/10/virtual-power-plant-in-south-australia.pdf
24. Project EDGE (2023) Project EDGE: final report. Technical Report. https://aemo.com.au/-/media/files/initiatives/der/2023/project-edge-final-report.pdf?la=en
25. AEMO (2023) Project symphony test and learn results—industry briefing—10 October 2023. Presentation. https://aemo.com.au/-/media/files/stakeholder_consultation/working_groups/der-program/wa-der-market-participation-forum/project-symphony_-industry-briefing-on-test-and-learn-results.pdf?la=en
26. Government of South Australia (2023) Technical regulator guideline—remote communications, capabilities for inverters, declared components of electricity infrastructure or an electrical installation associated with an electricity generating plant, deemed methodologies for remote disconnection and reconnection of electricity generating plants, remote updating methods & export limiting methods. Guideline, v1.5. https://www.energymining.sa.gov.au/__data/assets/pdf_file/0007/808225/2022D066388-Technical-Regulator-Guidelines-Distributed-Energy-Resources-Version-1.5-1.pdf
27. Kuiper G (2022) What is the state of virtual power plants in Australia?. Institute for energy economics and financial analysis
28. Rahman S, Sales-Ortiz J, Ardakanian O (2023) Making a virtual power plant out of privately owned electric vehicles: from contract design to scheduling. In: Proceedings of the 14th ACM international conference on future energy systems (e-Energy '23). Association for Computing Machinery, New York

Interconnected Microgrid Clusters Through Various Provisional Power Exchange Links

S. M. Ferdous, Farhad Shahnia, and G. M. Shafiullah

Abstract A standalone microgrid in a remote area may frequently experience overloading due to lack of sufficient power generation and/or renewable-based over generation causing unacceptable voltage and frequency deviation, which in turn lead the microgrid to operate with less resiliency and reliability. Conventionally, such problems are alleviated by load shedding or renewable curtailment. Alternatively, such autonomously operating microgrid clusters in a certain geographical area can be provisionally connected to each other to enable power exchange among them to address the problems of overloading or overgeneration more efficiently and cost-effective way. The power exchange link among the microgrids can be of different types such as a three-phase ac, a single-phase ac, or a dc-link. Power electronic converters are required to interconnect such power exchange networks to the three-phase ac microgrids and control the power-sharing amongst them. Such arrangement is also essential to interconnect microgrid clusters to each other with proper isolation while maintaining autonomy if they are operating in different standards. In this chapter, the topologies, and structures of various forms of power exchange links are investigated and an appropriate framework is established under which power exchange will take place. This approach is a decentralized control mechanism to facilitate power-sharing amongst the converters of the neighboring microgrids without any data communication, that can be implemented at the primary level based on the localized measurements. The dynamic performance of the control mechanism for all the topologies is illustrated through simulation studies in PSIM® to verify that such overloading or overgeneration situations can be effectively alleviated through proper frequency regulation. The chapter also presents a comparative analysis of the topologies in terms of stability and sensitivity.

Keywords Coupled microgrids · Power-sharing · Power exchange link · Back-to-Back converter · Frequency regulation

---

S. M. Ferdous (✉) · F. Shahnia · G. M. Shafiullah
School of Engineering and Energy, Murdoch University, Perth, Australia
e-mail: SM.Ferdous@Murdoch.edu.au

F. Shahnia and J. M. Guerrero (eds.), *Microgrids and Virtual Power Plants*, Power Systems, https://doi.org/10.1007/978-981-97-6623-9_14

1 Introduction

A microgrid is usually referred to as a small-scale interconnected network of multiple distributed generators that are predominantly renewable energy source-based and power electronic converter-interfaced, distributed loads, and energy storage [1–3]. Such a system can operate under a centralized, semi-centralized, or decentralized control scheme [2, 4–6], that may have two modes of grid-connected and autonomous (islanded). Where a utility grid is not feasible or unavailable due to cost and economic constraint, e.g., the edge-of-grid or remote/regional areas, the microgrids can operate under a decentralized control scheme and permanently autonomous, to minimize the cost by saving the requirement of communication technologies and thus facilitating to integrate a large number of renewable sources [4–7]. One or some distributed generators in such a scheme usually operates based on droop control to realize the desired power-sharing amongst them while also regulating the desired frequency and voltage [8–13]. Thus, the active and reactive powers are regulated at the output of each DG operating in droop control, and the microgrid's voltage and frequency are regulated within the acceptable limits and thus, the system's stability will be maintained [14–18].

To electrifying remote and regional areas, autonomous microgrids are a cost-effective solution and preferred over commissioning long transmission and distribution lines. Hence, microgrids operating permanently in the autonomous mode can be commissioned for such areas [19–24]. However, seasonal variation, intermittency, and unpredictability in renewable sources and fuel transportation cost impose further challenges for the operation of autonomous microgrids, and hence the microgrid operators are more interested in employing renewable energy resources in places where wind or solar-based resources are plentiful and thus, reducing the dependency of the network on diesel or gas [25–30]. Consequently, for remote locations like this, microgrid projects are gaining attention and popularity [4, 22, 31–34]. Such microgrid projects can significantly reduce the levelized cost of energy as demonstrated in several studies [4–8]. Therefore, standalone autonomous microgrids are getting prioritized and preferred for the electrification of remote off-grid towns. Power Electronic converter interfaced renewable energy sources along with diesel/gas-driven synchronous generators mostly dominate such microgrids' power supply [13–15, 20, 31–34].

Such microgrids can always experience over-generation and overloading despite the design considerations due to the high intermittency and uncertainty of the renewable energy sources [10, 32, 35–39]. Overloading is defined as a temporary shortage in the microgrid's generation capability against its total load demand. Load-shedding is the simplest mechanism to alleviate the problem in such a scenario and return the frequency and voltage to the acceptable limits [40, 41]. Conversely, over-generation is referred to as a state when the microgrid experiences excess power production through its non-dispatchable Distributed generators [35, 36, 42–45]. An easy solution in such a condition is to curtailing renewable energy source-based distributed generators. Though, both solutions are uneconomic, undesirable, reduces the resiliency and reliability of the microgrid. If not alleviated, such overloading or overgeneration scenario can cause unacceptable frequency and voltage deviation from its desired limits and eventually lead to system instability [46, 47].

Careful planning and considerations are required for such autonomous microgrids to account for the seasonal dependency, unpredictability, and intermittency associated with renewable energy sources [4–7]. Risk-based planning can decrease the probability of facing power shortage in the system; however, such system planning will result in a further increase in the projects' initial capital costs by either overdesigning and oversizing the distributed generators or installation of large-scale energy storage systems [4–6, 43, 44, 48]. The problems of over-generation or overloading can be effectively alleviated using oversized energy storage systems, but it will increase the installation and operational costs of the microgrids [44–48]. Regardless of its cost, floating battery energy storage units are used conventionally and are considered the most practical approach to support microgrids [49]. Such battery units, controller through power electronics-based converters, can play an essential role during peak demand, power deficiency, or overloading as they can respond faster than microturbines and diesel generators. Nonetheless, using large battery energy storage units to alleviate the problem of overloading or overgeneration, is cost-ineffective as overloading or peak power demand or overgeneration through renewable sources are either temporary or intermittent, causing oversizing and overdesign of both battery energy storage unit and its converter [31, 32, 44–49].

A third operation mode can be assumed to address these problems in which the neighboring autonomous microgrids interconnect temporarily to exchange power with one another in such situations [50–53]. The purpose of such an interconnection is that the microgrids support each other to retain the voltage and frequency within the acceptable limits and thus, system stability will be maintained. Such a provisional interconnection of neighboring microgrids to facilitate power exchange amongst them also enhances and improves system reliability and resiliency [35, 36, 49, 54–57].

With the rapid development of modern renewable sources along with power electronic converters and different types of energy storage systems, it can be anticipated that a remote area will comprise several neighboring microgrids in a close geographical area. Based on such prediction, a suitable power deficiency or surplus mitigation strategy can be instigated by suitably coupling two or more of those microgrids [36, 54, 58–60]. In the remainder of this chapter, a system of coupled microgrids (CMGs) refers to s group of interconnected neighboring autonomous microgrids. The CMG model and its operation based on the availability of communication infrastructure and the concept of the power market is discussed in ref [50, 56, 57]. According to this concept, a common power exchange link (i.e., distribution lines) that enables a physical connection among the microgrids will be introduced to interconnect the neighboring microgrids. Such an interconnecting link can be in either form of ac or dc [49, 55, 61–64]. A direct ac-ac connection between a microgrid and the power exchange line can be easily realized through a conventional circuit breaker or an interconnecting static switch which is very economical to a large extent [60]. However, the structure and control mechanism of each microgrid can be different from its neighboring microgrids and such a direct ac-ac connection will reduce the autonomy of their operations. In order to add proper isolation amongst the microgrids to retain the autonomy in their operation, the power electronic converter-based interlinking structure [65, 66] will enable every microgrid to function independently and autonomously, as well as facilitate power exchange with the neighboring microgrids [36, 60–62]. Power-sharing between two islanded microgrids during mutual contingency is studied in [54, 59], whereas an autonomous control approach to sharing power with neighboring microgrids through a back-to-back converter in the presence of a utility grid is proposed in [36, 61]. Neighboring microgrids can also interconnect to support each other during the occurrence of fault leading to the outage of a particular section [67] or during normal conditions to minimize the levelized cost of electricity [68]. If one of the microgrids temporarily observes a high-power generation by one of its DERs, the other neighboring microgrid(s) can get interconnected to one another to import power and thus offers electricity at a lower price. A transformative architecture is proposed for coupling the nearby microgrids is to improve the system resiliency during faults [69]. Ref [70, 71] analyse the reliability aspects of a CMG, while the voltage and current controllability in CMGs is analyzed in [72]. The dynamic operation of DERs within CMGs is investigated in [73] whereas [74] examines the dynamic security of the CMGs.

Detailed analysis of different topologies and architectures of forming CMGs is shown in [56, 75]. Optimal control of a utility grid-connected CMG is studied in [36, 66], while an interactive control method to share the load in a CMG ensuring a wide range of system stability is shown in [47, 76–81]. An optimization-based technique is developed in [82–84] to ensure optimal power exchange between the microgrids. A decision-making-based approach is proposed to determine the most suitable microgrid(s) to connect with an overloaded microgrid, based on different criteria such as available surplus power, electricity cost, reliability and the distance of the neighboring microgrids as well as the voltage/frequency deviation in the CMG [85–87]. The ultimate objective is that a microgrid can be coupled to any other microgrid (and not necessarily an adjacent microgrid) if a general link is available to act as a power exchange highway [55, 56, 75, 88]. At the same time, back-to-back power electronic converter-based interface has been proposed in [35, 54, 89, 90] that can be used between each microgrid and the interconnecting three-phase ac line to ensure autonomy in its operation. Such a scheme is comparatively expensive as two voltage source converters (VSCs) are employed for every microgrid. However, the main objective of such a connection is to provide isolation between the microgrids and thus, microgrids with different operational regulations can exchange power while being isolated through the dc-links within the back-to-back converters [35, 90]. Furthermore, such an interconnecting link can also be realized through a single-phase ac power line instead of a three-phase ac power line [90] to reduce the cost while fulfilling the requirements of the power transfer during overloading and/or overgeneration. It should be noted that such an interconnecting line is only operational during contingency of overloading or overgeneration and the total power exchanged through the links is comparatively smaller (around 10–20% of the microgrid maximum capacity). Therefore, such an alternative single-phase ac line over the three-phase ac lines can be more economical as the number of lines and the power electronic filter components will be less. However, the use of the back-to-back converter is limited to couple of two microgrids and not scalable to couple more than two microgrids. To couple multiple microgrids, the control technique and coordination among the converters will be complex as it involves 2*N* number of converters to interconnect *N* microgrids and a common interlinking link that operates in a predefined frequency and voltage [35, 90]. At the same time, a communication network is necessary to coordinate such ac power exchange link as proper synchronization among the microgrids is necessary for power exchange which is not possible unless a suitable data communication link is established [91–94]. Hence, commissioning a dc-link rather than the ac links proves to be a better and effective operational strategy and, in such topology, only one converter is adequate for connecting each microgrid to the dc power exchange link [49, 88, 89]. Thus, such a dc interconnecting link practically costs half of the price of the two previous ac topologies in terms of the power electronic converters. Even though the power loss in the interconnecting lines appears to be smaller; however, most of the commercial power line infrastructure, equipment, and accessories are based on ac systems. Besides, the network operators are usually more familiar with ac distribution systems as compare to dc systems. Moreover, the control and protection equipment and schemes for ac systems are more economical than dc systems to some extent. Also, dc power transmission is only economic and viable over ac power transmission if the distance between the sending and receiving end is over a certain level for a given voltage range [90].

Properly coordinating and efficiently managing the power transaction from one microgrid to another through the VSCs and lines, is the main challenge that requires to be addressed. This chapter has focused on developing an appropriate power transaction management strategy among the microgrids within the CMG along with the proper structure and control techniques for the power electronic converters. The main objective is to control the power flow from a microgrid with sufficient generation to a neighboring microgrid when it experiences power deficiency and hence, considerable frequency and/or voltage drop. In this regard, the VSC does not require to continuously exchange power but would only do so when a microgrid needs support. Similarly, if a renewable source-dominated microgrid suddenly experiences excessive generation, and thereby, leads to unacceptable voltage or frequency rise, a neighboring microgrid can absorb this surplus power. In this chapter an appropriate energy transaction management scheme is discussed with three different structures for CMG formation and control principle for the converters to realize such operations. The exchange of power through an interconnected system of microgrid networks requires a suitable protocol and electrical framework along with a definite power exchange strategy. A generic arrangement of CMG forming such a structure and the three interlinking topologies are shown in Figs. 1 and 2 respectively.

This technique is a fully decentralized approach to ease both over-generation and overload problems by coupling the neighboring microgrids which is fully scalable to couple *N* autonomous microgrids in order to exchange power among them using any one of the above-mentioned three topologies. This approach does not need any data communication among the microgrids and their controller; thus, operates only

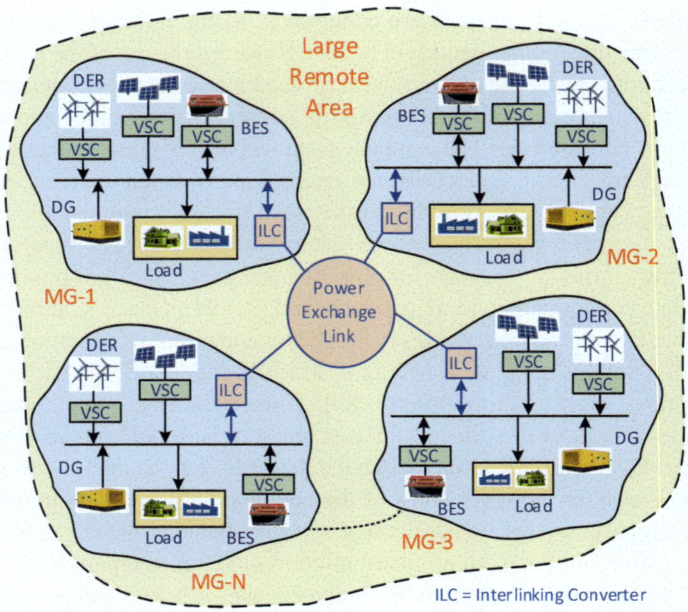

Fig. 1 CMG network interconnection through a common power exchange link

Fig. 2 Topologies for Interlinking Converters used for power exchange links

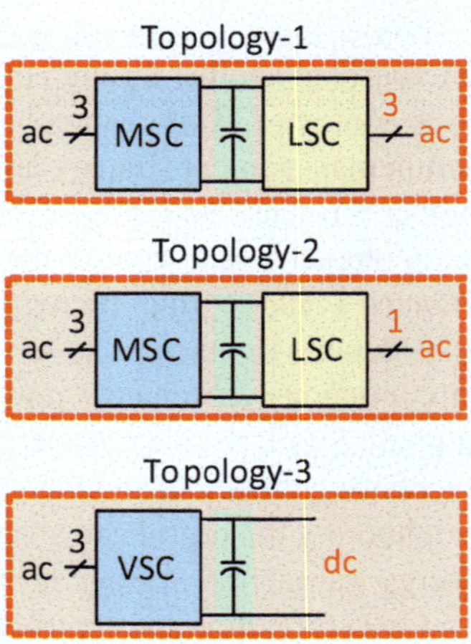

based on local measurements at the terminals of the converter at the point of common coupling which offers the additional advantage of eliminating the requirement as well as the cost of communication infrastructure. Thus, it enhances the resiliency and self-healing capability of each microgrid while ensuring autonomy in the operation of individual microgrids enabling them to operate at any frequency and voltage level. Besides, this mechanism can also work as a backup for CMG networks that are operating under a centralized control scheme, in the event of a communication link failure or any contingency, and hence, offering further reliability. The efficacy of this mechanism is evaluated by numerous examples with various scenarios in microgrids. To demonstrate its robustness and applicability for all three topologies, the technique is further validated with stability and sensitivity analysis of the CMG network against the key design and operational factors. A comparison of the implemented control technique and features of the three topologies against other approaches, available in the literature, has been summarized and provided in Table 1.

2 Power Exchange and CMG Operation

This provisional power exchange mechanism can be realized by interconnecting the neighboring microgrids to a common power exchange link to facilitate power exchange among the microgrids during an overload/excess generation. Thus, the neighboring microgrids can be coupled via a power electronic converter-based interface and ac/dc interconnecting power lines to form a CMG. Such a strategy is attained by monitoring and controlling the frequency of each microgrid. It is obvious that autonomous microgrids are conventionally operating under droop control to share power amongst their DERs and in such a control scheme, as the total load demand of a microgrid increases, its frequency decreases, and vice versa. Such a distinct behavior can be used as an indication of the microgrid's power shortage or surplus, and hence, the approach eliminates the necessity of any sort of data communication link or a centralized controller. The key benefit of this approach is that it eliminates the necessity of installing and operating a communication infrastructure to enable the microgrids to exchange power. It should be mentioned here, even in the presence of a central controller with communication links to monitor and regulate the power exchange, this approach can be implemented as the backup system that is activated in the event of communication failure or during any contingency. Thus, a fully decentralized mechanism to exchange power amongst the microgrids dynamically is introduced for effective overload/over-generation power management based on a predefined framework. Such provisional power exchange facility amongst the neighboring microgrids increases the reliability and resiliency of the overall system.

Based on the above-mentioned framework, the microgrids can participate in forming the CMG based on a predefined set of standards, as permitting all (or many) microgrids to participate in the CMG network formation may lead to system instability [47, 76–78, 80, 81, 95, 96]. Hence, the formation of a CMG would be allowed or denied based on the microgrids' existing loading level, which is identified by measuring their frequencies. Based on their loading level, every microgrid is classified and flagged in either of the three categories of a healthy, problem, or floating microgrid (respectively denoted by HMG, PMG, and FMG), as introduced below:

Table 1 Comparison of different CMG features

CMG features	[35]	[36]	[49]	[54]	[55]	[60]	[61]	[62]	[88]	Top-1	Top-2	Top-3
Communication Links	×	✓	×	✓	✓	✓	✓	×	✓	×	×	×
OL prevention	✓	✓	✓	✓	✓	✓	✓	×	×	✓	✓	✓
OG prevention	✓	✓	×	×	✓	×	×	×	×	✓	✓	✓
Multiple microgrids	✓	×	×	×	✓	×	×	×	✓	✓	✓	✓
Type of interconnecting link	CAC	CDC	CDC	CDC	CDC	DAC	CDC	CDC	CDC	CAC (3-ph)	CAC (1-ph)	CDC
Utility grid connection	×	✓	×	×	×	×	✓	✓	×	×	×	×
Energy storage	×	×	✓	×	×	✓	×	✓	×	×	×	×
Scalability	✓	×	✓	×	✓	×	×	×	✓	✓	✓	✓
Level of implementation	P	P	P	P	P/S	P/S	P	P	P/S	P	P	P
Control type	D	Dist	D	C	C	C/D	Dist	Dist	Dist	D	D	D
Unbalanced microgrid	×	×	×	×	×	×	×	×	×	✓	✓	✓
Desired power-sharing using droop control	✓	×	×	×	✓	×	×	×	×	✓	✓	✓

DAC = Direct ac coupling; **CDC =** Converter-based dc coupling; **CAC =** Converter-based ac coupling;
P = Primary level; **S =** Secondary Level; **Mod.** = Moderate; **Med.** = Medium; **D** = Decentralized; **C** = Centralized; **Dist.** = Distributed

- An HMG refers to the status of a microgrid when it is operating within its nominal frequency and voltage range and can support the other neighboring microgrids by exchanging power.
- A PMG refers to an operating condition when the microgrid experiences over-generation or overloading. Any PMG within the CMG network is expected to connect to one or more neighboring HMGs and exchange power to retain their frequency and voltage within acceptable limits.
- An FMG refers to a microgrid's loading condition that is in the boundary of becoming a PMG. In other words, the microgrid's frequency or voltage is in the boundary of maximum or minimum acceptable limits, and hence, should not be allowed to participate in power exchange within the CMG.

This classification of microgrids is shown in Fig. 3 and their corresponding mode of operation based on frequency level is tabulated in Table 2.

To explain the operation of such a system, let us consider Fig. 3 in which f_{MG} is the operating frequency of the microgrid while f_{FL}^{OL} and f_{FL}^{OG} are the two frequency limits, beyond which the overloading and over-generation begin in the microgrid, respectively. Parameters f_{LIM}^{OG} and f_{LIM}^{OL} are the microgrid's maximum and minimum

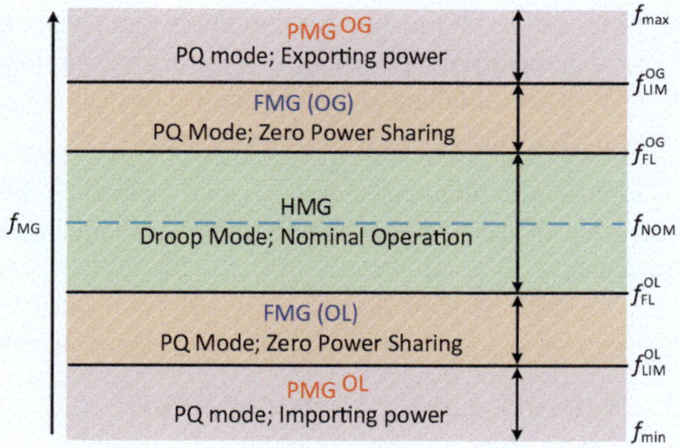

Fig. 3 Microgrid classification based on its frequency level

Table 2 Classification of microgrids based on their frequency

Frequency levels	MG Status	Mode of operation
$f_{FL}^{OL} < f_{MG} < f_{FL}^{OG}$	HMG (nominal operation)	Droop Control
$f_{LIM}^{OL} < f_{MG} < f_{FL}^{OL}$	FMG (boundary of overloading)	PQ mode (zero power exchange)
$f_{FL}^{OG} < f_{MG} < f_{LIM}^{OG}$	FMG (boundary of excess generation)	PQ mode (zero power exchange)
$f_{MG} \leq f_{LIM}^{OL}$	PMG (overloading)	PQ mode (importing power)
$f_{MG} \geq f_{LIM}^{OG}$	PMG (over-generation)	PQ mode (exporting power)

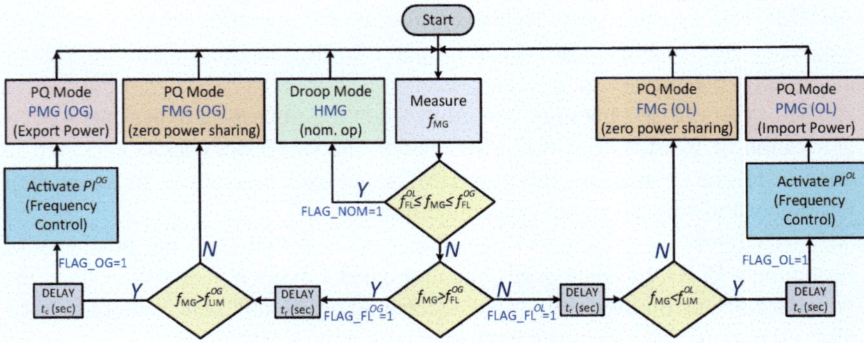

Fig. 4 Flowchart of the microgrid's different mode of operation for power exchange

permissible frequency to operate within the acceptable limit, while f_{max} is the no-load frequency and f_{min} is the frequency where the microgrid becomes unstable due to excessive loading. Referring to Table 2, the microgrid's operating status can be denoted as:

- The microgrid is defined as HMG when $f_{FL}^{OL} < f_{MG} < f_{FL}^{OG}$.
- FMG can be attained in two conditions; i.e., when $f_{LIM}^{OL} < f_{MG} < f_{FL}^{OL}$ (increasing load demand) or $f_{FL}^{OG} < f_{MG} < f_{LIM}^{OG}$ (decreasing load demand).
- PMG can be attained in two conditions; i.e., when $f_{min} < f_{MG} < f_{LIM}^{OL}$ (overloading) or $f_{LIM}^{OG} < f_{MG} < f_{max}$ (over-generation).

Figure 4 demonstrates the steps of detecting the microgrid's status and the operation of the CMG to enable power exchange amongst the microgrids using a flowchart.

3 Network Under Consideration

Consider a network of Fig. 1 where N autonomous microgrids, each consisting of several DERs, distributed generators, loads, and one interlinking converter coupled to a power exchange link. The interlinking converter can be any one of the three topologies shown in Fig. 2. Topologies for Interlinking Converters used for power exchange links. The DERs are connected to the microgrid through VSCs which are controlled in different operating modes to regulate the power and frequency of each microgrid. The DERs are assumed to be operating in the $P - f$ and $Q - V$ droop control mode, where the voltage and frequency at the output of each DER is determined from [73]

$$f_{MG} = f_{max} - m_i P_i \tag{1a}$$

$$V = V_{\text{rated}} - n_i Q_i \tag{1b}$$

where m_i and n_i are the droop coefficients of the ith DER of the microgrid, and can be obtained from

$$m_i = \frac{f_{\max} - f_{\min}}{P_i^{\max}} \tag{2a}$$

$$n_i = \frac{V_{\max} - V_{\min}}{2Q_i^{\max}} \tag{2b}$$

where subscript $_{\max}$, $_{\min}$, and $_{\text{rated}}$ represent the maximum, minimum, and rated limits for frequency and voltage of the microgrids while P_i and Q_i are active and reactive powers injected by each DER to the microgrid. All DERs within a microgrid are assumed to have the same $\Delta f = f_{\max} - f_{\min}$ and $\Delta V = V_{\max} - V_{\min}$; V_{rated} is the rated voltage of the DER's point of common coupling, i.e., one per unit. However, these parameters can be different in each neighboring microgrid. Each DER is provided with its own local energy storage and thus it can be assumed as a dispatchable DER. The structure and operation of the DERs are discussed in [16, 54, 73]. Lastly, each microgrid is connected to a common power exchange link via its interlinking converter to exchange power with the other microgrids in the event of power deficiency or over-generation. Each microgrid can operate independently at any predefined frequency and voltage level without affecting the operation of other neighboring microgrids.

4 Various CMG Structure and Their Control

The CMG formation among the neighboring microgrids is achieved through power electronics converters and interconnecting power lines using the three different topologies. For topology-1 and 2, a back-to-back converter structure is employed whereas, a single-stage converter is enough for topology-3. The converter connected to the microgrid is labeled as the microgrid-side converter (MSC) while the other one is labeled as the line-side converter (LSC). In topology-1 and 2, both MSC and LSC are VSCs connected back-to-back with each other via a common dc-link to provide the isolation which enables the microgrids to operate with full autonomy and no synchronization of MSC with LSC is required (see Fig. 2). In such arrangement, MSC enables bi-directional power flow between the microgrid and dc-link whereas, LSC enables bidirectional power flow between dc-link and the interconnecting lines. In this approach, the purpose of MSC is to regulate only the dc-link voltage; however, the mode of operation of LSC depends on the frequency (i.e., loading level) of the microgrid to which it is connected. On the other hand, in topology-3, a single-stage voltage source converter is used to connect the dc power lines to the microgrid and labeled as VSC (see Fig. 2), which can operate either in dc droop control mode or constant PQ control mode depending on the status of the microgrid.

4.1 Three-Phase AC Interconnection (Topology-1)

Topology-1 is a three-phase ac interconnecting line with two back-to-back connected power electronic converters. In this topology, both the MSC and LSC are assumed to be three-phase, three-leg VSC using IGBTs or MOSFETs. Each VSC is connected to its point of common coupling through a three-phase LCL filter. The structures of MSC and LSC are shown schematically in Fig. 5. However, any other three-phase topologies of VSCs can also be used instead of the considered topologies here.

4.1.1 Control of MSC

The main function of MSC is to regulate and maintain the dc-link voltage of V_{dc} at the desired reference level of V_{dc}^{ref}. This is valid for any status of the microgrid (i.e., HMG, PMG, FMG). The dc-link will exchange the appropriate amount of power with the microgrid through proper voltage regulation, such that the dc-link voltage remains constant. This can be realized through a closed-loop control mechanism (i.e., the outermost loop in the MSC part of Fig. 5) to determine the reference active power (P_{ref}) that should be drawn from or injected into the microgrid to avoid a deviation in V_{dc}. The MSC is not likely to exchange any reactive power with the microgrid and hence, the reference for reactive power (Q_{ref}) is assumed to be zero. Based on these two reference parameters, the PQ controller determines the three-phase reference voltage (v_c^{ref}) across the capacitor C_f of LCL filter of the MSC. The

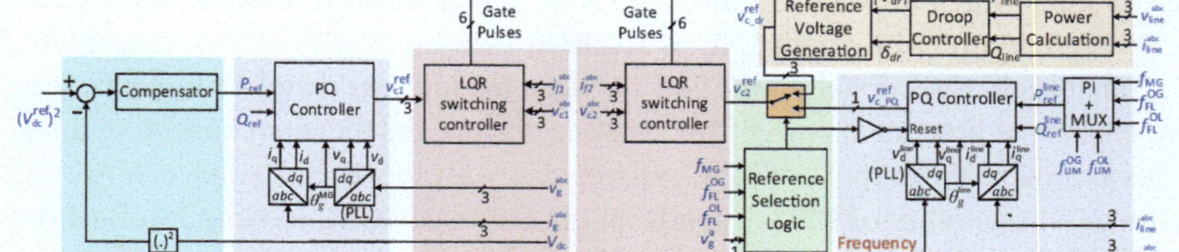

Fig. 5 Assumed structure and close loop control mechanism of the MSC and LSC for topology-1

most inner control loop of MSC in Fig. 5 is the voltage tracking and switching control block. To achieve robust and optimum performance, the linear quadratic regulator is used to track the sinusoidal reference voltage, as discussed in [73].

Voltage Tracking and Switching Control

The voltage tracking and switching control of the MSC is designed based on the dynamics of the filter inductance, L_f and filter capacitor, C_f of the LCL filter of the MSC that is connected at the point of common coupling on the microgrid side, which operates as the inner-most loop of the closed-loop control mechanism of MSC. The dynamics of the LCL filter can be represented in the state-space form by

$$\dot{\boldsymbol{x}} = \boldsymbol{A}\boldsymbol{x} + \boldsymbol{B}_1 u_c + \boldsymbol{B}_2 i_g \tag{3}$$

where
$\boldsymbol{A} = \begin{bmatrix} 0 & 1/C_f \\ -1/L_f & -R_f/L_f \end{bmatrix}, \boldsymbol{B}_1 = \begin{bmatrix} 0 \\ V_{dc}/2L_f \end{bmatrix}, \boldsymbol{B}_2 = \begin{bmatrix} -1/C_f \\ 0 \end{bmatrix}$ and $\boldsymbol{x} = [v_c i_f]^T$ while u_c is the control switching function and i_g is the current flowing through the coupling inductor and ^{T} is the transpose operator. The current, i_f is desired to have only a low-frequency component (i.e., 50 Hz) and all high-frequency switching components are needed to be filtered out. Hence, a high-pass filter (with a cut-off frequency of ω_c) is introduced in the feedback control structure to extract the high-frequency components of i_f as denoted by $\tilde{i}_f$. Hence, (3) can be rewritten as

$$\dot{\boldsymbol{x}}_c = \boldsymbol{A}_c \boldsymbol{x}_c + \boldsymbol{B}_c u_c \tag{4}$$

where $\boldsymbol{A}_c = \begin{bmatrix} 0 & 1/C_f & 0 \\ -1/L_f & -R_f/L_f & 0 \\ -1/L_f & -R_f/L_f & -\omega_c \end{bmatrix}, \boldsymbol{B}_c = \begin{bmatrix} 0 \\ V_{dc}/2L_f \\ V_{dc}/2L_f \end{bmatrix}$ and $\boldsymbol{x}_c = \begin{bmatrix} v_c i_f \ \tilde{i}_f \end{bmatrix}^T$.

An appropriate full state-feedback control law along with a reference input (which, in this case, is the reference filter capacitor voltage, v_c^{ref}) can be expressed in the form of

$$u_c = k_1 \left(v_c^{ref} - v_c\right) - k_2 i_f - k_3 \tilde{i}_f = -\boldsymbol{K}\dot{\boldsymbol{x}}_c + k_1 v_c^{ref} \tag{5}$$

where $\boldsymbol{K} = \begin{bmatrix} k_1 \ k_2 \ k_2 \end{bmatrix}$ is the state-feedback gain matrix in which k_1, k_2 and k_3 are optimally determined by minimizing an appropriate cost function using a linear quadratic regulator. Based on this optimization method, a highly robust switching controller can be designed with a bandwidth large enough to accurately track the low-frequency sinusoidal signal (i.e., 50 Hz) with zero phase delay and steady-state error. Therefore, (4) can be rewritten as

$$\dot{\boldsymbol{x}}_c = \boldsymbol{A}_{cl} \boldsymbol{x}_c + \boldsymbol{B}_{cl} v_c^{ref} \tag{6}$$

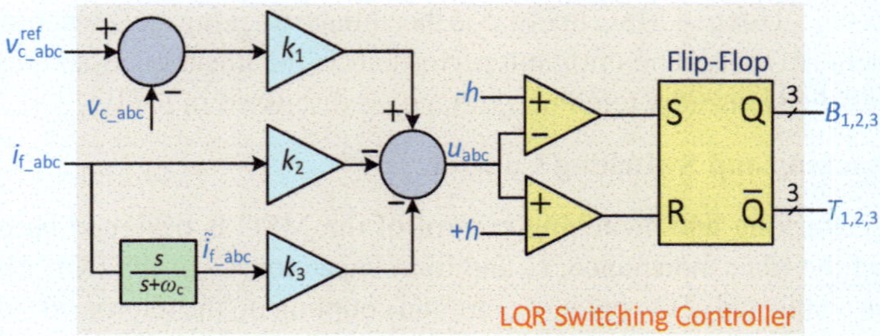

Fig. 6 Linear Quadratic Regulator based switching and voltage tracking

where $\boldsymbol{A}_{cl} = \boldsymbol{A}_c - \boldsymbol{B}_c\boldsymbol{K}$ and $\boldsymbol{B}_{cl} = k_1\boldsymbol{B}_c$. The implemented controller offers infinite gain and phase margin, and thus, robustly stable. The switching control of the MOSFET/IGBT can be implemented through a hysteresis function [97] with an adequately small band (e.g., $h = \pm 10^{-4}$). The control block diagram of such an LQR switching controller is shown in Fig. 6.

Power Control

The next control loop is to determine the appropriate reference voltage across the capacitor, C_f of LCL filter by regulating the active and reactive power. For this purpose, the conventional direct-quadrature (dq) axis close loop power controller (i.e., PQ controller) is implemented. The PQ controller is the first of two outer control loops. The benefit of the PQ controller is that the active and reactive powers are decoupled and can be controlled independently. The active and reactive power are regulated by controlling the d-axis current i_d and q-axis current i_q, respectively. Hence, the PQ controller is also known as the current controller. The controller is made of two proportional-integral (PI) controllers that are responsible for producing the required d and q axis reference voltages of v_d^{ref} and v_q^{ref}. Then, these two voltages are converted into three-phase ac reference voltage, v_c^{ref} across the filter capacitor, C_f using the grid angle θ_g^{MG} as obtained through a synchronous reference frame phase-locked loop (SRF-PLL) system. The dynamics of the output R-L branch of the LCL filter, comprised of coupling inductance, L_g, and resistance, R_g, are considered to design and tune the two PI controllers using the pole-zero cancellation method [98, 99]. The transfer function of the PI controller is given by

$$C_{PI}(s) = \frac{K_c(sT_c + 1)}{sT_c} \tag{7}$$

where $T_c = L_g/R_g$ and $K_c = 2\pi f_{BW} L_g$ where f_{BW} is the closed-loop bandwidth of the controller that conventionally, requires to be at least 10 times smaller than the bandwidth of the inner voltage tracking loop, and hence, the inner loop can be assumed as a unity gain block while tuning the PI controllers of the PQ controller. In this design, f_{BW} is chosen to be 1000 times faster than the voltage tracking loop to ensure no overlap in the bandwidth of the control loops and proper sinusoidal voltage tracking at 50 Hz. The closed-loop transfer function of the PQ controller is then given by

$$G_c(s) = \frac{i_{dq}^{ref}}{i_{dq}} = \frac{1}{s\tau_c + 1} = \frac{P_{ref}}{P_s} \tag{8}$$

where $\tau_c = L_g/K_c$ is the time constant of the current controller, P_{ref} is the reference power and P_s is the active power output of the converter. It should be noted here, P_{ref} is generated through the outermost control loop i.e., the dc-link voltage controller. The assumed close loop control block diagram of d and q axis current controller is shown in Fig. 7 while the PQ controller for MSC is shown in Fig. 8.

DC-Link Voltage Control

The outermost loop of the control structure is the dc-link voltage controller. The dynamics of the voltage across the dc-link capacitor, V_{dc}, with respect to converter output power, P_s, can be described in terms of a transfer function as [98, 100]

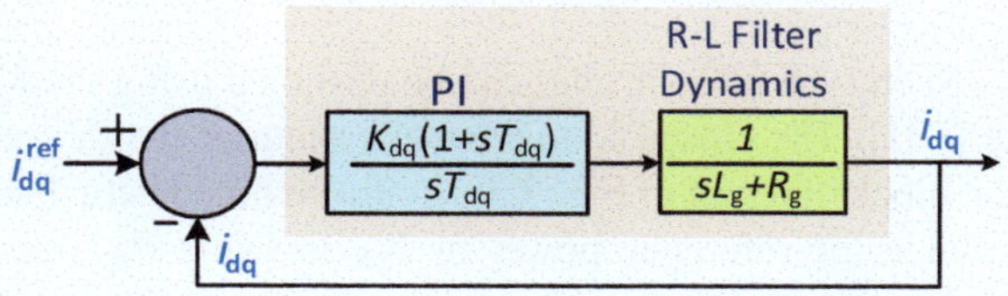

Fig. 7 Block diagram of d and q axis current controller

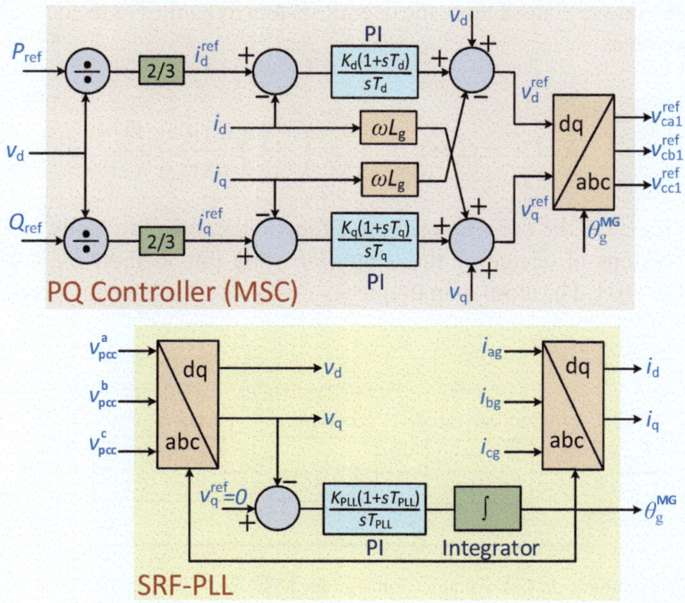

Fig. 8 Block diagram of PQ controller for MSC

$$G_{dc}(s) = \frac{V_{dc}^2}{P_s} = \left(\frac{-2}{C_{dc}}\right)\frac{(s\tau_{dc} + 1)}{s} \tag{9}$$

where $\tau_{dc} = 2(L_f + L_g)P_{so}/3V_{do}^2$. The transfer function of $G_{dc}(s)$ is a linearized small-signal model where P_{so} is the steady-state power output of the converter and V_{do} is the d-axis steady-state voltage at the point of common coupling of the converter to the microgrid. It is a non-minimum phase system with a right half plane zero, whose location depends on the steady-state value of P_{so} and V_{do}. If these two parameters are varied, then the controller needs to be redesigned to ensure the required stability margin. In other words, the controllers are designed to exchange rated power at a specific voltage level, and hence the exchanged power through the dc-link should not exceed its maximum limit. In order to design a controller that regulates the dc-link voltage with the required gain and phase margin to ensure satisfactory performance, a type-2 controller can be designed. The generic transfer function of a type- 2 controller can be written as

$$C_{dc}(s) = \frac{K_{dc}(s + \omega_z)}{s(s + \omega_p)} \tag{10}$$

Such type of controller can be designed using the K-factor method, as described in [101]. The closed-loop control system of the dc-link voltage can be realized by the block diagram of Fig. 9. It should be noted here that, while regulating the dc-link voltage the reference power of P_{ref} in (8) is P_{dc} (i.e., the power that needs to be exchanged bi-directionally to keep the dc-link voltage fixed to its reference value) as can be seen from Fig. 9. As the voltage tracking loop is made almost 1000 times faster than the power control loop, the overall system dynamics of the dc-link voltage can be assumed as

$$G_{OL}(s) = G_c(s)G_{dc}(s) = \left(\frac{-2}{C_{dc}}\right)\frac{(s\tau_{dc} + 1)}{s(s\tau_c + 1)} \tag{11}$$

Hence, $G_{OL}(s)$ is the open-loop transfer function assumed to apply the K-factor method. The steps of designing the controller using this method are discussed in detail in [100, 101]. The close loop frequency responses of all the three control loops

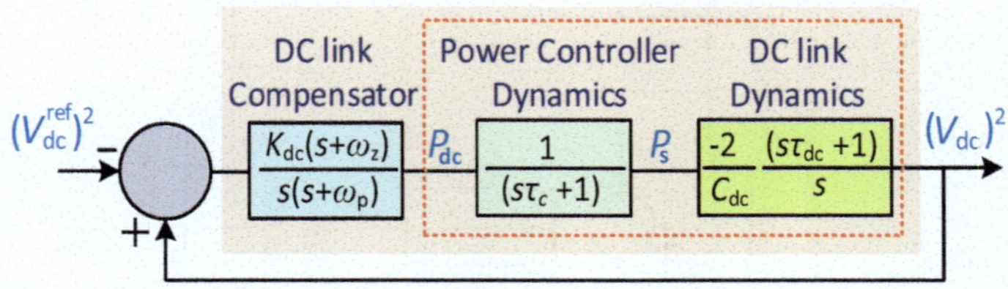

Fig. 9 Block diagram of dc-link voltage controller for MSC

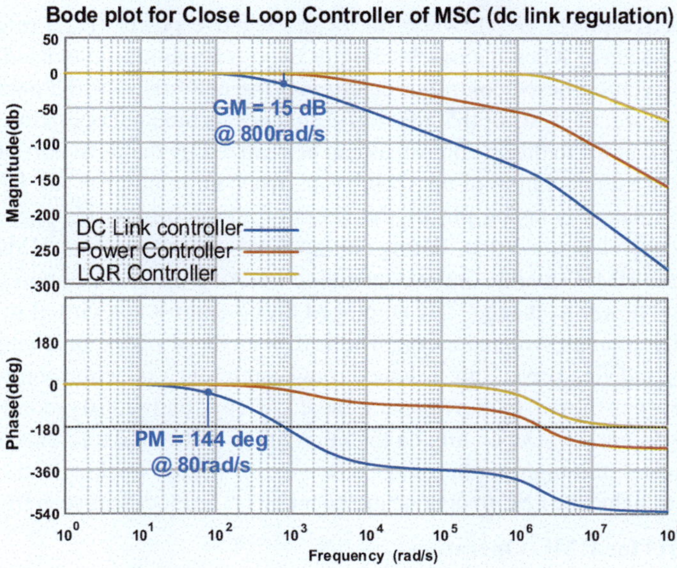

Fig. 10 Frequency response plots of the control loops for the MSC

are shown in Fig. 10. The close loop bandwidths of voltage tracking, power, and dc-link controller are 318 k, 300, and 20 Hz, respectively.

4.1.2 Control of LSC

Two separate modes of operation for the LSC are considered that depend on the status of the microgrid. It has been proposed that to form a CMG, the LSC of all HMGs should operate under droop control mode to enable power-sharing, whereas the LSC of the FMGs and PMGs should operate in constant power (PQ) control mode (see the flowchart of Fig. 4). The LSCs connected to HMGs, form the CMG and provide the required droop voltage and angle references that enable power-sharing among the HMGs within a CMG, whereas it will absorb or inject the desired amount of power from the interconnecting power lines to alleviate the overloading or over-generation situations respectively when a microgrid becomes PMG.

To realize the above objective, the frequencies of microgrids need to be continuously monitored. If the microgrid is overloaded, then a control mechanism should determine the amount of power that needs to be absorbed from the interconnecting link by the PMG to return the frequency of the microgrid to the acceptable limit. The concept is equally applicable when the microgrid has surplus power generation and its frequency rises beyond the acceptable limit. Thus, the LSC of an overloaded or over-generating microgrid will start to operate in the constant PQ mode (see the flowchart of Fig. 4). A frequency deviation-based control system determines the reference active power (P_{ref}), while the of the reactive power (Q_{ref}) reference is assumed to be zero. When the overloading or over-generation is alleviated, another frequency deviation-based controller determines the new status of the microgrid as HMG, and hence, the LSC of that microgrid changes its mode of operation from the constant PQ control to droop control, as can be realized by the flowchart of Fig. 4.

During the idle operation of the CMG, (i.e., when all microgrids are HMGs and no power exchange is taking place), the LSCs of all the microgrids operate under the droop control mode. Also, during the nominal operation of an HMG within the CMG, the HMG should continue operating in droop mode to support the PMG and facilitate power exchange. The employed droop control is a virtual impedance technique-based angle-voltage droop. The LSC's mode of operation changes from droop to PQ and vice versa. This is done through a reference selection controller that switches between the two modes of operation of LSC, by continuously monitoring the microgrid's frequency. Depending on the microgrid's status, the reference selection controller selects the appropriate voltage reference for voltage tracking and switching controller to enable the LSC to operate in either one of the modes, as seen from Fig. 5.

LSC Control for HMG Operation

The LSC of an HMG operates under the angle-voltage droop control mode (on contrary to the conventional frequency-voltage droop that has been employed for the DERs within the microgrids). Employing such an angle-voltage droop instead of the frequency-voltage droop will set the frequency of the interconnecting lines at a fixed value (e.g., 50 Hz). The angle and voltage droop equations for the LSC of kth microgrid can be written as [102–105]

$$\delta_k = \delta_o - m_{MG-k} P_{line} \tag{12a}$$

$$V_k = V_{max} - n_{MG-k} Q_{line} \tag{12b}$$

where m_{MG-k} and n_{MG-k} are the droop coefficients of the kth microgrid's LSC, and derived from

$$m_{MG-k} = (\delta_{max} - \delta_{min})/P_{line}^{max} \tag{13a}$$

$$n_{MG-k} = (V_{max} - V_{min})/2Q_{line}^{max} \tag{13b}$$

in which subscript $_{min}$ and $_{max}$ represents the minimum and maximum allowable limits for angle and voltage while P_{line}^{max} and Q_{line}^{max} are the maximum active and reactive powers injected/absorbed by each microgrid to the interconnecting lines. All the LSCs have similar $\Delta\delta = \delta_{max} - \delta_{min}$ and $\Delta V = V_{max} - V_{min}$. It should be explained here, that the transition from PQ to droop mode cannot be realized using the frequency-voltage droop because, when all the microgrids become HMG with no exchange of power through the interconnecting lines, the voltages at the output of the LSCs will be operating at the same frequency. Thus, the synchronization among the

LSCs is not possible in this situation without having a centralized controller backed up by a data communication link. The decentralized control scheme would work only under the angle-voltage droop control technique. However, the conventional angle-voltage droop control technique cannot ensure the desired power-sharing among the HMGs unless a modified angle droop [102] mechanism is implemented using the virtual impedance method. To implement the virtual impedance-based modified angle droop, (12) needs to be transformed into the $\alpha\beta$ coordinates and the voltage across the virtual impedance needs to be subtracted. Hence, the reference angle and voltage equation can be modified and rewritten $\alpha\beta$ coordinates as

$$V_{\alpha-k}^{\text{ref}} = V_k \cos\delta_k + X_{\text{v}}^k I_{\text{g}}^k \sin\theta_{\text{g}}^k \tag{14a}$$

$$V_{\beta-k}^{\text{ref}} = V_k \sin\delta_k - X_{\text{v}}^k I_{\text{g}}^k \cos\theta_{\text{g}}^k \tag{14b}$$

where $X_{\text{v}}^k = \omega_{\text{ref}} L_{\text{v}}^k$ and $i_{\text{g}}^k = I_{\text{g}}^k \angle\theta_{\text{g}}^k$ are respectively the virtual impedance and current of the coupling inductance. Then, the reference voltage and phase angle for the LSC of microgrid-k operating under the modified droop control technique, are calculated as

$$V_{\text{dr}}^k = \sqrt{\left(V_{\alpha-k}^{\text{ref}}\right)^2 + \left(V_{\beta-k}^{\text{ref}}\right)^2} \tag{15a}$$

$$\delta_{\text{dr}}^k = \tan^{-1}\left(\frac{V_{\beta-k}^{\text{ref}}}{V_{\alpha-k}^{\text{ref}}}\right) \tag{15b}$$

From (15), the instantaneous reference voltage for the LQR switching control block of the LSC in microgrid-k can be determined as

$$V_{\text{a}-k}^{\text{ref}} = V_{\text{dr}}^k \sin(\omega_{\text{ref}} t + \delta_{\text{dr}}^k) \tag{16a}$$

$$V_{\text{b}-k}^{\text{ref}} = V_{\text{dr}}^k \sin(\omega_{\text{ref}} t + \delta_{\text{dr}}^k - 120^\circ) \tag{16b}$$

$$V_{\text{c}-k}^{\text{ref}} = V_{\text{dr}}^k \sin(\omega_{\text{ref}} t + \delta_{\text{dr}}^k - 240^\circ) \tag{16c}$$

where ω_{ref} is selected based on the desired frequency level in the three-phase ac power lines (e.g., 50 Hz). The virtual impedances to share power between microgrid-i and j can be calculated based on their droop coefficients as [102, 104, 105]

$$\frac{m_{\text{MG}-i}}{m_{\text{MG}-j}} = \frac{L_{\text{g}} + L_{\text{v}}^i}{L_{\text{g}} + L_{\text{v}}^j} \tag{17}$$

Figure 11 shows the block diagram of the modified angle-voltage droop control technique. It may be noted here that, the central controller of a microgrid determines

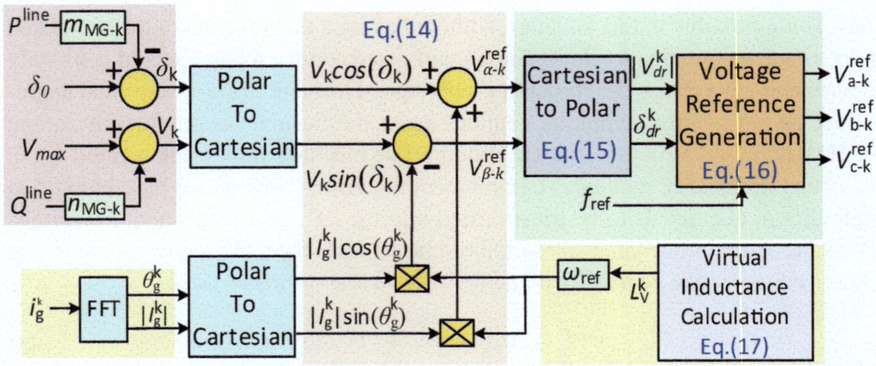

Fig. 11 Modified angle-voltage droop control employed for the LSC during HMG operation

the amount of power that is going to be shared by each HMG and it is governed by numerous factors such as distance among the microgrids, cost of power generation, line loss, willingness to participate in power exchange and so on [106–108]. Different types of optimization and decision-making techniques are proposed in ref [109–112] that determine the optimum power-sharing and evaluates the desired droop coefficients. Thus, the droop coefficient of m_{MG} and n_{MG} will be modified based on (17) to achieve the desired power-sharing.

The frequency response plot of LSCs of two HMGs operating under modified angle droop control, while supporting one PMG, is shown in Fig. 12. The plot shows both the angle and voltage droop response of LSCs.

It may be mentioned here that, the power-sharing ratio of the two LSCs operating under droop mode is set to 2:1 in this example. Hence it can be seen from Fig. 12 that, the LSC with the larger droop coefficients has a slightly smaller Gain and phase margin compared to the other LSC with smaller droop coefficients.

LSC Control for PMG Operation

When a microgrid becomes PMG, the LSC of that particular microgrid switches its operation from droop control to frequency regulation mode, which is a constant PQ control mode where the active power reference is generated through an outer frequency control loop. This mode of operation of PMG through its LSC can be termed as frequency regulation or control mode. The PMG alleviates its overloading and/or over-generation situation through this mode of operation. If a microgrid becomes PMG by overloading, a frequency deviation-based control mechanism termed as overload frequency controller (OLFC) gets activated and determines the amount of power that should be absorbed from the neighboring microgrids to retain the microgrid frequency within the acceptable limit. Likewise, during over-generation, the over-generation frequency controller (OGFC) gets activated to determine the required amount of power to inject into the neighboring microgrids. Thus, the LSC will start to operate in the constant PQ mode when a microgrid becomes a PMG. Figure 13 shows an arrangement of such a control mechanism along with

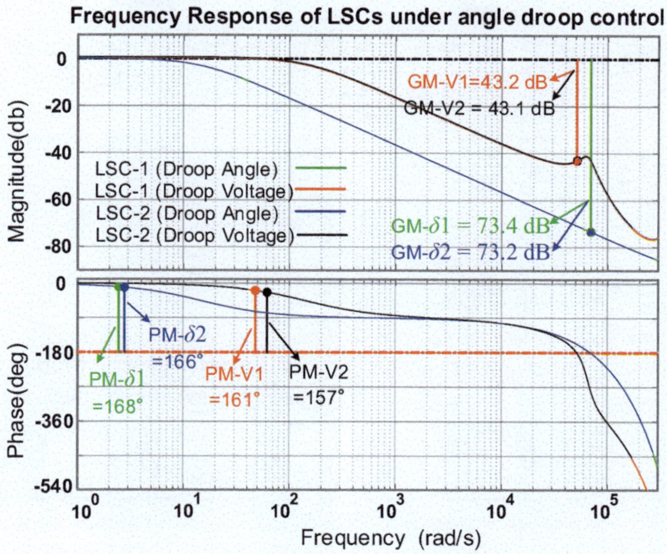

Fig. 12 Frequency response plots of LSC operating under modified angle-voltage droop control

its control activation logic. The reference of the active power (P_{ref}) is determined by either OLFC or OGFC block while the reference of the reactive power (Q_{ref}) is assumed to be zero. Once the overloading or over-generation problem is alleviated, another frequency deviation-based control system determines the new status of the microgrid as an HMG that can resume its normal operation, and hence, the LSC of that microgrid should change its operation from the constant PQ mode to the droop control mode.

The desired active power reference is determined by the frequency controller that employs two separate PI controllers and the necessary control logic and circuitry. Then, the desired reference power during an overload or over-generation is obtained either by activating the OLFC or OGFC, respectively, and is passed through a 4 × 1 multiplexer that selects the proper reference. The overall control block diagram of the frequency controller is shown in Fig. 14.

To design the frequency controller, P_{MG} and f_{max} can be considered as a disturbance to the system. The PI controller employed for the PLL is very fast compared to the overall system dynamics and LPF; hence, the PLL block can be assumed to be a unity gain block, and thus, the overall control block diagram of Fig. 14 can be represented as shown in Fig. 15. Based on the closed-loop control system of Fig. 15, the PI controllers can be designed based on the symmetrical optimum method as described in [98]. The bode plot of the closed-loop frequency regulation controller is shown in Fig. 16.

A proper reference selection logic is required to facilitate a smooth transition between constant PQ control mode (for FMG and PMG) and droop control mode (for HMG) without causing unwanted transients, oscillations, and instability. Such an

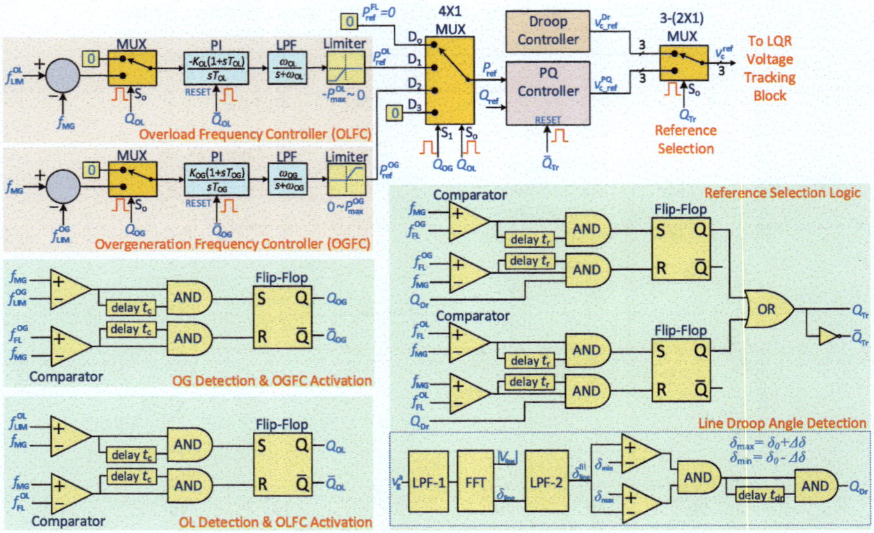

Fig. 13 Frequency regulation controller of LSC for PMG operation in topology-1

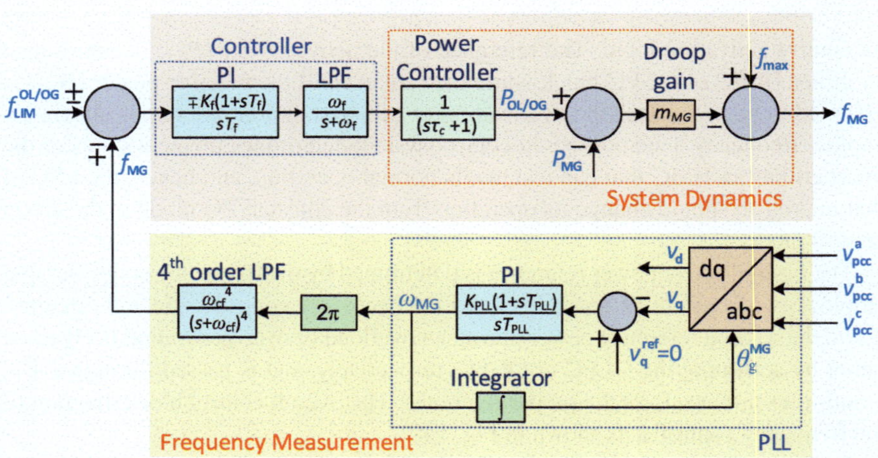

Fig. 14 Frequency controller for overload and over-generation prevention

arrangement to change over between the operating modes of LSC is shown in Fig. 13. The three-phase reference voltages are either selected from the droop control block or PQ controller block through three sets of 2×1 multiplexers using the reference selection logic employed in the mode transition controller. Then the appropriate reference voltages are passed to the LQR voltage tracking and switching control block. When the voltage references are selected from the droop control block, the PI controllers of the PQ control block should be held into reset mode to avoid integrator

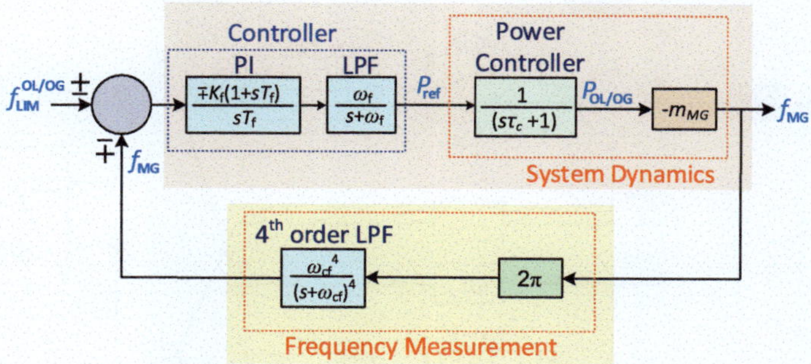

Fig. 15 Assumed close loop control mechanism of frequency controller

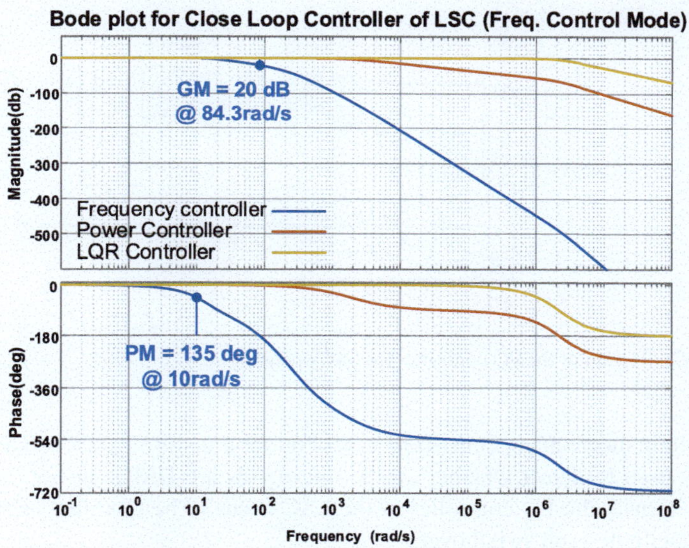

Fig. 16 Frequency response plots of LSC operating under frequency regulation mode

drift, windup, and unwanted overshoot upon activation. The switchover from droop to constant PQ mode (i.e., an HMG becoming an FMG or PMG) is straightforward as only the reset signals need to be withdrawn and the PI controllers will be made operational. However, during switchover from constant PQ to droop control mode (i.e., an FMG or PMG becoming an HMG), an additional line angle detection logic is required. When overloading or over-generation problem is over, a PMG or FMG should not be allowed to switch back to droop mode unless all the remaining microgrids within the CMG network are HMG (i.e., no operational PMG) and operating in the same phase angle, δ_0, where δ_0 is the phase angle of the line voltage when there is zero or no power-sharing among the microgrids through the interconnecting lines (i.e., $P_{\text{line}} = 0$). For this purpose and to facilitate such mode transition, a line

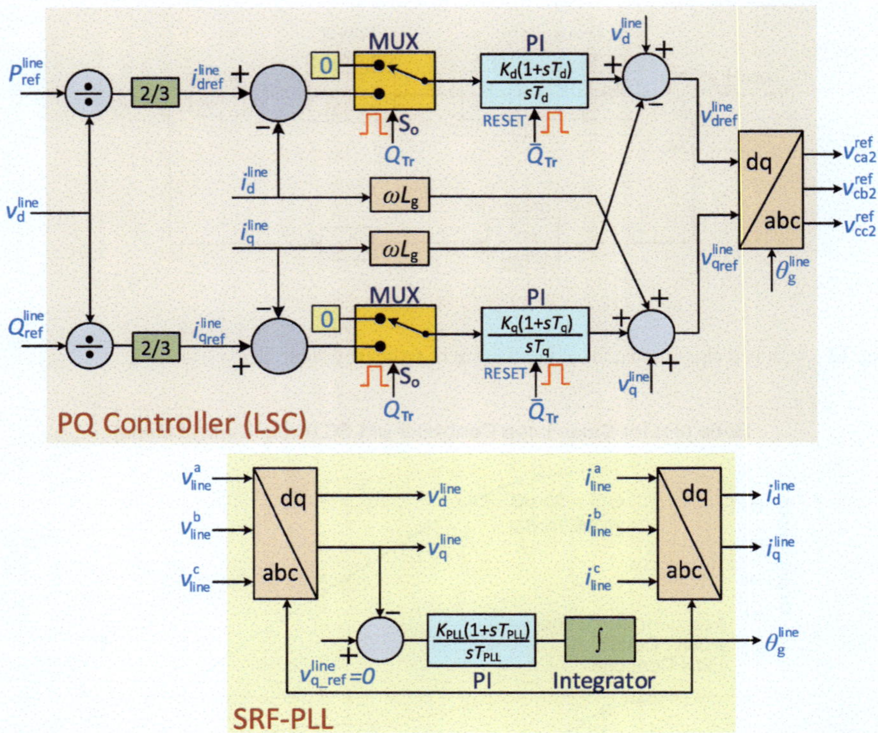

Fig. 17 Block diagram of the PQ controller for LSC operation

angle detection mechanism, consists of a fast Fourier transform block and a window detector, is used to detect whether all the microgrids are HMGs or not, as shown in Fig. 13. Hence, the PQ controller of the LSC needs to be modified as shown in Fig. 17 to facilitate such switchover.

4.1.3 Example-1

To evaluate the dynamic performance of the control technique, let us consider the network of Fig. 1 with 3 microgrids that have formed a CMG through back-to-back converters and a three-phase ac interconnecting link. In this example, the performance of such a system has been evaluated under various conditions, i.e., the normal, overloaded, and over-generating. Each microgrid is operating under a frequency and voltage droop control, while the technical parameters of the network under consideration, and the controllers are provided in Table 4. The study results are presented in Fig. 18 which shows the frequency of each microgrid and the exchanged power among the microgrids via the interlinking lines (measured at the point of common coupling of their corresponding LSCs). The figures also illustrate the dc-link voltage of the back-to-back converter, as well as the reference power and the droop angle at the output of the LSCs. The transition in the control modes of the LSCs from droop to constant PQ and vice versa is also illustrated.

Assume that, initially, the system is at steady-state condition, and all microgrids are healthy and the LSCs operating in droop control. At $t = t_1$, microgrid-1 becomes overloaded by 20% (see Fig. 18-microgrid power), and hence, its frequency decreases to 49.3 Hz (lower than the minimum frequency limit of 49.5 Hz). Thus, after a predefined delay of 0.6s, microgrid-1 changes its mode from droop to constant PQ control and at $t = t_2$, microgrid-2 and 3 start to transfer power to microgrid-1 to retain its frequency back to the acceptable limit (see Fig. 18—Frequency). Note that, the power delivered by microgrid-2 is twice that of microgrid-3. This is determined by the employed 1: 2 droop ratios. At $t = t_3$,microgrid-3 experiences over-generation after an internal demand decrease.

As such, it turns into an FMG and later a PMG after 0.6s of delay at $t = t_4$. Consequently, microgrid-3 changes its mode from droop control to constant PQ control with zero power-sharing (i.e., the floating mode) as it is not participating in supporting microgrid-1. During this interval, microgrid-2 supplies all the surplus demand of microgrid-1 alone. At $t = t_5$, microgrid-3 changes its mode of operation to constant PQ and feeds its excess power of 0.12 pu to microgrid-1. Hence, microgrid-2 only supplies the rest of the microgrid-1's excess demand. At $t = t_6$, microgrid-1's overloading is eased following the reduction in its internal demand by 20%. Thus, microgrid-1 becomes an HMG. However, it cannot be allowed to switch back to droop mode as it cannot be synchronized with the interconnecting link. So, it continues its operation under the constant PQ mode but with zero output power reference (as FMG). In this condition, the excess power of microgrid-3 is consumed by microgrid-2. Finally, at $t = t_7$, all the microgrids become HMGs. After a delay of 2s (determined by the mode transition controller), both microgrid-1 and 3 change back to droop mode and resumes operating under their normal condition. Table 3 summarizes the events applied to the microgrids and the corresponding actions taken according to the control mechanism for example- 1 (Table 4).

4.2 *Single-Phase AC Interconnection (Topology-2)*

Topology-2 is a single-phase ac interconnecting line with back-to-back power electronic converters. In this topology, the MSC is assumed to be a three-phase, three-leg VSC whereas, the LSC is assumed to be a single-phase H-bridge VSC using IGBTs or MOSFETs. Each VSC is connected to their point of common coupling through LCL filters (i.e., a three-phase LCL for the MSC and a single-phase LCL for the LSC). The structures of MSC and LSC are shown schematically in Fig. 19. Like topology- 1, any other type of three-phase and single-phase VSC topologies can also be used instead of the assumed topologies for the MSC and LSC, respectively.

Fig. 18 Simulation results of the CMG network of example- 1

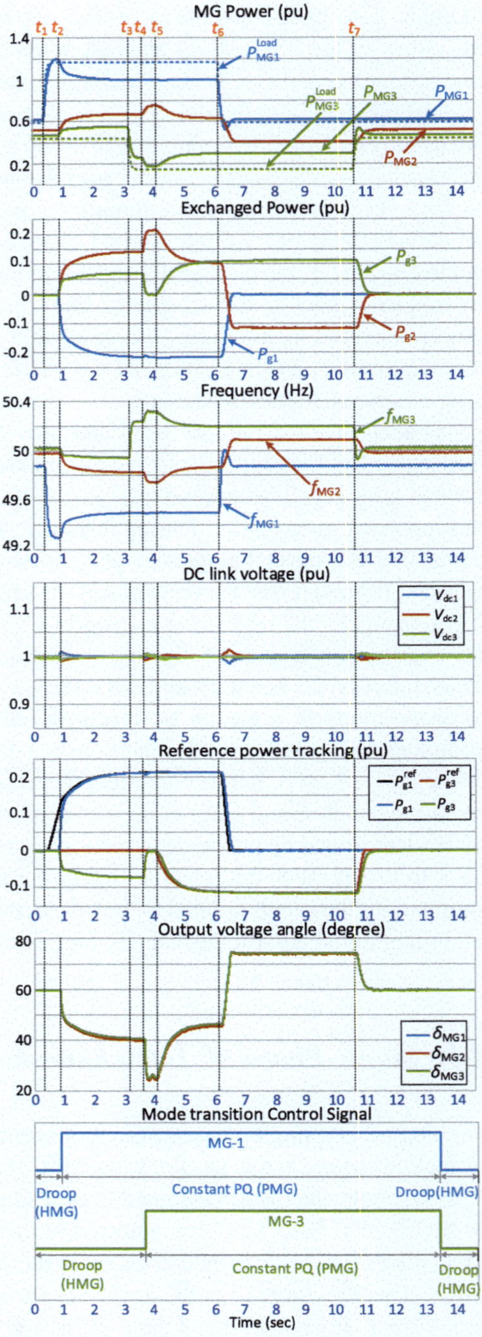

Table 3 Events applied to microgrids in example- 1 at different intervals

Time	Microgrid status			Action
	Microgrid-1	Microgrid-2	Microgrid-3	
t_1	**OL**	HMG	HMG	Normal operation for microgrid-2 and microgrid-3 while microgrid-1 is overloaded
t_2	PMG	HMG	HMG	Microgrid-1 becomes overloaded, and hence, microgrid-2 and 3 support it
t_3	PMG	HMG	**OG**	Microgrid-3 is over-generating
t_4	PMG	HMG	FMG	Microgrid-3 changes its operation mode from droop to PQ but shares zero power. Microgrid-2 is supporting microgrid-1
t_5	PMG	HMG	PMG	Both microgrid-1 and 3 are supported by microgrid-2
t_6	FMG	HMG	PMG	Microgrid-1 overload condition is removed while microgrid-2 is supporting microgrid-3
t_7	HMG	HMG	HMG	All the microgrids become HMG and all microgrids switch to droop mode

Table 4 Technical data of network under consideration

Network	$V_{\text{rated}} = 415\text{V}, f_{\text{nom}} = 50\text{Hz}, f_{\min} = 49.5\text{Hz}$, $f_{\max} = 50.2\text{Hz}$, $P_{\max} = 100\text{kW}$, $V_{\text{dc}} = 1.5\text{kV}$, $Z_{\text{MG}} = 0.1 + j0.5\Omega$, $Z_{\text{line}} = 0.1 + j0.08\Omega$
DER droop	$m = 0.01\text{Hz/kW}, n = 0.1\text{V/kVAR}$
Angle droop	$m_{\text{d}} = 0.002\text{rad/kW}, n_{\text{d}} = 0.004\text{V/kVAR}$
DC Droop Parameters	$m_{\text{MG}-1} = 2.5\text{V/kW}$, $m_{\text{MG}-2} = 5\frac{\text{V}}{\text{kW}}, m_{\text{MG}-3} = 5\text{V/kW}$
VSC and filters	$R_{\text{f}} = 0.1\Omega$, $L_{\text{f}} = 2\text{mH}$, $R_{\text{g}}(\text{MSC}) = 0.04\Omega$, $L_g(\text{MSC}) = 1.36\text{mH}$, $R_{\text{g}}(\text{LSC}) = 0.12\Omega$, $L_g(\text{LSC}) = 10\text{mH}$, $C_{\text{f}} = 25\mu F$, $C_{\text{dc}} = 4700\mu F$
Linear quadratic regulator gains	$k_1 = 58.63, k_2 = 97.18, k_3 = 1.2, \omega_c = 6283.2\text{rad/s}$
PI—PQ controller (MSC)	Proportional Gain, $K_{\text{c}} = 0.628$, Time Constant, $T_{\text{c}} = 0.034\text{s}, f_{\text{BW}} = 73.5\text{Hz}$
PI—PQ controller (LSC)	Proportional Gain, $K_{\text{c}} = 0.628$, Time Constant, $T_{\text{c}} = 0.0834\text{s}, f_{\text{BW}} = 10\text{Hz}$
Frequency controller	$K_{\text{OL}} = 15\text{k}$, $T_{\text{OL}} = 0.025\text{s}$, $\omega_{\text{OL}} = 314$ rad/s, $K_{\text{OG}} = 20\text{k}$, $T_{\text{OG}} = 0.05\text{s}$, $\omega_{\text{OG}} = 37.7$ krad/s
Dc-link voltage controller	$K_{\text{dc}} = 222, \omega_z = 2.2\text{rad/s}, \omega_{\text{p}} = 1000\text{rad/s}$, $\tau_{\text{dc}} = 392.2\text{ms}, P_{\text{so}} = 100\text{kW}, V_{\text{d}} = 380\text{V}$

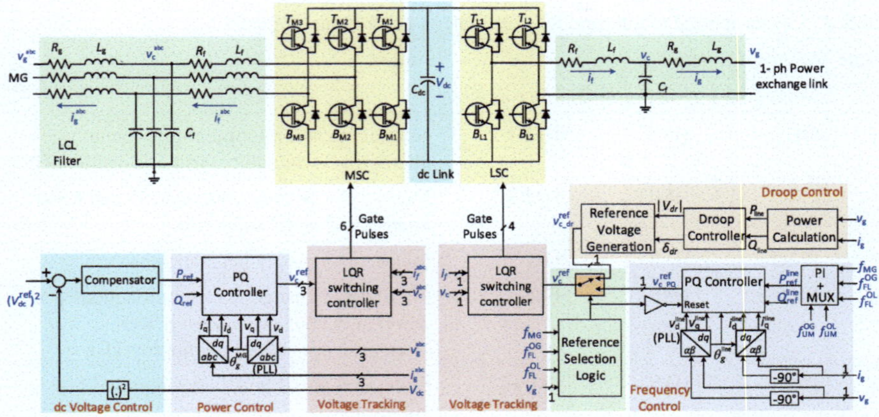

Fig. 19 Assumed structure and close loop control mechanism of the MSC and LSC for topology-2

4.2.1 Control of MSC

The control of MSC for this topology is identical to the technique as discussed in Sect. 4.1.1 for three-phase ac interconnection lines (topology-1). It should be noted here that, in this topology, the dc-link of the back-to-back converter is comprised of only one capacitor and should not be grounded, unlike the topology-1 where the dc-link is comprised of two split capacitors with the midpoint grounded.

4.2.2 Control of LSC

The control technique for LSC in this topology is similar to the methods discussed for topology-1. However, some modifications in the control technique are required to implement them for single-phase VSC. Similar to topology-1, the LSCs in this topology are operating in two modes—droop control and constant PQ mode, depending on the microgrid's status. The conditions of their operation and transition between the modes based on the microgrid's status are identical to topology-1 and will not be repeated here. Only the modifications adopted to incorporate the control mechanism will be discussed here.

The first key difference is reference voltage generation for the single-phase VSC while operating in droop mode (i.e., HMG operation). In this regard, Eq. (16) can be rewritten as

$$V_{\mathrm{a}-k}^{\mathrm{ref}} = V_{\mathrm{dr}}^{k}\sin(\omega_{\mathrm{ref}}t + \delta_{\mathrm{dr}}^{k}) \tag{18a}$$

$$V_{\mathrm{b}-k}^{\mathrm{ref}} = V_{\mathrm{dr}}^{k}\sin(\omega_{\mathrm{ref}}t + \delta_{\mathrm{dr}}^{k} - 180^{\circ}) \tag{18b}$$

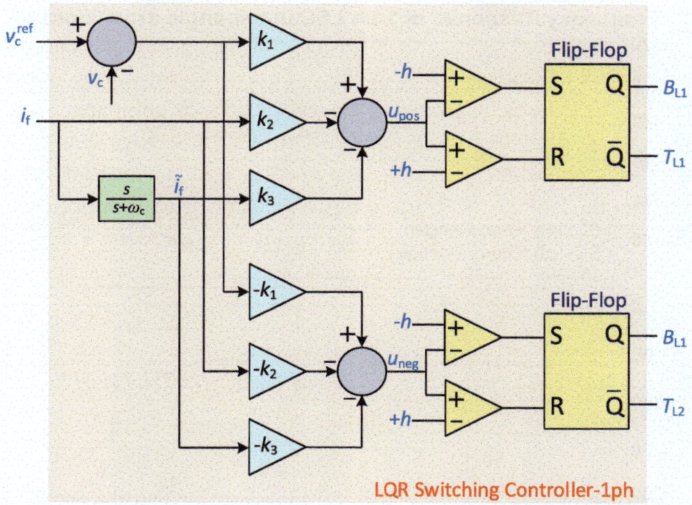

Fig. 20 Linear Quadratic Regulator based switching and voltage tracking for single-phase VSC

which are the reference voltages, v_c^{ref} for the single-phase LSC and applied to the LQR voltage tracking and switching block. Thus, the LQR switching controller needs to be modified as shown in Fig. 20. The bode plots for single-phase LSC operating under droop control are shown in Fig. 21. It may be noted here that, the single-phase LSC has a relatively smaller phase and gain margin as compared to the three-phase LSC. In other words, the three-phase LSC operating in droop mode appears to be more stable than the single-phase LSC.

Finally, the PQ controller for the LSC needs to be modified and implemented for single-phase topology. The single-phase PQ controller implemented in dq frame for LSC is shown in Fig. 22 along with its reset and control signals. To implement the PQ controller in dq frame for a single-phase converter, $\alpha\beta$ to dq transformation (Clark transformation) is used where the measured line voltage v_g and current i_g are the α components for voltage and current respectively, whereas a 90° phase-shifted version of v_g and i_g are the β components for voltage and current. The phase shift in voltage and current are introduced by a critically damped second order low pass filter that operates with a gain of 0.5 and phase shift of 90° at its cross over frequency, ω_{ps} (which is 50 Hz or 314.16 rad/s in this case). To compensate for this gain of 0.5, the signal needs to be amplified with a gain of 2. To determine the angle of the interconnection line, θ_g^{line},a synchronous reference frame PLL is used as before.

4.2.3 Example-2

Consider the network of Fig. 1 with 3 microgrids that formed a CMG using a single-phase ac interconnecting lines and back-to-back converters and is operating at steady-state condition. In this example, the performance of the CMG has been evaluated

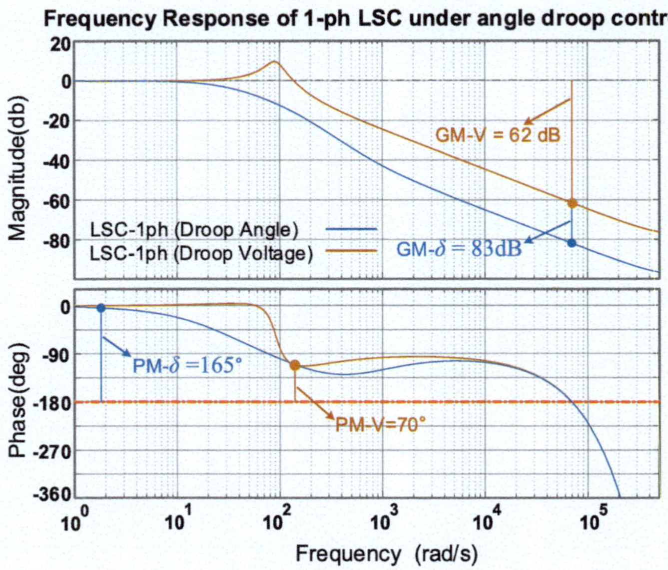

Fig. 21 Frequency response of single-phase LSC operating under modified angle-voltage droop control

under various conditions, i.e., the normal, overloaded, and over-generating, same as example- 1. The results of the study are presented in Fig. 23.

Initially, all the microgrids are HMGs and the respective LSCs are operating in droop mode. At $t = t_1$, microgrid-3 starts to experience over-generation by 8% after a decrease in its internal load demand (see Fig. 23- microgrid Power), and hence, microgrid-3's frequency rises to 50.28 Hz (higher than the maximum frequency limit of 50.2 Hz). Thus, after a predefined delay of 0.6s, microgrid-3 changes its mode of operation from droop to constant PQ control and at $t = t_2$, microgrid-1 and 2 start to absorb power from microgrid-3, so that it can retain its frequency back to the acceptable limit of 50.2 Hz (see Fig. 23). It can be seen that the power delivered by microgrid-2 ($P_{g2} = -0.0474$pu) is twice than that of microgrid-1 ($P_{g1} = -0.024$pu). This is determined by the employed 1: 2 droop ratios (see Fig. 23- Exchanged Power). At $t = t_3$, microgrid-1 becomes overloaded by 10% and frequency falls to 49.4 Hz. Therefore, it turns into an FMG and later a PMG after 0.6s of delay at $t = t_4$. In such a condition, microgrid-1 changes its mode from droop control to constant PQ control with zero power-sharing (i.e., the floating mode) as it is not participating in absorbing power from microgrid-3. During this interval, microgrid-2 absorbs all the excess generating power of microgrid-3 alone. At $t = t_5$, microgrid-1 starts to take power from both over-generating PMG of microgrid-3 and HMG of microgrid-2. Thus, microgrid-1 first receives the total over-generating power ($P_{g3} = 0.075$pu) of microgrid-3 and then the rest of the overload demand was supplied by microgrid-2 ($P_{g2} = 0.025$pu) to retain microgrid-1's frequency back to minimum limit of 49.5 Hz.

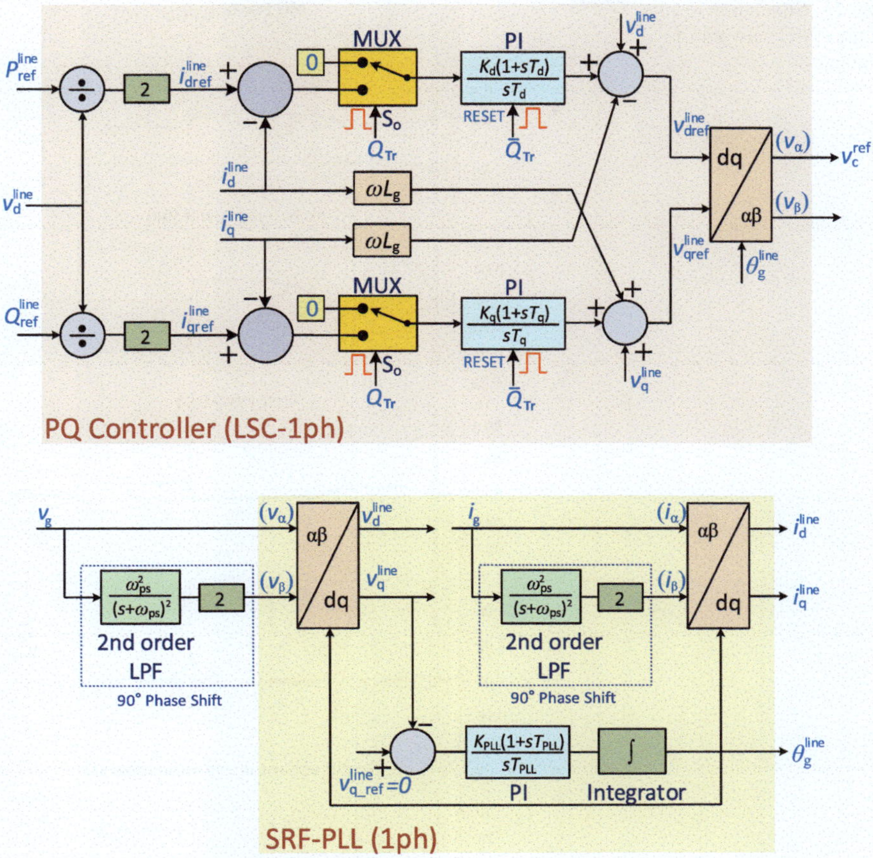

Fig. 22 Block diagram of the PQ controller for single-phase LSC

At $t = t_6$, overloading of microgrid-1 is alleviated due to a reduction in its internal demand by 10%. Therefore, microgrid-1 becomes an HMG. But it cannot be allowed to switch back to droop mode as it is not possible to synchronize the LSC with the interconnecting link. So, it continues its operation under the constant PQ mode but with zero output power reference (as FMG). In this situation, the surplus power of microgrid-3 is absorbed by microgrid-2 only. Finally, at $t = t_7$, the over-generation of microgrid-3 is over and all the microgrids become HMGs. However, both microgrid-1 and microgrid-3 will continue to operate as FMG until the droop angle of the interconnecting lines returns to δ_o (which is 60 degrees in this case). Once the line voltage angle is attained as δ_o, indicating that the exchange of power has ceased (i.e., all the microgrids are HMGs), both microgrid-1 and 3 change back to droop mode after a preset delay of 2s (determined by the mode transition controller). All the microgrids resume operating under droop mode in normal condition with no power flow through the power exchange link. Table 5 summarizes the events applied to the

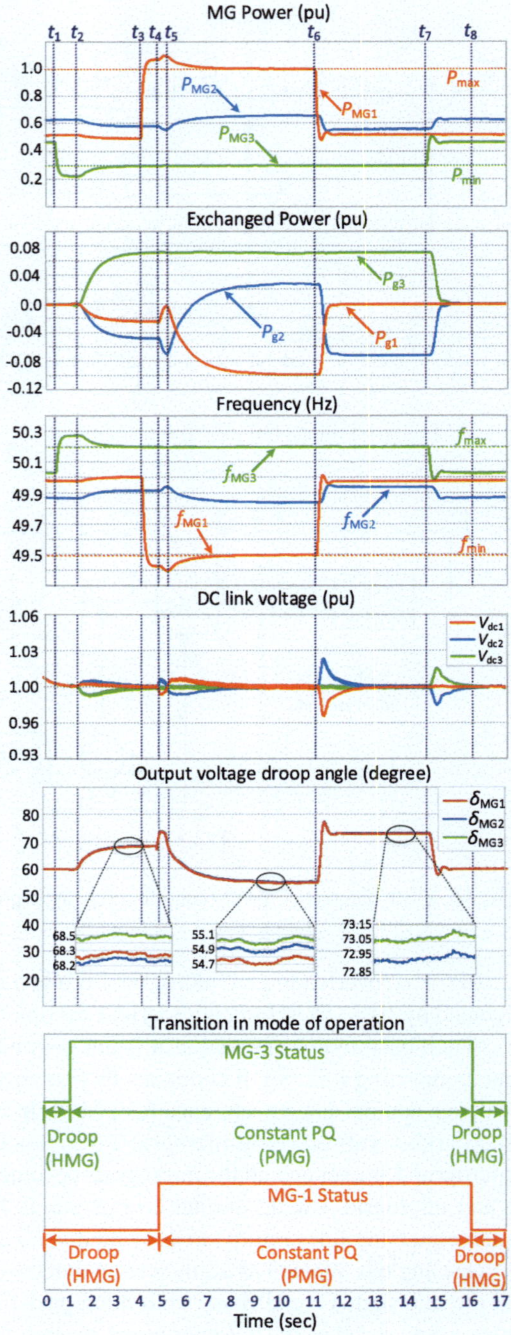

Fig. 23 Simulation results of the CMG network of example-2

Table 5 Events applied to microgrids in example-2 at different intervals

Time	Microgrid status			Action
	Microgrid-1	Microgrid-2	Microgrid-3	
t_1	HMG	HMG	**OG**	Normal operation for microgrid-1 and microgrid-2 while microgrid-3 is over-generating
t_2	HMG	HMG	PMG	Microgrid-3 becomes over-generating, and hence, microgrid-1 and 2 support it
t_3	**OL**	HMG	PMG	Microgrid-1 is overloaded
t_4	FMG	HMG	PMG	Microgrid-1 changes its operation mode from droop to PQ but shares zero power. Microgrid-2 is supporting Microgrid-3
t_5	PMG	HMG	PMG	Both microgrid-1 and 3 are supported by microgrid-2
t_6	FMG	HMG	PMG	Microgrid-1's overload condition is removed while microgrid-2 is supporting microgrid-3
t_7	FMG	HMG	FMG	Microgrid-3's over-generation is over and both microgrid-1 and microgrid-3 are in PQ mode with zero power-sharing
t_8	HMG	HMG	HMG	All the microgrids become HMG and all microgrids switch to droop mode

microgrids and the corresponding actions taken according to the control mechanism. The network and controller parameters are as shown in Table 4.

4.3 DC Interconnection (Topology-3)

Topology-3 is a dc interconnecting line with a single VSC. It is connected to the microgrid's three-phase ac lines and facilitates bi-directional power flow between the microgrid and the dc-link. The dc-links are directly connected to the dc interconnecting lines that form the CMG network. In this topology, a single VSC serves the functions of both the MSC and LSC of the previous two topologies. The VSC connected to each HMG within the CMG regulates the dc voltage across the dc-link capacitor. As the HMGs will share the power injected/absorbed by the PMGs of the CMG, an active power-dc voltage droop technique is employed that determines the reference dc voltage such that the power shared by the HMGs is as defined by the ratios of the droop coefficients. On the other hand, the VSC connected to each PMG operates in constant PQ mode and facilitates absorption or injection of the required active power to retain and regulate the frequency of the PMG within its acceptable limits. The structure and control mechanism of the VSC are discussed below:

4.3.1 VSC Structure

The assumed VSC structure is a three-phase, three-leg topology using MOSFETs or IGBTs. The VSC is connected to the microgrid through an LCL filter while the dc-link is formed with two series-connected dc capacitors with midpoint grounded. The schematic of the VSC structure is shown in Fig. 24a. It may be noted that any other three-phase VSC topology can also be used.

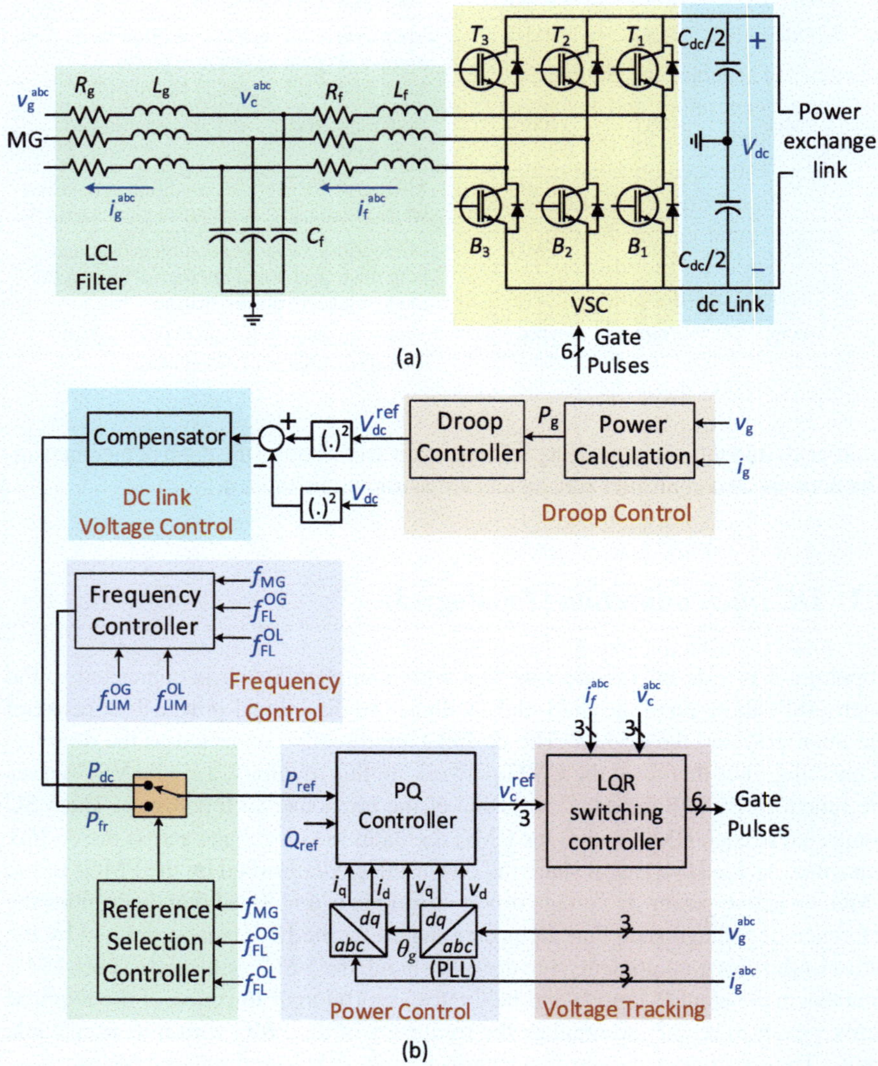

Fig. 24 VSC operation for topology-3 (a) VSC structure, (b) Closed-loop control system

4.3.2 VSC Control

The overall close loop control system of the VSC is shown in Fig. 24b. The detailed operation and design steps of the LQR Switching controller, PQ controller, and dc-link voltage controller are identical to the MSC control of topology-1 as discussed in Sect. 4.1.1. The only difference is that, in this topology, the reference voltage of the dc power exchange link (i.e., the voltage across the dc capacitor) should be generated through dc droop control block whereas, in topology-1 the dc-link voltage was fixed and predefined. The VSC operates in two different modes, depending on the status of the microgrid, i.e., when it is connected to an HMG or a PMG. Each of these modes is discussed separately below.

VSC of an HMG

The VSC connected to an HMG is operating in droop mode and primarily responsible for regulating the dc-link voltage across the dc capacitor V_{dc}, to the desired level of V_{dc}^{ref}. The desired voltage level is not constant as the value depends on the total power that the HMG should absorb from the link or supply to that. If the CMG network includes multiple HMGs, then they can share the power supplied to the PMGs or received from them at any desired ratio. This can be easily realized by using an active power—dc voltage droop control mechanism for the output dc-link voltage of each VSC. This is schematically illustrated by the dc droop control block in Fig. 24b and can be mathematically represented as

$$\left(V_{dc}^{ref}\right)_{MG-i} = V_{nom} - m_{MG-i}P_{MG-i} \tag{19}$$

where m_{MG} and P_{MG} denote the dc droop coefficient and the active power absorbed/injected by the VSC respectively and V_{nom} is the nominal voltage of the dc-link (e.g., 1 pu). It should be noted that an active power absorption will be considered as negative whereas, active power injection by the VSC is considered as positive. According to (19), the ratio of the power absorbed (or injected) by HMG i and j will be the reciprocal of the ratio of their droop coefficients, i.e.,

$$m_{MG-i}P_{MG-i} = m_{MG-j}P_{MG-j} \tag{20}$$

SC of a PMG

The VSCs of a PMG should absorb the required amount of power from the dc-link into the PMG when it is overloaded. Likewise, if the PMG experiences over-generation, the VSC should inject the required amount of power from the PMG to its dc-link. Thus, the frequency of the PMG will be retained to the acceptable range through such exchange of power. To facilitate such power exchange, the VSC is suggested to operate in the constant PQ mode. As the key objective of power exchange mechanism in CMG network is to control the active power flow, the reference of the reactive power (Q_{ref}) is set to be zero. The reference for the active power (P_{ref}) is determined

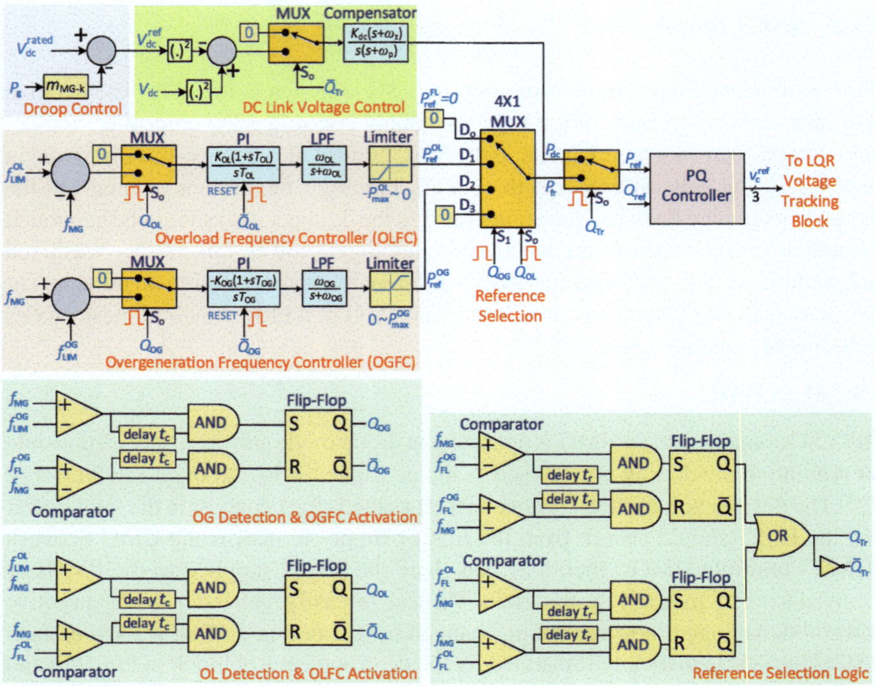

Fig. 25 Frequency regulation controller of VSC for PMG operation in topology-3

through the deviation of the PMG's frequency from the acceptable limit. Such a frequency regulation controller is schematically shown in Fig. 25.

The switchover in the mode of operation from droop to PQ and vice versa is achieved through a reference selection controller which monitors the microgrid's frequency continuously. Depending on the status of the microgrid, the appropriate P_{ref} will be selected, as schematically illustrated in Fig. 25.

When a microgrid becomes a PMG, the frequency controller gets activated. During overloading, the OLFC determines the total amount of power that should be received by that particular PMG from the neighboring microgrids to return the microgrid frequency to the acceptable limits. Similarly, during over-generation, OGFC gets activated and determines the required power that should be delivered to the neighboring microgrids. Thus, when a microgrid becomes a PMG, its VSC should change its operation from droop control to the constant PQ mode. Such a control mechanism can be realized using a hysteresis controller implemented with the reference selection logic and schematically shown in Fig. 25. The active power reference (P_{ref}) is determined by a frequency deviation-based overload or over-generation control mechanism while the reactive power reference (Q_{ref}) is set to be zero. Once the overloading or over-generation situation is mitigated, another frequency deviation-based control mechanism will determine the new status of the microgrid. After the detection of the microgrid's status and certain predefined time delay as set by the controller, the VSC of that particular microgrid will change from the constant PQ mode to the droop control mode and hence, deactivates the frequency controller.

The frequency controller determines the desired power reference through two different PI-based closed-loop control systems along with its necessary control logic and circuitry. Then, the desired power reference is selected through a 4 × 1 Multiplexer circuit that selects the proper power reference during the overload or excess-generation situation respectively by activating the OLFC or OGFC. To coordinate between droop control mode (for HMG) and constant PQ control mode (for the FMG and PMG), a proper reference selection logic is needed to switchover between the modes smoothly without causing unwanted transients, oscillations, and instability. The reference selector will select proper active power reference either from the droop control-based dc-link controller block or frequency controller block through a 2 × 1 multiplexer and pass it to the PQ control block. Once a PMG becomes an HMG, the reference selector will enable the VSC to switch back to droop-based dc-link voltage control mode.

4.3.3 Example-3

Consider the system of Fig. 1 with three CMGs. Figure 26 illustrates the study results and shows each microgrid's total demand and frequency, along with the exchanged power amongst the microgrids, the dc-link voltage at the output of the VSCs, and the transition in the mode of operation of VSCs from droop to PQ and vice versa. Unlike, the previous two examples, all the microgrids are heavily unbalanced as can be seen from Fig. 26. All the three microgrid's individual per phase load power and their variation over the simulation period are shown.

Let us assume all the microgrids are HMG operating under a voltage and frequency droop and hence, the VSC of all HMGs are operating under the dc voltage droop control mode with no power flow in the power exchange dc-link. At $t = t_1$, microgrid-2 starts over-generation by 10% by withdrawing some of the connected loads (note that all the phases are underloaded nonuniformly and the underloaded microgrid is heavily unbalanced), and hence, the frequency rises to 50.4 Hz (i.e., over the maximum frequency limit of 50.3 Hz) while the other two microgrids are operating at nominal condition. Therefore, the VSC of microgrid-2 changes its mode of operation from droop control to constant PQ control and starts to deliver an active power of 0.1 pu to alleviate the over-generation situation. As a result, at $t = t_2$, the other two HMGs start to support microgrid-2 by absorbing the excess power produced by it after a pre-defined delay of 0.6 s. microgrid-1 absorbs 0.062 pu of power while microgrid-3 absorbs 0.03 pu which is half of that of microgrid-1 because of the preset ratio of droop coefficients ($m_{\text{MG-3}}/m_{\text{MG-1}} = 2:1$) and thus, the frequency of microgrid-2 is regulated to the desired maximum limit of 50.3 Hz. It may be noted here that, the choice of droop coefficients to share power depends on the microgrid loading condition. As the connected load in microgrid-1 is higher than microgrid-3, the former is allowed to receive more power than the latter.

Fig. 26 Simulation results of the CMG network of example-3

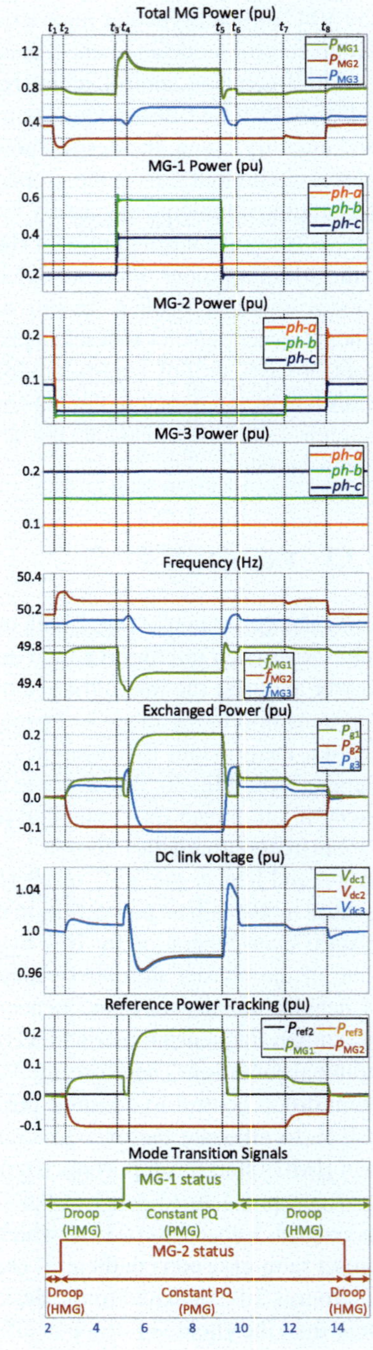

This continues until $t = t_3$ at which microgrid-1 experiences overloading of 20% due to the addition of loads in phase-b and c, and thus, its frequency falls and settles at 49.35 Hz (beyond the acceptable minimum limit of 49.5 Hz). During this interval, the frequency controller of microgrid-1 is triggered and determines the amount of power that needs to be imported from the dc-link to regulate its frequency to the desired level. It changes its mode of operation from dc voltage droop control to constant PQ mode and at $t = t_4$, the VSC of microgrid-1 starts absorbing power of 0.2 pu. This power is provided by the over-generating PMG of microgrid-2, as well as the HMG of microgrid-3. microgrid-1 only absorbs the amount of power needed to improve its overloading as a result microgrid-3 starts to deliver a power of 0.12 pu in addition to the excess power of 0.1 pu as provided by microgrid-2 which is also operating as a PMG. This scenario continues until $t = t_5$ at which the overloading of microgrid-1 is removed and does not need external support. As a result, the VSC of microgrid-1 switches its mode of operation from constant PQ mode to the dc voltage droop control mode with zero power exchange. However, as the CMG still includes an over-generating PMG (i.e., microgrid-2), the HMG of microgrid-3 starts to absorb the excess power of microgrid-2. At $t = t_6$ microgrid-1 switches back in dc droop mode and starts absorbing excess power in addition to microgrid-3, based on their preset droop coefficient value as before. At $t = t_7$, some load is added back to phase-b of microgrid-2 and hence reduces the excess generation from 10 to 6%. The frequency controller of microgrid-2 adjusts the power reference to 0.06 pu and thus the power absorbed by microgrid-1 and microgrid-3 are 0.038 pu and 0.018 pu respectively. At $t = t_8$, the situation of over generation in microgrid-2 is alleviated and goes back to its nominal operating condition to become an HMG. Thus, the exchange of power is ceased and VSC of microgrid-2 switches back to dc voltage droop.

Table 6 lists the different events occurring in this example over time and shows the status of the microgrids, as well as the measures taken by the power exchange strategy after each event. The network and controller parameters are as shown in Table 4.

5 Stability

The stability of the power exchange strategy for all three CMG topologies is evaluated. The stability analysis is performed by varying numerous key design parameters such as the active and reactive power droop coefficients, interconnecting line's impedance, and PMG's power demand. The eigenvalue plots obtained from the stability analysis are shown in Figs. 27, 28, 29, 30, 31, 32, 33, 34 and 35.

Among all these parameters, the active power droop coefficient, m_{MG}, (for topology-1 and 2) exhibits the most dominating effect over the CMG stability and can drive the system towards instability; whereas the CMG's dynamics and response are affected by the other parameters but have relatively less influence on the system's stability. For topology-3, parametric variations affect the network dynamics, but it

Table 6 Events applied to microgrids in example-3 at different intervals

Time	Microgrid status			Action
	Microgrid-1	Microgrid -2	Microgrid -3	
t_1	HMG	**OG**	HMG	Normal operation for microgrid-1 and 3 while microgrid -2 is over-generating
t_2	HMG	PMG	HMG	Microgrid-2 switches to PQ mode to deliver power and microgrid -1 and 3 are operating in droop to support it
t_3	**OL**	PMG	HMG	Microgrid-1 is over-loaded
t_4	PMG	PMG	HMG	Microgrid-1 changes its operation mode from droop to PQ to receive power. Both microgrid-1 and 2 are supported by microgrid-3
t_5	FMG	PMG	HMG	Microgrid-1's overload is removed while microgrid-3 was supporting microgrid-2. Microgrid-1 exchanges zero power at this instant
t_6	HMG	PMG	HMG	Microgrid-1 switches to droop and joins microgrid-3 in supporting microgrid-2
t_7	HMG	PMG	HMG	Microgrid-2's over-generation is reduced but remains overloaded. The absorbed power by HMGs is adjusted
t_8	HMG	HMG	HMG	Microgrid-2's overgeneration is over. All the microgrids become HMG and all microgrids switch to droop mode

is not as dominant as it is in topology-1 and 2 and more importantly, do not lead the system to instability.

5.1 Variation of Droop Coefficients

The active power droop coefficient of m_{MG} is changed in steps from its nominal value of 0.002rad/kW for both topology-1 and 2, while the reactive power droop coefficient of n_{MG} is kept constant. The coefficient m_{MG} and n_{MG} are defined in (13a) and (13b), respectively. The critical value of $m_{\mathrm{MG}}^{3\varnothing}$ for topology-1 is obtained as $m_{3\varnothing}^{\mathrm{crit}} = 0.08\mathrm{rad/kW}$ which can be seen from Fig. 27a, whereas, the critical value of $m_{\mathrm{MG}}^{1\varnothing}$ for topology-2 is obtained as $m_{1\varnothing}^{\mathrm{crit}} = 0.125\mathrm{rad/kW}$ that can be seen from Fig. 28a. Any further increase in the value of m_{MG} in both topologies will make the CMG unstable. This is also verified by time-domain simulation in PSIM and the critical values are obtained as, $m_{3\varnothing}^{\mathrm{crit}} = 0.076\mathrm{rad/kW}$ and $m_{1\varnothing}^{\mathrm{crit}} = 0.122\mathrm{rad/kW}$ which are almost equal to the values obtained from MATLAB simulation. On the other hand, the variation of n_{MG} for both topology-1 and 2, does not significantly affect the stability of the CMG as the reactive power flowing through the CMG link

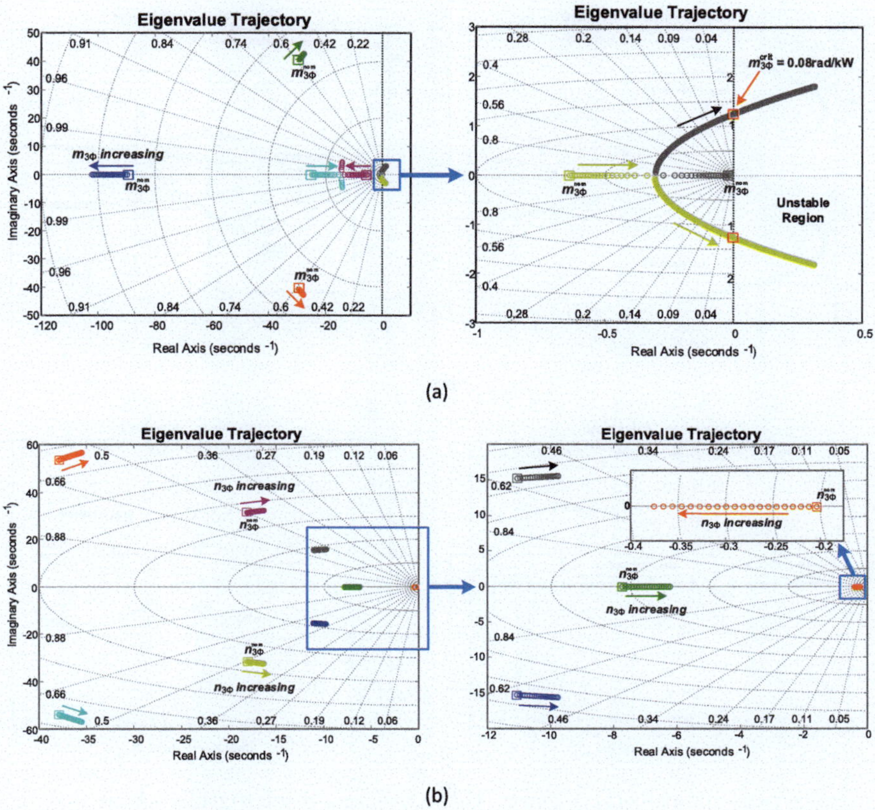

Fig. 27 CMG Stability analysis for topology-1: Eigenvalue trajectory for variation of droop coefficients (a)$m_{MG}^{3\varnothing}$, (b)$n_{MG}^{3\varnothing}$

is very small, and hence the variation of n_{MG} does not have a noticeable effect on the system stability, as seen from Fig. 27b and Fig. 28b, respectively. The value of n_{MG} was varied from its nominal value of 0.004rad/kVAR up to 0.4rad/kVAR (i.e.,100 times of its nominal value) to obtain the eigenvalue trajectory plots.

At the nominal value of droop coefficients $m_{\mathrm{MG}}^{\mathrm{nom}}$ and $n_{\mathrm{MG}}^{\mathrm{nom}}$, the response of topology-2 is more oscillatory (less damping) compare to the topology-1 which can also be justified from the eigenvalue location for nominal values of droop coefficients as shown in Figs. 27 and 28, and also from the frequency response plots of Figs. 12 and 21. However, topology-1 is more sensitive to variation of droop coefficients and becomes unstable with a relatively smaller value ($m_{3\varnothing}^{\mathrm{crit}} = 0.08\mathrm{rad/kW}$) as compared to topology-2 ($m_{1\varnothing}^{\mathrm{crit}} = 0.125\mathrm{rad/kW}$). In other words, topology-2 has better tolerance to droop coefficient variations and a larger stability range even though the nominal response is more oscillatory compared to topology-1. If the CMG network requires to operate with a larger value of droop coefficients, topology-2 is more suitable than topology-1.

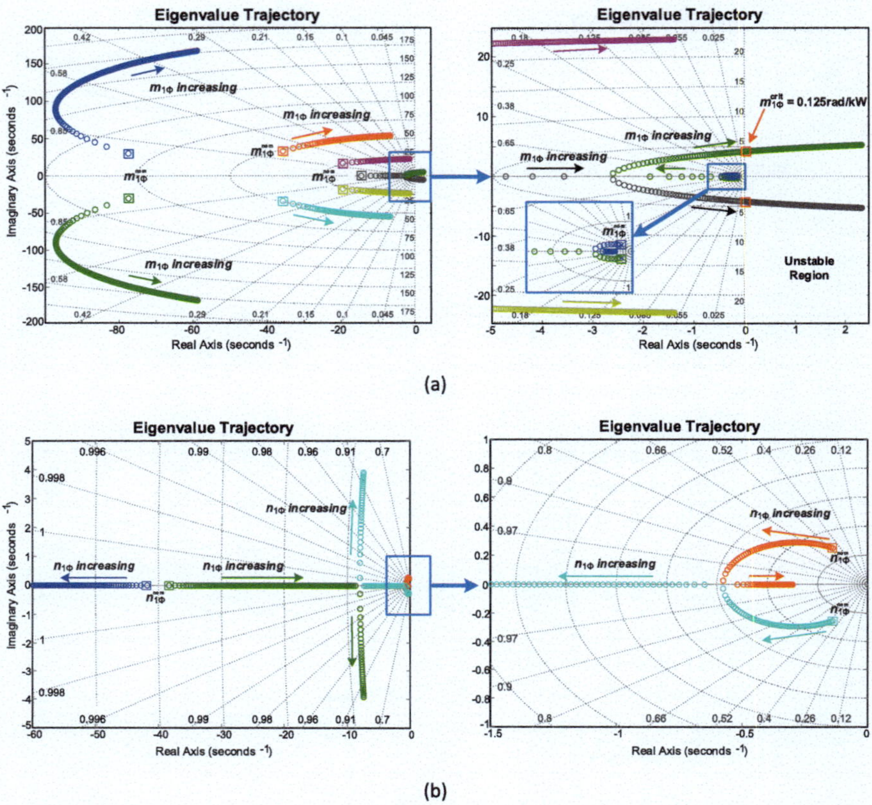

Fig. 28 CMG Stability analysis for topology-2: Eigenvalue trajectory for variation of droop coefficients (a)$m_{MG}^{1\varnothing}$, (b) $n_{MG}^{1\varnothing}$

For topology- 3, the eigenvalue trajectory plot is obtained by varying only the active power droop coefficient m_{MG} as the exchange of reactive power is not possible in dc droop control and the associated dc power exchange link. The value of $m_{\text{MG}}^{\text{dc}}$ is incremented from its nominal value of $1V/\text{kW}$ up to $50V/\text{kW}$ to obtain the eigenvalue trajectory plot as shown in Fig. 29. It can be observed that, increasing the value of $m_{\text{MG}}^{\text{dc}}$ does not make the CMG network unstable like topology-1 and 2 and hence no critical value of $m_{\text{MG}}^{\text{dc}}$ was obtained. However, the response of the dc power exchange link is affected by such variation of m_{MG}. The response of the dc power exchange link becomes over-damped and sluggish as the value of m_{MG} is increased. It can be seen from the eigenvalue plot of Fig. 29 that, the dominant pole moves further towards the origin as m_{MG} is increased and hence make the system response very slow and thus the dc power exchange links relatively longer time to settle to its steady-state value when power exchange occurs among the microgrids. A smaller value of m_{MG} makes the dc-link to respond faster. However, there is a limiting value of m_{MG} that depends on the value of line resistance. To share the desired amount of power among the

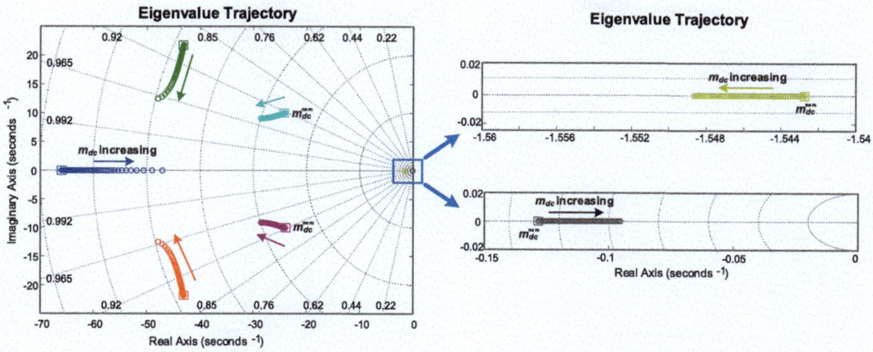

Fig. 29 CMG Stability analysis for topology-3: Eigenvalue trajectory for variation of droop coefficients m_{MG}^{dc}

microgrids according to their droop ratio, m_{MG} should be large enough as compared to the line resistance, otherwise, the assumption based on which conventional dc droop method is developed does not remain valid, and hence the property of the desired power-sharing among the microgrids according to their droop coefficients is lost. This phenomenon is further discussed in Sect. 6.1 of sensitivity analysis.

5.2 *Variation of PMG Power*

The dynamic response of the CMG is also affected by the amount of power absorbed (overload) from or injected (over-generation) to the CMG network through the PMG. To observe this effect, the PMG power was varied from 0.02 pu to 0.15 pu for all the three topologies and the corresponding eigenvalues are plotted. This is shown in Figs. 30, 31 and 32. In all the three topologies, as the PMG power increases, the system damping increases and the response becomes slightly sluggish. In other words, the CMG network performs better when rated power is exchanged among the microgrids forming the CMG. In topology-1, the dominant eigenvalue moves towards the right as the PMG power increases and hence makes the system response sluggish; however, such variation of PMG power level does not lead the system to instability as can be seen from the eigenvalue plot of Fig. 30. A similar scenario can be observed in topology-2 also, but the dominant eigenvalues in this topology are complex conjugate pair that moves towards left as the PMG power increases and thus the CMG stability remains unaffected as shown in Fig. 31. Finally, in topology-3, the dominant eigenvalue moves towards the right with PMG power variation and the effect is similar to the effect as discussed for topology-1. The eigenvalue plot for topology-3 is shown in Fig. 32.

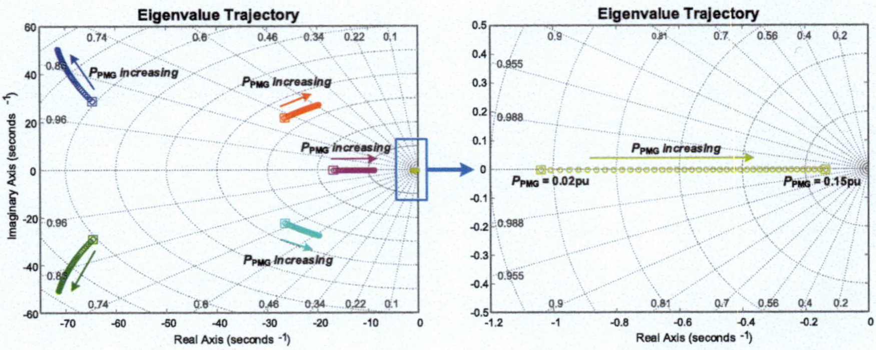

Fig. 30 CMG Stability analysis for topology-1: Eigenvalue trajectory for PMG Power variation

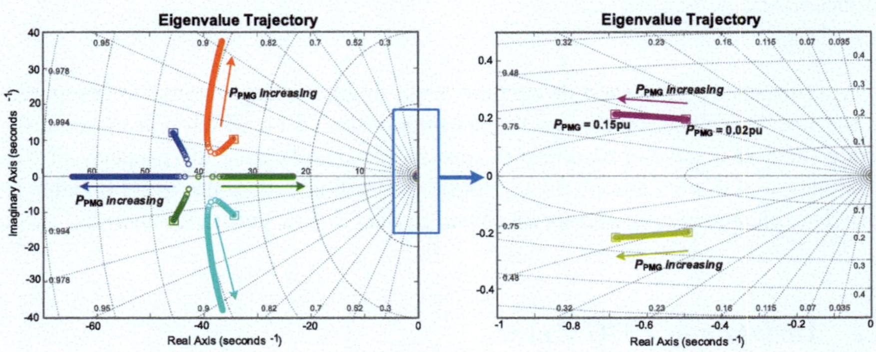

Fig. 31 CMG Stability analysis for topology-2: Eigenvalue trajectory for PMG Power variation

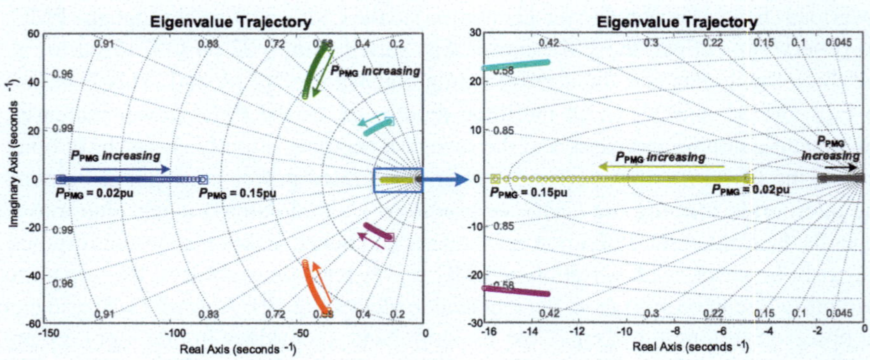

Fig. 32 CMG Stability analysis for topology-3: Eigenvalue trajectory for PMG Power variation

5.3 *Variation of Line Parameters*

The impedance of the line (i.e., line reactance to resistance ratio, termed as X/R) was varied for topology-1 and 2 whereas, the line resistance, R_{line} is varied for topology-3, to see the effect on stability. It may be noted here that, as topology- 3 is a dc interconnection link, the power exchange lines are assumed to be purely resistive with zero inductance. In topology-1 and 2, the line inductance is increased from 0.16 mH to 3.2 mH while the resistance of the line is held constant at 0.1 Ω. The evaluated eigenvalues are shown in Figs. 33, 34 and 35. It can be observed that increase in X/R ratio i.e., increase in line inductance, reduces the system damping and makes the system response oscillatory as the complex conjugate pair of eigenvalues moves towards the right (see Figs. 33 and 34), whereas in topology-3, the system damping increases and makes the response very sluggish as the dominant eigenvalue moves further towards the right with increasing value of line resistance (see Fig. 35). Such increase in line resistance deteriorates the property of power-sharing based on the microgrid's droop coefficient but, does not lead the system towards instability.

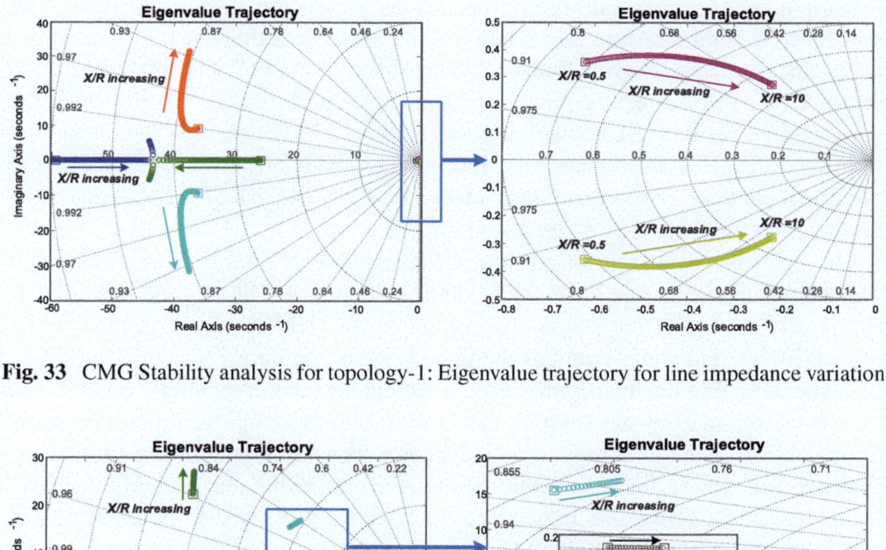

Fig. 33 CMG Stability analysis for topology-1: Eigenvalue trajectory for line impedance variation

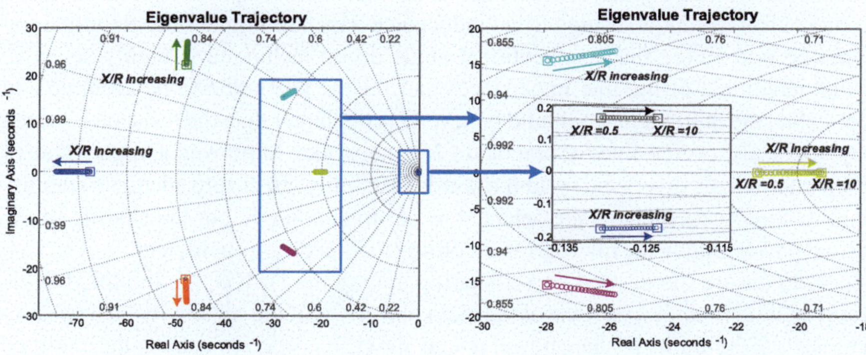

Fig. 34 CMG Stability analysis for topology-2: Eigenvalue trajectory for line impedance variation

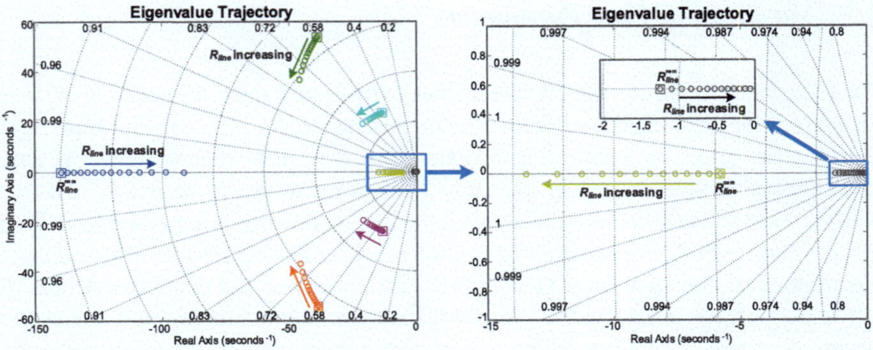

Fig. 35 CMG Stability analysis for topology-3: Eigenvalue trajectory for line impedance variation

6 Sensitivity

The sensitivity of the power exchange link and employed control strategy for all the three topologies against the design and operational parameters has been analyzed. The objective is to determine how the variation of the interconnecting line impedance (i.e., the distance of the microgrids from each other) and the amount of power delivered or absorbed by a PMG affect the power exchange link's performance. The line loss, the amount of power delivered/absorbed from HMGs to PMG, the LSC's voltage and droop angle, the HMG's frequency, dc-link voltage, and finally error in power-sharing based on their droop ratio are evaluated in the sensitivity analysis. The obtained results are shown in Figs. 36, 37, 38, 39, 40, 41 and 42.

The study is carried out for all three CMG topologies for a scenario where two HMGs are supporting one PMG while sharing the power with a droop ratio of 2:1. The results show that the power-sharing among the HMGs based on their droop ratios is affected to some extent as the impedance of the interconnecting lines (i.e., the distance among the microgrids) is increased in the case of topology-1 and 2. The error in power-sharing stays within the limit due to the employed modified angle droop control technique using virtual inductance which minimizes the error margin. It can be noted here that, conventional angle droop technique is highly sensitive to the magnitude of the line impedances and for the technique to be accurate and applicable, the line reactance should be several times larger than the resistance, which may not be always feasible for practical CMG networks. Moreover, a larger value of droop coefficients drives the system towards instability. To ensure adequate stability margins, droop coefficients are selected to be small enough, but the sharing of the desired power among the microgrids according to their droop coefficients will be lost. Hence, to ensure the desired power-sharing among the HMGs, the modified angle droop method is employed. One of the main objectives of the sensitivity analysis is to demonstrate the ability to share power among the HMGs operating under the modified angle droop while line impedances are varied, and the ratio of line resistances to reactances is larger than unity.

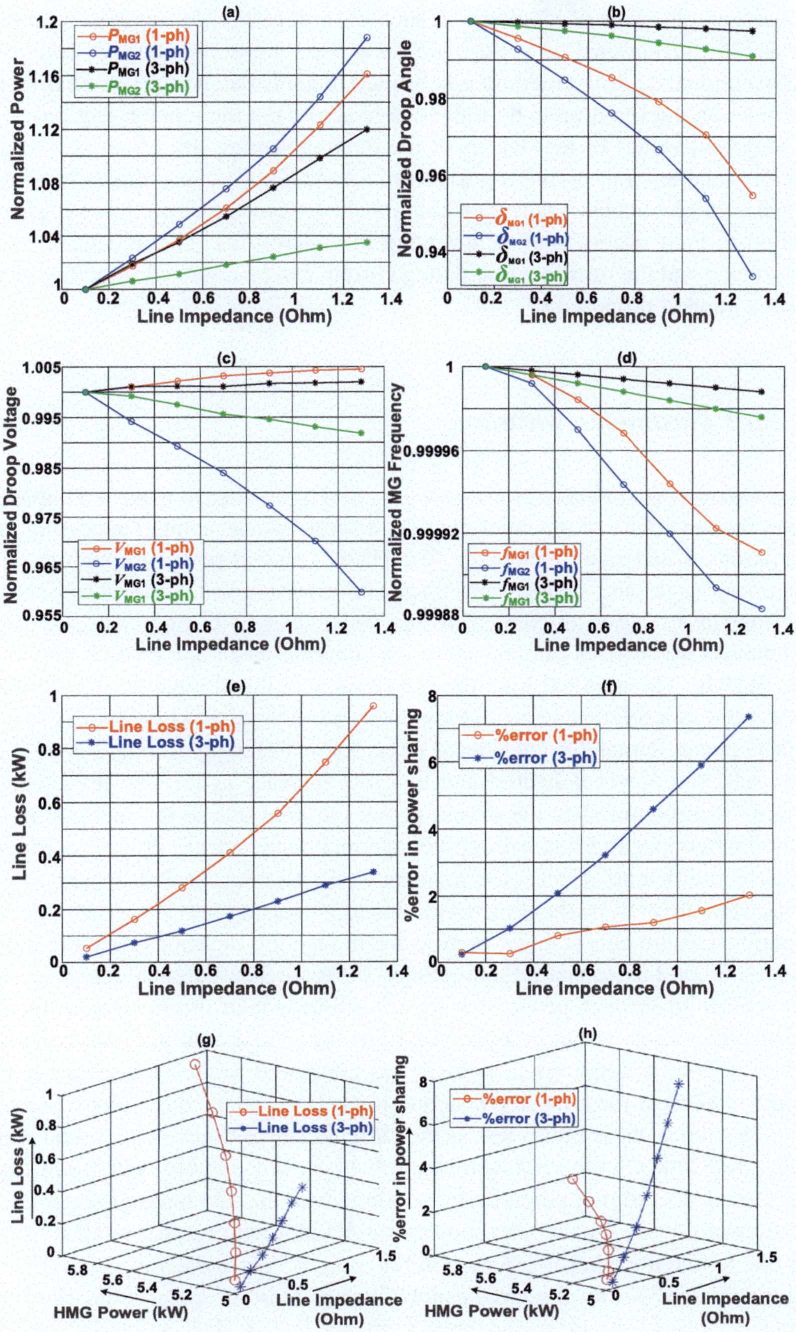

Fig. 36 Comparative CMG sensitivity analysis of topology-1 and 2 for line impedance variation

However, in the case of topology- 3, such error in power-sharing is substantially larger as the conventional dc voltage–power droop control technique is employed. The conventional dc droop method is subjected to a mismatch in power-sharing and experiences an uneven drop in dc-link voltage across the interconnecting link [113, 114]. Different values of resistances of the interconnecting lines (that is governed by the distance among the microgrids forming the CMG) cause this voltage drop and such voltage mismatch is responsible for the loss of the desired power-sharing according to their droop coefficient among the microgrids. This mismatch in the droop voltage and the resulting circulating current can be removed using the virtual resistance method proposed in [114].

6.1 Line Parameter Variation

The first analysis is conducted for topology-1 and 2 together to make a comparison between the sensitivity of the three-phase and single-phase ac interconnecting links for variations of the same parameter. The HMG's shared power through the link, LSC's droop angle and voltage, line loss, and error percentage in power-sharing as the impedance varies are shown in Fig. 36a–h. For both topologies, the sensitivity analysis was carried out considering a situation when one PMG is supported by two HMGs. The nominal line impedance used in the simulation of example-1 and 2, is used as a reference and all parameters are normalized based on this value. The PMG power demand is considered to be 5 kW. Referring to Fig. 36a–h, it can be seen that, the power delivered and line loss increase as the line impedance is increased. To accommodate the excess power required due to the increase in line loss, the interconnecting line's droop voltage and angle start to drop, according to their droop coefficients. Hence, the frequency of each microgrid also starts to drop as the power delivered by them increases due to increasing line loss. The variations in all parameters are almost linear as expected though the variations are more prominent in topology-2 as compared to topology-1. Figure 36e, f show the line loss and the percentage of error in power-sharing according to their droop ratio, as the line impedance increases. It can be observed from Fig. 36e–h that the line loss in the single-phase link of topology-2, is higher as compared to the three-phase link of topology-1, whereas the error in power-sharing is larger in the three-phase link. The line loss in a three-phase link is less as the current is three times smaller than that of a single-phase link. On the other hand, a single-phase link exhibits better accuracy in power-sharing according to their droop coefficient among the microgrids. The Line loss and error percentage are also shown as a 3D plot with respect to HMG's total power due to line impedance variation in Fig. 36g, h, respectively.

Figures 37 and 38 show the surface plot of line loss, droop angle, voltage and error percentage in power-sharing as the power delivered by two HMGs is varied due to an increase in the line impedance in topology-1 and topology-2, respectively. It can be observed that the droop angle plots of both HMG follow a similar pattern, whereas the droop voltage plots are not similar. As the impedance of the interconnecting lines

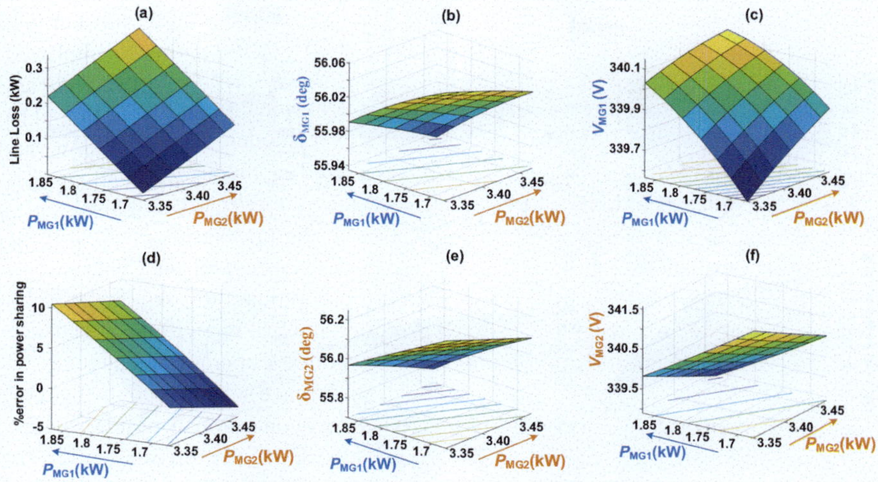

Fig. 37 Effect of line impedance variation on CMG network for topology-1

is increased, the sharing of desired power, according to their droop coefficients, starts to deviate slightly and produces an error. The reason behind such behavior is that, as the impedances of the lines become larger, the assumption of droop coefficients being greater than the line reactance does not remain valid any longer, and thus, the property of power-sharing, as the reciprocal of the droop coefficients, is lost. As can be seen from Fig. 37d, the variation of impedance causes an error of –5 to + 10% in topology-1 whereas, the error ranges from –8 to + 4% in topology-2, as can be seen from Fig. 38d. In both topologies, the error stays in an acceptable range because of the implemented modified angle droop, which is an effective way to ensure accurate power-sharing among the microgrids.

The sensitivity analysis for topology-3 due to line resistance variation is shown in Fig. 39. It can be observed from Fig. 39a–d that the line loss and power delivered by each microgrid are increased as the line resistance (i.e., the distance of the microgrids from each other) increases, whereas to accommodate the excess power required due to line loss, the frequency and dc-link voltages start to drop according to their droop coefficients values. The nominal PMG power demand is assumed to be 10 kW. As the resistance of the interconnecting lines is increased, the property of power-sharing based on their droop coefficients starts to become erroneous. This is due to the fact that as the resistances of the lines become larger than droop coefficients values, the property of power-sharing as the reciprocal of the droop coefficients, is not valid any longer.

This is can be seen in Fig. 39e. The Line loss and percentage of error are also shown as a 3D plot with respect to total HMG power variation due to the increased line resistance, in Fig. 39f, g, respectively. It can be observed from the above analysis that, the error in power-sharing based on the HMG's droop coefficient, is noticeably larger in topology-3 as compared to topology-1 and 2.

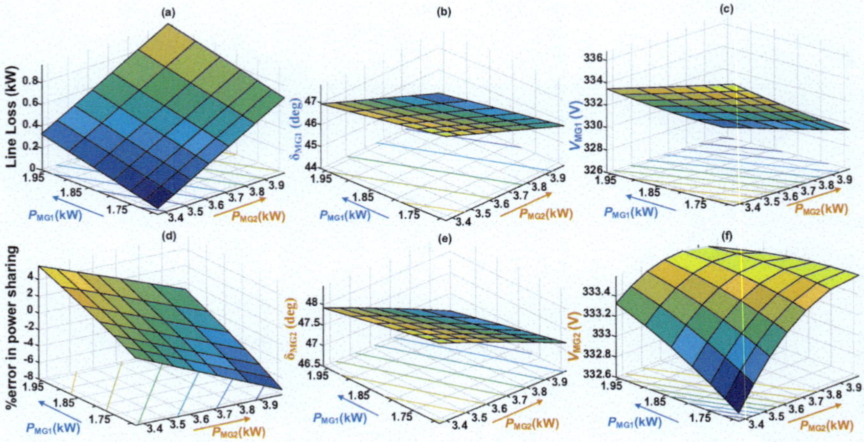

Fig. 38 Effect of line impedance variation on CMG network for topology-2

Figure 40 shows the surface plot for line loss, frequency, the dc-link voltage of each HMG, and the percentage of error in power-sharing as a function of power delivered by both HMGs due to the increase in line resistance. The surface plots of frequency and dc-link voltage of individual HMG follow a similar pattern. From Fig. 40, it can be observed that the power, frequency and dc-link voltage of microgrid-1 (with larger droop value, provides 33% of the total PMG power demand and line loss) drops more abruptly where for microgrid-2 (with a smaller droop value, provides 67% of the total PMG power demand and line loss) the parameters remain fairly constant. This scenario can also be observed from the surface plots.

6.2 PMG Power Variation

The second sensitivity analysis was performed by varying the power demand of the PMG in the CMG network. The parameters of CMG in both topology-1 and 2 are shown in Fig. 41 as the power consumed by the PMG varies from 2 to 10 kW. The delivered HMG power, line loss, the HMG's droop angle, voltage, and percentage of error in power-sharing are plotted as a function of the PMG's power demand. The plots are normalized based on a value of 6 kW.

It can be seen from Fig. 41 that for both topologies, the power-sharing between the two HMGs remains constant irrespective of the PMG's power demand. Hence, it can be observed in this analysis that, the error in power-sharing as per their droop coefficient is very negligible. In other words, the modified angle droop control-based power-sharing approach is insensitive to the PMG's power variation.

Finally, the sensitivity analysis for topology-3 is conducted by varying the PMG's power from 2 to 20 kW. Figure 42a–f show the normalized line loss, HMG power,

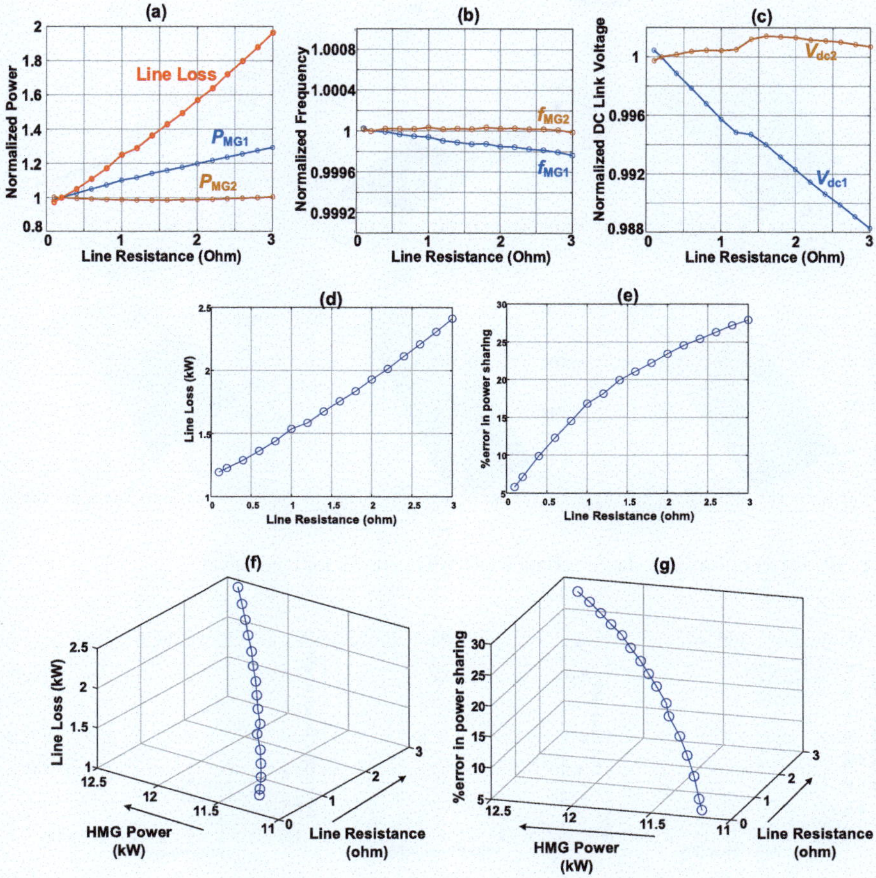

Fig. 39 CMG sensitivity analysis of topology-3 for line resistance variation

frequency, respective HMG's dc-link voltages, actual line loss, and percentage of error in power-sharing as the PMG's power demand are varied. The values are normalized based on the PMG power demand of 10 kW.

Figure 42f–g show the 3D plot for line loss and error percentage in power-sharing as the PMG power demand and the total supplied power of the HMGs are varied. It can be seen from Fig. 42a, b that the sharing of power between the two HMGs remains fairly constant irrespective of the PMG's power demand. Hence, the error in power-sharing according to their droop ratio is very negligible when the PMG's power demand varies. Thus, the sharing of power amongst the HMGs is insensitive to the PMG's power demand.

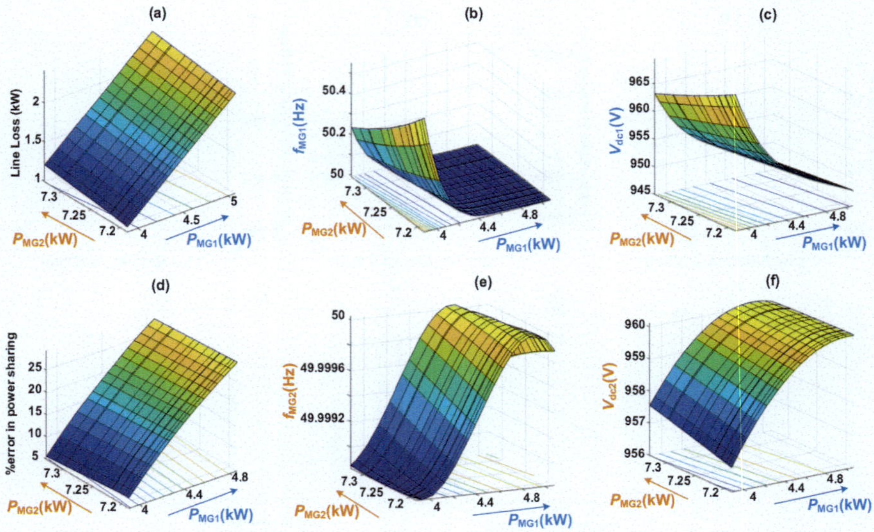

Fig. 40 Effect of line impedance variation on CMG network for topology-3

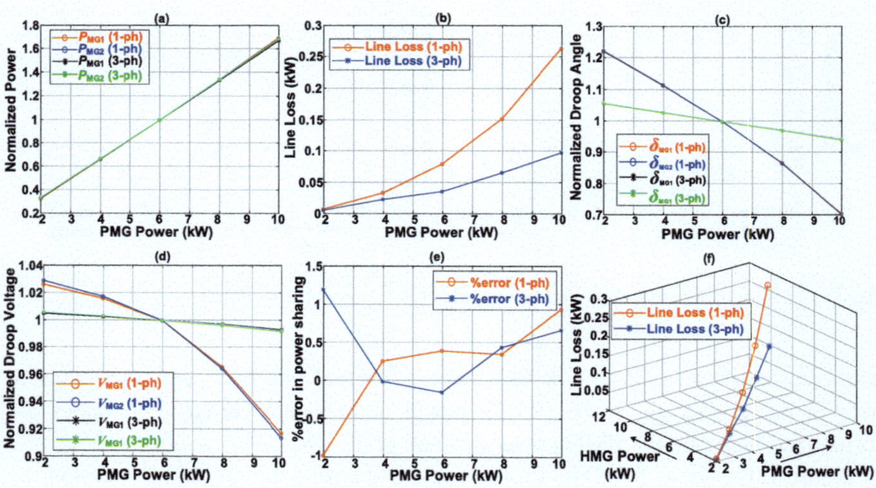

Fig. 41 Comparative CMG sensitivity analysis of topology-1 and 2 for PMG power variation

7 Comparison Among the Three Topologies

The chapter has discussed that the power exchange amongst the microgrids can be realized through any of the three topologies where each topology has its own advantages or drawbacks. However, it should be noted that such links only enable power exchange provisionally and during any contingency that requires to exchange

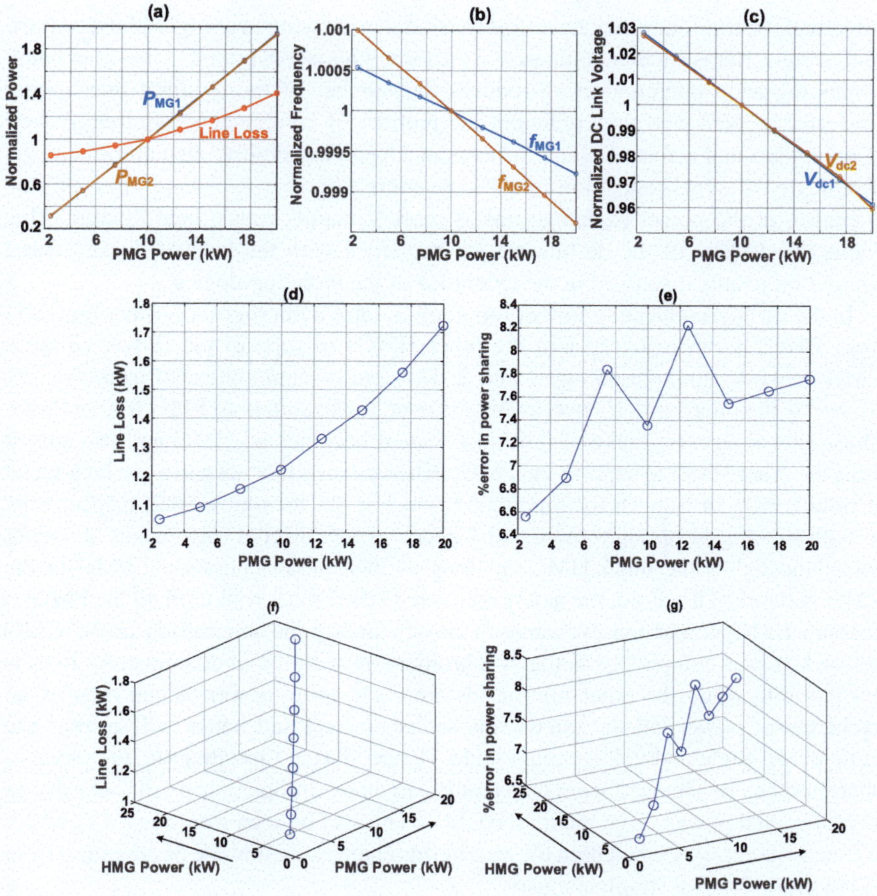

Fig. 42 Sensitivity analysis of the CMG network in topology-3 with PMG power demand variation

10–20% of the total generation capacity of each microgrid. To realize such interconnecting links that only carry a small amount of power for short periods, proper economical, feasible and most optimal option needs to be selected. It is obvious that based on the distance among the microgrids and the amount of power that is going to be shared will play the most important role in selecting the most optimal topology for such interconnection. For microgrid clusters located in close proximity, ac interconnection may be more suitable and economical whereas, if the distance is large, then dc interconnection will be preferred. Furthermore, when selecting an ac interconnecting link based on the above-mentioned criteria, a single-phase link can be more suitable over a three-phase link or vice versa. A single-phase ac link of topology-2, and single-phase power electronics converter requires fewer components than a three-phase ac link of topology-3. Also, the required number of current and voltage sensors, closed-loop controllers, and signal processing circuitry are less.

Nevertheless, the computational effort for signal processing in a single-phase voltage source converter is smaller compared to the three-phase converter. On the other hand, employing a dc interconnecting reduces the number of the required interlinking converter by 50% and thus reducing the number of components required as well as computational effort of signal processing. Applying dc-link also eliminates the complexity of synchronization of the converters to power exchange lines and the operation of the power exchange link is more versatile, robust, and flexible. This distinctive feature of the dc-link can be explained with the help of the presented simulation results discussed in the examples of the three topologies.

In the dc interconnecting link of topology-3, once a microgrid becomes an HMG from PMG, it can instantly join the other HMGs to support the PMGs within a CMG network, unlike topology-1 and 2. This can be comprehended from Fig. 26. At, $t = t_5$ the microgrid-1's overloading is over, the becomes an FMG from a PMG. Thus, after the preset delay of 0.6s at, $t = t_6$ it becomes an HMG and instantly it joins the other HMG to support the PMG. Whereas, a similar scenario can be noticed at time $t = t_6$ in Fig. 18 for topology-1 and Fig. 23 for topology-2, respectively. In both topologies, though microgrid-1 becomes an HMG at this instant, it cannot get connected to the other HMGs as long as there is an operational PMG in the CMG network. Therefore, the newly recovered HMG must wait until all the PMG(s) become HMG(s) and the exchange of power among the microgrids in the CMG network ceases completely as the synchronization with the interconnecting lines is not possible while the other microgrids are exchanging power. When there is no exchange of power, all the microgrids are at no load and hence will assume the same phase angle and voltage magnitude. Hence, during this situation, the mode of operation can be changed from constant PQ to angle droop control without causing any unwanted power surge within the CMG. This can be seen from Figs. 18 and 23, at time instant $t = t_8$ when both the microgrids changes their mode of operation from PQ to droop control simultaneously.

A comparison among the three topologies is given in Table 7. In topology-3, as the number of switching devices, LCL filters, and conductors is less, the power losses in these components will also be the minimum among all the three topologies, and a better converter efficiency can be achieved. At the same time, the number of sensors and LCL filters (the most expensive components of the converter) required for a dc-link is 50% less than that of the three-phase link and 40% less than that of the single-phase link, and thus offers a substantial reduction in cost and size of the converter.

Finally, the number of conductors required in the single-phase link is one (assuming earth neutral), whereas two and three conductors are required for dc and three-phase ac link, respectively. However, the total power that can be exchanged in a single-phase link is one-third of the three-phase link, whereas the maximum power that can be exchanged for a given voltage level, is through the dc-link. If the same amount of power needs to be transferred through a single-phase link, either the reference voltage across the filter capacitor of C_f needs to be increased or the size of the conductors needs to be increased to minimize the line impedance (approximately, one-third than that of three-phase lines). Also, the coupling inductance L_g needs to

Table 7 Comparison of employing different topologies for power exchange link

Parameters	Single-phase link	three-phase link	dc-link
No. of switching devices	10	12	6
No. of LCL filters	4	6	3
No. of current sensors	8	12	6
No. of voltage sensors	8	12	6
Number of lines	1	3	2

be reduced to accommodate the desired power flow. Each one of these solutions has its own limitation. First, the voltage reference across C_f cannot be made very large as it may produce excess stress over the switching devices and the other components and thus may incur an additional cost. Also, the voltage across C_f is related to the dc-link voltage of the converter as its value should be less than half of the dc-link voltage to avoid saturation. Second, using conductors with a larger cross-section area would make the system and its transmission structures more expensive. Third, L_g can not be made very small as the switching harmonics may not get properly filtered out and can cause harmonic distortion in the output voltage waveform. At the same time, the modified angle droop with virtual impedance is employed to ensure the desired power-sharing among the HMGs. To implement such a droop control scheme, L_g needs to be large enough to decouple the active and reactive power-sharing and needs to be larger than a certain minimum value. Otherwise, it may deteriorate the property of power-sharing based on droop coefficients.

The main purpose of using a single-phase ac link over a three-phase link to couple microgrids, is to reduce the cost as the number of equipment and interconnecting lines will be less. However, this is valid in the case of autonomous microgrids which are located close to each other in a certain geographical area. If the distance among the microgrids is large, then the benefits of the single-phase links over three-phase links may not be applicable anymore. Using dc-link to couple microgrids can be more feasible but, the protection equipment for dc system are expensive compare to ac system up to some extent. Also, the majority of the power distribution line infrastructure and commercially available equipment are based on ac systems. At the same time, the network operators are usually more accustomed to ac distribution systems as compared to dc systems. Moreover, dc systems would be more economical or cost-effective if the distance among the microgrids is long.

The stability of all the CMG topologies was examined for several parametric variations. It is demonstrated that the stability of the CMG is affected by the active power droop coefficient in topology-1 and 2, and hence the critical value of the parameter is evaluated. The reactive power droop coefficient cannot affect the CMG stability and its effect is almost negligible. However, in topology-3, variation of dc droop coefficient does not make the system unstable; however, a large value of dc droop coefficient makes the system highly overdamped. The power injected or consumed by PMG also affects the CMG dynamics. A larger amount of PMG power offers more damping and makes the system response sluggish whereas, a smaller value of PMG power reduces the system damping and makes the response oscillatory as can be seen in all the three topologies. Line resistance and reactance reasonably affect the CMG dynamics but not dominating enough to cause instability unless an extremely large value of reactance is used.

Finally, sensitivity analysis of all the three topologies shows that the minimum power loss has occurred in case of three-phase topology whereas, the dc topology is capable of exchanging the maximum power and the distance among the microgrids does not affect the exchanged power level. On the other hand, for three-phase and single-phase ac links, line impedance affects the level of power exchange. If the line impedance is very large (i.e., the distance among the microgrids is large) the CMG network will fail to exchange the desired amount of power or the amount of power that can be exchanged via the interconnecting link will be limited. The LSC's LCL filter also restricts the power level up to some extent.

Considering all the above points, a detailed feasibility study and network planning are required to select the most optimum topology to couple autonomous microgrids to exchange power among them and the selection of the most suitable topology among the three would depend on several key factors and parameters.

8 Conclusion

This chapter has analyzed and discussed three different approaches to forming a system of interconnected microgrid networks that will facilitate power exchange through a power electronic converter-based interface only when one or more of those microgrids are experiencing either overloading due to lack of adequate power generation or over-generation from its renewable-based sources. The studied topologies provide the necessary isolation of the microgrids so that each microgrid can operate autonomously according to its own internal standards and control techniques while enabling power-sharing amongst them. The control mechanism of each structure is based on a decentralized local frequency monitoring and regulation technique for the individual microgrids that eliminates the requirement for a centralized control or data communication infrastructure. Also, a suitable framework has been discussed that ensures adequate and proper power-sharing among the microgrids based on droop control mechanism when more than one microgrid will support overloaded or over-generating microgrid(s). Thus, the approach reduces or prevents load-shedding or curtailment of renewable sources when microgrid experiences such overloading or excessive generation. The mechanism includes an approach to realize the desired power-sharing ratio among multiple HMGs and various control techniques to detect the microgrid's status. At the same time, it also determines the required level of active power that needs to be exchanged to support the PMG which will operate in the constant PQ control mode. Such transitions between the various modes of operation of the microgrids are achieved based on measuring the local parameters only. The simulation-based case studies are carried out in PSIM® to validate the efficacy of the discussed approaches in addressing the problems of the unacceptable frequency deviation in the microgrids, following a power excess or shortfall. The stability analysis identifies the system parameters that are responsible to affect the dynamics of the CMG network. Finally, the performed sensitivity analysis of the CMG topologies demonstrates a nearly linear variation of the different system parameters and quantities with respect to the variation of design and operational factors, thus the technique offers a considerably robust performance against variations of those parameters.

References

1. Lasseter RH (2002) 'Microgrids', IEEE power engineering society winter meeting. Conference 1:305–308
2. Lasseter RH, Piagi P (2004) Microgrid: a conceptual solution. IEEE Power Electron Specialists Conf 6:4285–4291
3. Yoldaş Y, Önen A, Muyeen SM, Vasilakos AV, Alan İ (2017) Enhancing smart grid with microgrids: challenges and opportunities. Renew Sustain Energy Rev 72:205–214
4. Ali L, Shahnia F (2017) Determination of an economically-suitable and sustainable standalone power system for an off-grid town in Western Australia. Renew Energy 106:243–254
5. Ali L, Shahnia F (2016) Selection of a suitable standalone power supply system for an off-grid town in Western Australia. 2016 IEEE 16th international conference on environment and electrical engineering (EEEIC), pp 1–6, https://doi.org/10.1109/EEEIC.2016.7555735
6. Ali L, Shahnia F (2016) Impact of annual load growth on selecting the suitable sustainable standalone system for an off-grid town in Western Australia. IEEE Int Conf Power Syst Technol (POWERCON) 2016:1–5. https://doi.org/10.1109/POWERCON.2016.7754043
7. Ali L, Shahnia F, Arefi A, Iu H, Fernando T (2017) Feasibility analysis of a sustainable system for an Australian remote town. 2017 3rd international conference on power generation systems and renewable energy technologies (PGSRET), pp 118–122, https://doi.org/10.1109/PGSRET.2017.8251812
8. de Matos JG, de Silva FSF, de LA, Ribeiro S (2015) "Power control in ac isolated microgrids with renewable energy sources and energy storage systems. IEEE Trans Indus Electron 62(6):3490–3498
9. Eghtedarpour N, Farjah E (2014) Power control and management in a hybrid AC/DC microgrid. IEEE Trans Smart Grid 5(3):1494–1505. https://doi.org/10.1109/TSG.2013.2294275
10. Katiraei F, Iravani MR (2006) Power management strategies for a microgrid with multiple distributed generation units. IEEE Trans Power Syst 21(4):1821–1831. https://doi.org/10.1109/TPWRS.2006.879260
11. Chandrasena RPS, Shahnia F, Ghosh A, Rajakaruna S (2014) Operation and control of a hybrid AC-DC nanogrid for future community houses. Australasian Univ Power Eng Conf (AUPEC) 2014:1–6. https://doi.org/10.1109/AUPEC.2014.6966617
12. Chandrasena RPS, Shahnia F, Rajakaruna S, Ghosh A (2013) Operation and control of three phase microgrids consisting of single-phase DERs. 2013 IEEE 8th international conference on industrial and information systems, pp 599–604, https://doi.org/10.1109/ICIInfS.2013.6732052
13. Shahnia F et al (2017) A three-phase community microgrid comprised of single-phase energy resources with an uneven scattering amongst phases. Int J Electr Power Energy Syst 84:267–283
14. Zia MF, Elbouchikhi E, Benbouzid M (2018) Microgrids energy management systems: a critical review on methods, solutions, and prospects. Appl Energy 22:1033–1055
15. Chandrasena RPS, Shahnia F, Rajakaruna S, Ghosh A (2013) Control, operation and power sharing among parallel converter-interfaced DERs in a microgrid in the presence of unbalanced

and harmonic loads. Australasian Univ Power Eng Conf (AUPEC) 2013:1–6. https://doi.org/10.1109/AUPEC.2013.6725382
16. Arefifar SA, Mohamed YARI (2014) DG mix, reactive sources and energy storage units for optimizing microgrid reliability and supply security. IEEE Trans Smart Grid 5(4):1835–1844
17. Nejabatkhah F, Li YW (2015) Overview of power management strategies of hybrid AC/DC microgrid. IEEE Trans Power Electron 30(12):7072–7089. https://doi.org/10.1109/TPEL.2014.2384999
18. Shoeb MA, Shahnia F, Shafiullah G (2018) A multilayer preventive control to regulate voltage and frequency in autonomous microgrids. Australasian Univ Power Eng Conf (AUPEC) 2018:1–6. https://doi.org/10.1109/AUPEC.2018.8758055
19. Shoeb MA, Shahnia F, Shafiullah GM, Su X (2019) Sensitivity of prediction error on the performance of a preventive controller for microgrids. IEEE Int Conf Indus Technol (ICIT) 2019:1717–1722. https://doi.org/10.1109/ICIT.2019.8755157
20. Kani AP et al (2019) Microgrid control overview. Scalability and stability of microgrid. IET Press, Variability
21. Shoeb MA, Shahnia F, Shafiullah G (2019) A two-stage optimal generation units dispatch for standalone microgrids. 2019 9th international conference on power and energy systems (ICPES), pp 1–6, https://doi.org/10.1109/ICPES47639.2019.9105544
22. Shahnia F (2021) Microgrids and their control. Hybrid renewable energy systems and microgrids. Academic Press, 399–462
23. Shoeb MA et al (2020) Peer-to-peer load allocation using potential field concept for optimal operation of standalone microgrids. IET generation, transmission & distribution
24. Shahnia F et al (2014) Application of DSTATCOM for surplus power circulation in MV and LV distribution networks with single-phase distributed energy resources. Electric Power Syst Res 117:104–114
25. Fornarelli R et al (2018) Selecting an economically suitable and sustainable solution for a renewable energy-powered water desalination system: a rural Australian case study. Desalination 435:128–139
26. Shahnia F, Moghbel M, Arefi A, Shafiullah GM, Anda M, Vahidnia A (2017) Levelized cost of energy and cash flow for a hybrid solar-wind-diesel microgrid on Rottnest island. Australasian Univ Power Eng Conf (AUPEC) 2017:1–6. https://doi.org/10.1109/AUPEC.2017.8282413
27. Jamal T, Shafiullah GM, Carter C, Ferdous SM, Rahman M (2018) Benefits of short-term PV forecasting in a remote area standalone off-grid power supply system. IEEE Power Energy Soc General Meeting (PESGM) 2018:1–5. https://doi.org/10.1109/PESGM.2018.8586416
28. Mansur AA, Ferdous SM, Shams ZB, Islam MR, Rokonuzzaman M, Hoque MA (2011) An experimental investigation of the real time electrical characteristics of a PV panel for different atmospheric conditions in Islamic University of Technology (OIC), Gazipur, Bangladesh. 2011 International conference & utility exhibition on power and energy systems: issues and prospects for Asia (ICUE), Pattaya, Thailand, pp 1–8https://doi.org/10.1109/ICUEPES.2011.6497767
29. Susanto J, Shahnia F (2019) Smoothing batteries for PV-diesel microgrids. 2019 IEEE 28th international symposium on industrial electronics (ISIE), pp 2609–2614, https://doi.org/10.1109/ISIE.2019.8781461
30. Moghbel M et al (2020) Output power fluctuations of distributed photovoltaic systems across an isolated power system: insights from high-resolution data. IET Renew Power Gen
31. Chandrasena RPS, Shahnia F, Ghosh A, Rajakaruna S (2014) Secondary control in microgrids for dynamic power sharing and voltage/frequency adjustment. Australasian Univ Power Eng Conf (AUPEC) 2014:1–8. https://doi.org/10.1109/AUPEC.2014.6966619
32. Pashajavid E, Shahnia F, Ghosh A (2015) Overload management of autonomous microgrids. 2015 IEEE 11th international conference on power electronics and drive systems, pp 73–78, https://doi.org/10.1109/PEDS.2015.7203515
33. Shahnia F (2017) Hybrid nanogrid systems for future small communities. Sustainable Development in energy systems. Springer, Cham, 19–38

34. Pashajavid E, Shahnia F, Ghosh A (2015) A decentralized strategy to remedy the power deficiency in remote area microgrids. 2015 50th international universities power engineering conference (UPEC), pp 1–6, https://doi.org/10.1109/UPEC.2015.7339865
35. Ferdous SM, Shahnia F, Shafiullah G (2019) Power sharing and control strategy for microgrid clusters. 9th international conference on power and energy systems (ICPES), Perth
36. Nutkani IU, Loh PC, Wang P et al (2015) Intertied ac–ac microgrids with autonomous power import and export. Int J Electr Power Energy Syst 6:385–393
37. Goyal M, Ghosh A, Shahnia F (2014) Overload prevention in an autonomous microgrid using battery storage units. 2014 IEEE PES general meeting | conference & exposition, pp 1–5, https://doi.org/10.1109/PESGM.2014.6939459.
38. Goyal M, Ghosh A, Shahnia F (2014) Distributed battery storage units for overload prevention in an islanded microgrid. Australasian Univ Power Eng Conf (AUPEC) 2014:1–6. https://doi.org/10.1109/AUPEC.2014.6966556
39. Shahnia F (2016) Semi-decentralized charging and discharging control of floating batteries in microgrids. 2016 IEEE 2nd annual southern power electronics conference (SPEC), pp 1–6, https://doi.org/10.1109/SPEC.2016.7846139.
40. Hong Y, Hsiao M, Chang Y, Lee Y, Huang H (2013) Multiscenario underfrequency load shedding in a microgrid consisting of intermittent renewables. IEEE Trans Power Delivery 28(3):1610–1617. https://doi.org/10.1109/TPWRD.2013.2254502
41. Hong Y, Nguyen M (2020) Multiobjective multiscenario under-frequency load shedding in a standalone power system. IEEE Syst J 14(2):2759–2769. https://doi.org/10.1109/JSYST.2019.2931934
42. Marwali MKC, Haili M, Shahidehpour SM, Abdul-Rahman KH (1998) Short term generation scheduling in photovoltaic-utility grid with battery storage. IEEE Trans Power Syst 13(3):1057–1062. https://doi.org/10.1109/59.709099
43. Jabalameli N, Deilami S, Masoum MAS, Shahnia F (2013) Rooftop PV with battery storage for constant output power production considering load characteristics. 2013 8th international conference on electrical and electronics engineering (ELECO), pp 142–146, https://doi.org/10.1109/ELECO.2013.6713820
44. Wang XY, Mahinda Vilathgamuwa D, Choi SS (2008) Determination of battery storage capacity in energy buffer for wind farm. IEEE Trans Energy Conver 23(3):868–878, https://doi.org/10.1109/TEC.2008.921556
45. Venu C, Riffonneau Y, Bacha S, Baghzouz Y (2009) Battery storage system sizing in distribution feeders with distributed photovoltaic systems. 2009 IEEE Bucharest PowerTech, Bucharest, Romania, pp 1–5https://doi.org/10.1109/PTC.2009.5282093
46. Shahnia F, Arefi A (2017) Eigenanalysis-based small signal stability of the system of coupled sustainable microgrids. Int J Electr Power Energy Syst 91:42–60
47. Shahnia F (2016) Stability and eigenanalysis of a sustainable remote area microgrid with a transforming structure. Sustain EnergyGrids Netw 8:37–50
48. Bahramirad S, Reder W, Khodaei A (2012) Reliability-constrained optimal sizing of energy storage system in a microgrid. IEEE Trans Smart Grid 3(4):2056–2062. https://doi.org/10.1109/TSG.2012.2217991
49. Ferdous SM, Shafiullah GM, Shahnia F, Elavarasan RM, Subramaniam U (2020) Dynamic frequency and overload management in autonomous coupled microgrids for self-healing and resiliency improvement. IEEE Access 8:116796–116811
50. Lasseter RH (2011) Smart distribution: coupled microgrids. Proc IEEE 99(6):1074–1082
51. Shahnia F (2017) Interconnected microgrid systems for remote areas. Sustainable development in energy systems. Springer, Cham, 39–63
52. Shahnia F et al (2014) Interconnected autonomous microgrids in smart grids with self-healing capability. Renewable energy integration. Springer, Singapore, 347–381
53. Shahnia F, Chandrasena RPS, Rajakaruna S, Ghosh A (2013) Autonomous operation of multiple interconnected microgrids with self-healing capability. IEEE Power Energy Soc Gen Meeting 2013:1–5. https://doi.org/10.1109/PESMG.2013.6672794

54. Goyal M, Ghosh A (2016) Microgrids interconnection to support mutually during any contingency. Sustain Energy Grids Netw 6:100–108
55. John B, Ghosh A, Goyal M, Zare F (2019) A dc power exchange highway based power flow management for interconnected microgrid clusters. IEEE Syst J 13(3):3347–3357
56. Bullich-Massague E, Díaz-González F, Aragüés-Peñalba M et al (2018) Microgrid clustering architectures. Appl Energy 212:340–361
57. Zou H, Mao S, Wang Y, Zhang F, Chen X, Cheng L (2019) A survey of energy management in interconnected multi-microgrids. IEEE Access 7:72158–72169
58. Shahnia F, Ghosh A (2016) Coupling of neighbouring low voltage residential distribution feeders for voltage profile improvement using power electronics converters. IET Gener Transm Distrib 10(2):535–547
59. Pashajavid E, Shahnia F, Ghosh A (2015) Overloading conditions management in remote networks by coupling neighboring microgrids. 2015 50th international universities power engineering conference (UPEC), pp 1–6, https://doi.org/10.1109/UPEC.2015.7339874
60. Pashajavid E, Shahnia F, Ghosh A (2017) Provisional internal and external power exchange to support remote sustainable microgrids in the course of power deficiency. IET Gener Transm Distrib 11(1):246–260
61. Nutkani IU, Loh PC, Blaabjerg F (2013) Distributed operation of interlinked AC microgrids with dynamic active and reactive power tuning. IEEE Trans Indus Appl 49(5):2188–2196
62. Loh PC, Li D, Chai YK, Blaabjerg F (2013) Autonomous operation of hybrid microgrid with ac and dc subgrids. IEEE Trans Power Electron 28(5):2214–2223
63. Ferdous SM, Shahnia F, Shafiullah G (2019) Provisional energy transaction management amongst neighboring microgrids through a DC power exchange link. 2019 9th International conference on power and energy systems (ICPES), pp 1–6, https://doi.org/10.1109/ICPES47639.2019.9105555
64. Ferdous SM, Shahnia F, Shafiullah G (2019) Realizing a system of coupled microgrids using a single-phase AC power exchange link. 2019 9th international conference on power and energy systems (ICPES), pp 1–6, https://doi.org/10.1109/ICPES47639.2019.9105363
65. Susanto J, Shahnia F, Ghosh A, Rajakaruna S (2014) Interconnected microgrids via back-to-back converters for dynamic frequency support. Australasian Univ Power Eng Conf (AUPEC) 2014:1–6. https://doi.org/10.1109/AUPEC.2014.6966616
66. Majumder R, Ghosh A, Ledwich G, Zare F (2010) Power management and power flow control with back-to-back converters in a utility connected microgrid. IEEE Trans Power Syst 25(2):821–834
67. Wang Z, Chen B, Wang J, Chen C (2016) Networked microgrids for self-healing power systems. IEEE Trans Smart Grid 7(1):310–319. https://doi.org/10.1109/TSG.2015.2427513
68. Wang Z, Chen B, Wang J, Begovic M, Chen C (2015) Coordinated energy management of networked Microgrids in distribution systems. 2015 IEEE Power & energy society general meeting, Denver, CO, USA, pp 1–1,https://doi.org/10.1109/PESGM.2015.7286148
69. Zhang Y, Xie L, Ding Q (2015) Interactive control of coupled microgrids for guaranteed system-wide small signal stability. IEEE Trans Smart Grid 7(2):1088–1096
70. Batool M, Islam SM, Shahnia F (2017) Master control unit based power exchange strategy for interconnected microgrids. Australasian Univ Power Eng Conf (AUPEC) 2017:1–6. https://doi.org/10.1109/AUPEC.2017.8282475
71. Batool M, Shahnia F, Islam SM (2019) Market model for clustered microgrids optimisation including distribution network operations. IET Gen Trans Distrib 13(22):5139–5150
72. Colson CM, Nehrir MH (2013) Comprehensive real-time microgrid power management and control with distributed agents. IEEE Trans Smart Grid 4(1):617–627. https://doi.org/10.1109/TSG.2012.2236368
73. Shahnia F, Chandrasena RPS, Rajakaruna S, Ghosh A (2014) Primary control level of parallel distributed energy resources converters in system of multiple interconnected autonomous microgrids within self-healing networks. IET Gener Transm Distrib 8(2):203–222
74. Zhang Y, Xie L (2015) Online dynamic security assessment of microgrid interconnections in smart distribution systems. IEEE Trans Power Syst 30(6):3246–3254. https://doi.org/10.1109/TPWRS.2014.2374876

75. Ferdous SM, Shahnia F, Shafiullah G (2020) Various structures and control strategies for provisional energy transaction management in coupled microgrid clusters. 2020 intermountain engineering, technology and computing (IETC), pp 1–6,https://doi.org/10.1109/IETC47856.2020.9249113
76. Susanto J, Shahnia F, Arefi A (2016) Small signal model load flow initialisation for converter-dominated microgrids. IEEE Int Conf Power Syst Technol (POWERCON) 2016:1–6. https://doi.org/10.1109/POWERCON.2016.7754050
77. Shahnia F (2016) Eigenvalues and their participation factors for the system of coupled adjacent microgrids. 2016 IEEE 2nd annual southern power electronics conference (SPEC), pp 1–6, https://doi.org/10.1109/SPEC.2016.7846138
78. Shahnia F (2016) Stability of a sustainable remote area microgrid. 2016 IEEE region 10 conference (TENCON), pp 1220–1223, https://doi.org/10.1109/TENCON.2016.7848205
79. Nikolakakos IP, Zeineldin HH, El-Moursi MS, Hatziargyriou ND (2016) Stability evaluation of interconnected multi-inverter microgrids through critical clusters. IEEE Trans Power Syst 31(4):3060–3072. https://doi.org/10.1109/TPWRS.2015.2476826
80. Kunwar A, Shahnia F, Bansal RC (2018) Impact of the capacity and number of inertial and non-inertial distributed energy resources within a microgrid on its stability margins. 2018 IEEE 27th international symposium on industrial electronics (ISIE), pp 55–60, https://doi.org/10.1109/ISIE.2018.8433587
81. Kunwar A, Shahnia F, Bansal RC (2020) Eigenvalue-oriented dynamic stability examination to enhance designing a microgrid hosting clusters of inertial and non-inertial distributed generators. IEEE Trans Smart Grid 11(3):1942–1955. https://doi.org/10.1109/TSG.2019.2945587
82. Minciardi R, Sacile R (2012) Optimal control in a cooperative network of smart power grids. IEEE Syst J 6(1):126–133
83. Shoeb MA, Shahnia F, Shafiullah GM (2019) Optimal coupling of multiple microgrid clusters. 2019 IEEE innovative smart grid technologies—Asia (ISGT Asia), pp 2809–2814, https://doi.org/10.1109/ISGT-Asia.2019.8881386
84. Batool M, Islam SM, Shahnia F (2017) Power transaction management amongst coupled microgrids in remote areas. 2017 IEEE innovative smart grid technologies—Asia (ISGT-Asia), pp 1–6, https://doi.org/10.1109/ISGT-Asia.2017.8378449.
85. Shahnia F, Bourbour S, Ghosh A (2017) Coupling neighboring microgrids for overload management based on dynamic multicriteria decision-making. IEEE Trans Smart Grid 8(2):969–983. https://doi.org/10.1109/TSG.2015.2477845
86. Bourbour S, Shahnia F, Ghosh A (2015) Selection of a suitable microgrid to couple with an overloaded neighboring microgrid based on decision making. North Amer Power Symp (NAPS) 2015:1–6. https://doi.org/10.1109/NAPS.2015.7335200
87. Jamal T, Urmee T, Shafiullah G, Shahnia F (2018) Using experts' opinions and multi-criteria decision analysis to determine the weighing of criteria employed in planning remote area microgrids. Int Conf Utility Exhib Green Energy Sustain Dev (ICUE) 2018:1–7. https://doi.org/10.23919/ICUE-GESD.2018.8635734
88. Majumder R (2014) Aggregation of microgrids with DC system. Electric Power Syst Res 108:134–143
89. Ferdous SM, Shahnia F, Shafiullah G (2020) Provisional energy transaction management amongst neighbouring microgrids through a DC power exchange link. IET Power Electron 13:4129–4139
90. Ferdous SM, Shahnia , F, Shafiullah G (2021) Realising a system of coupled microgrid networks using single-phase interconnection lines. IET Smart Grid
91. Setiawan MA, Shahnia F, Ghosh A, Rajakaruna S (2014) Proposing a new algorithm for defining the shortest distance among ZigBee-based communication devices in microgrids. Australasian Univ Power Eng Conf (AUPEC) 2014:1–6. https://doi.org/10.1109/AUPEC.2014.6966517
92. Setiawan MA, Shahnia F, Ghosh A, Rajakaruna S (2014) Developing the ZigBee based data payload coding for data communication in microgrids. Australasian Univ Power Eng Conf (AUPEC) 2014:1–6. https://doi.org/10.1109/AUPEC.2014.6966516

93. Setiawan MA, Shahnia F, Chandrasena RPS, Ghosh A (2014) Data communication network and its delay effect on the dynamic operation of distributed generation units in a microgrid. IEEE PES Asia-Pacific Power Energy Eng Conf (APPEEC) 2014:1–6. https://doi.org/10.1109/APPEEC.2014.7066197
94. Shahnia F, Bourbour S (2017) A practical and intelligent technique for coupling multiple neighboring microgrids at the synchronization stage. Sustain Energy Grids Netw 11:13–25
95. Susanto J, Shahnia F, Arefi A (2017) Effects of network characteristics and topology on the stability of converter-dominated microgrids. Australasian Univ Power Eng Conf (AUPEC) 2017:1–6. https://doi.org/10.1109/AUPEC.2017.8282512
96. Wang Z, Wang J (2015) Self-healing resilient distribution systems based on sectionalization into microgrids. IEEE Trans Power Syst 30(6):3139–3149. https://doi.org/10.1109/TPWRS.2015.2389753
97. Ferdous SM, Oninda MAM, Sarowar G, Islam KK, Hoque MA (2016) Non-isolated single stage PFC based LED driver with THD minimization using Cúk converter. 2016 9th international conference on electrical and computer engineering (ICECE), pp 471–474, https://doi.org/10.1109/ICECE.2016.7853959
98. Yazdani A, Iravani R (2010) Voltage-sourced converters in power systems: modeling, control, and applications. Wiley-Blackwell
99. Ferdous SM, Garcia P, Oninda MAM, Hoque MA (2016) MTPA and field weakening control of synchronous reluctance motor. 2016 9th international conference on electrical and computer engineering (ICECE), Dhaka, Bangladesh, pp 598–601, https://doi.org/10.1109/ICECE.2016.7853991.
100. Yazdani A, Iravani R (2005) An accurate model for the dc-side voltage control of the neutral point diode clamped converter. IEEE Trans Power Delivery 21(1):185–193
101. Ferdous SM, Shafiullah G, Oninda MAM et al (2017) Close loop compensation technique for high performance MPPT using ripple correlation control. 27th Australasian universities power engineering conference (AUPEC), pp 1–6, Melbourne, Australia
102. Hosseinimehr T, Ghosh A, Shahnia F (2017) Cooperative control of battery energy storage systems in microgrids. Int J Electr Power Energy Syst 87:109–120
103. Hosseinimehr T, Shahnia , F Ghosh A (2015) Power sharing control of batteries within autonomous microgrids based on their state of charge. 25th Australasian universities power engineering conference (AUPEC), pp 1–6, Wollongong, Australia
104. Hosseinimehr T, Shahnia F, Chandrasena RPS, Ghosh A (2014) Different power sharing techniques for converter-interfaced DERs in an autonomous microgrid. IEEE PES Asia-Pacific Power Energy Eng Conf (APPEEC) 2014:1–6. https://doi.org/10.1109/APPEEC.2014.7066196
105. Hosseinimehr T, Shahnia F, Ghosh A (2015) Dynamic power sharing control among converter-interfaced DERs in an autonomous microgrid. IEEE Eindhoven PowerTech 2015:1–6. https://doi.org/10.1109/PTC.2015.7232566
106. Bourbour S, Shahnia F (2016) Impact of the weightings of different criteria in selecting the suitable microgrids when forming a system of coupled microgrids. 2016 IEEE innovative smart grid technologies—Asia (ISGT-Asia), pp 1151–1156, https://doi.org/10.1109/ISGT-Asia.2016.7796548
107. Bourbour S, Shahnia F (2016) A suitable mechanism for the interconnection phase of temporary coupling of adjacent microgrids. 2016 IEEE innovative smart grid technologies—Asia (ISGT-Asia), pp 624–629, https://doi.org/10.1109/ISGT-Asia.2016.7796457
108. Shahnia F, Arefi A (2016) Defining the suitable adjacent microgrids to form a temporary system of coupled microgrids. 2016 IEEE region 10 conference (TENCON), pp 1216–1219, https://doi.org/10.1109/TENCON.2016.7848204
109. Shoeb MA, Shahnia F, Shafiullah G (2017) A multilayer optimization scheme to retain the voltage and frequency in standalone microgrids. 2017 IEEE innovative smart grid technologies—Asia (ISGT-Asia), pp 1–6, https://doi.org/10.1109/ISGT-Asia.2017.8378465
110. Shoeb MA, Shahnia F, Shafiullah GM (2019) A multilayer and event-triggered voltage and frequency management technique for microgrid's central controller considering operational

and sustainability aspects. IEEE Trans Smart Grid 10(5):5136–5151. https://doi.org/10.1109/TSG.2018.2877999

111. Batool M, Shahnia F, Islam SM (2019) Multi-level supervisory emergency control for operation of remote area microgrid clusters. J Modern Power Syst Clean Energy 7(5):1210–1228. https://doi.org/10.1007/s40565-018-0481-6
112. Batool M, Shahnia F, Islam SM (2019) Impact of scaled fitness functions on a floating-point genetic algorithm to optimise the operation of standalone microgrids. IET Renew Power Gener 13(8):1280–1290
113. Kim J-W (2002) Hang-seok choi and bo hyung cho, "a novel droop method for converter parallel operation,." IEEE Trans Power Electron 17(1):25–32
114. Augustine S, Mishra MK, Lakshminarasamma N (2015) Adaptive droop control strategy for load sharing and circulating current minimization in low-voltage standalone DC microgrid. IEEE Trans Sustain Energy 6(1):132–141

Cooperative and Transactive Integration of Multiple Microgrids

Nelson L. Díaz, Jan L. Díaz, Yuly V. García, and Adriana C. Luna

Abstract The constant changes in electrical grids, such as the incorporation of distributed generators and the integration of microgrids, have resulted in various alterations in the way the power system functions. To address these changes, recent years have witnessed ongoing research into the concept of transactive energy. Here, the conventional operation of the electrical power system, historically characterized by unidirectional energy flow from bulk generation through the transmission and distribution systems to end-users, has experienced shifts due to the evolving dynamics of generation and demand, in which energy can be shared or transacted among many energy prosumers. This chapter presents an overview, from management to hardware, of the operation and requirements for effective energy transactions between prosumers. Game theory is introduced and explored as one of the most promising market topologies used in Transactive Energy. A real-life optimization problem, along with its Python code, is provided in the chapter. Additionally, a perspective on hardware requirements for enabling energy transactions between multiple microgrids is presented by exploring the multi-terminal converter. The topologies, control, and configuration are discussed, highlighting the advantages of using Voltage Source Converters interconnected through a common DC link.

Keywords Transactive energy · Game theory · Multi-terminal converter

N. L. Díaz
Universidad Distrital Francisco José de Caldas, Bogotá, Colombia
e-mail: nldiaza@udistrital.edu.co

J. L. Díaz · Y. V. García · A. C. Luna (✉)
University of Puerto Rico, Mayaguez, Puerto Rico
e-mail: adriana.luna4@upr.edu

J. L. Díaz
e-mail: jan.diaz12@upr.edu

Y. V. García
e-mail: yuly.garcia@upr.edu

F. Shahnia and J. M. Guerrero (eds.), *Microgrids and Virtual Power Plants*,
Power Systems, https://doi.org/10.1007/978-981-97-6623-9_15

1 Introduction

The shift in the paradigm from conventional electrical systems, characterized by centralized generation, long transmission, and distribution lines, to a new concept where customers actively participate in the generation, becoming prosumers, signifies a redefinition and reshaping of how we perceive energy exchange and power flow. When energy storage units are added to the local power system, and all the energy resources are coordinated to ensure a reliable power supply to the local load, a microgrid is configured. In this context, new technical challenges emerge since load supply reliability should be ensured, both when the local power system is connected to and disconnected from the utility grid [1].

Due to the integration of local energy resources, microgrids have encountered a niche application in the electrification of remote communities not integrated into the utility grid or in providing energy resilience to communities without a reliable energy supply [2]. In this scenario, when communities actively engage in the operation of their electrical networks, microgrid-based energy communities emerge as an opportunity for integration with neighboring communities or even with the public grid, facilitating energy exchange [3]. The integration aims to enhance the overall reliability of the system by enabling energy exchange between prosumers within a cooperative environment, where system reliability remains the primary objective [4]. Within this framework, consideration of self-interest in obtaining the best profit should also align with the objectives of each community, and prosumer [3, 5].

To address these changes, recent years have witnessed ongoing research into the concept of transactive energy, defined as a mechanism that enables a dynamic balance between energy supply and demand while taking control and economic effects into account [6]. Here, the role of a distribution system operator (DSO) changes significantly. The DSO bears the responsibility of ensuring the reliable and cost-effective operation of the distribution system, as well as facilitating retail transactions between prosumers [7]. To effectively fulfill these responsibilities, the DSO requires a platform equipped with novel hardware and software applications, and user interfaces. For instance, it requires new performance metrics, controls, and analytics to uphold the dependable and economically efficient operation of the distribution system. Transactive energy strategies also entail the development of incentives to maximize advantages while minimizing costs for prosumers. Moreover, the platform's capabilities should encompass the transactional exchange of product information to align prosumer investment decisions with operational distribution requirements and to support the delivery of Distributed Energy Resources (DER) projects [8].

To deal with the transactive challenges between microgrids, new strategies, frameworks and approaches appear, mostly based on artificial intelligence and decision-making like Machine Learning [9], Fuzzy Logic [10, 11], Game Theory [12, 13], Consensus [14], which appears as a feasible solution to optimize energy consumption, aiming to the prosumers to maximize revenue of each autonomous energy system most of them based on minimized the cost, maximized the comfort, reducing peak loads or a combination of them. For that reason, we are going to talk about one of them, Game Theory, which is widely used to solve different type of problems find in transactive, such as energy management, energy exchange between prosumers, reduce costs, increase system reliability, increase revenues, among others.

On top of that, electrical networks should be properly conditioned to enable energy exchange, and trade between microgrid-based energy communities. As a feasible alternative for enabling the interconnection of different electrical systems appear the Multi-terminal Converters based on DC links. This technology provides additional flexibility as well as isolation between grid regulation, management, and power flow control for each microgrid compared to using simple rakers or static switches [15].

This chapter is organized as follows: Section two will introduce the main concepts related to Transactive energy. Section three will explore Game Theory approaches as one of the most promising market topologies used in Transactive Energy. Practical examples and python codes can be found. Finally, section four will explore the power exchange problem from a hardware point of view. Here multi-terminal converters based on Voltage source Converters with a common DC link are explored.

2 Transactive Energy

Transactive Energy (TE) is all about changing the way energy has been handled. Instead of the old-fashioned grid system, TE promotes a more decentralized network where different energy sources can talk to each other. This means everyone involved in making and using energy can communicate. When we talk about TE, interoperability is a big deal—it's about making sure all the systems can share energy info while being relieved to their rules.

The main idea behind TE is to make sure energy gets from the sellers to the buyers in a smart way. The market, where this smart sharing happens, needs to be decentralized. That means lots of different buyers and sellers working together, not just one big manager. Imagine it as a layered structure, where each layer has its own goals and only shares important information with the others [16].

2.1 Market Topologies Used in Transactive Energy

Markets have been around since power systems began. But with more decentralized energy sources growing up, we need a market that fits with how we generate and use power all over the place. The electricity market is tricky because there are a lot of different players and service providers interacting. To attempt these challenges, people are suggesting new ways to organize the energy market.

In this study [17], they talk about four market setups: peer-to-peer (P2P), prosumer-to-interconnected, prosumer-to-island, and organized prosumer groups. Another study [18] has three ideas for P2P markets: the full P2P market, the community-based market, and the hybrid P2P market. There's also [19], which gives a big picture of two market setups: the full P2P market and the community-based market. Meanwhile, [20] looks at three market setups: the community-based market, P2P, and hybrid.

2.1.1 Peer to Peer

The way this market works is by having direct talks between people selling and buying electricity. In a P2P market, they negotiate to decide how much electricity to exchange and at what price. What's different here from regular markets is that it's decentralized, meaning there are no middlemen involved. In a P2P market, each person producing and using electricity (prosumer) decides how to best use their local energy sources, like renewable energy and storage devices, after getting price info from others. The only thing they share is the price, keeping their details private [18, 20, 21].

2.1.2 Community Based Market

In this market, someone needs to manage electricity trading between people selling and buying. Each person using electricity talks to the manager, but there's no direct talking or sharing business info between them. The manager gathers info about how much electricity is wanted and available to make sure the market works smoothly. This kind of market is great for close communities with similar interests, where people can trade together in a friendly and competitive way [18, 20, 21]. The community-based market setup has good things, like making the community stronger and improving relationships among its members. But there are also challenges, like deciding what type of energy the community prefers and dealing with difficulties in putting the plan into action.

2.1.3 Hybrid Market

This market represents a fusion of P2P and community-based markets, amalgamating the strengths of both approaches. Notably, it delivers enhanced scalability, showcasing its ability to expand more effectively. The market is distinguished by its hierarchical structure, typically featuring P2P energy exchange at the top level and community-based energy exchange at another level [18, 20, 21].

2.1.4 One of the Most Applied Optimization Techniques Applied in Transactive Energy

According to the concept of transactive energy, the goal is to achieve an energy balance that provides economic benefits for all participants. For this reason, various optimization strategies have been developed to carry out this operation. One of the most prominent optimization techniques is game theory, due to the low consumption of computational resources and low complexity compared with other widely used strategies [22–24].

3 Game Theory Approach

Initially, the game theory approach was developed by John Von Neumann and Oskar Morgenstern in 1944 to solve economic problems. In the 1950s, John Nash made significant contributions that improved the mathematical model, marking a major step in its development [25, 26] Nowadays, this approach is one of the most widely used strategies for decision-making in many areas such as biology, economy, psychology, engineering, among others. Therefore, in this section, the basic definitions, types of games, and solution strategies are introduced to finally apply them in an example that allows for a better understanding of how this approach would be applied in a mathematical problem, specifically in optimization in a real case.

3.1 Definition

This section will discuss the definition of game theory and the main components of this approach, which are always present in every game.

Let's start by defining what game theory is. Based on the definition given by Oxford English Dictionary, game theory is: "the branch of mathematics concerned with the analysis of strategies for dealing with competitive situations where the outcome of a participant's choice of action depends critically on the actions of other participants" [27].

Now the parts of the game theory approach that will always find will be defined below.

- **Agents**: The agents are the players, the people who participate in the game and who must decide.
- **Game**: The game is a situation in which the agents are involved, which forces them to decide.
- **Payoff**: The payoff is the outcome, which can be good or bad, meaning that it can be beneficial (positive payoff) or detrimental (negative payoff) to the agent. These payoffs are usually represented in a matrix called normal form or matrix form, but in some cases, they are represented in the form of a tree, such as a decision tree called extensive form. The latter is generally used when there is an order in agent participation.
- **Strategy**: The strategy is the way or plan that the agents choose to solve the game. Depending on the strategy that is selected we can know which is the result for

Table 1 Payoff matrix for zero-sum game

Player 1/Player 2	A	B
C	$< -2,2 >$	$< 1, -1 >$
D	$< 5, -5 >$	$< -3,3 >$

each agent, likewise, depending on the strategy we can have one or more results for the game.

3.2 *Type of Games*

Now that the basic definitions about game theory approach are set, it is possible to continue with the types of games that we can find, considering that in this opportunity we are going to work with two-agent games. Among the types of games are.

3.2.1 Zero-Sum Games

The main characteristic of a zero-sum game for a two-agent game is that the sum of the payoffs of all agents always equals zero, this means that there is one player who wins and one who loses. For example, a poker game, or the matching pennies game.

A zero-sum game is represented in Table 1. Where there are two players or agents who can choose between two options.

As we can see in the example above, if we add the payoffs of any choice of both agents, the sum is equal to zero. To demonstrate it if player 1 chose C, and player 2 chose B, the payoffs for both agents are $< 1, -1 >$ and if we sum both, the result is zero. Is the same case if player 1 chose C and player 2, A with a payoff of $< -2,2 >$. This kind of game belongs to pure competition games, where one agent or player wins and the other loses, the winning or losing values can be equal such as zero-sum game or different.

3.2.2 Cooperative Games

The cooperative games, also called coalition games, are those in which the agents reach an agreement prior to the game so that everyone wins or loses; these gains or losses may be equal or different depending on their contribution or the agreement. Some examples of this are elections, funds, contest, these are examples where multiple agents must work together to obtain the maximum revenue and finally the prize was divided among the participants. In the next table we can see an example about a payoff matrix for a cooperative game (Table 2).

Table 2 Payoff matrix for a cooperative game

Player 1/Player 2	A	B
C	$< -5, -5 >$	$< 2,4 >$
D	$< 8,5 >$	$< -3, -1 >$

Table 3 Payoff matrix for a non-cooperative game

Player 1/Player 2	A	B
C	$< 2,5 >$	$< -3,8 >$
D	$< -2,0 >$	$< 0, -5 >$

From the example it is possible to see that, for each combination both agents get positive or negative payoffs, depending on the combination these values can be the same or different.

3.2.3 Non-Cooperative Games

Non-cooperative games are the most widely used in literature. These games are because each agent wants to maximize its profit, but as is characteristic of game theory, this profit depends on the decision of all agents. In the next table we can observe a payoff matrix for a non-cooperative game (Table 3).

As we can see in the table above, the payoffs for each agent are different and these payoffs, unlike those seen above, are not related to each other.

3.2.4 Repeated Games

As its name implies, it is a game that is repeated a finite number of times, and each time the agents can choose something different to change their profit. This kind of game has two variations, the first one is when the agent does not have a penalty for changing its decision while in the other for each change in the decision the agent has a penalty factor. These repeated games can be like the play the matching pennies game for five times in which if the player one chooses face, the player must keep his choice until the game ends, but if he wants to change his decision he will be penalized.

3.3 Solution Strategies

The strategies are like the solution method that the agent's choice to solve the game and get a result. Among the best known and most widely used, considering that we are evaluating two-agent games, we find the following.

Like in other areas there are some conditions to consider after applying these strategies, the first is that all agents behave rationally, the second one is that all possible solutions or payoffs are known, both their own and those of the opponent [25].

3.3.1 Pure Strategy Nash Equilibrium

The Nash Equilibrium was developed by John Nash in the 50 s [26], it was developed like a solution to apply this approach to solve problems or games with two or more solutions [25].

The purpose of this strategy is to seek the best solution or the best payoff of the game for each agent considering the influence of the decisions made by the other agent.

Let's apply this strategy as an example (Table 4).

An easy form to find the Nash Equilibrium is as follows.

- **First Step**: Choose one agent and one option. For example, Player 1 chooses C.
- **Second Step**: Evaluate the best response for the other player. For example, when Player 1 chooses C, the best payoff for Player 2 is when chooses A, because it earns 10 instead of 0 when chooses B.
- **Third Step**: Evaluate each option for each agent. In this example we obtain the following results.

First agent: P1(C) → P2(A), P1(D) → P2(A)
Second agent: P2(A) → P1(C), P2(B) → P1(D).

These combinations mean the following. For example, P1(D) → P2(A), means that when Player 1 chooses D, the best response for Player 2 is choose A.

- **Forth Step:** Evaluate all the results obtained in the third step and look for a common point. In this example the common point is when the Player 1 chooses C and Player 2 chooses A, it was the common point because when we play as Player 1, we get that point (P1(C) → P2(A)) and when we play as Player 2, we get the same point (P2(A) → P1(C)).

This common point is the Nash Equilibrium point, we have to consider that there can be multiple points in common and each of these is considered as a Nash Equilibrium point.

Table 4 Payoff matrix

Player 1/Player 2	A	B
C	< 1,10 >	< 2,0 >
D	< 0,8 >	< 8,5 >

3.3.2 Mixed-Strategies Nash Equilibrium

The mixed-strategies are based on the fact that each agent can make a random decision or there is a probability of choosing one of the options, these probabilities helps the agents to calculate their own payoff and best response based on the probabilities of the other agent, which can be the opponent.

In addition, this strategy is widely used to set the probabilities of one agent's choices to eliminate the differences between the other agent's choices. The set of probabilities that achieve the above is known as the Mixed-Strategies Nash Equilibrium.

Let's apply this strategy in the previous example. Instead of the previous example, here we will consider the probability "r" that Player 1 chooses C, and a probability "1-r" to choose D; the same for Player 2, "q" is the probability to choose A and "1-q" to choose B (Table 5).

In this example we are going to apply the two forms mentioned above.

First with a defined values of "r" and "q" equal to 0.3 and 0.6 respectively, where we will have to follow the following steps:

- **First Step**: Choose one player, and one option, in this example I will choose Player 1, option C.
- **Second Step**: Calculate the expected value of the choice made in the first step. This is done as follows, first multiply the obtained payoff by Player 1 when Player 2 chooses the first option (A) with the probability that Player2 chooses that option and added with the product obtained when Player 2 chooses the second option. Consider the data of the example, mathematically this is represented as follows.

$$EP(P1(C)) = v_{C,A} * p(P2(A)) + v_{C,B} * p(P2(B)) \tag{1}$$

where $EP(P1(C))$ is the representation of the expected payoff for Player 1 when chooses the option C, $v_{C,A}$ is the payoff of Player 1 chooses C and Player 2 chooses A, and finally the $p(P2(A))$ is the probability of Player 2 chooses A.

Given numerical values to the equation (1) based on the example we get the follow.

$$EP(P1(C)) = 1 * q + 2 * (1 - q) = 1 * 0.6 + 2 * 0.4 = 1.4$$

This means that the expected payoff for Player 1 when chooses C is 1.4.

- **Third Step**: Repeat it for all the players and options. When we do it, we get the following results.

Table 5 Payoff matrix for a mixed-strategy solution

	Probabilities	q	1-q
Probabilities	Player 1/Player 2	A	B
r	C	< 1,4 >	< 2,6 >
1-r	D	< 0,8 >	< 8,5 >

$$EP(P1(C)) = 1 * q + 2 * (1 - q) = 1 * 0.6 + 2 * 0.4 = 1.4$$

$$EP(P1(D)) = 0 * q + 8 * (1 - q) = 0 * 0.6 + 8 * 0.4 = 3.2$$

$$EP(P2(A)) = 4 * r + 8 * (1 - r) = 4 * 0.3 + 8 * 0.7 = 6.8$$

$$EP(P2(B)) = 6 * r + 5 * (1 - r) = 6 * 0.3 + 5 * 0.7 = 5.3$$

- **Forth Step**: One we get the expected payoff for each player and option; we know the best option for each player, which are the option D for Player 1 and option A for Player 2.

Additionally, we can know the expected payoff for each player in the full game. Which is calculated as follows.

$$EP(P1) = EP(P1(C)) * p(P1(C)) + EP(P1(D)) * p(P1(D)) \tag{2}$$

where $EP(P1)$ is the expected payoff for Player 1 in the game, $EP(P1(C))$ and $p(P1(C))$ are the expected payoff for Player 1 when chooses the option C and the probability for Player 1 to chooses the option C. Given numerical values to the Eq. (2) based on the example we get the follow.

$$\begin{aligned} EP(P1) &= EP(P1(C)) * p(P1(C)) + EP(P1(D)) * p(P1(D)) \\ &= 1.4 * 0.3 + 3.2 * 0.7 = 2.66 \end{aligned}$$

$$\begin{aligned} EP(P2) &= EP(P2(A)) * p(P2(A)) + EP(P2(B)) * p(P2(B)) \\ &= 6.8 * 0.6 + 5.3 * 0.4 = 6.20 \end{aligned}$$

These values mean that the expected value for Player 1 for participating is *2.66*, while for Player 2 is *6.20*.

Finally, we are going to calculate the Mixed-Strategy Nash Equilibrium, for the example shown above.

- **First Step**: Select one agent. In this case we are going to start with Player 1.
- **Second Step**: Equalize the expected values for both options, following the Eq. (3).

$$EP(P1(C)) = EP(P1(D)) \tag{3}$$

where each term in the equation (3) is calculated using the equation (1). Consider that in this case the probabilities are unknown, and it is necessary to calculate them.

Based on the example, the numerical solution of the equation (3) is:

$$EP(P1(C)) = EP(P1(D))$$

$$v_{C,A} * p(P2(A)) + v_{C,B} * p(P2(B)) = v_{D,A} * p(P2(A)) + v_{D,B} * p(P2(B))$$

$$1 * q + 2 * (1 - q) = 0 * q + 8 * (1 - q)$$

$$q = 6/7$$

- **Third Step**: Repeat the second step for the other agents. In this case, for the Player 2.

$$EP(P2(A)) = EP(P2(B))$$

$$v_{C,A} * p(P1(C)) + v_{D,A} * p(P1(D)) = v_{C,B} * p(P1(C)) + v_{D,B} * p(P1(D))$$

$$4 * r + 8 * (1 - r) = 6 * r + 5 * (1 - r)$$

$$r = 3/5$$

- **Forth Step**: Once we get all the probabilities values, we have to define the Nash Equilibrium. Which is for Player 1 the probabilities of $(3/5, 2/5)$ for the options C and D respectively, and for Player 2, the probabilities of $(6/7, 1/7)$ for options A and B respectively.

3.3.3 Dominated-Strategies

The dominated-strategies are based on the idea that one option is better than another regardless of the other agent's choice. This strategy is divided in two categories, the strictly dominated and the weakly dominated, the main difference between these two categories is that in the first, one option must always be better than the other, in other words, the payoffs of the second option will be higher than the first option; while in the second category, the payoffs of the second option will be higher or equal to those in the first option. Once we determine the dominated option, this option will be eliminated, and we continue with the next option. Let's start with an example for a better understanding of this strategy (Table 6).

To solve this problem with the dominated-strategies we have to follow the next steps.

- **First Step**: Select one agent and evaluate the options to determine if there is one option better than the other. We will start with Player 2, we can see that option A

Table 6 Original payoff matrix

Player 1/Player 2	A	B	C
X	< 1,10 >	< 2,0 >	< 5,6 >
Y	< 0,8 >	< 8,5 >	< 5,10 >
Z	< 5,7 >	< 6,6 >	< 1,4 >

is better than B, because for each option chosen by Player 1, Player 2 has a better payoff for choosing A instead of B. As we can see as follows.

If Player 1 chooses X, Player 2 should choose A instead of B, because he will earn 10 instead of 0.

If Player 1 chooses Y, Player 2 should choose A instead of B, because he will earn 8 instead of 5.

If Player 1 chooses Z, Player 2 should choose A instead of B, because he will earn 7 instead of 6.

This is known as strictly dominant strategy, and is expressed as B is strictly dominated by A (Table 7).

- **Second Step**: Erase the dominated option, in this case option B. The original table will become as follows.

- **Third Step**: Repeat the first and second steps until only one cell remains. Continuing with the example, we proceed to solve the problem step by step.

We can notice that for Player 1, option Y is weakly dominated by X, this is because for Player 2 option A, the payoff of choosing X is better than choosing Y, but for option C, payoffs are equals. Then we eliminate option Y for Player 1 (Table 8).

Now, we can see that for Player 2, option C is strictly dominated by A, for that reason C is eliminated (Table 9).

Finally, for Player 1, X is strictly dominated by Z (Table 10).

The best response for this game is when Player 1 chooses Z and Player 2 chooses A.

Table 7 Payoff matrix without option B

Player 1/Player 2	A	C
X	< 1,10 >	< 5,6 >
Y	< 0,8 >	< 5,10 >
Z	< 5,7 >	< 1,4 >

Table 8 Payoff matrix without options B and Y

Player 1/Player 2	A	C
X	< 1,10 >	< 5,6 >
Z	< 5,7 >	< 1,4 >

Table 9 Payoff matrix without options B, Y and C

Player 1/Player 2	A
X	< 1,10 >
Z	< 5,7 >

Table 10 Payoff matrix without options B, Y, C and X

Player 1/Player 2	A
Z	< 5,7 >

3.4 Example

The previous sections have been dedicated to show the different types of games and strategies to solve the game theory. In the section is oriented to solve, from scratch, an optimization exercise based in a real problem using game theory.

3.4.1 Study Case

Let's start with an example, imagine a small neighborhood in Mayagüez – Puerto Rico, where within in there are two people who have their own renewable generation, in this case, the first one has photovoltaic solar (PV) system and the second one has wind generation, also both have different load profiles and different batteries capacity.

3.4.2 Profiles Generation

Once we are situated in the example, we will begin to do the mathematical part of the example, we will start by obtaining the solar irradiance and wind speed for that area, which was obtained from. "PVWatts Calculator" [28], with this information the generation profiles will be calculated based on the following.

To convert the solar irradiation and the temperature into solar power, we use the Eq. (4) based on [29].

$$P_{PV}^{t} = \eta_{PV}.A_{PV}.I_{r}^{t}.\left(1 - 0.005\left(Temp_{out}^{t} - 25\right)\right) \tag{4}$$

where η_{PV} and A_{PV} are the efficiency and the area of the solar panels respectively. I_r^t is the solar irradiance in the step "t" and $Temp_{out}^t$ is the temperature in the step "t" in degrees Celsius.

In addition, to obtain the wind power, we used the Bornay datasheet [30], from which we used the Bornay 6000 to convert the wind speed into wind power based on the Performance Table shown on page 24.

Based on the above mentioned, we obtained both generation profiles, which are shown in Fig. 1.

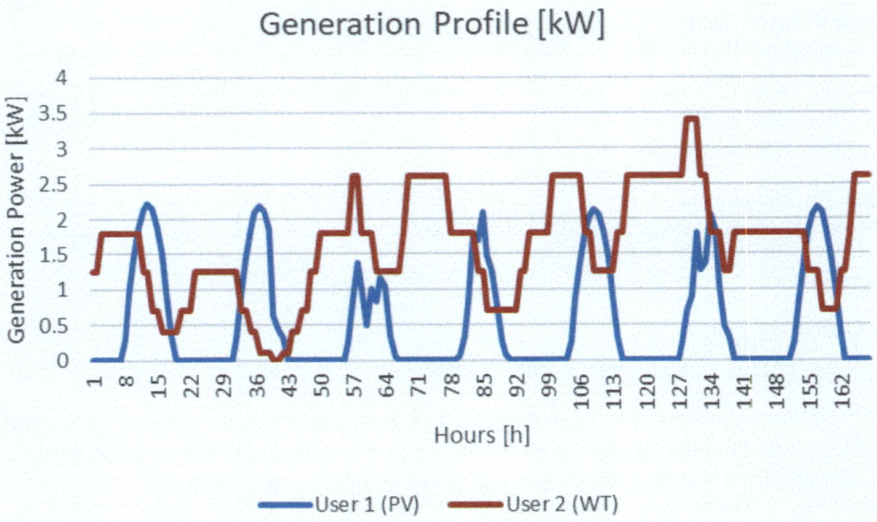

Fig. 1 Generation profiles for two users

Finally, the load profiles were obtained based on [31], where we select the configuration of a home with two rooms for the user 1, and one with five or more rooms for the user 2. The graphic representation of these load profiles is shown in Fig. 2.

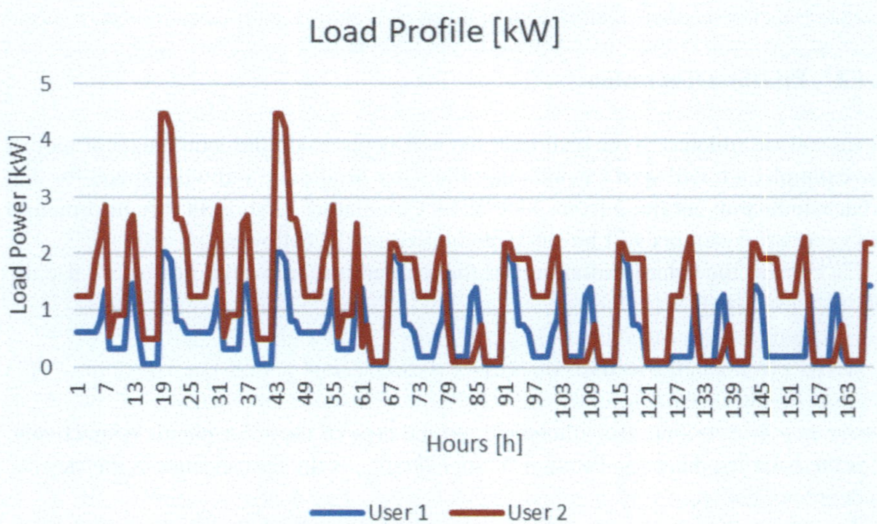

Fig. 2 Load profiles for two users based on [31]

3.4.3 Mathematical Modeling

After getting the profiles for both users, we must define how they will work, they have two alternatives, work alone, or work together. If they decided work alone, we must execute the optimization regardless the interaction between them, but if they decided work together, we must consider the interaction, which consist in the fact that part or all the injected power for one user can be used by the second user.

Based on the above mentioned the equations used to get the results for both alternatives are the following.

- **Objective Function:**

The objective function for this optimization will be the minimization of the absorbed energy cost, which depends mainly on the absorbed energy and the energy cost, but in this case, we are going to consider a constant energy cost, for that reason the minimization only affects the absorbed energy. So, the equation that describes the objective function is.

$$min\sum_{t=1}^{n}\left\{P_{g,\alpha}(t).C_g(t)+a.\left[P_{ji,\alpha}(t).C_{ji,\alpha}(t)-b.\left(P_{ij,\beta}(t).C_{ij,\beta}(t)\right)\right]\right\}.\Delta T \quad (5)$$

where $P_{g,\alpha}$ is the power absorbed by the user from the grid, C_g is the absorbed energy cost; $P_{21,\alpha}$ and $C_{ji,\alpha}(t)$ are the energy absorbed and the cost for absorb this energy by the user "*i*" from the user "*j*"; $P_{ij,\beta}(t)$ and $C_{ij,\beta}(t)$ are the injected energy and the cost for inject this energy by the user "*i*" to the user "*j*"; and ΔT is the time step in hours. Finally, we can find two binary terms which are very important and help with the activation or deactivation of some terms depending on the two works alternatives. The term "*a*" is equals to 0 when the non-cooperative alternative is evaluated but is 1 when the cooperative alternative is evaluated; but the term "*b*" is equals to 0 when the non-cooperative alternative is evaluated or when only one user is cooperating but is 1 when the cooperative alternative with the two users cooperates is evaluated.

- **Constraints**:

Like in the objective function, within the constraints there are some of them that will be activated or deactivated based on the selected alternative.

We will start with one of the most important constraints in this kind of example, which is the power balance equation, and it is expressed as follows.

$$P_L(t)=P_{g,\alpha}(t)-P_{g,\beta}(t)+a.\left[P_{ji,\alpha}(t)-b.P_{ij,\beta}(t)\right]+P_{ESS}(t)+P_{Gen}(t) \quad (6)$$

where P_L is the user load; $P_{g,\beta}$ is the power injected to the grid; P_{ESS} is the power given by the battery, that can be positive when is discharging or negative when is charging, and P_{Gen} is the generated power by the PV system or the wind turbine. The terms "*a*" and "*b*" are the same of the Eq. (5).

Another important constraint is which governs the battery behavior, which is represented based on the following equations.

$$SoC(t) = SoC(t-1) + \Delta SoC(t) \quad (7)$$

$$\Delta SoC(t) = \begin{cases} \frac{P_{ESS}(t).\eta_{ch}}{Cap}, where P_{ESS}(t) \geq 0, (charging) \\ \frac{P_{ESS}(t)}{\eta_{dch}.Cap}, where P_{ESS}(t) < 0, (discharging) \end{cases} \quad (8)$$

where the SoC is the Battery State of Charge, η_{ch} and η_{dch} are the battery charging and discharging efficiency respectively and finally Cap is the total capacity of the battery in Ah.

There are more constraints that help us to obtain coherent results. Such as the following.

$$X_{min} \leq X(t) \leq X_{max} \quad (9)$$

where X can be $P_{g,\propto}$, $P_{ji,\propto}$, $P_{ij,\beta}$, P_{ESS} and SoC, this constraint helps us to stablish the upper and lower boundaries for most of the terms used in the optimization.

3.4.4 Parameters and Values Used

In this section we are going to tell you all the parameters and values that we used for the optimization execution (Table 11).

And the boundaries values used for some parameters (Table 12).

As we can see in the last two parameters, the value of the upper boundaries are other parameters, this is because first we have to execute the optimization program for a non-cooperative scenario, when we get the values of the injected and absorbed power, these are the input for the cooperative scenarios.

Table 11 Used parameters

Parameter	User 1	User 2
Time Window [h]	168	168
Time Step (ΔT) [h]	1	1
Absorbed Energy Cost (C_g) [$]	0.3047	0.3047
Absorbed Energy Cost $(C_{ji,\propto})$ [$]	0.2133	0.2133
Injected Energy Cost $(C_{ij,\beta})$ [$]	0.2133	0.2133
Battery Charge Efficiency (η_{ch})	0.95	0.95
Battery Discharge Efficiency (η_{dch})	0.98	0.98
Battery Capacity (Cap) [kWh]	25.92	69.12
Initial State of Charge (SoC) [%]	50	50

Table 12 Boundaries values

Parameter	Lower Boundary	Upper Boundary
SoC [%]	20	80
Power Absorbed $(P_{g,\alpha})$ [kW]	0	10
Power Injected $(P_{g,\beta})$ [kW]	0	10
Batter Power (P_{ESS}) [kWh]	–7.2	7.2
Absorbed Power $(P_{ji,\alpha})$ [kW]	0	$P_{g,\beta}$
Injected Power $(P_{ij,\beta})$ [kW]	0	$P_{g,\alpha}$

3.4.5 Optimization Results

The results obtained for work alone or together were shown in the following Table.

		Grid		User 1		User 2	
		Absorbed [kWh]	Injected [kWh]	Absorbed [kWh]	Injected [kWh]	Absorbed [kWh]	Injected [kWh]
Non-Cooperative	User 1	20.2272	10.1036	–	–	–	–
	User 2	24.4984	105.7163	–	–	–	–
One Cooperate	User 1	5.8320	10.1036	–	–	15.0073	0.0000
	User 2	24.4984	105.7163	0.0000	0.0000	–	–
Both Cooperate	User 1	5.8320	10.1036	–	–	15.0073	0.0000
	User 2	24.4984	90.7090	0.0000	15.0073	–	–

In the Table we can see all the decisions for both users, which are non-cooperative, only one cooperates and both cooperate. We can see that in the first case, all the absorbed and injected power comes from the grid, but in the second and third cases, part of the absorbed and injected power comes from the other user.

As we can see in the Table there are the values of the absorbed or injected energy to the grid or to the user, that mean that based on the Eq. (5) this is the result of multiply the power with the time step $(P(t).\Delta T)$, regardless of cost. But to convert this Table into money we have to use the energy cost established in the previous point and multiply these energy values with the cost.

	Non-Cooperative	One Cooperate	Both Cooperate
User 1	6.16	4.98	4.98
User 2	7.46	7.46	4.26

3.4.6 Game Theory Approach

With the values obtained from the optimization problem, we can build the payoff matrix for a non-cooperative game theory. This is a non-cooperative type because each user or prosumer tries to maximize its own payoff, for this case both users try to minimize their costs.

Like the previous examples we are going to establish the agents and the options, in this case we have two agents, the User 1 and User 2, and both have two options, which are cooperates or no cooperates. With these parameters, we can build the matrix.

We are going to explain one cell of the payoff matrix, and the other ones can be calculated like the example.

For example, if the User 1 cooperates, but the User 2 does not cooperate. When User 1 cooperates, the savings are equals to $6.16 - 4.98 = 1.18$ and represent the 19.16% of savins, but for User 2 is equals to $7.46 - 7.46 = 0.00$, and represents the 0% of savings.

User 1/User 2	No Cooperate	Cooperate
No Cooperate	< 0.00,0.00 >	< 0.00,0.00 >
Cooperate	< 19.16,0.00 >	< 19.16,42.92 >

Once we get the payoffs matrix, we can select a strategy to get a result, for this example we are going to use the Pure Strategy Nash Equilibrium, and the best response for the game is when both cooperates. Also, this game can be solved using dominated strategies, and we get the same result.

3.5 *Python Code*

In this section we are going to bring you the Python code based on [32] to solve the example.

This code begins with the data acquisition from an Excel Sheet, then the values and boundaries for the different parameters are stablished, as well as the objective function and the constraints needed. And ends with the execution of the optimization and the visualization of the necessary parameters by graphics. Based on that the code was divided in eight sections, which are as follows:

- **First Section**: In the first section the libraries that are needed for the optimization execution were installed, like "pyomo" who help to the solution and the definition of the optimization, "pandas" which helps with the data acquisition from Excel and finally the "matplotlib" which helps with the graphic generation.

- **Second Section**: In this section the load and generation profiles are called from the Excel Sheet with the help of pandas library, then these profiles were saved into a variable called "profiles", where each column of this variable is called as same as they are called in the Excel Sheet.
- **Third Section**: In this section the initial values, as well as the maximum and minimum values for some parameters are stablish. In this section of the code, you can change the battery parameters and the limits of the interaction between the prosumer and the grid.
- **Forth Section**: In this section the lower and upper boundaries were stablish based on the values stablished in the third section, also the type of variable is defined, for example if some of them is a binary variable or only positive integer. Also in this section is defined the dependencies of the variable, for example in this specific case we can see that all the variables depends on the variable "T".
- **Fifth Section**: In this section the objective function of the optimization problem is defined, this function is based on the Eq. (5) defined above, is also necessary to define in the code whether it is a minimization or maximization problem.
- **Sixth Section**: In this section the constraint functions were defined, as can see, all the constraint were based on the Eqs. (6)–(8), like in the previous section is necessary to define the function and if this is a constraint. Additionally to the equations previous mentioned, a new constraint functions were added, which can help to prohibit the charging and discharging the battery and the injection and discharging battery at the same time. These constraints are used to avoid charging and discharging the battery and the same time in one time period and avoid the battery discharging to inject the energy to the grid or to other user.
- **Seventh Section**: In this section the solver for the optimization code was selected, also the solution of the optimization was saved into different variables like injected energy, absorbed energy, state of charge of the battery.
- **Eighth Section**: In this last section the variables saved in the previous section were plotted, in order to obtain a better way to analyse them and compare their behaviours.

```
###First Section
###Install libraries
from pyomo.environ import *
import matplotlib.pyplot as plt
import pandas as pd

###Second Section
###Get the load and generation profiles from Excel
model = ConcreteModel()
path = r"path"
profiles = pd.read_excel(path, sheet_name='Name', header=0, index_col=0)
T = profiles.index.tolist()
model.T = Set(initialize=T, ordered=True)

###Third Section
###Stablish the values of the parameters
Delta_t = 1  # Time Step
Predabs_max = 10  # Maximum Absorbed Power from the Grid kW
Predinj_max = 10  # Maximum Injected Power to the Grid kW
CapkWh = 25.92 # Total Battery Capacity Ah
PESScargamax = 7.2  # Maximum Battery Charging Power kWh
PESSdescmax = 7.2  # Maximum Battery Discharging Power kWh
SoC0 = 0.5  # Initial State of Charge
SoCmax = 0.90  # Maximum State of Charge
SoCmin = 0.20  # Minimum State of Charge
Eff_ch = 0.98  # Battery Discharge Efficiency
Eff_dch = 0.95  # Battery Charge Efficiency
Cg = 0.3047 #Absorbed Energy Cost $
Cji = 0.2133 #Absorbed Energy Cost $
Cij = 0.2133 #Injected Energy Cost $
a = 0 #Binary variable, we have to change it manually
b = 0 #Binary variable, we have to change it manually

###Forth Section
###Variables
# Injected Power
model.Predinj = Var(model.T, domain=NonNegativeReals, bounds=(0.0, Predinj_max))
# Abosrbed Power
model.Predabs = Var(model.T, domain=NonNegativeReals, bounds=(0.0, Predabs_max))
# Battery Discharge Power
model.PESSdch = Var(model.T, domain=NonNegativeReals, bounds=(0.0, PESSdescmax))
# Battery Charge Power
model.PESSch = Var(model.T, domain=NonNegativeReals, bounds=(0.0, PESScargamax))
# Battery State of Charge
model.SoC = Var(model.T, domain=NonNegativeReals, bounds=(SoCmin, SoCmax))
# Binary Variable for Charge=1 and Dischar=0
model.xESS = Var(model.T, within=Binary)
# Absorbed Power for Other User
model.Pji = Var(model.T, domain=NonNegativeReals, bounds=(0.0, Predabs_max))
# Injected Power to Other User
model.Pij = Var(model.T, domain=NonNegativeReals, bounds=(0.0, Predinj_max))

### Fifth Section
###Objective Function
def Obj_rule(model, t):
    return sum(Cg * model.Predabs[t] + a * (Cji * model.Pji[t] -
b*model.Pij[t]*Cij) for t in T)

model.Obj = Objective(rule=Obj_rule,sense=minimize)
```

```
###Sixth Section
###Constraints
def Elec_rule(model, t):
    return profiles.Generation[t] + model.Predabs[t] - model.Predinj[t] +
model.PESSdch[t] - model.PESSch[t] + model.Pji[t] \
        - model.Pij[t] == profiles.PLoad[t]

model.balance = Constraint(model.T, rule=Elec_rule)

def Pow_in(model, t):
    return model.Pji[t] <= profiles.P21[t] #This means that go from 2 to 1,
because we are evaluating user 1

model.Potencia_21 = Constraint(model.T, rule=Pow_in)

def Pow_out(model, t):
    return model.Pij[t] == profiles.P12[t]

model.Potencia_12 = Constraint(model.T, rule=Pow_out)

def SoC_rule(model,t):
    if t == 1:
        return model.SoC[t] == SoC0 + Eff_dch * model.PESSch[t] * Delta_t /
CapkWh \
              - (model.PESSdch[t] * Delta_t) / (Eff_ch * CapkWh)
      else:
          return model.SoC[t] == model.SoC[t-1] + Eff_dch * model.PESSch[t] *
  Delta_t / CapkWh \
              - (model.PESSdch[t] * Delta_t) / (Eff_ch * CapkWh)

  model.CSoC = Constraint(model.T, rule=SoC_rule)

  def c_lim_rule(model,t):
      return model.PESSch[t] <= PESScargamax * model.xESS[t]

  model.ESScar_lim = Constraint(model.T, rule=c_lim_rule)

  def d_lim_rule(model,t):
      return model.PESSdch[t] <= PESSdescmax * (1 - model.xESS[t])

  model.ESSdes_lim = Constraint(model.T, rule=d_lim_rule)

  def no_bat_inj(model,t):
      return model.Pij[t] <= Predinj_max * model.xESS[t]

  model.Pinj = Constraint(model.T, rule=no_bat_inj)

  ###Seventh Section
  ###Solver
  def pyomo_postprocess(options=None, instance=None, results=None):
      results.write()
      model.Obj.display()
      model.Predabs.display()


  opt = SolverFactory("glpk")
  results = opt.solve(model)

  pyomo_postprocess(None, None, results)
  print("\nSolucion optima encontrada\n" + '-' * 80)
```

```
E_Buy = [model.Predabs[t].value for t in T]
E_Sell = [model.Predinj[t].value for t in T]
PESSc = [model.PESSch[t].value for t in T]
PESSdc = [model.PESSdch[t].value for t in T]
SoCp = [model.SoC[t].value for t in T]
xESSp = [model.xESS[t].value for t in T]

###Eighth Section
###Plot the Variables

x = list(range(0, 24*7))
fig1 = plt.figure(figsize=(7, 5))
plt.step(x, E_Buy, 'b', label='P Absorbed From the Grid')
plt.step(x, E_Sell, 'r', label='P Inject To the Grid')
plt.xlabel('Time [h]')
plt.ylabel('Power [kW]')
plt.xticks(x)
plt.legend()
plt.grid()
plt.show()
plt.close(fig1)
fig2 = plt.figure(figsize=(7, 5))
plt.step(x, PESSc, 'b', label='Charging')
plt.step(x, PESSdc, 'r', label='Discharging')
plt.xlabel('Time [h]')
plt.ylabel('Power [kW]')
plt.xticks(x)
plt.legend()
plt.grid()
plt.show()
plt.close(fig2)

fig3 = plt.figure(figsize=(7, 5))
plt.step(x, SoCp, 'b', label='SoC')
plt.xlabel('Time [h]')
plt.ylabel('SoC')
plt.xticks(x)
plt.legend()
plt.grid()
plt.show()
plt.close(fig3)

fig4 = plt.figure(figsize=(7, 5))
plt.step(x, xESSp, 'b', label='xESS')
plt.xlabel('Time [h]')
plt.ylabel('xESS')
plt.xticks(x)
plt.legend()
plt.grid()
plt.show()
plt.close(fig4)
```

4 Power Exchange Through Multi-terminal Converters

To enable power exchange between microgrids, apart from the coordination and management strategy, a reliable channel for interconnecting independent microgrids is required. In general, three main methods for interconnecting independent AC power systems could be mentioned. The first one is based on direct interconnection through electromechanical or static switches, as shown in Fig. 3 [33]. The second option is the use Variable Frequency Transformers (VFT) as shown in Fig. 4 [34, 35]. Finaly, the third option is by means of an intermediate DC link in which bidirectional power electronics converters are used between the AC grid and the DC link in form of Back-to-Back conversion system as shown in Fig. 5 [36].

The use of switches for direct interconnection between microgrids requires that all the connected grids have the same rated voltage and frequency. Therefore, additional synchronization processes are required to homogenize the grids. This means that, in the end, independent and asynchronous power grids will be unified into a larger one, with different security and stability issues that should be reconsidered when integrating a new system [37].

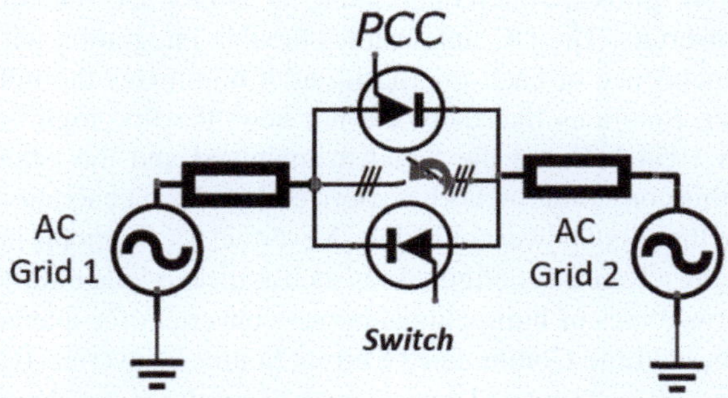

Fig. 3 Direct interconnection through switches

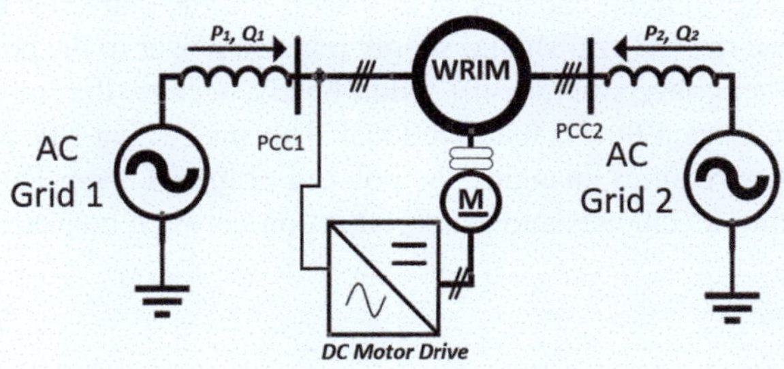

Fig. 4 VFT system configuration [35]

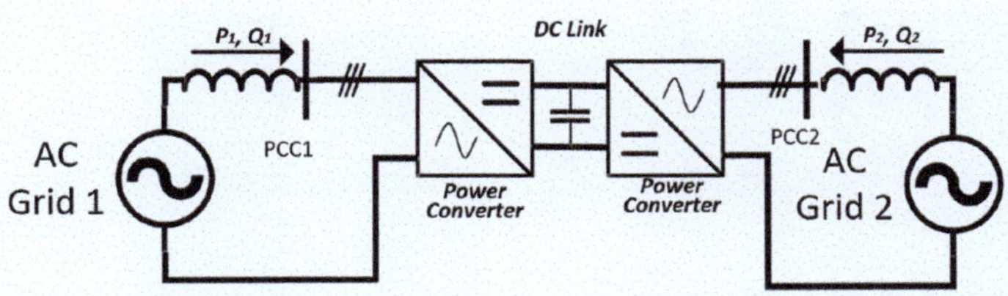

Fig. 5 Back-to-Back interconnection

The VFT uses a doubly fed induction machine to interconnect two separate electrical networks: one connected to the stator and the other connected to the rotor, as shown in Fig. 4 [34]. This technology has proven to be more cost-effective for the integration of large-scale asynchronous power grids compared to other technologies, such as high voltage direct current systems [37]. Additionally, among the main advantages of a VFT, it is important to mention its high efficiency, low harmonic generation, and easy maintenance [37]. Nevertheless, this technology is limited to enable the power flow between two peers which means that no more grids can be added to the power trade and exchange. This fact is an important limitation by considering the scalability required for a power exchange strategy between multiple microgrids.

Figure 5 shows the schematic of a back-to-back connection between two AC microgrids. The AC grids are interconnected to a common DC link through bidirectional power converters. The DC link enables flexible integration without compromising the independence of each microgrid, as it minimizes the influence of one grid on the other. Solutions like these are not new; the first high voltage dc link was installed in 1954 between the Swedish mainland and the island of Gotland [38]. However, this configuration is characterized by a high generation of switching harmonics due to the use of power converters. Nevertheless, harmonic injection can be minimized by using advanced configurations such as the modular multilevel converter [39]. There are two types of bidirectional power converters for interconnecting AC grids: thyristor-based Line-Commutated Current Source Converters (CSCs) (Fig. 6), and those that use self-commutated devices such as insulated-gate bipolar transistors (IGBTs) in Voltage Source Converters (VSCs) (Fig. 7) [36].

In the case of a CSCs, it is required that the converter is connected to a strong AC grid with a short-circuit capacity of at least twice the converter rating. This is necessary because the AC grid should supply reactive power to the power converter to enable the power conversion[36, 40]. Additionally, because the converter is based on thyristors, the current flow is unidirectional. This means that changing the direction of power flow requires changing the polarity of the DC bus. This fact implies complex switching techniques and synchronization between converters. On top of

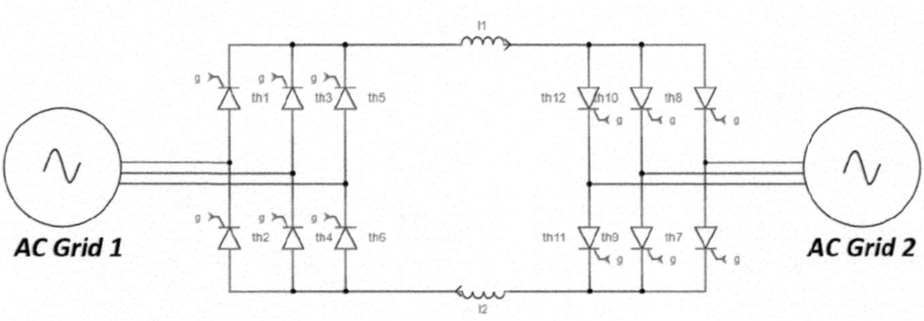

Fig. 6 CSCs Back-to-Back system

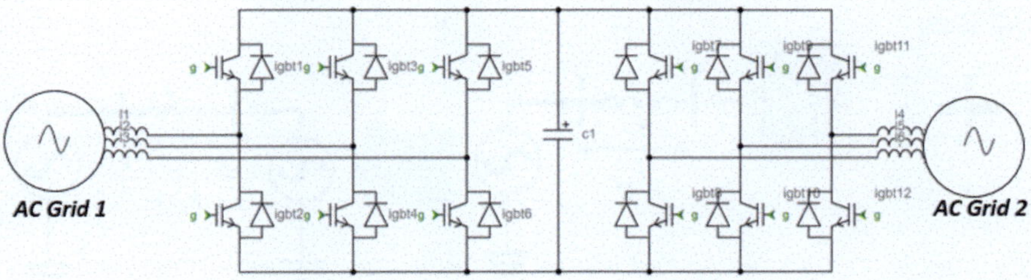

Fig. 7 Back-to-Back interconection

that, the complexity increases significantly when more than two converters need to be connected to the DC link [41, 42].

On the other hand, VSCs can rapidly and independently control the active and reactive power flow at each terminal. This capability arises from their ability to generate an AC sinusoidal voltage at any magnitude and phase, thanks to the use of self-commutated devices. Because of this capability, VSCs can be connected to weak AC networks, provide reactive power support, and even perform black start functions [36, 40]. Additionally, the current direction at the DC link can be adjusted without changing its polarity. Because the DC link is regulated, DC-DC converters can be directly connected to link DC microgrids. These characteristics make VSCs more suitable for integrating multiple independent microgrids, enabling power transactions between them.

4.1 VSC Operation and Control Modes

Because a VSC can generate a sinusoidal voltage with any amplitude and phase when connected to an AC grid, the system can be seen as two generators connected through an inductive reactor, as shown in Fig. 8. Based on the assumption that the two voltages, the fundamental component of the converter voltage ($V_c(t)$), and the grid voltage ($V_G(t)$), are completely synchronized, and neglecting losses, the active and reactive power flow between the AC grid and the converter can be simplified as shown in Eqs. (10) and (11) respectively. Synchronization between the AC grid voltage (V_G) and the voltage generated by the converter (V_C) is always required for a reliable interaction between the VSC and the power grid. Phase-Locked Loops (PLLs) should always be used in the controllers to ensure this synchronization [43].

$$P = \frac{V_C V_G}{X} sin(\varphi) \tag{10}$$

$$Q = \frac{V_G(V_G - V_C cos(\varphi))}{X} \tag{11}$$

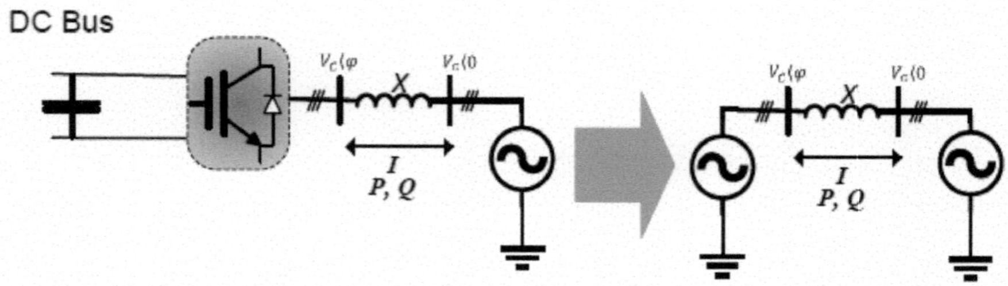

Fig. 8 Equivalence between VSC connected to a power grid and two power grids interconnected

In Eq. (10) and (11) P and Q are the active and reactive power exchanged between the converter and the grid, respectively, φ is the phase shift between the sinusoidal voltage of the grid and the sinusoidal voltage generated by the converter, V_C and V_G are the RMS values for the converter and grid voltage, respectively, and X is the reactance of the equivalent inductor between the converter and the connection point with the main grid.

4.1.1 Power Flow Control

From Eq. (10), it is possible to see that the active power flow can be changed by varying the value of φ. In this case, a leading value for φ means that the inverter is injecting power into the grid, whereas a lagging value for φ means that the inverter is absorbing power from the AC grid. Alternatively, from Eq. (11), it is possible to infer that the reactive power flow can be adjusted by increasing or decreasing the voltage generated by the converter V_C with respect to the grid voltage V_G.

In addition, from Eqs. (10) and (11), it is possible to see that the active and reactive power are not completely decoupled, since a small difference between the converter and grid voltage may also cause a variation in the active power.

However, the active and reactive power flow control can be decoupled if the VSC is controlled in the synchronous rotating d-q frame by means of the Clarke-Park transformation. In this case, the active and reactive power in a balanced system can be written as:

$$P = \frac{1}{2}\left(v_{dG}i_d + v_{qG}i_q\right) \tag{12}$$

$$Q = \frac{1}{2}\left(v_{dG}i_q - v_{qG}i_d\right) \tag{13}$$

where v_{dG} and v_{qG} are the d-q components of the voltage at the point where the inverter is connected to the grid, i_d and i_q are the d-q components of the inductor current. Since independent microgrids define their own voltage amplitude and frequency they impose the reference for the synchronization, performed by the PLL. The voltage at the grid is then considered to have a phase shift equal to zero. Therefore, $v_{qG} = 0$.

Under this condition Eqs. (12) and (13) can be rewritten as.

$$P = \frac{v_{dG} i_d}{2} \tag{14}$$

$$Q = \frac{v_{dG} i_q}{2} \tag{15}$$

From Eqs. (14) and (15), it is possible to see that the active and reactive power flow can be controlled simply by adjusting the inductor current. Additionally, if independent controllers are defined for the d and q components of the inductor current, active, and reactive power control are decoupled. Figure 9 shows the control scheme for the inverter connected to an AC-formed grid. Similarly, Fig. 10 illustrates the control block diagram for the VSC behaving as a power controller.

In Fig. 9, the current reference generator will calculate the reference values (I_d^*) and (I_q^*) for the inductor current as a function of the active power reference P^* and reactive power reference Q^* as:

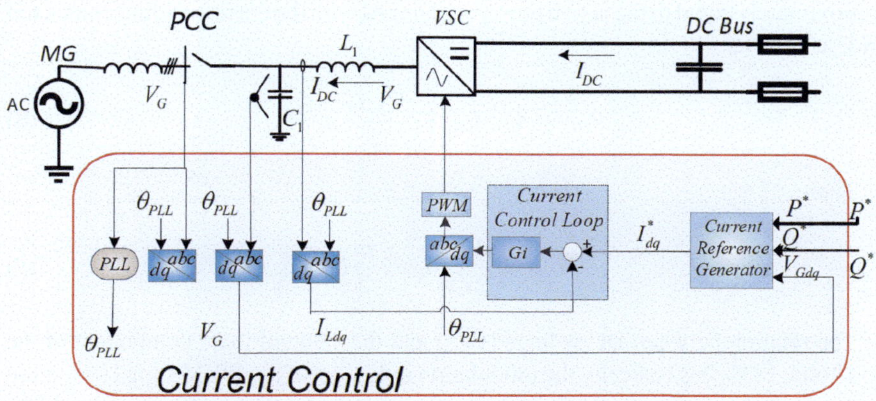

Fig. 9 Control scheme of a VSC in powe flow control mode

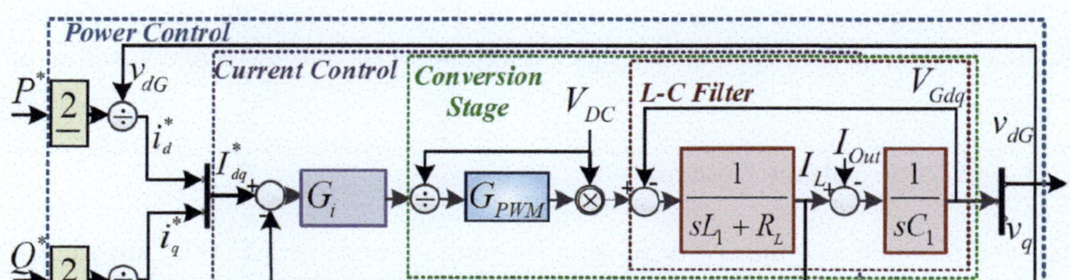

Fig. 10 Control block diagram for the VSC under power flow control

$$I_d^* = 2\frac{P^* v_{dG} - Q^* v_{qG}}{v_d^2 + v_q^2} \tag{16}$$

$$I_q^* = 2\frac{P^* v_{qG} + Q^* v_{dG}}{v_d^2 + v_q^2} \tag{17}$$

When $v_{qG} = 0$ the current reference generator is simplified and can be implemented as shown in Fig. 10. The reference values mainly for the reactive power are expected to be scheduled by the transactive strategy between microgrids.

4.1.2 DC Link Voltage Control

A crucial part of the system's operation is to form and regulate the DC link voltage. To ensure this, it is required that at least one of the VSCs assumes control of the DC bus. The units taking control of the DC bus will operate as forming units and will also manage the power balance in the system. Therefore, DC voltage control is related to power flow control, considering that the power at the DC side will be approximately equal to the active power exchanged with the corresponding connected grid, as follows:

$$P_{DC} = V_{DC} I_{DC} \cong \frac{V_{dG} I_d}{2} = P \tag{18}$$

Then,

$$I_{DC} \cong \frac{V_{DC}}{2V_{dG}} I_d \cong \frac{m}{2} I_d \tag{19}$$

The relation between DC voltage (V_{DC}) and the value of the d component of the grid voltage (V_{dG}) is basically the modulation index (m) of the PWM signal switching the converter. It is important to mention that the modulation index (m) should be kept smaller than 1 to avoid overmodulation of the sinusoidal signal. Therefore, it is required that the value of the DC bus voltage be larger than the amplitude of all the AC grids interconnected in the system. Figures 11 and 12 show the scheme for the DC link control of the VSC and the block diagram for the DC voltage control loop.

4.1.3 AC Side Voltage Control

It is also possible to control a VSC as a forming unit on the AC side, allowing it to regulate the voltage of the interconnected AC grid. In this case, it is necessary for the DC bus to be regulated to a voltage level higher than the desired voltage amplitude of the AC grid. When the inverter operates as a forming unit on the AC side, it will draw or inject the amount of power required to maintain power balance

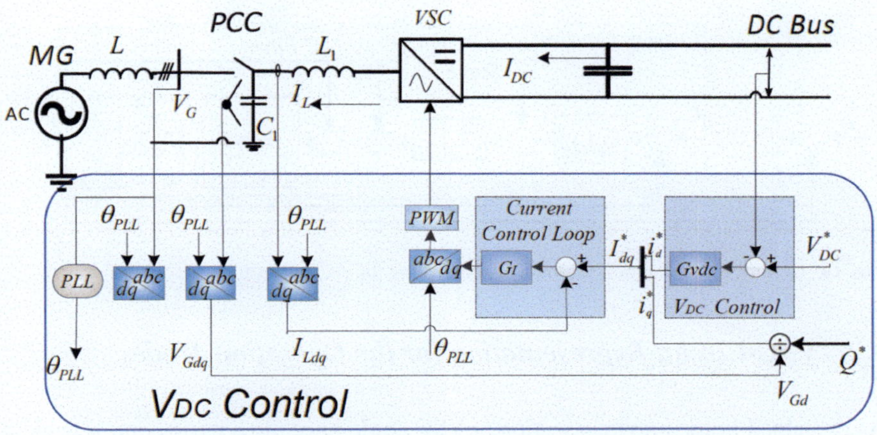

Fig. 11 Control scheme of a VSC under DC link voltage control

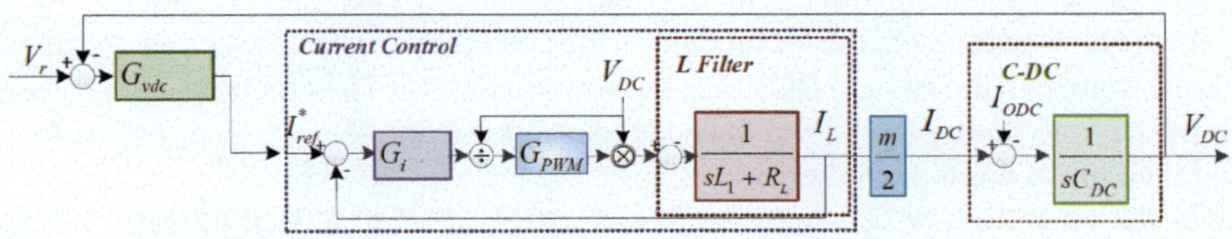

Fig. 12 Control block diagram for the VSC under DC link voltage control

in the interconnected AC microgrid. Figures 13 and 14 show the block diagram for the AC voltage control loop and the scheme for controlling the AC voltage for the VSC, respectively.

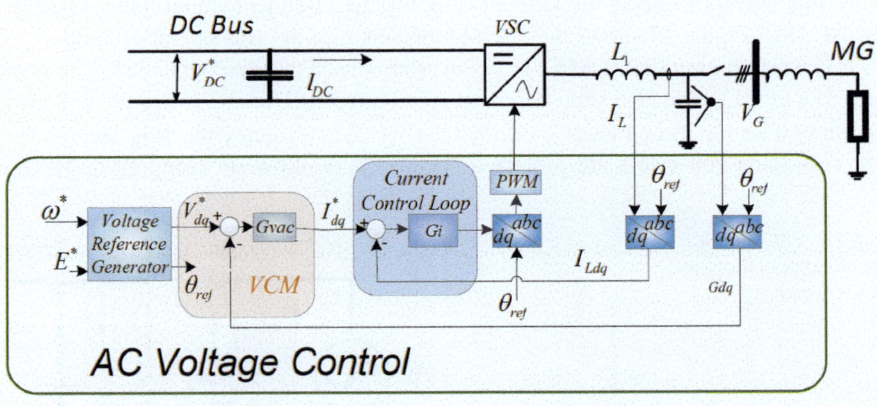

Fig. 13 Control scheme of a VSC under AC side voltage control

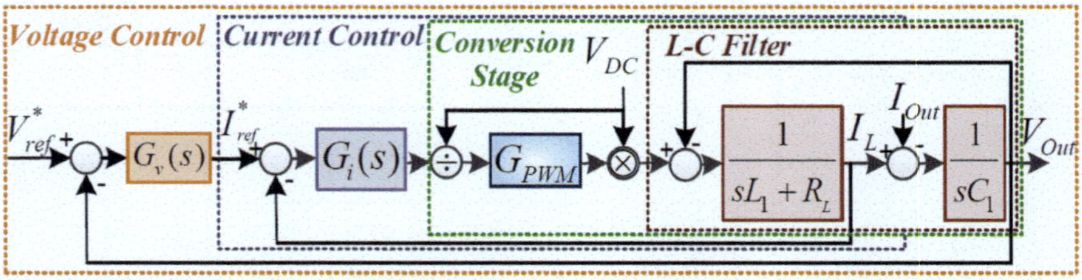

Fig. 14 Control block diagram for the VSC controling the AC side voltage

4.2 VSC-Circuit Representation for the Operation Modes

As explained in the previous section, a VSC could have three basic control modes: DC link voltage control, AC voltage control, and power flow control. One way to better understand the operation of the converter under these different control modes is by representing them with circuit diagrams for both the AC and DC sides.

In the first case, where the VSC controls the power flow, it can be seen as a current source from both the AC and DC sides, as shown in Fig. 15. The function of the power converter is to draw power from one side, convert it from AC to DC, or DC to AC, and then inject it into the other side.

In the operation mode where the VSC controls the DC link, it operates like a voltage source from the DC side and like a current source from the AC side. In this mode, the power converter draws or injects the amount of power required to maintain the DC link voltage regulated. Figure 16 shows the circuit scheme for this operation mode.

Finally, in the last operation mode where the power converter controls the voltage at the AC side, the VSC is seen as a current source on the DC side and as a voltage source on the AC side. In contrast to the previous mode, in this case, the power converter draws or injects the amount of power required to regulate the voltage at the AC side. Figure 17 shows the circuit representation for this operation mode.

It is worth noting that if a DC-DC converter is used to interconnect a DC micro-grid to a multi-terminal system based on a DC link, the DC-DC converter will also operate under the three modes described earlier. When interconnecting DC micro-grids, isolated converters are preferred because they allow for stepping down the

Fig. 15 VSC-circuit representation for power flow control

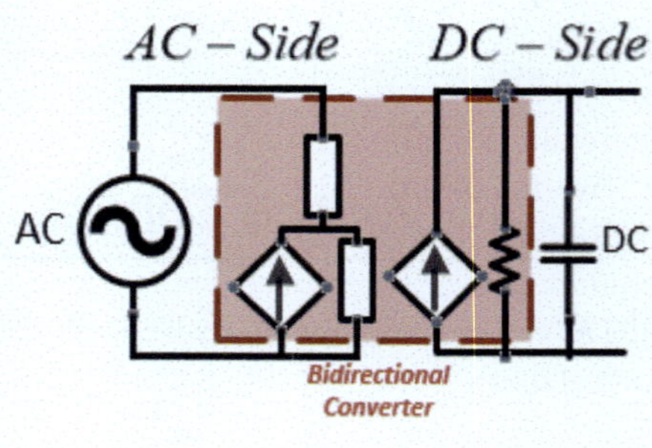

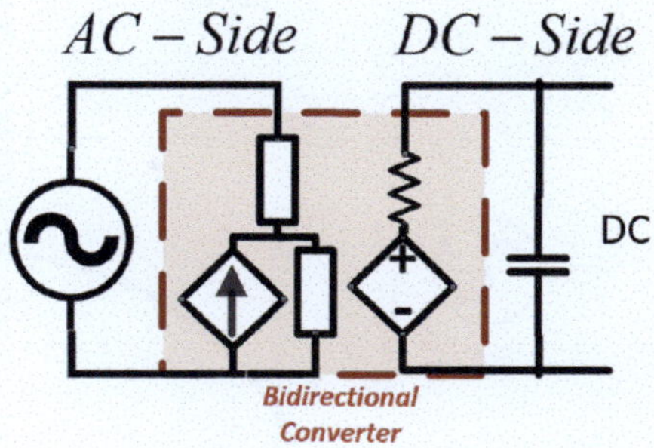

Fig. 16 VSC-circuit representation for DC link control

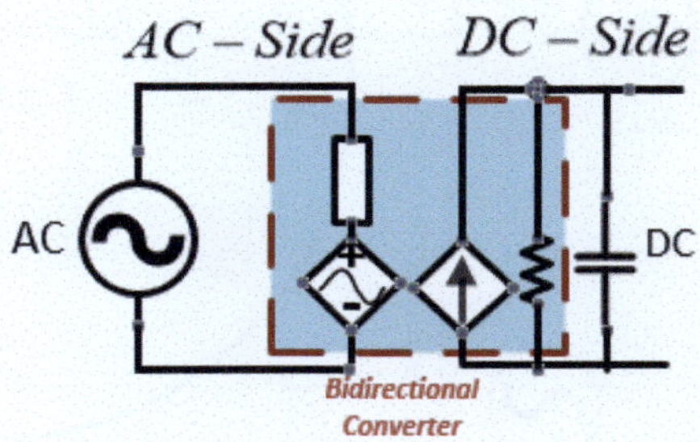

Fig. 17 VSC-circuit representation AC side voltage control

voltage level at the DC microgrid side. For instance, when connecting a 48 V DC microgrid, while maintaining higher voltage levels at the multi-terminal DC link. The dual active bridge is the most used isolated converter for microgrids and interlink applications [44, 45].

4.3 Multi-terminal Architectures

The main characteristic of a multi-terminal converter is its ability to integrate several different AC or even DC microgrids into a common electrical system through a specific connection port for each one, as shown in Fig. 18. Each connection port should have a bidirectional converter: a VSC for interconnecting AC microgrids or a dual active bridge for DC microgrids. These converters regulate the power exchange between each microgrid and the common electrical system. Internally, the electrical system of the multi-terminal converter should be able to auto-regulate its power balance inside the system, enabling proper energy exchange between all the interconnected users.

Two main configurations, series and parallel connections can be used at the DC link of a Multi-terminal converter [40]. The series connection, illustrated in Fig. 19, requires that one of the converters control a common rated current flowing through the DC circuit. Meanwhile, the other converters connected at the other terminals will control the power transfer. This configuration is only possible for CSCs because they

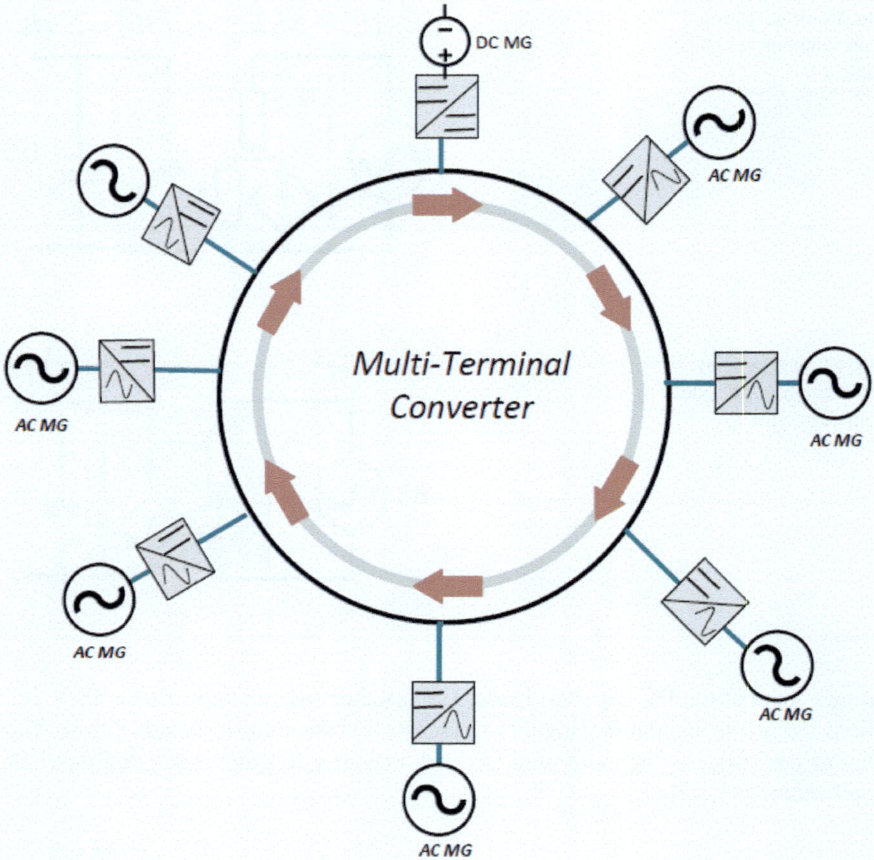

Fig. 18 Representation of a Multi-terminal converter

essentially function as voltage sources at the DC side, allowing them to control their DC voltage over a wider range, from positive to negative values, and even to zero. This behavior is not possible for VSCs because they always require a DC side voltage value higher than the peak voltage of the AC grid. Additionally, other drawbacks for the series connection can be mentioned, such as higher losses and multiple points of failure, because If one unit fails, the common circuit is broken.

The other option is parallel connection, in which all the converters are connected to the same DC bus, as shown in Fig. 20. This configuration is more suitable for VSCs and DC-DC converters because it establishes a regulated DC link, enabling the required conditions for the operation of VSCs. Additionally, this configuration offers plug-and-play capability, as a failed converter at any terminal can be disconnected without affecting the operation of the others.

For the proper operation of the Multiterminal converter, it is required that at least one of the converter units assumes the role of forming the common DC bus.

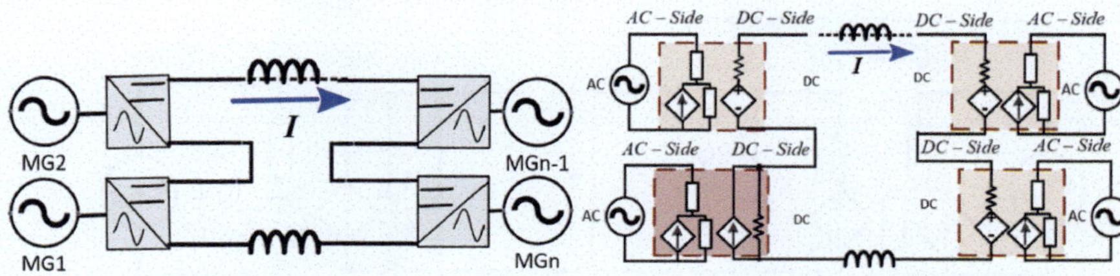

Fig. 19 Multi-terminal converter in series connection

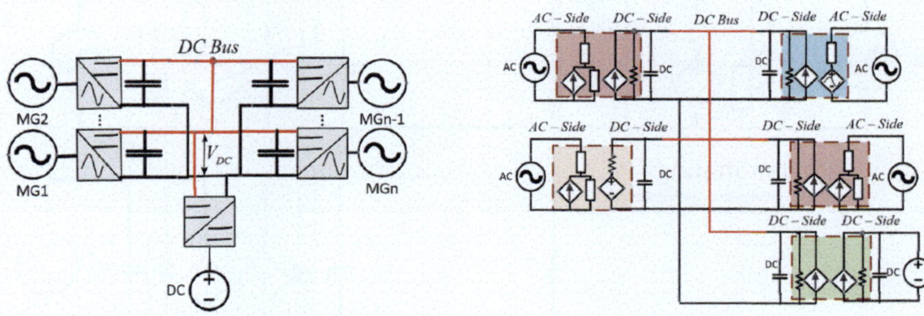

Fig. 20 Multi-terminal converter in parallel connection

This means that at least one converter should be controlled in DC voltage mode. Meanwhile, the other converters at the other terminals may be controlled as current sources at the DC side, as shown in Fig. 20.

However, it is important to consider the power rating of the converters during the operation of the system, especially for the converter regulating the DC link, as it will be responsible for ensuring power balance. It may occur that the amount of power to be compensated by the converter regulating the DC link exceeds its limits, for instance, if all the ports are injecting or drawing power at the same time. This issue should be considered and properly managed within the transactive energy strategy, at the upper control levels, where the power dispatch of all the units is performed [4]. Another option to solve this issue is to configure more than one converter as a forming unit at the DC side (DC link control), as shown in Fig. 21. The VSCs can easily move from one control mode to another during system operation. This fact will require coordination functions that can also be defined within the management strategy [46]. When the VSCs are configured in DC link control, the power balance can be shared between two or more units, increasing the power regulation capability of the integrated power system.

To enable power sharing between units, droop control loops can be configured for the converters participating in power balance regulation, following Eq. (20), where R_{vir} represents the virtual resistance, V_{DC}^* the rated DC voltage operation, and V_{DC} the voltage measured at the common connection point of the multi-terminal converter. Figure 22 shows the DC droop curves representing how the current, and consequently the power, can be shared between two forming units when the virtual resistances at

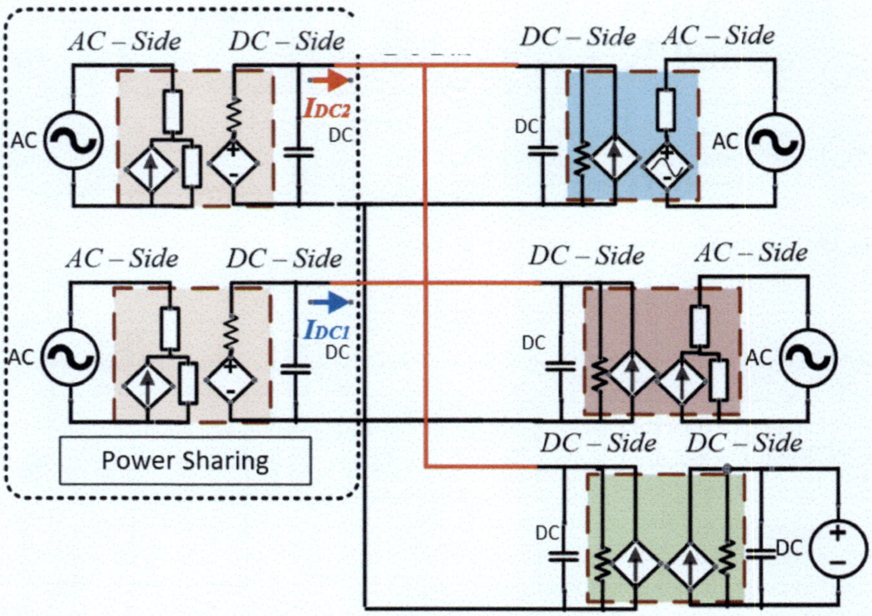

Fig. 21 Multi-terminal converter in parallel connection with two forming units at the DC link

each forming unit have the same or different values. The droop coefficients, or virtual resistances (R_{vir}), can also be adjusted by the upper control levels, the 'transactive energy and coordination strategy,' based on the scheduled power dispatch. However, stability and voltage deviation should be considered when selecting the appropriate values for the virtual resistances [47]. Figure 23 shows the control scheme for a VSC under DC link control with the droop control loops.

$$V_{DC} = V_{DC}^{*} - I_{DC}R_{vir} \tag{20}$$

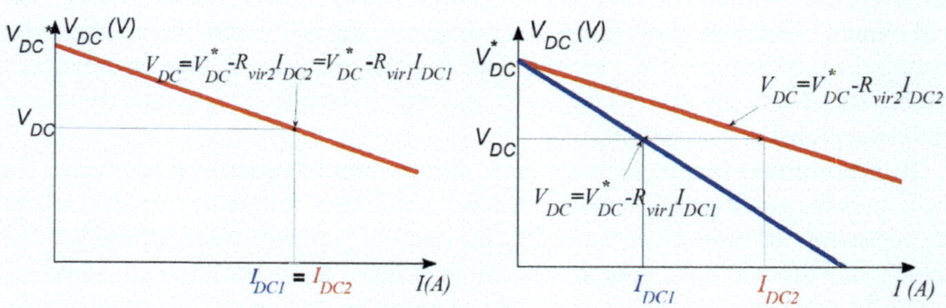

Fig. 22 Current sharing curves for two VSC operating as formin units at the DC side

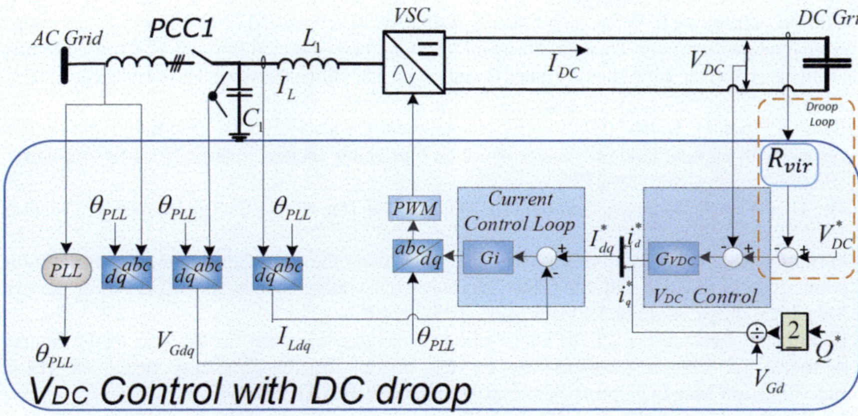

Fig. 23 Control scheme for a VSC under DC link control with the droop control loops

5 Conclusion

This chapter presents perspectives from different viewpoints on energy transactions between microgrids. The problems and challenges are mainly addressed from two perspectives. The first is from an upper-level management strategy, presenting perspectives and possibilities of using game theory as the basis of the transactive energy strategy and as can be seen along the chapter, this strategy give a lot of possibilities, because it allows the evaluation and prediction of the results depending on the choices made by the users. In this specific case a problem with two users trying to exchange energy between them was developed, given as result that the best strategy to obtain the best revenue is when both cooperate. This means that the excess of energy of one user is use by the second one and vice versa. The other perspective is explored at lower levels, examining the hardware requirements and topologies that will enable transactive energy to become a feasible reality between independent microgrids and prosumers.

References

1. Joseph E, Lasseter R, Schenkman B, et al. (2008) CERTS microgrid laboratory Testbed. Consortium for Electric Reliability Technology Solutions (CERTS). Columbus
2. Valencia F, Billi M, Urquiza A (2021) Overcoming energy poverty through micro-grids: An integrated framework for resilient, participatory sociotechnical transitions. Energy Res Soc Sci 75. https://doi.org/10.1016/j.erss.2021.102030
3. Caramizaru A, Uihlein A (2019) Energy communities : an overview of energy and social innovation

4. Díaz NL, Guinjoan F, Velasco-Quesada G, Luna AC, Guerrero JM (2023) Fuzzy-based cooperative interaction between stand-alone microgrids interconnected through VSC-based multi-terminal converter. Int J Electr Power Energy Syst 152. https://doi.org/10.1016/j.ijepes.2023.109226
5. Jiang A, Yuan H, Li D (2021) Energy management for a community-level integrated energy system with photovoltaic prosumers based on bargaining theory. Energy 225. https://doi.org/10.1016/j.energy.2021.120272
6. The GridWise Architecture Council (2015) GridWise Transactive Energy Framework Version 1.0
7. Forfia D, Knight M, Melton R (2016) The view from the top of the mountain: Building a community of practice with the GridWise transactive energy framework. IEEE Power Energy Mag 14(3). https://doi.org/10.1109/MPE.2016.2524961
8. Rahimi FA, Mokhtari S (2018) Distribution management system for the grid of the future: A transactive system compensating for the rise in distributed energy resources. IEEE Electrification Mag 6(2). https://doi.org/10.1109/MELE.2018.2816846
9. Iringan RA, Janer AMS, Tria LAR (2022) A machine-learning based energy management system for microgrids with distributed energy resources and storage. In: 2022 International Conference on Electrical Machines and Systems, ICEMS 2022. https://doi.org/10.1109/ICEMS56177.2022.9983198
10. Ali M, Hossain MI, Shafiullah M (2022) Fuzzy logic for energy management in hybrid energy storage systems integrated DC microgrid. In: 2022 International Conference on Power Energy Systems and Applications, ICoPESA 2022. https://doi.org/10.1109/ICoPESA54515.2022.9754406
11. Patil A, Patil SR (2023) Energy management system for smart microgrid systems using adaptive fuzzy logic. In: 2023 3rd International Conference on Intelligent Technologies, CONIT 2023. https://doi.org/10.1109/CONIT59222.2023.10205643
12. Zhang Y, Ma Y, Zhang S, Chen L, Liu H (2022) Building-level demand-side energy management based on game theory. In: 2022 IEEE International Conference on Mechatronics and Automation, ICMA 2022. https://doi.org/10.1109/ICMA54519.2022.9856118
13. Tatari FR, Bizhani H, Muyeen SM, Ali L, Sanjeevikumar P (2022) A game theory based optimal planning for a hybrid energy system considering time of use tariffs. In: Proceedings - 2022 IEEE 4th Global Power, Energy and Communication Conference, GPECOM 2022. https://doi.org/10.1109/GPECOM55404.2022.9815742
14. Attia EM, Abdelsalam HA, Rashad EEM (2022) Consensus-based tracking control of energy storage for microgrid energy management. In: 2022 23rd International Middle East Power Systems Conference, MEPCON 2022. https://doi.org/10.1109/MEPCON55441.2022.10021697
15. Ferdous SM, Shahnia F, Shafiullah GM (2021) Stability and robustness of a coupled microgrid cluster formed by various coupling structures. Chin J Electr Eng 7(4). https://doi.org/10.23919/CJEE.2021.000038
16. Garcia YV, et al. (2023) Overview on transactive energy—Advantages and challenges for weak power grids. Energies 16(12). MDPI. https://doi.org/10.3390/en16124607
17. Parag YSBK (2016) Electricity market design for the prosumer era. Nat Energy 1:1–6
18. Sousa T, Soares T, Pinson P, Moret F, Baroche T, Sorin E (2019) Peer-to-peer and community-based markets: A comprehensive review. Renew Sustain Energy Rev 104. Elsevier Ltd, pp. 367–378. https://doi.org/10.1016/j.rser.2019.01.036
19. Khorasany M (2020) Market design for peer-to-peer energy trading in a distribution network with high penetration of distributed energy resources. Queensland University of Technology
20. Nizami S, Tushar W, Hossain MJ, Yuen C, Saha T, Poor HV (2022) Transactive energy for low voltage residential networks: A review. Appl Energy 323. Elsevier Ltd. https://doi.org/10.1016/j.apenergy.2022.119556
21. Song M, Cai Y, Gao C, Chen T, Yao Y, Ming H (2022) Transactive energy in power distribution systems: Paving the path towards cyber-physical-social system. Int J Electr Power Energy Syst 142. Elsevier Ltd. https://doi.org/10.1016/j.ijepes.2022.108289

22. Tan LN, Gupta N, Derawi M (2023) Adaptive dynamic programming and zero-sum game-based distributed control for energy management systems with internet of things. IEEE Internet Things J 10(24):22371–22385. https://doi.org/10.1109/JIOT.2023.3303448
23. Fang F, Zhu Z, Jin S, Hu S (Jul.2020) Two-layer game theoretic microgrid capacity optimization considering uncertainty of renewable energy. IEEE Syst J 15(3):4260–4271. https://doi.org/10.1109/jsyst.2020.3008316
24. Marzband M, Ardeshiri RR, Moafi M, Uppal H (2017) Distributed generation for economic benefit maximization through coalition formation–based game theory concept. Int Trans Electr Energy Syst 27(6). https://doi.org/10.1002/etep.2313
25. López-Dóriga Ostolaza J, "El legado de John Nash: la teoría de juegos," EL ECONOMISTA. Accessed: Feb. 06, 2024. [Online]. Available: https://www.eleconomista.com.mx/opinion/El-legado-de-John-Nash-la-teoria-de-juegos-20150525-0003.html
26. Davis MD, Brams SJ, "game theory," Encyclopedia Britannica. Accessed: Feb. 06, 2024. [Online]. Available: https://www.britannica.com/science/game-theory
27. Oxford University, "Oxford English Dictionary [OED]." Accessed: Feb. 06, 2024. [Online]. Available: www.oed.com
28. NREL, "NREL's PVWatts Calculator." Accessed: Feb. 12, 2024. [Online]. Available: https://pvwatts.nrel.gov/
29. Lokeshgupta B, Sivasubramani S (Mar.2019) Cooperative game theory approach for multi-objective home energy management with renewable energy integration. IET Smart Grid 2(1):34–41. https://doi.org/10.1049/iet-stg.2018.0094
30. Bornay, "Join the Bornay experience." [Online]. Available: bornay.com
31. Campo-Ossa DD, Vega Penagos CA, Garzon, OD, Andrade F (2023) Modeling, load profile validation, and assessment of solar-rooftop energy potential for low-and-moderate-income communities in the Caribbean. Appl Sci (Switz) 13(2). https://doi.org/10.3390/app13021184
32. Díaz Aldana NL, Luna Hernández AC, Trujillo Rodríguez CL (2022) Control y Gestion para Microrredes Eléctricas Basadas en Convertidores de Potencia, Primera Edición. Bogota
33. Park KW, Kim CH (2021) Bi-directional power flow in switchgear with static transfer switch applied at various renewable energies. Energies (Basel) 14(11). https://doi.org/10.3390/en14113187
34. Merkhouf A, Doyon P, Upadhyay S (2008) Variable frequency transformer—concept and electromagnetic design evaluation. IEEE Trans Energy Convers 23(4):989–996. https://doi.org/10.1109/TEC.2008.2001460
35. Ambati BB, Khadkikar V (2016) Variable frequency transformer configuration for decoupled active-reactive powers transfer control. IEEE Trans Energy Convers 31(3). https://doi.org/10.1109/TEC.2016.2550558
36. Bahrman MP, Johnson BK (2007) The ABCs of HVDC transmission technologies. IEEE Power Energy Mag 5(2). https://doi.org/10.1109/MPAE.2007.329194
37. An B, Jeon J, Lee M, Park T (2019) Economical comparisons of interconnection technology for asynchronous AC power grids. J Phys: Conf Ser. https://doi.org/10.1088/1742-6596/1304/1/012009
38. Long W, Nilsson S (2007) HVDC transmission: Yesterday and today. IEEE Power Energy Mag 5(2). https://doi.org/10.1109/MPAE.2007.329175
39. Nami A, Liang J, Dijkhuizen F, Demetriades GD (2015) Modular multilevel converters for HVDC applications: Review on converter cells and functionalities. IEEE Trans Power Electron 30(1). https://doi.org/10.1109/TPEL.2014.2327641
40. Baradar M (2013) Modeling of multi terminal HVDC systems in power flow and optimal power flow formulations. KTH, Stockholm
41. Lescale VF, Kumar A, Juhlin LE, Björklund H, Nyberg K (2008) Challenges with multi-terminal UHVDC transmissions. In: 2008 Joint International Conference on Power System Technology POWERCON and IEEE Power India Conference, POWERCON 2008. https://doi.org/10.1109/ICPST.2008.4745171
42. He S, Hao L, Liu C, He J, Chen Z (Aug.2023) Master-slave control strategy of the cascaded multi-terminal ultra-high voltage direct current transmission system. IET Gener Transm Distrib 17(16):3638–3647. https://doi.org/10.1049/gtd2.12919

43. Rocabert J, Luna A, Blaabjerg F, Rodríguez P (2012) Control of power converters in AC microgrids. IEEE Trans Power Electron 27(11):4734–4749
44. Ordono A, Unamuno E, Barrena JA, Paniagua J (2019) Interlinking converters and their contribution to primary regulation: a review. Int J Electr Power Energy Syst 111. https://doi.org/10.1016/j.ijepes.2019.03.057
45. Shao S, et al. (2022) Modeling and advanced control of dual-active-bridge DC-DC converters: A review. IEEE Trans Power Electron 37(2). https://doi.org/10.1109/TPEL.2021.3108157
46. Díaz NL, Vasquez JC, Guerrero JM (Mar.2018) A communication-less distributed control architecture for Islanded microgrids with renewable generation and storage. IEEE Trans Power Electron 33(3):1922–1939. https://doi.org/10.1109/TPEL.2017.2698023
47. Diaz NL, Dragicevic T, Vasquez JC, Guerrero JM (Sep.2014) Intelligent distributed generation and storage units for DC microgrids - A new concept on cooperative control without communications beyond droop control. IEEE Trans Smart Grid 5(5):2476–2485. https://doi.org/10.1109/TSG.2014.2341740

Practical Aspects of Pre-engineering Design of Clustered Microgrids

Alexandre Nassif

Abstract Although microgrids are a popular research topic, design objectives of a real-life project are probably the largest uncertainty faced by planning engineers. Most real microgrid systems, i.e., those not built exclusively as a demonstration project, have as a primary goal increasing the reliability to be afforded to customers within the microgrid boundary, as well as resilience to ensure the microgrid can maintain its supply even for very long duration outages caused by high-impact-low-probability events. A secondary goal is oftentimes increasing the share of renewable energy produced and consumed by the microgrid loads, both under blue sky and black sky operation. As a result, the load serving capability of the microgrid operating under different fractions of renewable energy is another aspect of interest. Whether operating in blue sky or black sky mode, each microgrid operation must result in a stable state and the microgrid must maintain safe and reliable operation. These objectives apply to single microgrids as well as to networked (clustered) microgrids, where more than one microgrid form a cluster that can have each microgrid operating individually or a microgrid cluster operating as a networked system. Design aspects go beyond the planning phase, and consider designing the operation modes, the clustering approach, and impacts therein. Several aspects of power quality must be analysed to ensure design parameters are addressed. Only with this foundation, a clustered microgrid is expected to engage seamless transition and continue operating in both grid-tied as well as in off-grid condition uneventfully. Oftentimes this creates a focal point on protection and control. It is expected the microgrid can either have a single scheme that enables the dual operation mode, including transition, or to have practical means to quickly switch settings groups if the philosophy calls for more than one. This chapter provides an overview of several engineering aspects of two networked island microgrids.

Keywords Load serving capability · Power quality · Seamless transition · Blackstart · Protection

---

A. Nassif (✉)
LUMA Energy, San Juan, PR, USA
e-mail: nassif@ieee.org

F. Shahnia and J. M. Guerrero (eds.), *Microgrids and Virtual Power Plants*, Power Systems, https://doi.org/10.1007/978-981-97-6623-9_16

1 Introduction

The term microgrid, coined decades ago, denotes a system containing a group of interconnected loads and DERs within clearly defined electrical boundaries that acts as a single controllable entity with respect to the grid. A microgrid can connect and disconnect from the grid to enable it to operate in both grid-connected or island mode. This definition, verbatim from the Department of Energy [1] does not specify use cases, nature of the DER fuel source, reliability or resilience metrics, or specific requirements for safe and reliable operation. Some of these items are addressed by IEEE Std 1547.4 [2], which highlights the basic requirements for equipment capabilities, system coordination, and operation. This standard also identifies the studies recommended to assess the system capabilities and derive recommendations on operation modes, setpoints, or equipment augmentation. This background provides a foundation on individual microgrids that can be extrapolated to a case of networked, or clustered microgrids that are subject to the exact same requirements. Important to also understand is that clustered operation may become more complex due to multiple sources that are, jointly, required to meet certain functionality such as synchronization among themselves as well as with the main grid, as well as load sharing mechanisms (e.g., droop control of several generators of different sizes and in different locations). This can become especially challenging if electrical distances among the various microgrids and/or the main grid are long, making synchronization schemes more challenging.

This chapter features a real case of two interconnected microgrids required to achieve clustered operation. Among the several aspects of individual microgrids that are somewhat understood by utility engineers [2], the clustered requirements are beyond those and represent new and emerging challenges that required novel, albeit practical approaches, that can be adopted in real systems. The chapter will focus on the following aspects that must be addressed during the pre-engineering stage of a microgrid project:

1. Topology analysis.
2. Definition of microgrid boundary.
3. Load serving capability.
4. Equipment functionality specification.
5. Inverter capabilities specification.
6. Protection system dependability and effective grounding.

The findings and recommendations were based on a real system case of resilience augmentation.

2 A General Clustered Microgrid

Consider a network shown Fig. 1 where two microgrids are represented to show two interconnected systems. To better understand functionalities and requirements, the system under analysis is illustrated as two island-able smaller systems. This also represents a real case the utility has developed and is currently undergoing development. This case also illustrates the vulnerability over which some systems are exposed to, as this series connection has the potential to expose both microgrids, and in particular Microgrid 2, to disconnection from the main grid source. It is also highlighted that at the pre-development stages, these microgrids might have comprised backup generation plants. In fact, there are many use cases for microgrids and one of them is to modernize the off-grid sources, which in the past could have been diesel generators and they may be modernized with more efficient fuel supplies, such as biodiesel or other form of biofuel. Even more common is the addition of renewable sources such as solar and wind, as well as energy storage. For all these cases, during outages, a backup generation plant is often treated by regulating bodies as a proxy for transmission, or in this case, sub-transmission.

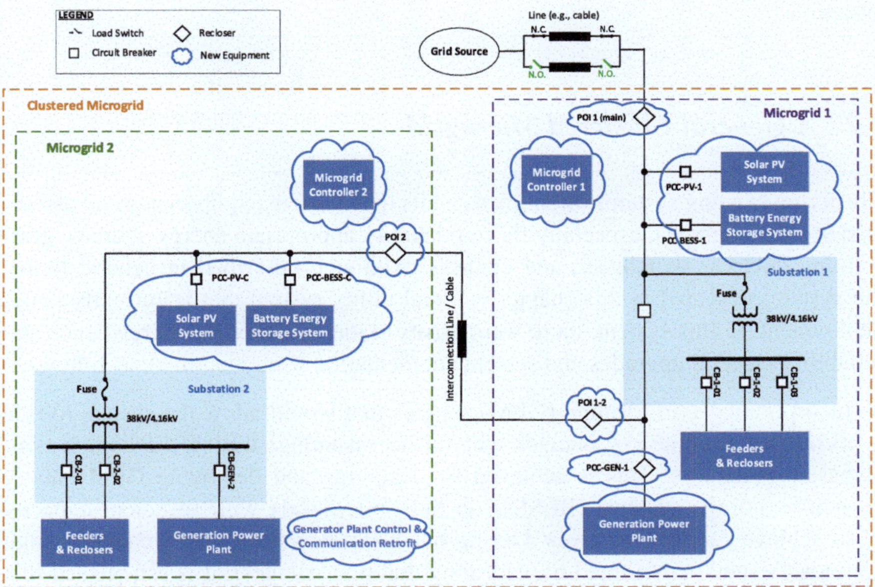

Fig. 1 A generic networked (clustered) microgrids

2.1 Microgrids Are not Just Backup Systems

Simple operation of each individual power system island as a backup system does not constitute microgrids for several reasons. First, backup combustion units have permission to operate only under a limited number of hours per year, due to Environmental Protection Agency, or EPA, requirements. Restrictions apply to any fossil-fueled generation plant and the system under analysis currently relies on diesel-fired generators. Second, backup generators are almost always not prime rated, also meaning they do not have continuous operation capability and need to be cycled down very constantly to avoid overload and subsequent accelerated aging and damage. As is well known to utility planners, prime rated engine sets have much greater CAPEX and OPEX expenditures at the deployment and lifecycle stages. Third, backup generators are designed and operated in such a way that they do not parallel with the grid. This is also done to avoid having to comply with interconnection requirements that every distributed energy resource (DER) is subject to, including potentially costly requirements such as direct transfer trip, possible system upgrades, protection upgrades, various control modes, etc., for the simple fact they are to be allowed to operate in grid-connected mode. This is a key differentiation inherent to conventional backup systems.

2.2 A General Clustered Microgrid

Clustering existing systems oftentimes results in extensive requirements and associated system upgrades, especially those related to appropriate energy sources, grid-forming devices, switchgear, and controls. Although described in general terms, the system presented in this chapter is a real utility system that is currently under deployment. In this system, there were drives to increase the system resilience and reliability. Several upgrades and system augmentation were identified, as follows:

- Incorporating grid forming (GFM) sources that would allow the system to ride through most blue-sky outages without necessitating starting the combustion-based generators. This is achieved by designing and deploying GFM battery energy storage systems (BESSs) on both microgrids with sufficient capacity to achieve certain objectives. During black-sky operation (low frequency, high impact events represented by very long duration outages), it is established that the conventional generators would start and participate in the generation mix to sustain the grid accordingly. A GFM BESS would allow cycling these generators out if required.
- Incorporating generation, in this case photovoltaic (PV) systems on both islands to offset their energy consumption during blue-sky operation, as well as prolong the load serving capability of the BESSs during black-sky operation, allowing reducing the usage of the conventional plants.

- Retrofitting conventional generation plants to allow integration in advanced microgrid control systems (MGCS).
- Deployment of MGCS with specific requirements to allow each microgrid to operate individually or allow them to engage clustering as required, allowing resource sharing among them.
- Equipment to integrate the microgrids, such as new breakers, reclosers, load-breaking switches that can all be controlled by the MGCS. New telecommunication media, such as fiber, is also required among the various microgrid sites.
- An illustration of this generic system is presented in Fig. 1.

3 Microgrid Use Cases and Sizing Objectives

The first step of microgrid planning is to clearly define its objectives. Common objectives of microgrid planning include:

1. Reduction in outage duration: This is likely the most common objective in most real microgrid projects. An entire concerted planning would typically remove all outages, including those resulting from high-impact-low-probability events. However, each microgrid element (i.e., generation or storage) could be sized individually (but not independently).
   a. For example, an objective may be to size the BESS so that it is just large enough to reduce the total customer-minute-interruptions by 50%, 70%, or 90% under blue sky operation. This means that excluding major day events, the total customer-minute-interruptions of a power system area would be reduced by to a targeted amount. Figure 2 displays an example of how this could be quantified. This figure essentially predicts how long the BESS within one microgrid or the clustered microgrid (where more than one microgrid shares resources with the others) could supply, without active sources generation (e.g., an outage occurs during the night where generation sources are predominantly PV), the entirety of the load it was designed for.
   b. Generation, for example, solar PV or wind turbines, will certainly extend this duration and this should also be quantified. Visual displays such as the one in Fig. 2 can be generated under different scenarios for easy comparison. This figure illustrates the average load serving capability of a BESS in case of an outage at a given hour of a given month and was derived from the average hourly loading over 30 days, which can then be used as a benchmark for the expected duration of the energy storage under its operating objectives.
   c. Different load profiles need to be captured. It is customary to characterize the load serving capability for existing load duration characteristics, as well as for a future forecasted state, which could be a 10-year forecast. The forecast needs to include load forecast, behind-the-meter generation such as

Hour	January	February	March	April	May	June	July	August	September	October	November	December
0	2.4	2.5	2.4	2.1	1.9	1.8	1.8	1.9	2.2	2.2	2.4	2.4
1	2.5	2.5	2.5	2.2	2.0	1.9	1.8	2.0	2.3	2.3	2.5	2.5
2	2.5	2.6	2.6	2.3	2.1	1.9	1.9	2.0	2.3	2.3	2.5	2.5
3	2.6	2.7	2.6	2.3	2.1	2.0	1.9	2.0	2.3	2.4	2.5	2.6
4	2.6	2.7	2.7	2.4	2.1	2.0	1.9	2.1	2.3	2.4	2.6	2.6
5	2.6	2.7	2.7	2.4	2.1	2.0	1.9	2.1	2.3	2.4	2.6	2.6
6	2.6	2.7	2.7	2.5	2.3	2.1	2.0	2.2	2.4	2.4	2.6	2.6
7	2.8	2.8	3.0	2.7	2.4	2.2	2.1	2.3	2.7	2.8	2.9	2.9
8	3.0	3.3	3.3	3.0	2.7	2.3	2.2	2.5	3.0	3.2	3.3	3.3
9	3.6	3.8	4.1	3.5	2.9	2.5	2.4	2.8	3.4	3.6	3.8	3.9
10	4.3	4.9	4.6	4.0	3.4	2.7	2.6	3.1	3.7	4.1	4.2	4.5
11	4.8	5.5	5.4	4.4	3.4	2.9	2.8	3.3	3.8	4.3	4.5	4.7
12	5.0	6.0	5.7	4.7	3.4	2.9	2.9	3.3	3.7	4.3	4.4	5.1
13	4.9	5.9	5.3	4.4	3.4	2.7	2.8	3.1	3.7	4.0	4.1	4.7
14	4.4	5.1	4.6	3.8	3.1	2.6	2.6	2.9	3.4	3.5	3.6	4.1
15	3.8	4.3	3.8	3.3	2.7	2.4	2.5	2.6	3.0	3.0	3.1	3.4
16	3.0	3.4	3.1	2.8	2.3	2.1	2.2	2.3	2.6	2.6	2.6	2.8
17	2.4	2.5	2.5	2.2	2.0	1.9	1.9	2.0	2.3	2.2	2.2	2.2
18	2.2	2.2	2.1	2.0	1.8	1.7	1.7	1.8	2.1	2.1	2.1	2.2
19	2.1	2.1	2.0	1.9	1.8	1.7	1.7	1.7	2.0	2.0	2.1	2.0
20	2.1	2.1	2.1	1.9	1.7	1.6	1.6	1.7	2.0	2.0	2.1	2.1
21	2.2	2.2	2.1	1.9	1.8	1.7	1.6	1.7	2.1	2.1	2.2	2.1
22	2.2	2.3	2.2	2.0	1.8	1.7	1.6	1.8	2.1	2.1	2.2	2.2
23	2.3	2.3	2.3	2.0	1.9	1.8	1.7	1.9	2.2	2.2	2.3	2.3

Fig. 2 Example of load serving capability illustration of microgrid resources

customer rooftop PV systems, electric vehicles, and retirement of conventional generation units where applicable. Methods to forecast these DER and prosumer-driven devices are complex and a subject of its own.

2. Reduction in sustained outage frequency: the definition of sustained outage is those that are not seamless, i.e., those that may result in a brief interruption of a few seconds to a minute (typically). A renewable microgrid containing a BESS is typically sufficient to greatly improve outage duration. As one can see, this objective is interrelated to the previous objective to reduce outage duration, as sizing must be done by considering both objectives simultaneously.
3. Increasing renewable energy share: this objective normally applies to sizing renewable generation units such as a utility-scale PV plant or a wind farm. The common method to incorporate this requirement is to annualize the energy production and account it as a percentage of the total energy consumed.

    a. For example, an objective could be to offset the total energy consumption by 50, 70, or 90% by using these renewable resources that require low maintenance.
    b. A secondary objective could be to include "renewable peakers" comprising thermal generation that is based on renewable resources, such as biodiesel, ethanol, or other biofuels that require higher maintenance than those of PV and wind, but that can play crucial role in resilience planning.

4. Reduction in project OPEX costs of maintaining and operating conventional thermal generation, such as engine overhauls, engine replacement, oil changes, and of course, the cost of fuel delivery and fuel system maintenance. This is always an engineering objective and always needs to be one of the main aspects of microgrid planning. On the other hand, and related to this task, it is also important to consider the costs and complexities of developing new renewable generation as these relate to obtaining permitting, environmental and historic preservation assessments, easements in rights-of-way, and other related complications.
5. These objectives do not overlap and are part of the same objectives.

4 Defining Conceptual and Physical Boundaries

Defining the boundary and loads to be encompassed by a microgrid is a critical step in meeting the objectives defined in the previous section. These are high impact decisions and strategies that will dictate the equipment to be procured.

4.1 Definition of Critical Facilities and Monitoring Requirements

Every electric utility will have their own definition of critical facilities and will typically prioritize their restoration as much as possible. Utilities oftentimes adopt a tiered approach. Hence, the critical load identification could be based on an area restoration that follows some form of an Emergency Response Plan. As a side note, emergency response plans need to be approved by the system Regulatory body or Utilities Commission. This plan needs to categorize critical loads into levels. One example of high tier level could contain loads that provide critical services for the health and safety of the public, such as wastewater treatment plants and hospitals with operation centers. A mid-tie level could contain loads that provide significant public services, such as telecommunication towers and detention centers. Lower tiered critical facilities could contain loads that provide public services less impactful to emergency response, such as grocery stores and schools.

Tiering critical infrastructure provides flexibility to allow segmentation of a portion of the distribution system inside the microgrid boundary. The microgrid designer and operator may choose to design and deploy switches or fault interrupters to realize this segmentation or may choose to not expose any part of a feeder by removing it from the microgrid boundary. The decision is based on technical implementation as well as economics.

Another aspect of dealing with critical facilities in the context of a microgrid is that these facilities should be given priority for monitoring and control. These facilities should be monitored at a level that allows the microgrid secondary layer to establish control commands that will prioritize these loads to stay connected and not be shed during times of insufficient resources or disturbance recovery. As of the date of writing this manuscript, many utilities have begun pilot projects of distribution phasor measurement units deployment in strategic distribution systems. Recently, [3] has addressed this subject very well and has issued a report created for the Department of Energy Oakridge National Laboratory. This report introduces many use cases for the technology in distribution systems.

Once critical loads are identified, it is possible to prioritize which of these loads should be always kept online, which can be curtailed a portion of the time, and which may be disconnected for the entire outage duration if worst case scenarios of high-impact-low-probability events. These strategies can then be subsequently programmed into the microgrid control system.

4.2 Protection Equipment and Microgrid Boundary

The microgrid conceptual diagram presented in Fig. 1 suggests three main points of interconnection. Microgrid boundary creation is popularly done by using switches, as exemplified in several publications [4, 5]. However, the lower cost of pole-mounted or padmounted reclosers has declined significantly over the years. In fact, nowadays the cost of a pole-mounted SCADA-enabled switch is not significantly lower than that of a pole-mounted recloser. The continuous rating of both would be similar, the recloser would likely have higher load breaking capability, including higher capacitive current rating, and the recloser has fault interrupting capability, which an air-insulated switch would not have.

Hence, the point of interconnection devices should have the following capabilities:

- Load breaking capability, including breaking capacitive current. Most air-insulated switches, even those with arc snuffers, cannot break capacitive current without failure (the capacitive current rating is very small as compared to the load breaking rating).
- Fault interrupting capability. Fault current rating must be specified as per system conditions. To note, most distribution voltage (35 kV and below) overhead reclosers have fault current interrupting capability of 16 kA or 12 kA, some even less.
- Protection elements that enable protection philosophy. These could be directional overcurrent (67/67N), overcurrent with voltage supervision (67V), line distance (21/21N), or line current differential (87/87N). Other protection schemes, such as under or overvoltage (27 and 59), or frequency or rate-of-change-of-frequency (81/81R) may also be required.
- Sychronization and synchronization check capability, required to initiate synchronization procedures and subsequently supervise closing operations.
- Communication capability with industry-accepted protocols to allow interoperability among the several devices as well as with the microgrid controller.

As a result, a recloser is the preferred method of isolation. The protective relaying of a recloser can be very sophisticated with features such as synchronization, synchronization check, voltage and frequency protective elements, rate-of-change-of-frequency, fault location, high-impedance fault detection, etc. Where fault interruption is required, however, reclosers may not suffice. In areas of very high fault currents, a breaker may be needed. Distribution level reclosers have typical fault interrupting capabilities of 12 kA or 16 kA. Sub-transmission level reclosers may have higher interrupting capability such as 32 kA, but anything higher than this would require a substation grade breaker. For example, refer to Fig. 1 again, where the three POI reclosers or breakers are to be used to separate each microgrid from the main grid or from each other. These reclosers must be able to drive synchronization, must contain synchronization check elements, and preferably can be integrated with other protection schemes such as line current differential or line distance elements.

Interoperability is key as well. Telecommunication and operation compatibility among the switches, reclosers and breakers with the microgrid controller is of prime importance and a pre-requisite for the proper operation of a microgrid. For the microgrid controller, there are many references available such as [25], and for this reason this portion will not be expanded in this chapter. The requirements contained therein include basic functionalities, dispatch and transition functions, and interoperability of microgrid control systems. These also cover generation dispatch and load management. Generation capability is another topic well covered and there are many standards, such as [16] and [26], that cover this topic adequately.

4.3 *Quasi Static Power Flow Considerations*

Power flow studies are bread-and-butter for system planning engineers. Load flow studies must obviously reflect the system load and be used to size and allocate generation and storage resources. Hourly or 15-min load-flows are intended to reflect the load duration characteristics of monitored point, which in real systems is often at feeder head locations. Hence, the time-series load flow analysis can be a good predictor of future microgrid generation dispatch cases and generation mixes. These studies can be used as a cursory indicator on how often energy storage would need to be charged or discharged and by which amounts, as well as frequency at which a thermal resource may be called upon.

Hourly load flow studies are used to characterize how parameters vary over time. Taking Fig. 3 as an example, the planner can visualize the voltage behavior and corresponding active power exchange at a microgrid point of interconnection while the microgrid is connected to the main supply. This is different from a simple hourly load flow study often used to interconnect renewable generation in that it will inform the microgrid controller and its algorithm on the expected system characteristics. Voltage mitigation equipment, for example, can be prescribed during the microgrid planning process. In this example, one can correlate the active power export from

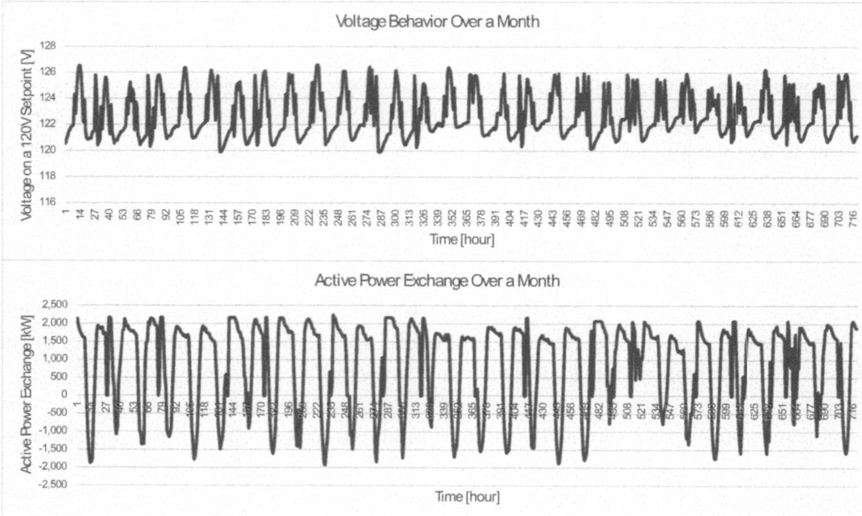

Fig. 3 Voltage and active power exchange measured at the microgrid point of interconnection

renewable sources with voltage peaks and understand the voltage regulation required at the microgrid POI.

Another benefit of these hourly load flow studies is the ability to visualize inverter controls, such as Volt-VAR settings adopted into the energy storage or renewable generation inverters. Figure 4 shows a histogram of voltages at a microgrid POI as a base case and with inverter Volt-VAR activated. This figure is the result of a real microgrid simulation and highlights how one can quantify the impact of such functions versus adopting a constant power factor mode. One of the observations of this example is that although Volt-VAR settings were custom-tuned, they may not be an ultimate solution to all cases as there are some voltage values that still exceed the target range.

5 Required Specifications for Inverter-Based Resources

Some capabilities are required for a successful operation of a microgrid, and they apply to each generation resource, load control, and interaction and interoperability of all these resources.

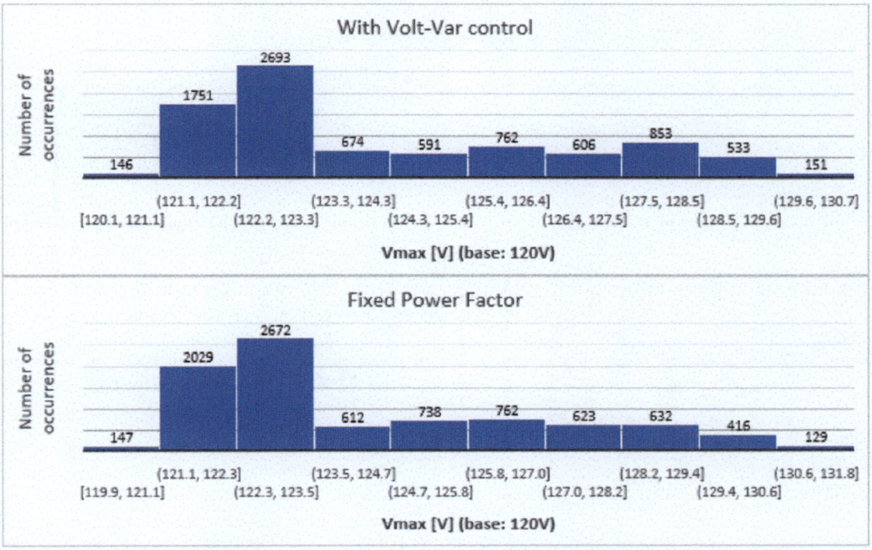

Fig. 4 Voltage measured at the microgrid point of interconnection under two inverter control modes

5.1 *Energy Storage and Inverter Controls*

In most microgrids, energy storage becomes the key component to establish the microgrid operation and capabilities. Where it is desirable to operate without thermal generation during periods of time, e.g., operation in diesel-off mode in hybrid microgrids, energy storage is expected to be the grid-forming device. When operating in diesel-off mode, the energy storage shall be coupled with GFM inverters to allow system operation when not tied to the main grid supply.

5.2 *GFM Inverter-Based Resources Capabilities and Functions*

GFM, an essential control system in power networks and electrical grids, plays a crucial role in preserving stability, regulating voltage and frequency, and ensuring the reliable operation of these systems. The growing integration of renewable energy sources underscores the necessity of GFM technologies to effectively manage intermittent resources and their impact on grid stability [6, 7]. GFM controls possess the capability to adjust grid voltage by manipulating the output power and reactive power devices while also establishing the frequency. In response to fluctuations in power supply and demand, GFM diligently maintains frequency close to its nominal value [8].

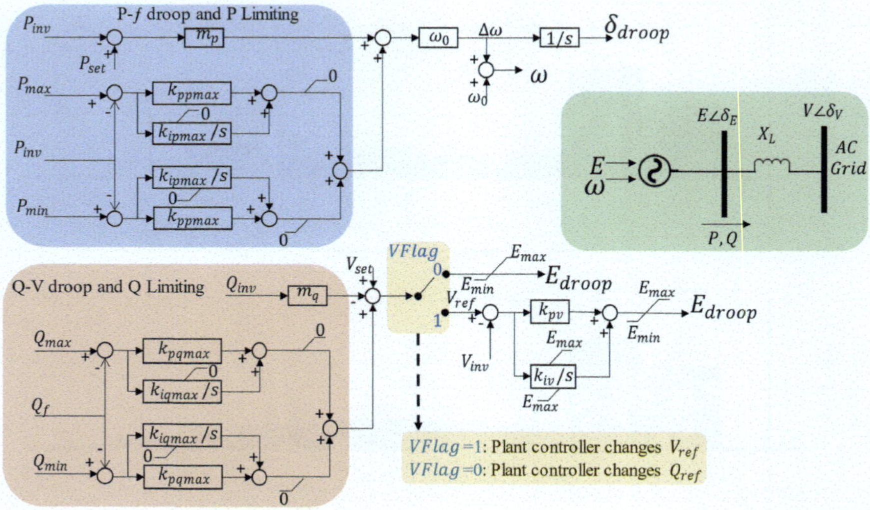

Fig. 5 Structure of a GFM control as proposed in [14, 15]

Designed to address grid faults and disturbances, such as short circuits or unforeseen load changes, GFM controls can dynamically adjust power generation and other parameters to prevent grid instability. The adaptability of GFM control systems arises from the implementation of sophisticated control algorithms, including droop control, voltage control loops, and frequency control loops [9]. A GFM model that has gained popularity was created by the Pacific Northwest National Laboratory and integrated into the commercial software package PSSE for BESS studies and is depicted in Fig. 5. It encompasses components such as voltage source, P-f and Q–V droop control, P/Q limiting, and fault current limiting, with technical parameters detailed in Table 1.

Within the model, the P-f droop ensures synchronization of phase angles among multiple voltage sources, while Q–V droop prevents substantial circulation of Vars between these sources. The adaptability and versatility of GFM control make it adept at responding to dynamic changes in the grid, including additions or removals of generation sources, fluctuations in demand, or the loss of a significant generator.

The general structure above reflects the operation of a droop controlled GFM inverter. As it can be seen in Fig. 5, one of the major advantages of this mode is that it does not rely on a phase-locked-loop (PLL) to measure voltage and extract its signal for synchronization. The absence of a PLL ensures not only a faster response to disturbances, but a more stable response as well. This is because the droop controlled GFM will not be susceptible to misoperation in case it is connected to a weak grid: when a disturbance occurs in a weak grid, it is possible that the voltage will be distorted enough that the PLL is not able to reliably measure it, causing the GFM control commands to drive the inverter into oscillatory instability.

Table 1 Possible GFM settings

Variables	Description	Value
m_q	Q–V droop gain	0.02
k_{pv}	Proportional gain for voltage controller	0.0
k_{iv}	Integral gain for voltage controller (>0)	5.86
k_{ppmax}	Proportional gain of the P_{max} and P_{min} controller	0.01
k_{ipmax}	Integral gain of the P_{max} and P_{min} controller	0.1
P_{max}	Upper limit of the inverter active power control	1.0
P_{min}	Lower limit of the inverter active power control	−1.0
P_{set}	Power Set Point	–
E_{max}	Upper limit of the output of voltage controller	1.25
E_{min}	Lower limit of the output of voltage controller	0
V_{set}	Voltage set point	–
Q_{max}	Upper limit of the inverter reactive power control	1
Q_{min}	Lower limit of the inverter reactive power control	−1
k_{pqmax}	Proportional gain of the Q_{max} and Q_{min} controller	3
k_{iqmax}	Integral gain of the Q_{max} and Q_{min} controller	20
ω	Frequency	50/60

However, GFM inverters can operate in other operational modes such as virtual synchronous machine (VSM), which typically relies on the PLL. The VSM mode has the advantage of following the same algebraic equations that characterize the behavior of synchronous generators, including load sharing schemes and programmable inertia constant, a capability that the droop control above does not have. This suggests that the VSM mode is more suitable to cases where there is a blend of rotating synchronous generators and inverter based resources (IBRs), and the main grid supply is strong (such as short-circuit ratios, or SCRs, exceeding 5) [10, 11]. Another operation mode that is gradually garnering more attention is a virtual oscillator, which has many of the advantages of both operation modes mentioned above. This is a promising technology that still has to mature to gain as much acceptance [12].

5.3 Control Function Response

When dealing with an IBR dominated system, control functions are the most impacting elements in system stability. When grid-tied, the battery system inverter is following the main grid voltage and will generally not have abrupt responses to disturbances, unless these disturbances are significantly exposed to the BESS GFM inverter directly. The control functions become of greater impact when there are

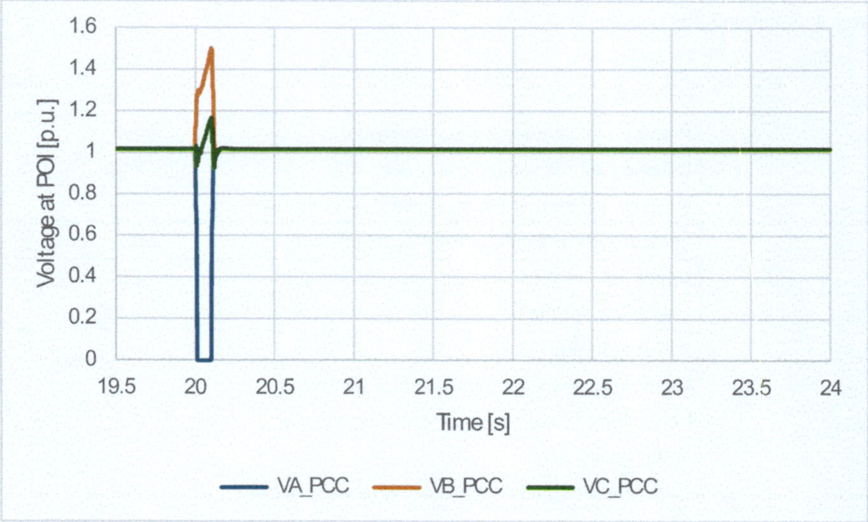

Fig. 6 Example of a seamless transition following a SLG fault near the GFM inverter terminals

transitions (i.e., grid-tied to off-grid) required. An illustration of such an event is shown in Fig. 6. This figure shows a simulation case of the generic microgrid of Fig. 1 for a fault at the main supply, and subsequent transition to an off-grid mode. In this simulation, a single-line-to-ground unimpeded fault is applied at t = 20 s and the supply line breakers sense and isolate the fault in 200 ms. The results observed for this particular GFM inverter model, which has a control philosophy based on VSM, is that it is able to ride through the low voltage it is exposed to and is also able to recover from the disturbance and maintain system stability. This simulation was obtained with a real GFM model and highlights the importance of a well-designed GFM control system.

5.4 *Short-Circuit Requirements*

IBRs derive their design and operational features from the voltage source converter topology. Despite this foundational connection, adherence to grid codes requires them to function as constant current sources. This alignment with constant current behavior is crucial for facilitating voltage regulation and ensuring optimal controllability for system operators. Consequently, IBRs are conventionally depicted and modeled as current sources within the power system framework. When examining a typical power system integrated with an IBR and exposed to a ground fault characterized by impedance R_f, an illustration is presented in Fig. 7. The power system under consideration draws its supply from an equivalent source (S) located in a substation, serves a designated load (L), and is interconnected with the IBR. Subscripts 0, 1,

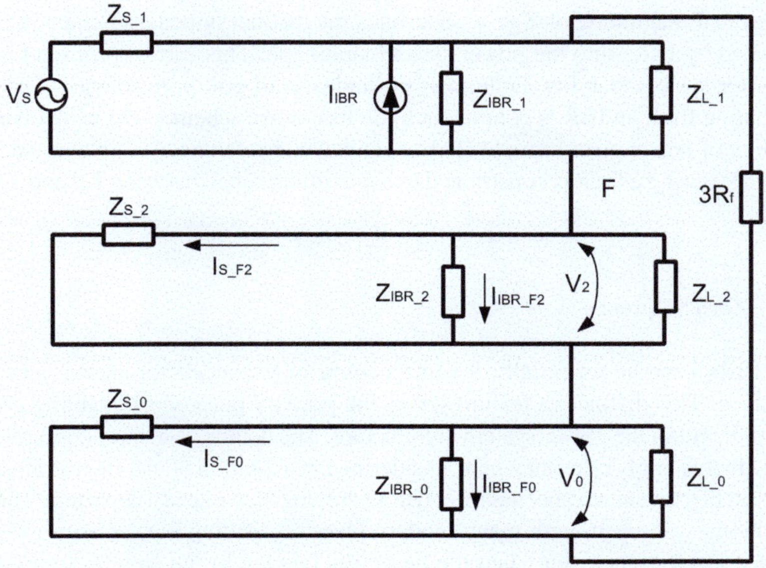

Fig. 7 Sequence network of a distribution system experiencing a ground fault while supplying an IBR

and 2 are utilized to signify the zero-, positive-, and negative-sequence networks, contributing to a comprehensive representation of the system dynamics. This representation enhances our understanding of the interaction between power system components and the IBR, particularly in challenging scenarios like ground faults. The emphasis on these technical nuances underscores the importance of precise modeling and representation for the effective analysis and operation of power systems incorporating IBRs.

5.4.1 Positive Sequence

An IBR exhibits a sophisticated control strategy focused on regulating its output current to uphold a constant active power generation profile. This distinctive characteristic positions the IBR as a controlled current source within the power system. The dynamics of modern inverters manifest in their high-speed switching capabilities, enabling rapid adjustments to the current control within a fraction of a cycle. This swift response is embedded in contemporary inverter designs, allowing for enhanced precision and adaptability in dynamic power system scenarios.

Delving into the IBR's behavior, consider the equivalent positive sequence, denoted as Z_{IBR_1}. Remarkably, Z_{IBR_1} exhibits a substantially higher magnitude when compared to traditional synchronous generators. This discrepancy is a direct result of the unique characteristics inherent in the inverter-based architecture, which diverges from the conventional attributes associated with synchronous generators.

The elevated magnitude of Z_{IBR_1} underlines the distinct impedance characteristics introduced by IBRs into the power system landscape. This carries profound implications for system stability, fault analysis, and overall grid performance. The fault contribution from an IBR is contingent upon its control scheme, which involves the switching of power electronics [13]. As elucidated in reference [14], the impedance of the IBR during a fault is constrained to a maximum range between 1.0 and 1.5 per unit (pu).

5.4.2 Zero Sequence

Older IBRs have an essentially floating neutral or grounded through a very high impedance. This distinctive feature serves the primary purpose of enabling ground protection within the power system architecture. The details of neutral configuration become increasingly relevant when considering the evolution of inverter designs over time. Notably, recent advancements in IBR technology have seen a departure from the conventional approach, with many modern inverters lacking the capacity for zero-sequence generation. Even when the neutral is bonded to the step-up transformer neutral, stringent control mechanisms within these inverters prevent any manifestation of zero-sequence current imbalance. This restriction is characteristic of control strategies implemented in state-of-the-art IBRs.

A key attribute characterizing modern IBRs is their tendency to exhibit a zero-sequence impedance that effectively appears infinite. This control objective prevents the emergence of zero-sequence currents. However, this seemingly infinite zero-sequence impedance introduces specific challenges and necessitates mitigation strategies. One effective method for mitigating the infinite zero-sequence impedance challenge involves the incorporation of grounding transformers. By strategically integrating grounding transformers into the system, engineers can tailor the neutral configuration and address potential issues related to zero-sequence impedance. This technical approach underscores the importance of planning and mitigation strategies to ensure the seamless integration and performance of modern IBRs.

In summary, IBRs are engineered to abstain from contributing any zero-sequence current to the system, resulting in $Z_0 = \infty$. Mitigating this issue involves managing zero sequence source equivalents through appropriate step-up transformer configurations, such as low-side delta and high-side grounded wye configuration or connecting grounding sources (transformers can effectively circumvent this limitation, ensuring an effectively grounded system [13–15].

5.4.3 Negative Sequence

The determination of the negative-sequence equivalent is important and intertwined with the inverter control algorithm. The sophistication of modern control algorithms allows for the intentional disabling of negative-sequence current generation, resulting in an equivalent infinite negative-sequence impedance. Conversely, a different facet of inverter designs unveils a scenario where inverters lack dedicated regulation for negative-sequence current. Instead, the positive-sequence current regulator takes precedence, boasting a wide bandwidth that inherently dominates the system dynamics. Consequently, this design approach yields an equivalent negative-sequence impedance that can be controlled to be of very large magnitude. The interplay between positive- and negative-sequence regulators underscores the diverse strategies employed by inverter manufacturers to tailor IBR characteristics.

Further complexity arises from the variation in positive-sequence regulator responsiveness among different inverters. In instances where positive-sequence regulators exhibit sluggish responses and fail to address current imbalances promptly, a very small equivalent negative-sequence impedance ensues. This dynamic characteristic poses challenges in accurately modeling IBRs. In response, many manufacturers are now proactively supplying negative-sequence impedance characteristics. This trend highlights the industry's need to provide engineers with data for more accurate modeling and analysis. It is imperative to acknowledge the consequence posed by an infinite equivalent negative-sequence impedance. The negative sequence impedance of the IBR is heavily dependent on the implemented control strategy, resulting in the suppression of I_2. This can lead to protection issues during unbalanced faults, as negative sequence current cannot be inherently generated by passive components, unlike in the case of zero-sequence, where transformer grounding configurations can create I_0 during a ground fault. To address this, grid codes mandate a certain level of negative sequence current injection by IBRs to mitigate ground fault overvoltage (GFOV) and obviate the need for protection replacement [15, 16]. IEEE Standard 2800 [16], which based off developments in other parts of the world such as Germany [15], has suggested that the equivalent negative-sequence impedance should not exceed twice that of its positive-sequence counterpart. This is a guideline that utilities may follow for new utility-scale interconnection, and it is the premise adopted in this manuscript: $Z_2 = 2 \times Z_1$. Field demonstrations have underscored the problem and its associated risks [17].

6 Protection Philosophy, Ground Fault Overvoltage and Overcurrent

Established and reliable protection and grounding schemes have long been in place for power systems connected to the Bulk Electric System (BES). However, with the emergence and widespread adoption of microgrids, it becomes imperative to reassess these approaches, as traditional philosophies prove inadequate in dealing with the reduced short circuit levels of IBRs. In the context of an IBR dominated microgrid, the conventional overcurrent elements face limitations, affecting protection coordination and selectivity. This necessitates the development of a nonstandard philosophy tailored to microgrid distribution systems. Moreover, given that most IBRs operate as constant current sources and do not naturally supply ground current, addressing performance grounding becomes a fundamental challenge for microgrids. To overcome this challenge, additional ground sources are required, and their sizing must align with the specific needs of the microgrid. Practical solutions to this problem are required, as fine-tuning in the field cannot rely on complicated algorithms. The theoretical framework presented in this section has been crafted based on experiences with an actual off grid microgrid, emphasizing the importance of adopting new philosophies that account for the practical aspects of these systems.

6.1 Challenges of IBR Operation in Microgrids

Protection and grounding form an intricately connected duo, and comprehending grounding requirements necessitates an initial exploration into traditional protection practices. In distribution system protection, the prevailing approach involves non-directional definite-time and timed overcurrent protection. These elements effectively safeguard systems supplied by a single source, such as a substation or a generation plant. However, in multi-sourced systems, where the protection sensing element may encounter reverse current in both steady-state and short-circuit scenarios, the implementation of overcurrent protection with torque control becomes imperative. These adaptations were successful in an era dominated by rotating synchronous machines, producing substantial short-circuit currents during system faults.

The advent of IBR, however, presents a new challenge. These topologies generate small amounts of short-circuit current compared to their nameplate ratings, rendering static overcurrent protection less effective. Even adaptive protection schemes, such as the one detailed in [18], may not guarantee fault detection due to the consistent current output of most inverters. While GFM inverters can supply imbalanced currents, their fault currents often align in magnitude with their steady-state ratings. Therefore, microgrids with a high penetration of IBRs demand protection philosophies that don't solely rely on current sensing for fault detection. Despite advancements in adaptive protection schemes, their complexity has hindered widespread commercial adoption, especially in critical applications like remote communities.

Undervoltage, a mature relaying element, has found application in microgrids, as discussed in [19], and is a common feature in generator protection suites. However, relying on undervoltage elements signifies a paradigm shift for several reasons:

- This element is not conventionally employed to protect distribution systems, making it challenging to determine appropriate settings.
- Achieving required trip times proves challenging as voltage behavior is less predictable than current, relying heavily on modeling assumptions in Electromagnetic Transient (EMT) simulations.
- Coordinating protection between undervoltage and overcurrent schemes is intricate due to their independent time characteristics.

In the current technological landscape, islanded microgrids may find themselves relying on undervoltage elements, particularly in mission-critical applications with low risk tolerance. IBRs offer an advantage as these inverters exhibit greater resilience to abnormal voltage and frequency conditions, allowing for extended voltage pickup settings. To enhance reliability, underfrequency elements can also be employed. Yet, characterizing the voltage and frequency behaviors of BESS proves challenging, requiring vendor-provided models for voltage studies and accurate BESS models coupled with EMT simulations for frequency evaluations.

6.2 Protection Philosophy

The incorporation of PV and BESS installations into a microgrid necessitates the adoption of new protective devices and a revised protection philosophy. Despite the prominence of the microgrid concept for over three decades, most implementations have been either proof-of-concept or demonstration projects. An illustrative instance is presented in [20], proposing the utilization of a separate grounding transformer and restricting DER step-up transformers from introducing ground sources. While a few isolated microgrid implementations have followed similar approaches, there is currently no unanimous recommendation on how to address performance grounding effectively, resulting in the absence of a "traditional" or "baseline" topology.

6.2.1 BESS Transformer Configuration

The BESS step up transformer winding configuration has great impact on the system operation. A Δ-Yg (Δ low, Yg high side) has been used on the GFM device very often for the following reasons:

- The step-up transformer is well-suited for supplying single-phase-to-ground loads, introducing a ground source and enabling feeder restoration even if no other GFM sources are present.
- This configuration could eliminate the need for a grounding transformer installation in the microgrid when operating in islanded mode.
- The transformer configuration reduces the degree of grounding, GFOV, and continuous operating voltage, thereby enhancing performance grounding [21].

However, there are notable disadvantages to this configuration:

- The additional ground source desensitizes the main supply overcurrent protection elements. This desensitization effect was analytically characterized in [22].
- There is a significant infeed from the BESS for LG faults. For proper operation, the BESS overcurrent protection must be torque-controlled by voltage (polarized).

In summary, despite the drawbacks, the decision to opt for a GFM BESS transformer Δ-Yg configuration is still common for most real microgrid systems driven by the flexibility it provides, allowing operation in isolation from the main supply.

6.2.2 Voltage Polarization and Auto-Reclosing Logic

Faul interrupter relay elements and settings in all IBRs (e.g., PV as well as GFM BESS) must be determined based on their respective fault contributions and intended functions. The fault contribution of PV systems is approximately equivalent to their steady-state current output, specifically observed in cases of a LLL faults. Consequently, voltage polarization is often unnecessary at the AC side of PV fault interrupters, as the inverter cannot supply fault current. Similarly, BESS typically have short-circuit contributions only slightly higher than the rated output for an LLL fault, remaining below typical overcurrent pick-up values. But in contrast to the case of PV inverters, the contribution significantly increases for a SLG faults because of the transformer configuration. This discrepancy is attributed to the transformer winding configuration, as elaborated on in the preceding subsection. Consequently, the settings chosen for the BESS relays could mirror those employed for the PV relays, but with the addition of voltage polarization. This precautionary measure prevents tripping on an upstream fault within the distribution system, which is intended to be cleared by the main feeder fault interrupter, while introducing protection dependability to detect real faults.

As mentioned previously, when a system is supplied by a BESS, the current supplied by all IBRs is negligible, rendering overcurrent elements ineffective in activating any protection element. In addition to voltage-supervised overcurrent (51 V), alternative elements may also need to be employed, a commonly applied being undervoltage protection (27). This element may have a longer time delay in favor of the 51 V element. The main reason for this time delay is that once a relay picks up on 27 and trips a breaker, there will be no information on fault location. Overcurrent element sensed currents can be used to calculate impedance to fault, however this is not the case of undervoltage elements. Hence, while element 27 is very dependable, it does degrade fault location capability.

Lastly, auto-reclosing functionality within the BESS relaying should be considered under the case of a microgrid. Considerations are:

- In grid-connected mode, reclosing remains deactivated since the interrupter's primary function is to protect the BESS step-up transformer. Any fault encountered in this mode is likely to be permanent, justifying the exclusion of reclosing.
- Auto-reclosing may trigger the need for two distinct setting groups, which increases complexity.
- In the hypothetical scenario where reclosing is enabled during islanded mode, the BESS could also face practical limitations. It might be unable to successfully ride through its trip and execute reclosing. This is attributed to the entire PV plant tripping in the event of an outage. Consequently, enabling reclosing in this context might prove futile and its need must be considered.

6.3 Ground Fault Overvoltage and Overcurrent Desensitization: A Conflicting Objective

The discussion in Sect. 6.2 revolved around the radial protection of a distribution feeder contained inside the microgrid. When a microgrid operates as a single entity in a networked environment, other challenges arise. This section addresses the case where a microgrid impacts the protection systems that cover the line at which its point of interconnection is tied into. Ensuring effective grounding is imperative for the secure and dependable functioning of power systems, as outlined by various standards, including [23]. This standard underscores the complexity of grounding applications, acknowledging that each potential solution may excel in certain aspects but may require trade-offs.

In the realm of microgrids and power systems, a challenge arises due to conflicting requirements between protection and grounding. GFOV often stem from inadequate ground currents during a fault. For instance, in a three-wire delta system, a ground fault can cause the healthy phases to rise to line-to-line values, while the faulty phase fails to provide any fault current to the ground. Conversely, an effectively grounded system (one that has a *Coefficient of Grounding*, or CoG < 0.8 as explained in [21]) attenuates excessive voltage rise in healthy phases but increases ground fault currents. Mitigating GFOV in a synchronous generator dominated system is well understood, and GOV can be mitigated with passive elements such as grounding transformers or adequate transformer winding configurations [24]. This reference also describes the complication arising when ground sources are distributed across the distribution system, leading to potentially severe desensitization of the source (equivalent sub-transmission bus, substation, or generation plant). This distribution of ground sources will desensitize the main feeder relay overcurrent elements, creating a challenge in maintaining effective protection. This desensitization effect, well-documented in [22] and illustrated in Fig. 8, showcases the impact of the current infeed generated by a microgrid. In such cases, the source fault current diminishes due to the contribution of fault current from the microgrid, further emphasizing the intricate balance required for optimal system performance. The microgrid may contain many ground sources, such as transformers with Δ-Yg configuration (Δ low, Yg high side), synchronous generators, or (more rarely) inverters designed to inject ground current.

Figure 8 describes this effect for both positive- and zero-sequence. Starting with the positive-sequence case, for a symmetrical fault, only the positive-sequence network exists, and the substation fault current contribution is governed by [22]:

$$i\prime_{f\_s} = i_{f\_s} - i_{f\_mg} \times \frac{Z_{L2}}{(Z_S + Z_{L1} + Z_{L2})}, \tag{1}$$

where Z_s is the positive-sequence substation source impedance, Z_{L1} is the positive-sequence impedance of the line segment between the substation and the microgrid POI, Z_{L2} is the positive-sequence impedance of the line segment between the microgrid POI location and the fault location, i_{f_s} is the original substation fault current

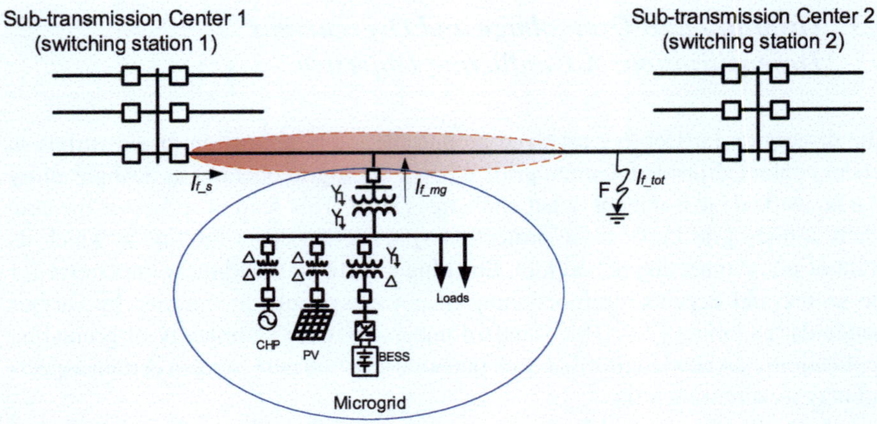

Fig. 8 Illustration of potential infeed from microgrid and its impact on terminal station protection

contribution, and i_{f_MG} is the totalized microgrid short-circuit current, and I_{f_tot} is the total fault current. Equation (1) governs how the microgrid infeed will cause a reduction in i_{f_s} upon fault inception.

Similarly, a single-line-to-ground fault results in the following zero-sequence infeed caused by the microgrid, also explained in [22]:

$$i'_{S\_f0} = \frac{V_0}{Z_{S\_0} + Z_{L1\_0} + Z_{L2\_0} + \left(\frac{[Z_{S\_0}+Z_{L1\_0}]\times Z_{L2\_0}}{Z_{G\_0}}\right)}, \tag{2}$$

where Z_{s_0} is the zero-sequence substation source impedance, Z_{L1_0} is the zero-sequence impedance of the line segment between the substation and the microgrid POI, Z_{L2_0} is the zero-sequence impedance of the line segment between the microgrid POI and the fault location, Z_{G_0} is the zero-sequence subtransient (or equivalent) source impedance of the microgrid and V_0 is the equivalent zero-sequence voltage. To note, determining a microgrid equivalent source is extremely difficult because a microgrid will contain several components that have their source determined by inverter control modes, control settings, and physical connections of transformers and (where applicable) rotating machines as well as loads.

During the planning of the microgrid shown in Fig. 1, consideration was given to these challenges. Basic analysis reveals that the system in its current state is not effectively grounded as per [21], since GFOV exceeds 1.38 pu. The microgrid construction, which includes a BESS, presents an opportunity to rectify this system problem at the same time mitigate problems that may be posed on the microgrid components due to overvoltage conditions.

Three options were considered:

1. A solidly grounded Δ-Yg BESS step up transformer.
2. A Δ-Yg BESS step up transformer with a 40 Ω reactor.

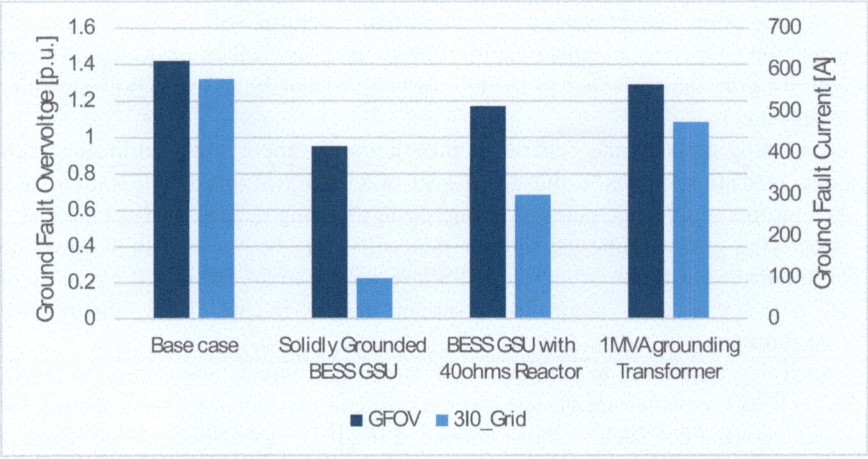

Fig. 9 Quantification of GFOV and source ground fault (from the sub-transmission center 1) with different BESS step-up transformer configurations

3. A Δ–Δ BESS step up transformer coupled with a 1MVA Δ-Yg bolted neutral grounding transformer.

Figure 9 shows a visual summary of the base case situation and the three mitigation options. As conceptually anticipated, under certain configurations of the BESS step-up transformer will cause the microgrid to inject large ground current into a fault, which will de-sensitize upstream protection elements. In this example, the sub-transmission line relays are wired via 1200:5 CTs and with this ratio, the line distance relay should sense at least 125 A to reliably pick up. The solidly grounded Δ-Yg BESS step up transformer, one of the cases, would result in non-recoverable desensitization. If this option is chosen, the best course of action is to replace line distance protection with line current differential protection.

7 Final Considerations

Clustered microgrids contain complex designs. Recently, many microgrids have been developed in laboratory or demonstration environments, but few have been built in real systems with real drives behind them. In these real systems, focal points must comprise:

- Resilience and reliability objectives, which must be verified by characterizing the load serving capability of a microgrid upon loss of normal supply.
- Renewable share targets, which are often integrated into the microgrid objectives and are part of all microgrid use cases.

- Grid operation and power quality, comprising careful voltage profile characterization to ensure adequate supply is provided, as well as ensuring inverters comply with standards such as IEEE Std 1547 [26] that dictate other power quality requirements.
- Developing an adequate generation mix that will support the grid during grid-connected mode, as well as during off-grid mode. Often a controllable source, such as a biofuel or CHP, is included in microgrid planning to achieve this objective.
- Specifying grid-forming capabilities that will likely be required in at least one of the inverter sources, often the BESS inverter (since the BESS has a consistent DC source that can be controlled to support the grid in case of large disturbance ride-through).
- Specifying grounding requirements for inverter resources, oftentimes the BESS as well as renewable-interfaced inverters. These may include requirements for negative-sequence such as those contained in [16]. Furthermore, zero-sequence requirements, possibly supplemented by an additional ground source, may also need to be dictated by requirements such as [23].
- Careful protection considerations as it comes to current infeed-caused desensitization, both for three-phase as well as single-line-to-ground faults, caused by equivalent positive-sequence and zero-sequence networks. Specification of transformer winding configurations and possibly other equipment should be considered.

References

1. Ton DT, Smith MA. The U.S. Department of Energy's Microgrid Initiative," available at https://www.energy.gov/sites/prod/files/2016/06/f32/The%20US%20Department%20of%20Energy%27s%20Microgrid%20Initiative.pdf
2. IEEE guide for design, operation, and integration of distributed resource island systems with electric power systems. IEEE Std 1547.4-2011, 20 July 2011. https://doi.org/10.1109/IEEESTD.2011.5960751
3. Distribution synchronized measurements roadmap. Prepared by Quanta Technology for the Department of Energy Oakridge National Laboratory, available at (last accessed 3 Mar 2024). https://www.naspi.org/sites/default/files/reference_documents/distribution_synch_measurement_roadmap_20210927.pdf
4. Elsayed WT, Farag H, Zeineldin HH, El-Saadany EF (2021) Dynamic transitional droops for seamless line-switching in islanded microgrids. IEEE Trans Power Syst 36(6):5590–5601. https://doi.org/10.1109/TPWRS.2021.3069852
5. Hamanah WM, Hossain MI, Shafiullah M, Abido MA (2023) AC microgrid protection schemes: a comprehensive review. IEEE Access 11:76842–76868. https://doi.org/10.1109/ACCESS.2023.3298306
6. Nassif A, Rahmatian M, Othman H, Paaso A (2022) Improvement of frequency nadir and RoCoF in Puerto Rico under high IBR scenarios by using grid-forming capabilities. CIGRE grid of the future symposium
7. Kenyon RW, Sajadi A, Hoke A, Hodge BM (2021) Open-source PSCAD grid-following and grid-forming inverters and a benchmark for zero-inertia power system simulations. In: 2021 IEEE Kansas power and energy conference (KPEC). IEEE, pp 1–6
8. Lasseter RH, Eto JH, Schenkman B, Stevens J, Vollkommer H, Klapp D, Linton E, Hurtado H, Roy J (2010) CERTS microgrid laboratory test bed. IEEE Trans Power Delivery 26(1):325–332

9. Du W, Lasseter RH, Khalsa AS (2018) Survivability of autonomous microgrid during overload events. IEEE Trans Smart Grid 10(4):3515–3524
10. Zhong Q-C (2016) Virtual synchronous machines: a unified interface for grid integration. IEEE Power Electron Mag 3(4):18–27. https://doi.org/10.1109/MPEL.2016.2614906
11. Sang W, Guo W, Dai S, Tian C, Yu S, Teng Y (2022) Virtual synchronous generator, a comprehensive overview. Energies 15:6148. https://doi.org/10.3390/en15176148
12. Johnson B, Rodriguez M, Sinha M, Dhople S (2017) Comparison of virtual oscillator and droop control. In: 2017 IEEE 18th workshop on control and modeling for power electronics (COMPEL), Stanford, CA, USA, pp 1–6. https://doi.org/10.1109/COMPEL.2017.8013298
13. Haddadi A, Farantatos E, Kocar I, Karaagac U (2021) Impact of inverter based resources on system protection. Energies 14(4):1050. https://doi.org/10.3390/en14041050
14. Mahamedi B, Fletcher JE (2020) The equivalent models of grid-forming inverters in the sequence domain for the steady-state analysis of power systems. IEEE Trans Power Syst 35(4):2876–2887. https://doi.org/10.1109/TPWRS.2020.2968114
15. VDE-AR-N 4110, 4120 and 4130 (German Grid Code). Technical connection rules: 10.2.3.3 dynamic network stability for type 2 power generating plants
16. IEEE Standard for Interconnection and Interoperability of Inverter Based Resources (IBRs) Interconnecting with Associated Transmission Electric Power Systems. IEEE Std 2800–2022, pp.1–180. https://doi.org/10.1109/IEEESTD.2022.9762253
17. Nassif AB, Loi E, Wheeler KA, Bahramirad S (2022) Impact of IBR negative-sequence current injection on ground fault temporary overvoltage and ground overcurrent protection. In: 2022 IEEE/PES transmission and distribution conference and exposition (T&D), New Orleans, LA, USA, pp 1–5. https://doi.org/10.1109/TD43745.2022.9816974
18. Brahma SM, Girgis AA (2004) Development of adaptive protection scheme for distribution systems with high penetration of distributed generation. IEEE Trans Power Deliv 19(1):56–63
19. Zamani MA, Sidhu TS, Yazdani A (2011) A protection strategy and microprocessor-based relay for low-voltage microgrids. IEEE Trans Power Delivery 26(3):1873–1883
20. Vukojevic A (2018) Lessons learned from microgrid implementation at electric utility. In: 2018 IEEE PES innovative smart grid technologies conference, Washington, DC, pp 1–5
21. EPRI Report 3002020130: effective grounding for inverter-connected DER: final report
22. Nassif AB (2018) An analytical assessment of feeder overcurrent protection with large penetration of distributed energy resources. IEEE Trans Indus Appl 54(5):5400–5407. https://doi.org/10.1109/TIA.2018.2810260
23. IEEE Std. C62.92.1–2019—IEEE guide for the application of neutral grounding in electrical utility systems—part I: introduction
24. Nassif AB, Aguero JR, Dong M (2022) Practical charts for sizing neutral grounding elements for machine-based distributed energy source step-up transformers. IEEE Trans Power Syst 37(3):2483–2486. https://doi.org/10.1109/TPWRS.2022.3156439
25. IEEE standard for the specification of microgrid controllers. IEEE Std 2030.7-2017, pp 1–43. https://doi.org/10.1109/IEEESTD.2018.8340204
26. IEEE standard for interconnection and interoperability of distributed energy resources with associated electric power systems interfaces. In: IEEE Std 1547-2018 (Revision of IEEE Std 1547-2003), pp 1–138. https://doi.org/10.1109/IEEESTD.2018.8332112

Military Microgrids with Renewable Energy and 5G Communication

Ali Mehrizi-Sani, Jeffery H. Reed, and Chen-Ching Liu

Abstract Military installations fit the widely accepted definition of an electrical microgrid very closely: they are geographically and electrically well-defined, need to be capable of operating in grid-connected and islanded modes, and can operate as a single entity when presented to the grid. The independent operation of a microgrid from the national grid can significantly enhance the resiliency, cybersecurity, and physical security of the nation's military bases. As a niche application of microgrids, several military base microgrids have been deployed in recent years. Renewable-based microgrids can help the military reduce its petroleum use, potentially saving \$8–\$20 billion over the next two decades. However, renewables-based microgrids require stringent control and protection facilities. Considering the large number of entities that need to be coordinated and controlled together, the control of assets within a microgrid typically becomes a distributed control problem—in distributed control, different assets communicate and coordinate with each other to optimally utilize the resources. This is where 5G communication networks can induce a paradigm shift. 5G has been identified by the United States Department of Defense as a critical strategic technology. A 5G-based technology platform is attractive from economic and security perspectives for military microgrids. 5G can address security while containing cost; 5G provides different layers of security, such as device-radio network security, device-core network security, device-service network security, and intra-network security. Even though 5G needs to be customized to meet power grid-specific requirements, baseline 5G features and designs are still significantly used. This chapter discusses the challenges and opportunities related to a renewable-based 5G-enabled microgrid for military installations.

Keywords Communication · Cybersecurity · Renewable energy · Military power system

Authors gratefully acknowledge support from the United States Army Research Laboratory (ARL) and State of Virginia's Commonwealth Cyber Initiative (CCI).

A. Mehrizi-Sani (✉) · J. H. Reed · C.-C. Liu
Bradley Department of Electrical and Computer Engineering, Virginia Tech, Blacksburg, VA, USA
e-mail: mehrizi@vt.edu

F. Shahnia and J. M. Guerrero (eds.), *Microgrids and Virtual Power Plants*, Power Systems, https://doi.org/10.1007/978-981-97-6623-9_17

1 Introduction

About one-third of the energy consumed in the United States industrial, commercial, and residential sectors is electrical energy. Since World War II (WWII), the number of soldiers in the army has decreased by a factor of 20. At the same time, fuel consumption per soldier has increased by a factor of 20; see Fig. 1 [1]. The military's need for energy is ever-increasing, necessitating the need to increase the efficiency of energy delivery. There are two major hurdles to accomplishing this goal: (i) Traditional form of delivery, i.e., deployment of fuel especially during an operation, is very costly and resource-intensive: it requires a carefully planned convoy raising severe limitations, particularly in active combat zones; and (ii) Increased need for energy requires additional delivery infrastructure; this is challenging even in civilian installations due to the high-quality (resilient and reliable) power requirements for installations and the dependency of an overseas base on a host country's goodwill.

Military entities are widely adopting electric vehicles. As with other forms of energy, petroleum availability depends on a robust supply chain. A classical military strategy against an enemy is to disrupt an adversary's supply chain. This supply chain disruption is much less likely if energy is generated and distributed locally. Additionally, environmental concerns have become very important, including to military organizations. Renewable energy sources, integrated through microgrids, provide a more environmentally friendly solution than petroleum.

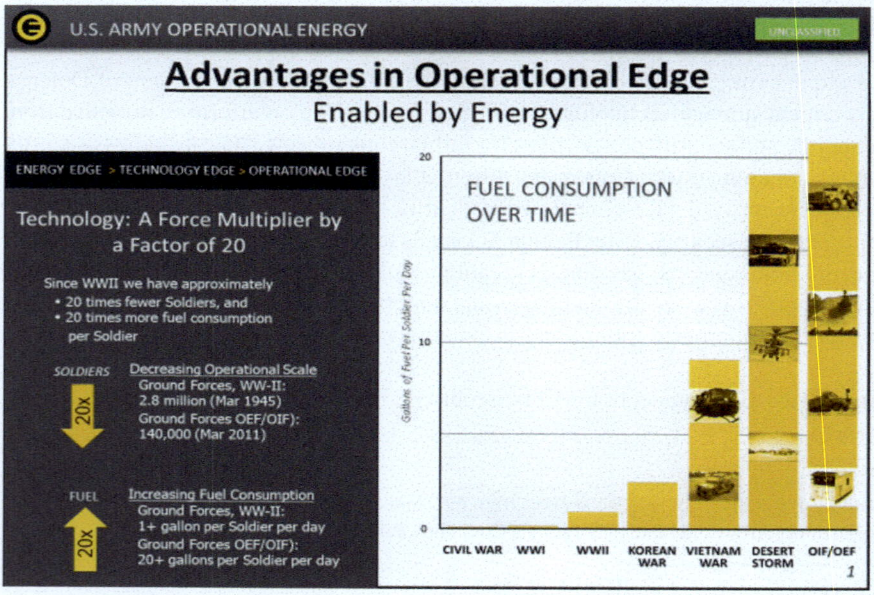

Fig. 1 Energy use increase in the US Army by a factor 20 since WWII [1]

Many of these problems also exist for the utility power grid. Therefore, with appropriate modifications, solutions proposed for the army (or military in general) will apply to the nation's power grid, increasing its reliability and resiliency. This is particularly important for national defense, as the power grid is a critical infrastructure component of national security.

2 Microgrids for Improved Security of Military Installations

Domestic army bases are typically connected to the grid.[1] The grid is one of the 20 greatest engineering achievements of the twenty-first century as selected by the United States National Academy of Engineering [2]. However, the grid is commonly impacted by various types of vulnerabilities. Figure 2 shows several examples of such vulnerabilities, including the 2013 sniper attack on a substation of the Pacific Gas & Electric Company, which caused $100 million in damages. Lloyd presented a hypothetical scenario in 2015 in which a cyberattack on only 10% of generators in the Eastern Interconnection could leave 93 million people without power. Thus, some priority mission bases should use backup generators to remain operational in case of outages, albeit at a lower efficiency. Therefore, even if bases are disconnected from the grid, the quality of power impacts services within the installation.

One attractive solution to increase the reliability and resiliency of the grid is to utilize the concept of microgrids. A microgrid is defined as a complete but miniature power system that is an aggregate of collocated resources (loads, generation units, and storage units or DERs [distributed energy resources]) that are interfaced to the main grid at the distribution level through a point of common coupling (PCC) and can operate in the grid-connected and islanded modes. Examples of microgrids are shown in Fig. 3a and b. The concept of microgrids is not new. The power system in Edison's time was effectively a collection of DC microgrids—each with its own set of loads and generators.

Military installations fit the widely accepted definition of an electrical *microgrid* very well [4]: they are geographically and electrically well-defined, they need to operate in both grid-connected and islanded modes (when the terrestrial grid is not present), and they can operate as a single entity when presented to the grid. The independent operation of a microgrid from the grid can significantly enhance the resiliency, cybersecurity, physical security, and reliability of the nation's military bases. Simultaneously, it is also the most challenging mode of operation. In this mode, the microgrid does not have generation or inertia support from the grid and can become unstable when subjected to large disturbances or large-scale cyberattacks.

---

[1] The grid is an informal term for the bulk power system. The bulk power system refers to facilities and control systems necessary for operating an interconnected electric energy transmission network (or any portion thereof).

Fig. 2 Common vulnerabilities of the grid: **a** extreme weather conditions, **b** physical damage, **c** cybersecurity, and **d** fossil fuel issues. Images taken from https://weather.com/news/news/sandy-power-outage-20121116, https://www.truecrimeneversleeps.com/san-jose-sniper-attack/, https://www.networkworld.com/article/947022/ukrainian-power-grid-hack-what-happened-and-what-needs-to-change.html, https://foreignpolicy.com/2021/10/05/energy-crisis-europe-electricity-gas-renewable-us/

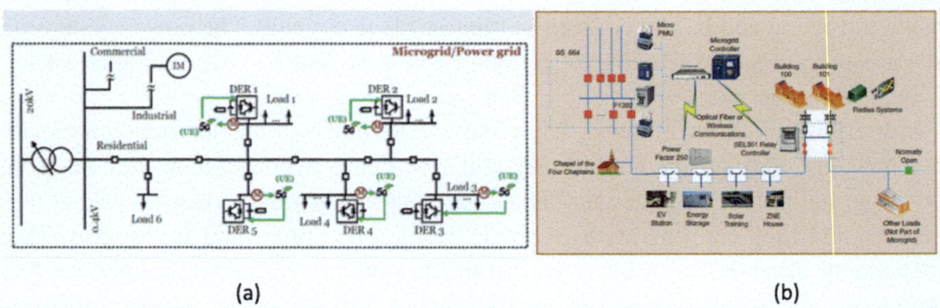

Fig. 3 **a** A sample microgrid topology and **b** the GridSTAR microgrid in Philadelphia [3]

One of the *niche* applications of microgrids has been military bases, and several such microgrids have been deployed in recent years. Examples are, see Fig. 4, (i) Marine Corps Air Station Miramar in San Diego, California, with 11 MW of combined solar photovoltaic (PV), landfill gas, and diesel/natural gas generation;

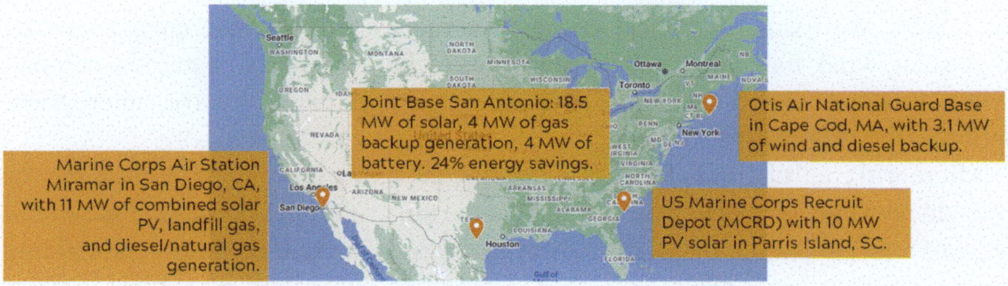

Fig. 4 Example of renewables-based bases in the continental United States

(ii) the Otis Air National Guard Base in Cape Cod, Massachusetts, with 3.1 MW of wind and diesel backup; and (iii) the United States Marine Corps Recruit Depot with 10 MW PV solar in Parris Island, South Carolina. Some Army bases rely on 5G, such as those near Ogden, Utah, and Albany, Georgia.

The United States Armed Forces have a growing trend of converting bases to microgrids. In February 2022, the Army announced it will build a microgrid at its 130 bases worldwide by 2035. In May 2022, the Navy announced it would build cyber-secure microgrids to protect from disasters while decarbonizing its energy supply. In November 2021, the Air Force announced leasing a solar microgrid system at Tyndall Air Force Base to compensate for the grid damage from Hurricane Michael in 2018. In summary, microgrids bring several advantages to the operation of Army installation power systems:

- Resiliency against power grid failures, e.g., due to weather
- Security: cybersecurity and physical security (smaller footprint; smaller attack surface)
- Reliability of electric energy delivery system

3 Power Needs of Installations

Unites States Department of Defense defines three levels of power sources:

- Tactical power: Low-voltage sources up to 600 V and 200 kW;
- Prime power: Medium-voltage sources from 600 V to 60 kV and more than 200 kW; and
- Utility power: High-voltage sources up to 230 kV connected to the national power grid.

In many cases, new operations start with tactical power with generation sources organic to the operational units, e.g., fuel cells, solar photovoltaics, and small gensets. They transition to prime and utility power as needed. The military also supports civil authorities through defense support of civil authorities (DSCA) tasks, which are normally coordinated by the United States Army Corps of Engineers. Providing temporary emergency power during domestic emergencies is another role for DSCA. We focus on tactical power as it relates to renewable energy sources (e.g., solar, wind, fuel cells), advanced storage technologies (e.g., batteries, pumped hydro, flywheels), and resilient operation (e.g., distributed control, microgrids, optimization) for installations.

Microgrid control differs from the control of the utility power grid because microgrids do not have the mechanical inertia that the utility grid has because of their bulk synchronous generation mass. Moreover, microgrids operate on a faster timeframe because of their power electronics-based inverter resources and are subject to more generation variability because of the possible change in renewable generation. Hence, microgrids experience more pronounced changes in their load because of their small size. These factors make microgrid control challenging; our past work in microgrid control has shown significant performance improvement in voltage and current control, power sharing, and renewables integration. This is discussed in more detail in the respective sections below.

We discuss microgrids because they can provide electric power where or when the terrestrial power grid is not available or reliable, thereby increasing resiliency and security (cybersecurity and physical security). Civilian bases, command centers, and other high-priority establishments benefit significantly from microgrids. For example, the United States Army is committed to microgrids on every installation by 2035 [5]; other services are following a similar path [6–9]. Army microgrids must be designed to control wider loading and power availability variations due to the requirements for all-electric non-tactical vehicles by 2027 and renewable energy sources by 2030.

Technological advances have reached a pivotal moment in the modernization efforts for the power grid. For the grid of the future, we need three components, as shown in Fig. 5: renewables as edge devices, communication and distributed control infrastructure, and performance guarantees. Recent advances in each of these areas have separately made them viable options. For example, in many parts of the country, solar energy is already cheaper than other forms of electricity. The United States Department of Energy has set a goal to cut the cost of utility-scale solar generation to \$0.02/kWh by 2030 [10]. These components are now at a critical stage of development, where their coordination can achieve the leap required for military applications. United States Department of Energy's Microgrid research program has the following vision [11]:

> *By 2035, microgrids are envisioned to be essential building blocks of the future electricity delivery system to support resilience, decarbonization, and affordability. Microgrids will be increasingly important for integration and aggregation of high penetration distributed energy resources. Microgrids will accelerate the transformation toward a more distributed and flexible architecture in a socially equitable and secure manner.*

The United States Government has similar priorities. United States Executive Order 14,057 directed several agencies, including the United States Department of Defense, to "increase its percentage use of carbon pollution-free electricity, so that it constitutes 100% of facility electrical energy use on an annual basis, and seek to match use on an hourly basis to achieve 50% 24/7 carbon pollution-free electricity,

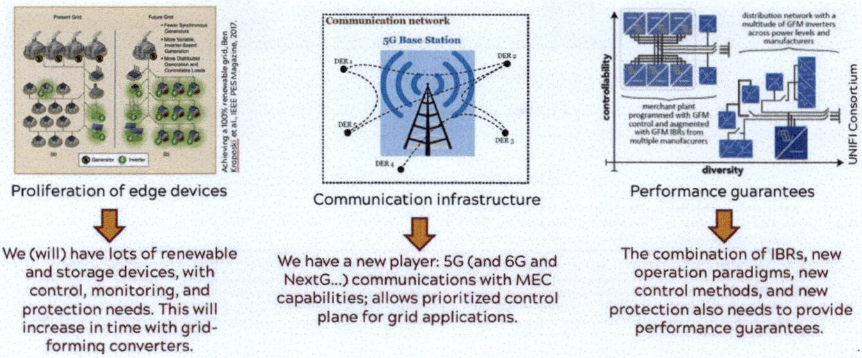

Fig. 5 The critical components of the power grid

by fiscal year 2030. In addition, agencies shall facilitate new carbon pollution-free electricity generation and energy storage capacity by authorizing the use of their real property assets, such as rooftops, parking structures, and adjoining land, for the development of new carbon pollution-free electricity generation and energy storage through leases, grants, permits, or other mechanisms, to the extent permitted by law" [13]. A similar priority is reflected in the Department of the Army's Climate Strategy as objectives 1.1, 1.2, 1.3, and 1.4 in Fig. 6 [12].

4 Renewables-Based, Communication-Enabled Military Installations

Renewable-based microgrids can reduce the military's use of petroleum, potentially saving \$8-\$20 billion over the next two decades [7]. However, such an operation also requires stringent control and protection facilities to be in place. Considering its large number of entities, control of assets within a microgrid typically becomes a distributed control problem—in distributed control, different assets communicate and coordinate with each other to optimally utilize the resources. In our view, this is where a communication-enabled distributed decision engine can induce a paradigm shift. Figure 7 shows our vision for the installations of the future.

5 Role of 5G Communication

5G communication plays an especially important role in microgrids and distributed energy resources (DER). The initial specifications for 5G were developed by the international standardization organization 3GPP. This organization defined the initial specifications for 5G in 2018, and such specifications will be enhanced well beyond

LINE OF EFFORT 1: INSTALLATIONS	
STRATEGIC OUTCOME:	
Enhance resilience and sustainability by adapting infrastructure and natural environments to climate change risks, securing access to training and testing lands into the future, and mitigating GHG emissions	
INTERMEDIATE OBJECTIVES:	
1.1	Install a microgrid on every installation by 2035
1.2	Achieve on-site carbon pollution-free power generation for Army critical missions on all installations by 2040
1.3	Provide 100% carbon-pollution-free electricity for Army installations' needs by 2030
1.4	Implement installation-wide building control systems by 2028
1.5	Achieve 50% reduction in GHG emissions from all Army buildings by 2032, from a 2005 baseline
1.6	Attain net-zero GHG emissions from Army installations by 2045
1.7	Field an all-electric light-duty non-tactical vehicle fleet by 2027
1.8	Field an all-electric non-tactical vehicle fleet by 2035
1.9	Continue to advocate for an expanded Army Compatible Use Buffer
1.10	Include climate change threat mitigation into Army land management decisions
1.11	Incorporate the latest climate and environmental science into stationing, construction, and fielding decisions

Fig. 6 Army installations climate strategy objective. Table taken from [12]

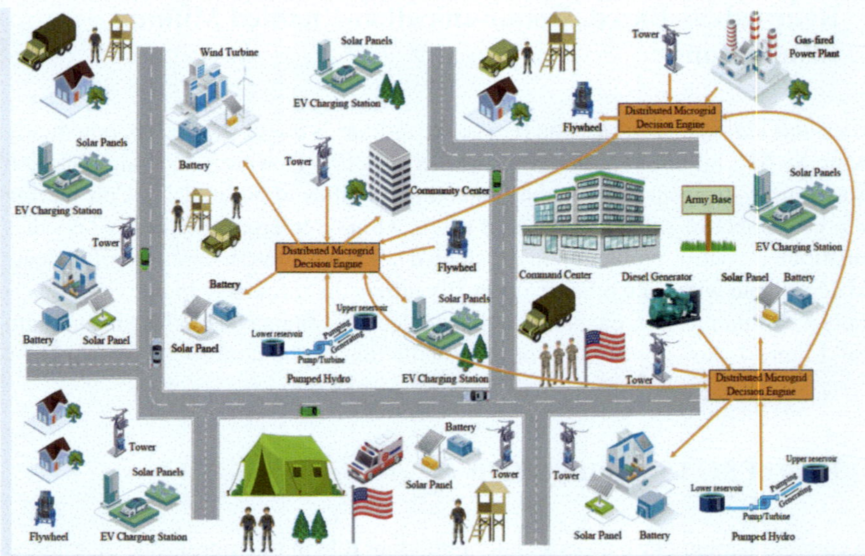

Fig. 7 Vision for the installations of the future: microgrids with renewables energy resources and storage controlled through a distributed architecture for increased resiliency and efficiency

the 2028 timeframe, the approximate time frame for 6G initial standardization. While standards from 1G to 4G have impacted various aspects of energy systems, 5G is expected to have a much more significant impact, with more diverse applications.

5G is a system-of-systems (Fig. 8), composed of User Equipment which may be cellular handsets or Internet-of-Things (IoT) devices; a 5G radio access network (RAN) or a next-generation RAN (NG-RAN) or gNB, which is the base station that does the transmission and scheduling of radio resources; and the 5G next generation core network (NGC) which performs functions such as authentication, quality of service management, routing and mobility management, billing, operations, and management, and interfacing with the Internet [14].

5G was created based on three pillar scenarios of applications:

- Enhanced Mobile Broad Band (eMBB) or high data rates
- Ultra-Reliable and Low Latency Communications (URLLC)
- Massive Machine-to-Machine Communications (mMTC) for the connectivity of IoT.

Figure 9 summarizes the applications for the three pillars. This flexibility to support applications with wide disparent requirements makes 5G a powerful standard.

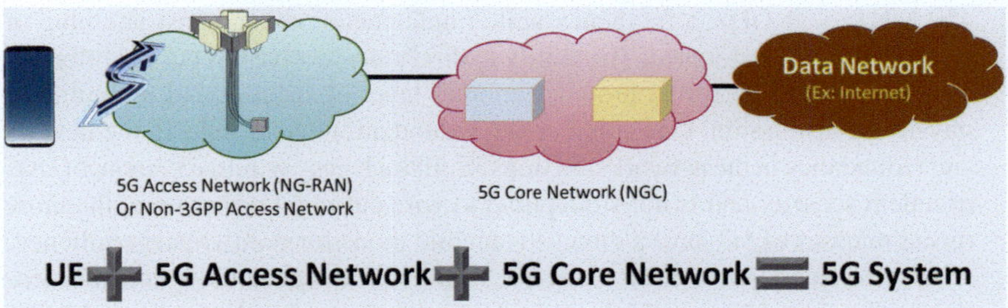

Fig. 8 A complete 5G system [15]

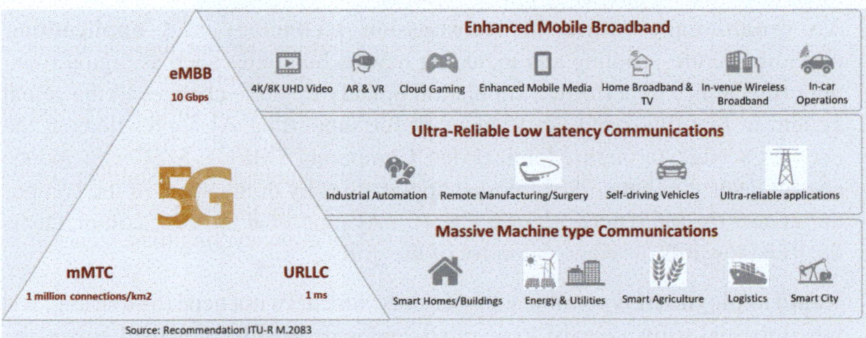

Fig. 9 Pillars of application scenarios for 5G

All the pillars have applications for microgrids and DERs. The specific advantages of 5G include the following:

1. **The ability to coordinate data collection and control with low latency.** 5G systems can be deployed quickly and inexpensively to perform this data collection and control function. The mMTC allows many objects to be controlled at a much greater scale and precision than before 5G. Massive amounts of data can be collected to understand grid status and improve performance.
2. **A unifying communication standard** allows communications for the grid and various data and voice services that can be implemented within a private cellular network. This ability to service multiple applications with disparate requirements is made feasible by the network slicing capabilities of 5G, where logical networks can be formed from a common physical network and provide the quality of service needed for disparent applications. Other advantages of slicing include giving the grid operator more control over devices, i.e., adding or deleting devices and negotiating service-level agreements between operators and service providers. A unified communications network is usually more economical than running several disparent networks.
3. **Secure communication with encryption resides at multiple levels within the network**. Furthermore, since 5G operates in dozens of bands spanning from 450 MHz to 71 GHz, a frequency-agile implementation can resist jamming or other forms of interference. The ability to direct and receive directional information through beamforming provides another layer of security and reliability to physical transmission. One could create redundant wireless links to the devices and redundancy in the network functions. Unlike a legacy wireline system, 5G has excellent security and is not susceptible to wire cutting. Furthermore, the more recent releases of 5G have a direct-to-satellite mode for even greater resiliency.
4. **Ability to easily incorporate electric vehicles through mobile communications, allowing electric vehicles to be both storage and sinks for the grid.** The 5G network could also provide cellular vehicle-to-anything with many applications to support safety and efficient transportation. Likewise, 5G can control and monitor portable emergency power systems.
5. **An evolutionary future in transmission technology and applications**, providing future-proofing and including new technologies such as digital twinning (the ability to perform a simulation quickly to guide changes to the actual system in real-time) and the framework for supporting AI applications at the network's edge through Mobile Edge Computing (MEC). MEC can reduce overall latency in network-supported applications by doing some of the computation near the radio network (see Fig. 10). AI support at the edge complements the trend towards distributed control of the grid.

6. **Rapid deployment** is possible with 5G since wire does not need to be strung, and functions and software exist to do rapid deployment with automated operations and management functions. Field hospitals and the emergence of portable energy-directed weapons that require significant power are examples of applications that could benefit from 5G microgrids.

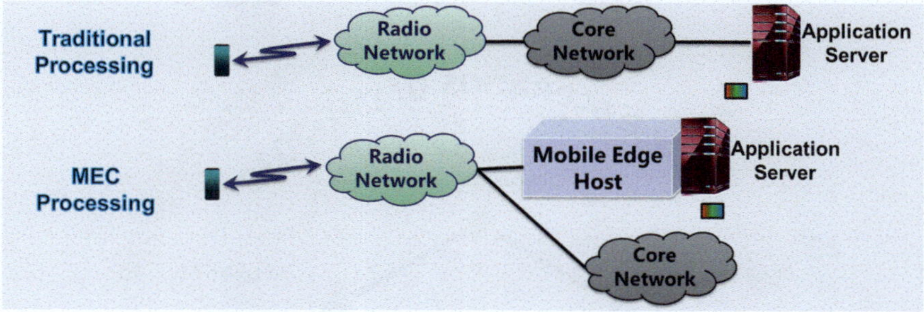

Fig. 10 Comparison of MEC enable network to traditional cellular network

Some of the applications for using 5G in smart grids and requirements are provided in Table 1.

While 5G does have many desirable properties, there are issues that, with further refinement, would make it even more closely suited for grid applications. First is a reduced latency, perhaps down to microseconds, which will open more applications. Next is the standardization of network slices as a common basis for grid management and control. For instance, specific slices could be created for.

- Grid configuration.
- Coordination of renewable sources feeding the grid.
- Coordination of battery storage (including making use of electric vehicles).
- Monitoring of grid information for analysis, optimization, and security.

Security is always a concern, and 5G is a complex system; ensuring the security of such a system is a continuing battle. Trading latency and security and latency and authentication certainty are trade-offs that need further exploration. A continuing challenge for these custom networks is the availability spectrum, especially for overseas military organizations. Spectrum sharing is one approach to managing these issues. Finally, for military applications, there needs to be a set of formal metrics for judging the performance and applicability of various elements that compose a 5G-enabled microgrid.

Table 1 5G Applications and their requirements for smart grids

Application	Requirements
Advanced Metering Infrastructure	Massive mMMC, extensive coverage, high-security
Distributed Automation for Fault Recovery and Dynamic Microgrid Fragmentation	URLLC, high-security, network slicing, edge AI
Power Grid Monitoring (i.e., line sensors, UAV control/cameras)	URLLC, eMBB, power efficient communications
Flexible and fine-grain Load Control	URLLS, mMMC, high security, network slicing with service level agreements

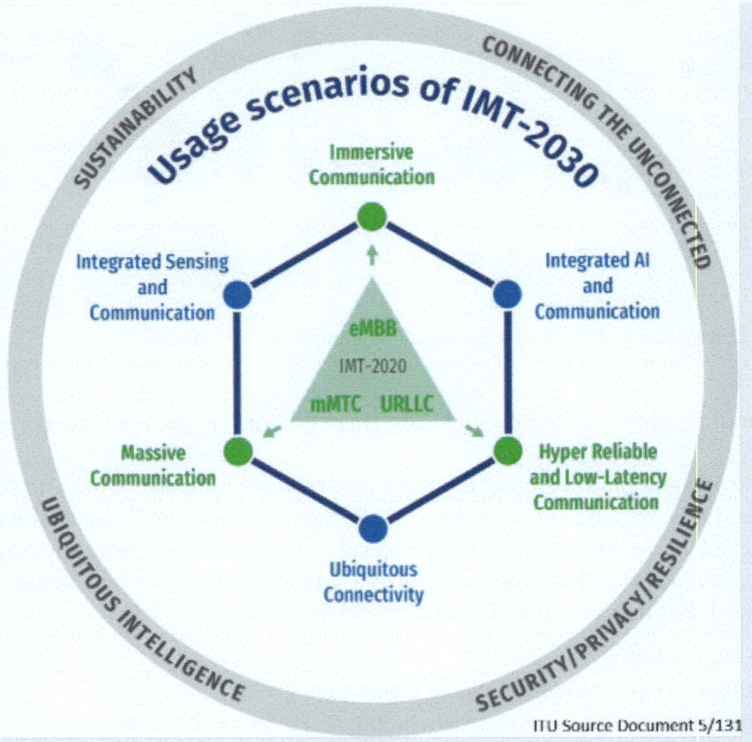

Fig. 11 Usage scenarios for 6G according to the International Telecommunications Union

Work is underway on 6G communications. The International Telecommunications Union (ITU) has set its vision for what 6G should address in its vision for IMT-2030. In addition to the three pillars of application scenarios defined in 5G, 6G will add ubiquitous connectivity, integrated AI and communications, and integrated sensing and communications, as illustrated in Fig. 11. All these application scenarios have applicability to microgrid operation.

6 Case Study: Inverter-Based Resources (IBR) Control Coordination in a 5G-Based Microgrid

This section discusses results for a coordinated set-point automatic adjustment with correction enabled (C-SPAACE) [16–19] framework that uses 5G communication for real-time control coordination between inverter-based resources (IBRs) in microgrids [20]. Utilizing slicing capability, 5G offers low-latency communication to C-SPAACE under normal conditions. However, given the multitude of power grid use cases, a certain 5G slice for C-SPAACE may have access only to limited radio spectrum resources, which, if not managed well, greatly undermines the communication needs of the C-SPAACE framework. Thus, optimally scheduling the available spectrum resources among IBRs in a sliced 5G network-based C-SPAACE framework becomes a critical problem. Time-domain simulation case studies are performed to evaluate the performance of C-SPAACE using 5G.

Figure 12 shows the CIGRE North American medium voltage (MV) distribution benchmark, used as the study system. It includes 14 IBRs with ratings shown in Table 2. The top two breakers in Fig. 12 are opened to establish islanded operation. The bottom three breakers are closed to create a meshed system. Each IBR communicates with its two nearest IBRs.

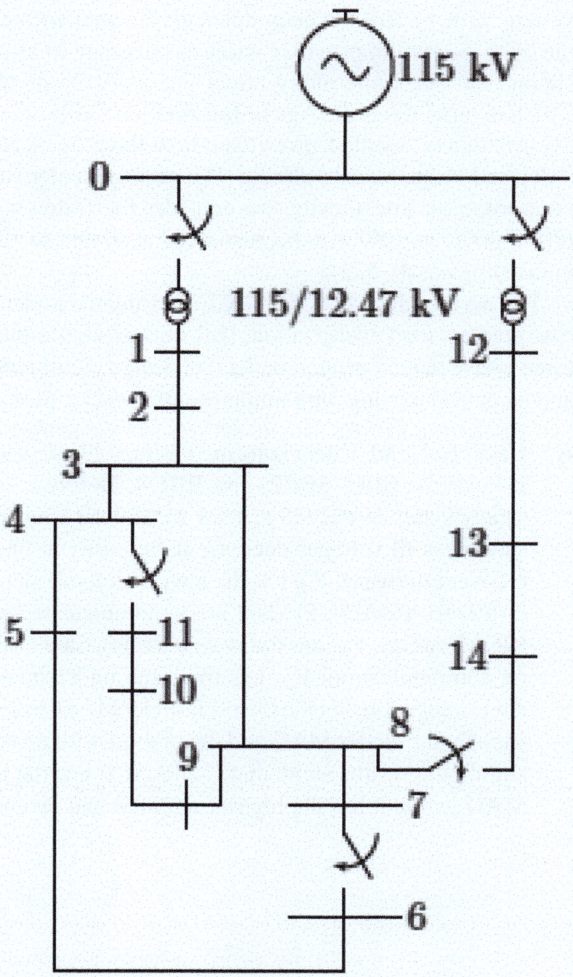

Fig. 12 CIGRE North American MV distribution benchmark system

Table 2 IBR ratings for the CIGRE North American MV distribution benchmark system

IBR Bus	Rating (MVA)	IBR Bus	Rating (MVA)
1	6	8	0.5
2	0.5	9	0.5
3	1	10	1
4	0.5	11	0.5
5	1	12	6
6	0.5	13	0.5
7	2	14	1

These IBRs and a 5G base station are located in a deployment region. The CIGRE system with 14 IBRs is geographically located within this deployment region, and the 5G base station provides wireless coverage to all 14 IBRs. Since both network slicing and 5G numerology affect the available number of radio resources (RB), which in turn affects the Age of Information (AoI) or latency performance of various 5G schedulers, we first investigate how three 5G schedulers of AoI, Round Robin (RR), and Proportional Fairness (PF) perform under varying network slicing and 5G numerologies. Specifically, we consider the following setting: (i) network slicing (20%, 40%, and 80% radio resources available to the power grid slice), and (ii) numerology μ (0, 1, 2).

The performance of C-SPAACE is evaluated under four time-domain simulation case studies: load energization, IBR outage, single-phase line-to-ground faults, and three-phase line-to-ground faults. 5G design parameters are set as shown in Table 3, and a network setting with numerology 2 with a slice size of 20% is considered.

A. **Case 1: Load Energization.** Figures 13a–c show the d-axis component of voltage for IBR1, IBR2, and IBR14. Before t = 0.5 s, the microgrid is in a steady state. At t = 0.5 s, a 5.1 MVA load is energized at bus 1. These figures show that all voltages decrease temporarily at the onset of the disturbance and recover afterward. V_{d1} has the lowest undershoot percentage under maximal age difference (MAD), PF, and no communication, while it has the highest under RR. Moreover, V_{d1} has the lowest settling time t_s with MAD and the highest with no communication. V_{d2} has the same undershoot under all schedulers and no communication, but the lowest t_s under MAD. Finally, V_{d14} has the lowest undershoot and t_s under MAD and the highest with no communication. In conclusion, simulation results show that C-SPAACE has the lowest undershoot and t_s with MAD, while it has the highest with RR and no communication.

Table 3 Communication parameters used in simulation case studies

Parameter	Value
Deployment region $l \times b$	1000×1000 m
Transmission power $p^{u}_{T,\mathrm{f}}$	30 dBm [15]
Noise power σ_o	10^{-10} W
Path loss exponent γ	3.76 [28]
Channel	Rayleigh fading channel
CSI error variance σ_e^2	0.01 [28]
Traffic model	Generate at will
Carrier Frequency f_D	2.63 GHz [37]
Uplink Bandwidth BW_u	40 MHz
Downlink Bandwidth BW_d	80 MHz
RBs needed by ith IBR B_i^{need}	$[1,\ 3]\ \forall i \in G$
BS location	(l/2, b/2)
Aggregated carriers J	8
Modulation order $Q_m^{(j)}$	4
Maximum layers for jth carrier $v_{\mathrm{Layers}}^{(j)}$	8
Scaling factor $f^{(j)}$	1
Overhead OH^j	0.14
Packer size L	150 Bytes [38]

B. **Case 2:** IBR Outage. Figure 14(a) shows V_{d1} under Case 2. Prior to t = 0.5 s, the microgrid is in a steady state. At t = 0.5 s, the 6 MVA IBR at bus 1 is suddenly disconnected. This figure shows that V_{d1} overshoots to 1.75 pu under all communication settings mainly because of droop control overcharging its filter capacitor. On the other hand, this figure also shows that MAD results in a faster response than the other schedulers and without communication. Figures 14(b) and (c) show V_{d2} and V_{d14} for Case 2. Both voltages decrease at the onset of the event as they pick up more power to maintain a balance between generation and load. However, both voltages stabilize after the disturbance. Table IV summarizes the performance results for V_{d2} and V_{d14}, with both voltages having the lowest undershoot and t_s with MAD and the highest with RR and no communication.

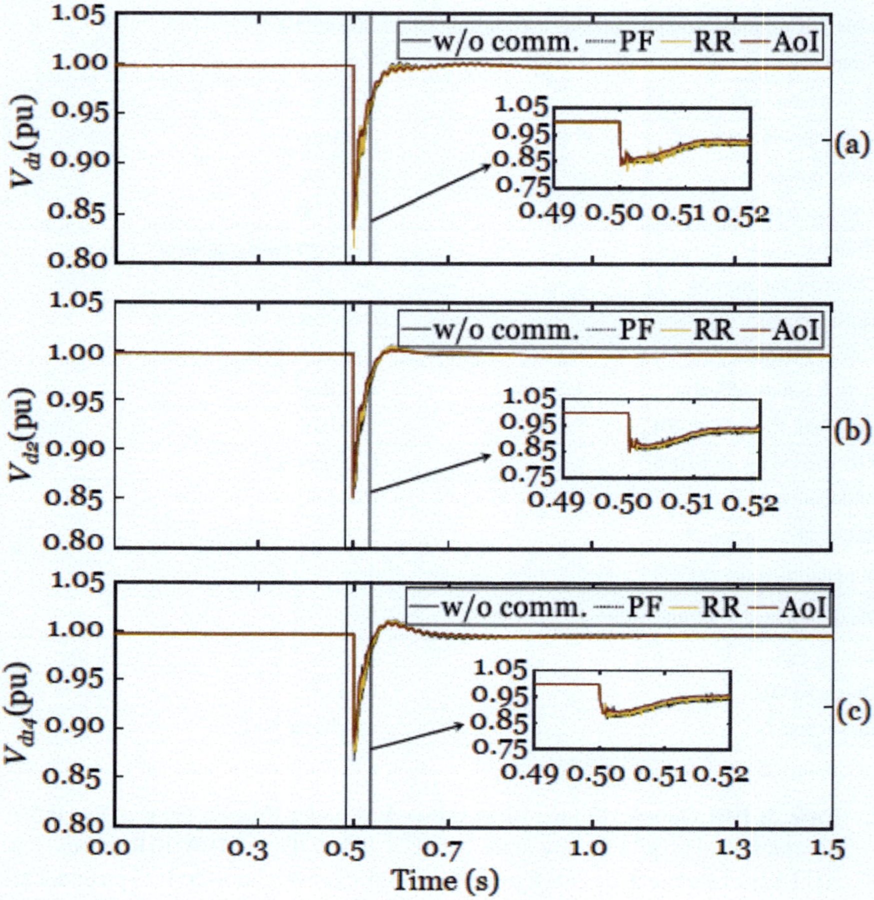

Fig. 13 Simulation results for case 1: (a) V_{d1}, (b) $V_{d2,}$ (c) V_{d14}

7 Testbed Facilities at Virginia Tech (VT)

Figure 15 shows the microgrid testbed at Virginia Tech. This testbed can incorporate technologies related to renewables-based microgrids and 5G communication. This testbed includes physical microgrid devices, inverters, solar panels, batteries, electronic loads, a four-quadrant amplifier, and a real-time RTDS simulator setup for power-hardware-in-the-loop simulation. This testbed also includes a digital twin (Figs. 16 and 17) of the Virginia Tech Electric Service, which is a full-service utility with new smart meters, PV inverters, and communications-enabled controllers partially enabled through Virginia Tech's Climate Action Commitment.[2] Virginia

[2] Virginia Tech, Climate Action Commitment, available at https://svpoa.vt.edu/content/dam/svpoa_vt_edu/vtcac/VT%202020%20Climate%20Action%20Committee%20Executive%20Report-FIN-111020.pdf

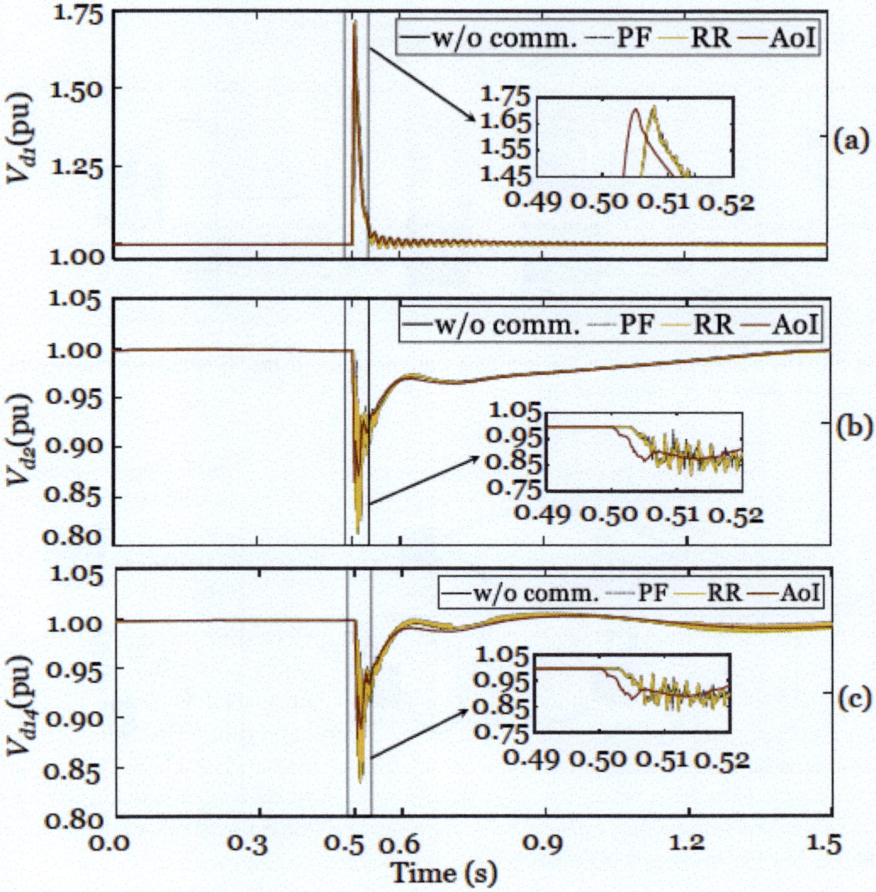

Fig. 14 Simulation results for case 2: (a) V_{d1}, (b) V_{d2}, (c) V_{d14}

Tech is one of the few universities in the United States that has acquired spectrum in the form of priority access licenses for Citizens Broadband Radio Service (CBRS). Two blocks (3610 and 3620 MHz) are reserved for research applications. These blocks are in Montgomery County (where Blacksburg resides) and can support 5G technology.

The microgrid testbed is also connected to the 5G testbed at the State of Virginia's Commonwealth Cyber Initiative (CCI) using a multi-connection TCP/IP server interface and an optical fiber connection to minimize the connection latency. The 5G testbed at the CCI is one of the largest university-based software-defined radio (SDR) testbeds in the United States, with 72 SDR base stations or user equipment. Figure 18 shows the portable in-lab 5G testbed at Virginia Tech. This portable testbed allows for network slicing in 5G, low roundtrip latency, and high reliability.

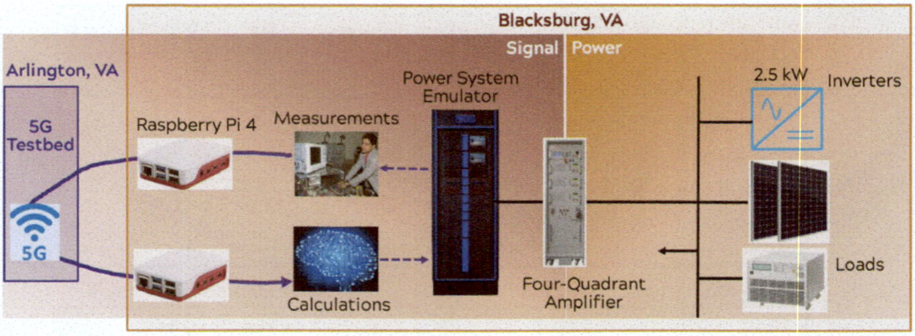

Fig. 15 The microgrid testbed at Virginia Tech with connection to the 5G testbed operated by the Commonwealth Cyber Initiative (CCI)

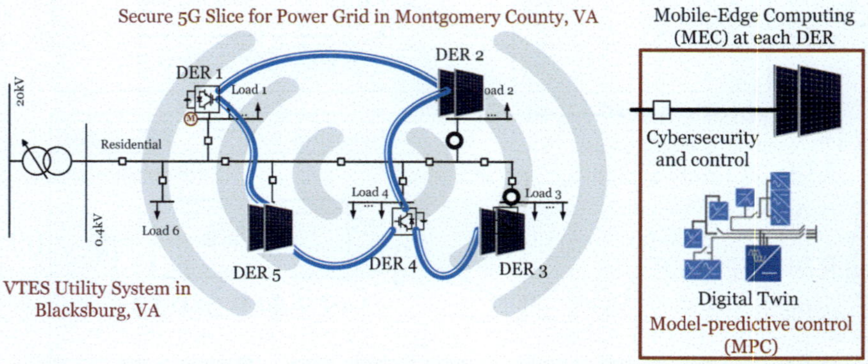

Fig. 16 Notional illustration of the technology

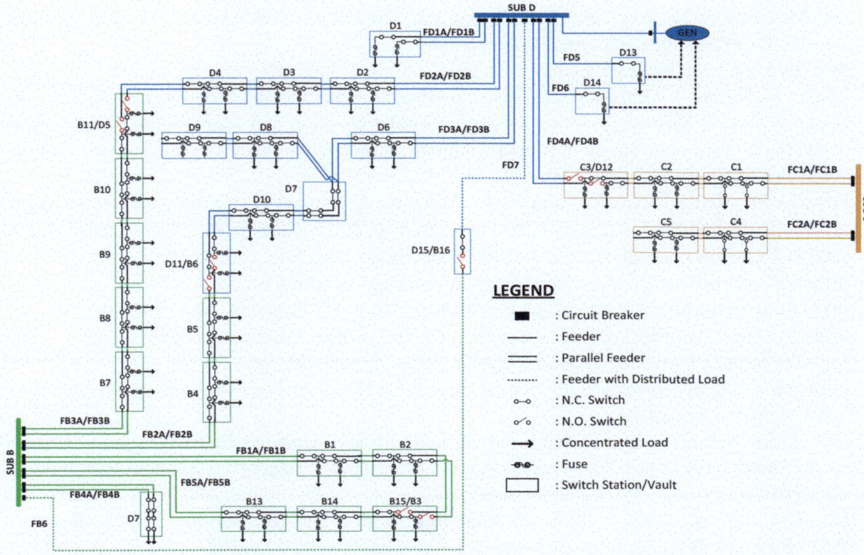

Fig. 17 A sanitized and simplified version of the Virginia Tech Electric Service network

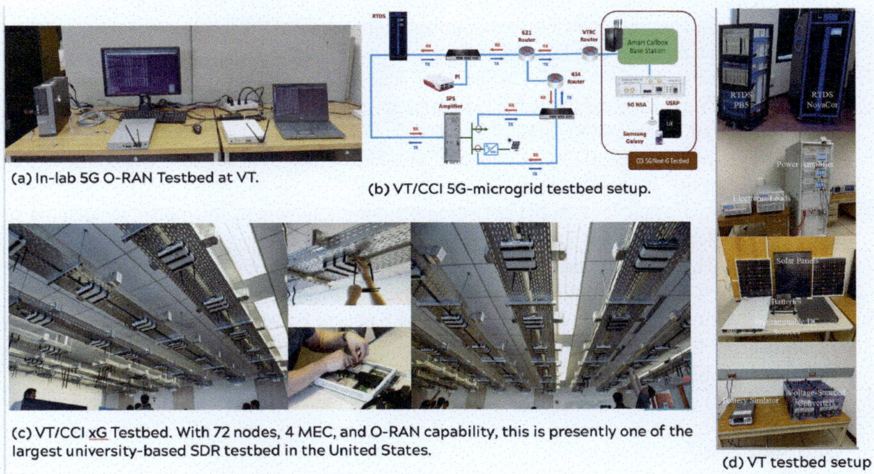

Fig. 18 5G testbed resources at Virginia Tech

References

1. Advantages in operational edge. R. Kidd, U.S. Army (2012) Army Energy and Sustainability Program. presentation, https://www.asaie. army.mil/Public/ES/doc/2-General%20Presentation.pdf accessed through National Academies of Sciences, Engineering,

and Medicine 2021. Powering the U.S. Army of the Future. Washington, DC: The National Academies Press. https://doi.org/10.17226/26052

2. http://www.greatachievements.org/
3. Uluski R, Kumar J, Venkata SSM, Vishwakarma D, Schneider K, Mehrizi-Sani A, Terry R, Agate W (2017) Microgrid controller design, implementation and deployment—Experiences at the Navy Yard community at philadelphia: A journey from conception to implementation. IEEE Power Energ Mag 15(4):50–62
4. Olivares D, Mehrizi-Sani A, Etemadi A, Canizares C, Iravani R, Kazerani M, Hajimiragha A, Gomis- Bellmunt O, Saeedifard M, Palma-Behnke R, Jimenez-Estevez G, Hatziargyriou N (2014) Trends in microgrid control. IEEE Trans Smart Grid 5(4):1905–1919
5. https://microgridknowledge.com/military-microgrids-spending/
6. https://www.cto.mil/wp-content/uploads/2020/05/DoD_5G_Strategy_May_2020.pdf
7. https://www.army.mil/e2/downloads/rv7/about/2022_army_climate_strategy.pdf
8. https://microgridknowledge.com/navy-microgrids-climate-strategy/
9. https://www.af.mil/News/Article-Display/Article/2837905/air-forces-first-energy-assurance-lease-signed-at-tyndall-afb/
10. U.S. DOE Solar Energy Technologies Office (SETO) Updated 2030 Goals for Utility-Scale Photovoltaics. https://www.energy.gov/eere/solar/solar-energy-technologies-office-updated-2030-goals-utility-scale-photovoltaics
11. Liu C-C, et al. (2022) Topic 3: Building blocks for microgrids," DOE Microgrid R&D Program. [Online] https://www.energy.gov/sites/default/files/2022-09/3-Building%20Blocks%20for%20Microgrids.pdf
12. Department of the Army, Office of the Assistant Secretary of the Army for Installations, Energy and Environment. February 2022. United States Army Climate Strategy. Washington, DC. https://www.army.mil/e2/downloads/rv7/about/2022_army_climate_strategy.pdf
13. https://www.whitehouse.gov/briefing-room/presidential-actions/2021/12/08/executive-order-on-catalyzing-clean-energy-industries-and-jobs-through-federal-sustainability/
14. Reed JH, Bernhard JT, Park J-M (2012) Spectrum access technologies: The past, the present, and the future. In: Proceedings of the IEEE, vol. 100, no. Special Centennial Issue, pp 1676–1684. https://doi.org/10.1109/JPROC.2012.2187140
15. Nishith D, Tripathi, Reed JH 5G Cellular Communications- Journey and Destination. The Wireless University. https://thewirelessuniversity.com/
16. Mehrizi-Sani A, Iravani R (Feb.2012) Online set point adjustment for trajectory shaping in microgrid applications. IEEE Trans Power Syst 27(1):216–223
17. Ghaffarzadeh H, Mehrizi-Sani A (Dec.2020) Adaptive set point modulation to mitigate transients in power systems. IET Gener Transm Distrib 14(23):5463–5470
18. Syed MH, Mehrizi-Sani A, Robowska M, Guillo-Sansano E, Wang D, Burt GM (Dec.2022) Dynamically robust coordinated set point tracking of distributed DERs at point of common coupling. Int J Electrical Power Energy Syst 143:108481
19. Ghaffarzadeh H, Stone C, Mehrizi-Sani A (Mar.2017) Predictive set point modulation to mitigate transients in lightly damped balanced and unbalanced systems. IEEE Trans Power Syst 32(2):1041–1049
20. Choudhury B, Mohammadhassani A, Iyer R, Alexander B, Mehrizi-Sani A, Shah VK, Reed JH (Feb.2024) Control coordination in inverter-based microgrids using AoI-based 5G schedulers. IET Smart Grid 7(1):38–50

SPRINGER NATURE

GPSR Compliance

The European Union's (EU) General Product Safety Regulation (GPSR) is a set of rules that requires consumer products to be safe and our obligations to ensure this.

If you have any concerns about our products, you can contact us on ProductSafety@springernature.com

In case Publisher is established outside the EU, the EU authorized representative is:

Springer Nature Customer Service Center GmbH
Europaplatz 3
69115 Heidelberg, Germany

The manufacturer's authorised representative in the EU is Springer Nature Customer Service Centre GmbH, Europaplatz 3, 69115 Heidelberg, Germany. If you have any concerns regarding our products, please contact ProductSafety@springernature.com

Printed and bound by CPI Group (UK) Ltd, Croydon, CR0 4YY

30/09/2026

02225253-0019